Lecture Notes in Electrical Engineering 277

For further volumes:
http://www.springer.com/series/7818

W. Eric Wong • Tingshao Zhu
Editors

Computer Engineering and Networking

Proceedings of the 2013 International Conference on Computer Engineering and Network (CENet2013)

Volume 2

Editors
W. Eric Wong
University of Texas at Dallas
Richardson, Texas
USA

Tingshao Zhu
Chinese Academy of Sciences
Beijing, China, People's Republic

ISSN 1876-1100
ISBN 978-3-319-01765-5
DOI 10.1007/978-3-319-01766-2
Springer Cham Heidelberg New York Dordrecht London

ISSN 1876-1119 (electronic)
ISBN 978-3-319-01766-2 (eBook)

Library of Congress Control Number: 2014931403

© Springer International Publishing Switzerland 2014

This work is subject to copyright. All rights are reserved by the Publisher, whether the whole or part of the material is concerned, specifically the rights of translation, reprinting, reuse of illustrations, recitation, broadcasting, reproduction on microfilms or in any other physical way, and transmission or information storage and retrieval, electronic adaptation, computer software, or by similar or dissimilar methodology now known or hereafter developed. Exempted from this legal reservation are brief excerpts in connection with reviews or scholarly analysis or material supplied specifically for the purpose of being entered and executed on a computer system, for exclusive use by the purchaser of the work. Duplication of this publication or parts thereof is permitted only under the provisions of the Copyright Law of the Publisher's location, in its current version, and permission for use must always be obtained from Springer. Permissions for use may be obtained through RightsLink at the Copyright Clearance Center. Violations are liable to prosecution under the respective Copyright Law.

The use of general descriptive names, registered names, trademarks, service marks, etc. in this publication does not imply, even in the absence of a specific statement, that such names are exempt from the relevant protective laws and regulations and therefore free for general use.

While the advice and information in this book are believed to be true and accurate at the date of publication, neither the authors nor the editors nor the publisher can accept any legal responsibility for any errors or omissions that may be made. The publisher makes no warranty, express or implied, with respect to the material contained herein.

Printed on acid-free paper

Springer is part of Springer Science+Business Media (www.springer.com)

Preface

This book is a collection of the papers accepted by CENet 2013—the third International Conference on *Computer Engineering and Network* (CENet), which was held from 20 to 21 July, 2013 in Shanghai, China. It has two volumes and three parts in each. Part I focuses on Algorithm Design with 29 papers over 232 pages; Part II emphasizes Data Processing containing 184 pages divided among 22 papers; Part III Pattern Recognition includes 29 papers in 234 pages; Part IV has 22 papers and 187 pages devoted to one of the most exciting technologies currently surging in popularity—Cloud Computing; Part V covers recent advances in Embedded Systems with 28 papers in 228 pages; and finally Part VI has 28 papers spanning 234 pages dedicated to Network Optimization.

Each part can be used as an excellent reference by industry practitioners, university faculty, and undergraduate as well as graduate students who need to build a knowledge base of the most current advances and state of practice in the topics covered by this book. This will enable them to produce, maintain, and manage systems with high levels of trustworthiness and complexity that provide critical services in a variety of applications.

Thanks go to the authors for their hard work and dedication as well as the reviewers for ensuring the selection of only the highest quality papers; their efforts made this book possible. Invaluable assistance with the publication was also provided by the editorial staff at Springer, especially Mr. Brett Kurzman and Miss Rebecca Hytowitz.

Richardson, Texas, USA	W. Eric Wong
Beijing, China	Tingshao Zhu

Contents

Volume 1

Part I Algorithm Design

1 Simulation Algorithm of Adaptive Scheduling in Missile-Borne Phased Array Radar 3
Qizhong Li, Shanshan Sun, and Jianfei Zhao

2 A Second-Order Algorithm for Curve Parallel Projection on Parametric Surfaces 9
Xiongbing Fang and Hai-Yin Xu

3 Computation Method of Processing Time Based on BP Neural Network and Genetic Algorithm 21
Danchen Zhou and Chao Guo

4 Integral Sliding Mode Controller for an Uncertain Network Control System with Delay 31
Zhenbin Gao

5 Synthesis of Linear Antenna Array Using Genetic Algorithm to Control Side Lobe Level 39
Zhigang Zhang, Ting Li, Feng Yuan, and Li Yin

6 Wavelet Analysis Combined with Artificial Neural Network for Predicting Protein–Protein Interactions 47
Juanjuan Li, Yuehui Chen, and Fenglin Wang

7 Application Analysis of Slot Allocation Algorithm for Link16 55
Hui Zeng, Qiang Chen, Xiaoqiang Li, and Jianguo Shen

8 An Improved Cluster Head Algorithm for Wireless Sensor Network 65
Feng Yu, Wei Liu, and Gang Li

9	**An Ant Colony System for Dynamic Voltage Scaling Problem in Heterogeneous System**... Yan Kang, Ying Lin, Yifan Zhang, and He Lu	73
10	**An Improved Ant Colony System for Task Scheduling Problem in Heterogeneous Distributed System**................. Yan Kang, Yifan Zhang, Ying Lin, and He Lu	83
11	**Optimization of Green Agri-Food Supply Chain Network Using Particle Swarm Optimization Algorithm**............... Qian Tao, Zhexue Huang, Chunqin Gu, and Chenxin Zhang	91
12	**A New Model for Short-Term Power System Load Forecasting Using Wavelet Transform Fuzzy RBF Neural Network**... Jingduan Dong, Changhao Xia, and Wei Zhang	99
13	**Energy-Effective Frequency-Based Adaptive Sampling Algorithm for Clustered Wireless Sensor Network**............. Meiyan Zhang, Wenyu Cai, Liping Zhou, and Jilai Liu	107
14	**An Indoor Three-Dimensional Positioning Algorithm Based on Difference Received Signal Strength in WiFi**................ Yibo Li and Xiting Liu	115
15	**The Universal Approximation Capability of Double Flexible Approximate Identity Neural Networks**..................... Saeed Panahian Fard and Zarita Zainuddin	125
16	**A Novel and Real-Time Hand Tracking Algorithm for Gesture Manipulation**.................................. Zhiqin Zhang	135
17	**A Transforming Quantum-Inspired Genetic Algorithm for Optimization of Green Agricultural Products Supply Chain Network**................................. Chunqin Gu and Qian Tao	145
18	**A Shortest Path Algorithm Suitable for Navigation Software**..... Peng Luo, Qizhi Qiu, Wenyan Zhou, and Pei Fang	153
19	**An Energy-Balanced Clustering Routing Algorithm for Wireless Sensor Networks**............................ Mingqiang Chen and Xianhai Tan	163
20	**Simulation and Analysis of Binary Frequency Shift Keying Noise Cancel Adaptive Filter Based on Least Mean Square Error Algorithm**............................ Zhongping Chen and Jinding Gao	171

21	Density-Sensitive Semi-supervised Affinity Propagation Clustering... Kunlun Li, Qi Meng, Shangzong Luo, Hexin Li, and Qian Wang	177
22	The Implementation of a Hybrid Particle Swarm Optimization Algorithm Based on Three-Level Parallel Model.............. Yi Xiao and Yu Liu	185
23	Optimization of Inverse Planning Based on an Improved Non-dominated Neighbor-Based Selection in Intensity Modulated Radiation Therapy........................... Xiao Zhang, Guoli Li, and Zhizhong Li	193
24	A Recommendation System for Paper Submission Based on Vertical Search Engine................................ Zhen Xu, Yi Yang, Fei Wang, Jiao Xu, Zhong Li, Fuqiang Mu, and Lian Li	201
25	Analysis and Improvement of SPRINT Algorithm Based on Hadoop.. Shanshan Fei, Qiaoyan Wen, and Zhengping Jin	209
26	Prediction Model for Trend of Web Sentiment Using Extension Neural Network and Nonparametric Auto-regression Method.... Haitao Zhang, Binjun Wang, and Guangxuan Chen	219
27	K-Optimal Chaos Ant Colony Algorithm and Its Application on Dynamic Route Guidance System...................... Hai Yang	227
28	A Certainty-Based Active Learning Framework of Meeting Speech Summarization.................................. Jian Zhang and Huaqiang Yuan	235
29	Application of Improved BP Neural Network in the Frequency Identification of Piano Tone.......................... Xu Chen and Jun Tang	243

Part II Data Processing

30	Implicit Factoring with Shared Middle Discrete Bits............ Meng Shi, Xianghui Liu, and Wenbao Han	255
31	Loading Data into HBase.................................... Juan Yang and Xiaopu Feng	265
32	Incomplete Decision-Theoretic Rough Set Model Based on Improved Complete Tolerance Relation.................. Xia Wang	273

33	A New Association Rule Mining Algorithm Based on Compression Matrix...............................	281
	Sihui Shu	
34	Decoupling Interrupts from the Internet in Markov Models.....	291
	Jinwen Ma, Jingchun Zhang, and Jinrong Guo	
35	Parallel Feature Selection Based on MapReduce...............	299
	Zhanquan Sun	
36	Initial State Modeling of Interlocking System Using Maude......	307
	Rui Ma, Zhongwei Xu, Zuxi Chen, and Shuqing Zhang	
37	Semi-supervised Learning Using Nonnegative Matrix Factorization and Harmonic Functions.....................	321
	Lin Li, Zhenyu Zhao, Chenping Hou, and Yi Wu	
38	Exploring Data Communication at System Level Through Reverse Engineering: A Case Study on USB Device Driver.......	329
	Leela Sedaghat, Brad Duerling, Xiaoxi Huang, and Ziying Tang	
39	Using Spatial Analysis to Identify Tuberculosis Transmission and Surveillance..	337
	Jinrong Bai, Guozhong Zou, Shiguang Mu, and Yu Ma	
40	Construction Method of Exception Control Flow Graph for Business Process Execution Language Process..............	345
	Caoqing Jiang, Shi Ying, Shanming Hu, and Hua Guan	
41	P300 Detection in Electroencephalographic Signals for Brain–Computer Interface Systems: A Neural Networks Approach....................................	355
	Seyed Aliakbar Mousavi, Muhammad Rafie Hj. Mohd. Arshad, Hasimah Hj. Mohamed, Putra Sumari, and Saeed Panahian Fard	
42	Web Content Extraction Technology.......................	365
	Zhenyu Jiao, Xiaoben Yan, Jinjin Sun, Yuchen Wang, and Jiangbin Chen	
43	A New Data-Intensive Parallel Processing Framework for Spatial Data...	375
	Dong Zhao, Yang Gu, and Zhenchun Huang	
44	The Approach of Graphical User Interface Testing Guided by Bayesian Model....................................	385
	Zhifang Yang, Zhongxing Yu, and Chenggang Bai	
45	A Model for Reverse Logistics with Collection Sites Based on Heuristic Algorithm.................................	395
	Xiaoqing Geng and Yu Wang	

46	The Storage of Wind Turbine Mass Data Based on MongoDB .... Qile Wang, Zhu Shen, Long Ma, and Shi Yin	403
47	Improvement of Extraction Method of Correlation Time Delay Based on Connected-Element Interferometry .... Fei Wang, Zhenfei Wang, Dun Li, and Bingjie Yang	411
48	Modeling and Evaluation of the Performance of Parallel/Distributed File System .... Tiezhu Zhao, Xin Ao, and Huaqiang Yuan	421
49	CoCell: A Low-Diameter, High-Performance Data Center Network Architecture .... Peng Wang, Huaxi Gu, Yan Zhao, and Xiaoshan Yu	429
50	Simulation Investigation of Counterwork Between Anti-radiation Missile and Active Decoy System .... Huaqiang Hu and Dandan Wen	437
51	Simulation Jamming Technique on Binary Phase-Coded Pulse Compression Radar .... Yulin Yang and Lijuan Qiu	445

Part III Pattern Recognition

52	Personalized Information Service Recommendation System Based on Clustering and Classification .... Yu Wang	455
53	Palmprint Recognition Based on Subclass Discriminant Analysis .... Pengfei Yu, Haiyan Li, Hao Zhou, and Dan Xu	465
54	A Process Quality Monitoring Approach of Automatic Aircraft Component Docking .... Guowei Yang, Chengjing Zhang, and Xiaofeng Zhang	473
55	Overhead Transmission Lines Sag Measurement Based on Image Processing .... Wengang Cheng and Long Chen	481
56	Chinese Domain Ontology Learning Based on Semantic Dependency and Formal Concept Analysis .... Lixin Hou, Shanhong Zheng, Haitao He, and Xinyi Peng	489
57	Text Classification Algorithm Based on Rough Set .... Zhiyong Hong	499
58	Robust Fragment-Based Tracking with Online Selection of Discriminative Features .... Yongqiang Huang and Long Zhao	507

59	Extraction Method of Gait Feature Based on Human Centroid Trajectory	515
	Xin Chen and Tianqi Yang	
60	An Algorithm for Bayesian Network Structure Learning Based on Simulated Annealing with Adaptive Selection Operator	525
	Ao Lin, Bing Xiao, and Yi Zhu	
61	Static Image Segmentation Using Polar Space Transformation Technique	533
	Xuan Luo, Tiancai Liang, and Weifeng Wang	
62	Image Restoration via Nonlocal P-Laplace Regularization	541
	Chen Yao, Lijuan Hong, and Yunfei Cheng	
63	Analysis and Application of Computer Technology on Architectural Space Lighting Visual Design	549
	Yiwen Cao	
64	Improving Online Gesture Recognition with WarpingLCSS by Multi-Sensor Fusion	559
	Chao Chen and Haibin Shen	
65	The Lane Mark Identifying and Tracking in Intense Illumination	567
	Yanyun Xing, Bo Yu, and Fangqun Yang	
66	Classification Modeling of Multi-Featured Remote Sensing Images Based on Sparse Representation	577
	Xiaoting Hao, Chunmei Zhang, Jing Bai, Mo Dai, Wenxing Bao, and Wei Feng	
67	A Parallel and Convergent Support Vector Machine Based on MapReduce	585
	Yingying Ma, Liming Wang, and Longpu Li	
68	Vehicle Classification Based on Hierarchical Support Vector Machine	593
	Mengwan Jiang and Haoliang Li	
69	Image Splicing Detection Based on Machine Learning Algorithm	601
	Yan Xiao	
70	A Lane Detection Algorithm Based on Hyperbola Model	609
	Chaobo Chen, Bofeng Zhang, and Song Gao	
71	Comparisons and Analyses of Image Softproofing Under Different Profile Rendering Intents	617
	Qingxue Yu, Yunhui Luo, Maohai Lin, and Quantao Liu	

72	**An Improved Dense Matching Algorithm for Face Based on Region Growing** .... Xin Xia and Shaoyan Gai	625
73	**An Improved Feature Selection Method for Chinese Short Texts Clustering Based on HowNet** .... Xin Chen, Yuqing Zhang, Long Cao, and Donghui Li	635
74	**Internet Worm Detection and Classification Based on Support Vector Machine** .... Huihui Liang, Min Li, and Jiwen Chai	643
75	**Real-Time Fall Detection Based on Global Orientation and Human Shape** .... Shuangcheng Wang, Yepeng Guan, and Ruiyue Xu	653
76	**The Classification of Synthetic Aperture Radar Oil Spill Images Based on the Texture Features and Deep Belief Network** .... Xixi Huang and Xiaofeng Wang	661
77	**The Ground Objects Identification for Digital Remote Sensing Image Based on the BP Neural Network** .... Shengkui Cao, Guangchao Cao, Kelong Chen, Chengyong Wu, Tao Zhang, and Jie Yuan	671
78	**Detection of Image Forgery Based on Improved PCA-SIFT** ...... Kunlun Li, Hexin Li, Bo Yang, Qi Meng, and Shangzong Luo	679
79	**A Thinning Model for Handwriting-Like Image Skeleton** .... Shijiao Zhu, Jun Yang, and Xue-fang Zhu	687
80	**Discrimination of the White Wine Based on Sparse Principal Component Analysis and Support Vector Machine** .... Rong Wang, Wu Zeng, and Jiao Ming	695

Volume 2

Part IV Cloud Computing

81	**Design of Mobile Electronic Payment System** .... Ting Huang	705
82	**Power Saving-Based Radio Resource Scheduling in Long-Term Evolution Advanced Network** .... Yen-Yin Chu, I-Hsuan Peng, Yen-Wen Chen, Chi-Fu Yi, and Addison Y.S. Su	713

83 Dispatching and Management Model Based on Safe
    Performance Interface for Improving Cloud Efficiency 723
    Bin Chen, Zhijian Wang, and Yu Wang

84 A Proposed Methodology for an E-Health Monitoring
    System Based on a Fault-Tolerant Smart Mobile 731
    Ahmed Alahmadi and Ben Soh

85 Design and Application of Indoor Geographical
    Information System 739
    Yongfeng Suo, Tianhe Chi, and Tianyue Liu

86 Constructing Cloud Computing Infrastructure Platform
    of the Digital Library Base on Virtualization Technology 747
    Tingbo Fu, Jinsheng Yang, Yu Gao, and Guang Yu

87 A New Single Sign-on Solution in Cloud 755
    Guangxuan Chen, Yanhui Du, Panke Qin, Lei Zhang, and Jin Du

88 A Collaborative Load Control Scheme for Hierarchical
    Mobile IPv6 Network 763
    Yi Yang, QingShan Man, and PingLiang Rui

89 A High Efficient Selective Content Encryption Method
    Suitable for Satellite Communication System 775
    Yanyan Xu, Bo Yang, Zhengquan Xu, and Tengyue Mao

90 Network Design of a Low-Power Parking Guidance System 783
    Ming Xia, Yabo Dong, Qingzhang Chen, Kai Wang,
    and Rongjie Wu

91 Strategy of Domain and Cross-Domain Access Control
    Based on Trust in Cloud Computing Environment 791
    Bo Li, Ming Tian, Yongsheng Zhang, and Shenjuan Lv

92 Detecting Unhealthy Cloud System Status 799
    Zhidong Chen, Buyang Cao, and Yuanyuan Liu

93 Scoring System of Simulation Training Platform
    Based on Expert System 809
    Wei Nie, Ying Wu, and Dabin Hu

94 Analysis of Distributed File Systems on Virtualized Cloud
    Computing Environment 817
    Tiezhu Zhao, Zusheng Zhang, and Huaqiang Yuan

95 A Decision Support System with Dynamic Probability
    Adjustment for Fault Diagnosis in Critical Systems 825
    Qiang Chen and Yun Xue

96	Design and Implementation of an SD Interface to Multiple-Target Interface Bridge............................ Guoyong Li, Leibo Liu, Shouyi Yin, Dajiang Liu, and Shaojun Wei	835
97	Cloud Storage Management Technology for Small File Based on Two-Dimensional Packing Algorithm.............. Zhiyun Zheng, Shaofeng Zhao, Xingjin Zhang, Zhenfei Wang, and Liping Lu	847
98	Advertising Media Selection and Delivery Decision-Making Using Influence Diagram................................. Xiaoxuan Hu and Fan Jiang	855
99	The Application of Trusted Computing Technology in the Cloud Security...................................... Bo Li, Shenjuan Lv, Yongsheng Zhang, and Ming Tian	865
100	The Application Level of E-commerce in Enterprises in China... Yinghan H. Tang	873
101	Toward a Trinity Model of Digital Education Resources Construction and Management............................ Yong Huang and Qingchun Hu	883
102	Geographic Information System in the Cloud Computing Environment... Yichun Peng and Yunpeng Wang	893

Part V Embedded Systems

103	Memory Controller Design Based on Quadruple Modular Redundant Architecture................................. Yuanyuan Cui, Wei Li, and Xunying Zhang	905
104	Computer Power Management System Based on the Face Detection...................................... Li Xie, Yong He, Yanfang Tian, and Tinghong Yang	913
105	Twist Rotation Deformation of Titanium Sheet Metal in Laser Curve Bending Based on Finite Element Analysis..... Peng Zhang, Qian Su, and Dong Luan	921
106	Voltage Transient Stability Analysis by Changing the Control Modes of the Wind Generator............................ Yu Shao, Feng Shi, and Xiang Li	929

107	The Generator Stator Fault Analysis Based on the Multi-loop Theory 939
	Yu Shao, Feng Shi, and Xiang Li
108	An Improved Edge Flag Algorithm Suitable for Hardware Implementation 947
	Lixiang Wang and Tiejun Xiao
109	A Handheld Controller with Embedded Real-Time Video Transmission Based on TCP/IP Protocol 955
	Mingjie Dong, Wusheng Chou, and Yihan Liu
110	Evaluating the Energy Consumption of InfiniBand Switch Based on Time Series 963
	Huifeng Wang, Zhanhuai Li, Xiaonan Zhao, Qinlu He, and Jian Sun
111	Real-Time Filtering Method Based on Neuron Filtering Mechanism and Its Application on Robot Speed Signals 971
	Wa Gao, Fusheng Zha, Baoyu Song, Mantian Li, Pengfei Wang, Zhenyu Jiang, and Wei Guo
112	Multiple-View Spectral Embedded Clustering Using a Co-training Approach 979
	Hong Tao, Chenping Hou, and Dongyun Yi
113	Feedback Earliest Deadline First Exploiting Hardware Assisted Voltage Scaling 989
	Chuansheng Wu
114	Design and Realization of General Interface Based on Object Linking and Embedding for Process Control 997
	Jiguang Liu, Jianbing Wu, and Zhiguo He
115	A Stateful and Stateless IPv4/IPv6 Translator Based on Embedded System 1007
	Yanlin Yin and Dalin Jiang
116	A Novel Collaborative Filtering Approach by Using Tags and Field Authorities 1017
	Zhi Xue, Yaoxue Zhang, Yuezhi Zhou, and Wei Hu
117	Characteristics of Impedance for Plasma Antenna 1027
	Bo Yin and Feng Yang
118	A Low-Voltage 5.8-GHz Complementary Metal Oxide Semiconductor Transceiver Front-End Chip Design for Dedicated Short-Range Communication Application 1035
	Jhin-Fang Huang, Jiun-Yu Wen, and Yong-Jhen Jiangn

119	A 5.8-GHz Frequency Synthesizer with Dynamic Current-Matching Charge Pump Linearization Technique and an Average Varactor Circuit .... Jhin-Fang Huang, Jia-Lun Yang, and Kuo-Lung Chen	1045
120	Full-Wave Design of Wireless Charging System for Electronic Vehicle .... Yongxiang Liu, Yi Ren, and Yi Wang	1055
121	A Hierarchical Local-Interconnection Structure for Reconfigurable Processing Unit .... Yujia Zou, Leibo Liu, Shouyi Yin, Min Zhu, and Shaojun Wei	1063
122	High Impedance Fault Location in Distribution System Based on Nonlinear Frequency Analysis .... Jinqian Zhai, Di Su, Wenjian Li, Feng Li, and Guohong Zhang	1073
123	Early Fault Detection of Distribution Network Based on High-Frequency Component of Residual Current .... Jinqian Zhai, Di Su, Wenjian Li, Feng Li, and Guohong Zhang	1083
124	A Complementary Metal Oxide Semiconductor D/A Converter with R-2R Ladder Based on T-Type Weighted Current Network .... Junshen Jiao	1091
125	Detecting Repackaged Android Applications .... Zhongyuan Qin, Zhongyun Yang, Yuxing Di, Qunfang Zhang, Xinshuai Zhang, and Zhiwei Zhang	1099
126	Design of Wireless Local Area Network Security Program Based on Near Field Communication Technology .... Pengfei Hu and Leizhen Wang	1109
127	A Mechanism of Transforming Architecture Analysis and Design Language into Modelica .... Shuguang Feng and Lichen Zhang	1117
128	Aspect-Oriented QoS Modeling of Cyber-Physical Systems by the Extension of Architecture Analysis and Design Language .... Lichen Zhang and Shuguang Feng	1125
129	Using RC4-BHF to Construct One-way Hash Chains .... Qian Yu and Chang N. Zhang	1133
130	Leakage Power Reduction of Instruction Cache Based on Tag Prediction and Drowsy Cache .... Wei Li and Jianqing Xiao	1143

Part VI Network Optimization

131 The Human Role Model of Cyber Counterwork 1155
Fang Zhou

**132 A Service Channel Assignment Scheme for IEEE 802.11p
Vehicular Ad Hoc Network** 1165
Yao Zhang, Licai Yang, Haiqing Liu, and Lei Wu

133 An Exception Handling Framework for Web Service 1173
Hua Guan, Shi Ying, and Caoqing Jiang

**134 Resource Congestion Based on SDH Network Static
Resource Allocation** 1181
Fuyong Liu, Jianghe Yao, Gang Wu, and Huanhuan Wu

**135 Multilayered Reinforcement Learning Approach for Radio
Resource Management** 1191
Kevin Collados, Juan-Luis Gorricho, Joan Serrat, and Hu Zheng

136 A Network Access Security Scheme for Virtual Machine 1201
Mingkun Xu, Wenyuan Dong, and Cheng Shuo

137 Light Protocols in Chain Network 1209
Ying Wang, Yifang Chen, and Lenan Wu

**138 Research and Implementation of a Peripheral Environment
Simulation Tool with Domain-Specific Languages** 1217
Maodi Zhang, Zili Wang, Ping Xu, and Yi Li

**139 Probability Model for Information Dissemination
on Complex Networks** 1225
Juan Li and Xueguang Zhou

140 Verification of UML Sequence Diagrams in Coq 1233
Liang Dou, Lunjin Lu, Ying Zuo, and Zongyuan Yang

**141 Quantitative Verification of the Bounded
Retransmission Protocol** 1245
Xu Guo, Ming Xu, and Zongyuan Yang

**142 A Cluster-Based and Range-Free Multidimensional
Scaling-MAP Localization Scheme in WSN** 1253
Ke Xu, Yuhua Liu, Cui Xu, and Kaihua Xu

**143 A Resource Information Organization Method Based
on Node Encoding for Resource Discovering** 1263
Zhuang Miao, Qianqian Zhang, Songqing Wang, Yang Li,
Weiguang Xu, and Jiang Xiao

144	The Implementation of Electronic Product Code System Based on Internet of Things Applications for Trade Enterprises	1271
	Huiqun Zhao and Biao Shi	
145	The Characteristic and Verification of Length of Vertex-Degree Sequence in Scale-Free Network	1281
	Yanxia Liu, Wenjun Xiao, and Jianqing Xi	
146	A Preemptive Model for Asynchronous Persistent Carrier Sense Multiple Access	1289
	Lin Gao and Zhijun Wu	
147	Extended Petri Net-Based Advanced Persistent Threat Analysis Model	1297
	Wentao Zhao, Pengfei Wang, and Fan Zhang	
148	Energy-Efficient Routing Protocol Based on Probability of Wireless Sensor Network	1307
	Kaiguo Qian	
149	A Dynamic Routing Protocols Switching Scheme in Wireless Sensor Networks	1315
	Zusheng Zhang, Tiezhu Zhao, and Huaqiang Yuan	
150	Incipient Fault Diagnosis in the Distribution Network Based on S-Transform and Polarity of Magnitude Difference	1323
	Jinqian Zhai and Xin Chen	
151	Network Communication Forming Coalition S4n-Knowledge Model Case	1331
	Takashi Matsuhisa	
152	An Optimization Model of the Layout of Public Bike Rental Stations Based on B+R Mode	1341
	Liu He, Xuhong Li, and Dawei Chen	
153	Modeling of Train Control Systems Using Formal Techniques	1349
	Bingqing Xu and Lichen Zhang	
154	A Clock-Based Specification of Cyber-Physical Systems	1357
	Bingqing Xu and Lichen Zhang	
155	Polymorphic Worm Detection Using Position-Relation Signature	1365
	Huihui Liang, Jiwen Chai, and Yong Tang	
156	Application of the Wavelet-ANFIS Model	1373
	Rijun Zhang, Caishui Hou, Hui Lin, Meiyan Zhuo, Meixin Zhang, Zhongsheng Li, Liwu Sun, and Fengqin Lin	

157	**Visualization of Clustered Network Graphs Based on Constrained Optimization Partition Layout**............... 1381
	Fang Huang, Wenjie Xiao, and Hao Zhang
158	**An Ultra-Wideband Cooperative Communication Method Based on Transmitted Cooperative Reference**................ 1395
	Tiefeng Li, Ou Li, and Zewen Zhou

Author Index for Volume 1................................... 1407

Subject Index for Volume 1................................... 1411

Author Index for Volume 2................................... 1417

Subject Index for Volume 2................................... 1421

CENet 2013 Committee

Advisory Chairs
Aniruddha Bhattacharjya   Amrita University, India

Program Chairs
C. E. Tapie Rohm	California State University San Bernardino, USA
W. Eric Wong	University of Texas at Dallas, USA
Hong Jiang	Hohai University, China
Jin Wang	Nanjing University of Information Science & Technology, China
Tingshao Zhu	Chinese Academy of Sciences, China

Program Committee
Abdalhossein Rezai	ACECR (Academic Center for Education, Culture and Research) and Semnan University, Iran
Akram Rashid	Air University, Islamabad, Pakistan
Amit Joshi	Sardar Vallabhbhai National Institute of Technology, India
Chen Hong	Beijing University of Aeronautics and Astronautics, China
Fatimah De'nan	Universiti Sains Malaysia, Malaysia
Feilong Liu	Hunan Institute of Science and Technology, China
Feng Xu	Hohai University, China
Fengjun Shang	Chongqing University of Posts and Telecommunications, China
Gyanendra Prasad Joshi	Yeungnam University, Korea
Hsieh Tzung-Yu	MingDao Univesity, Taiwan, China
Jesús C. Hernandez	University of Jaen, Spain
Jiandong Sun	Zhejiang University, China
Qian Yu	University of Regina, Canada

RADJEF Mohammed Said	University of Bejaia, Algeria
Rui Chen	Xi'an University of Electronic Science and Technology, China
S. EL-Rabaie	Faculty of Electronic Engineering, 32952 Menouf, Egypt
Stephan Chalup	Heidelberg University, Germany
Sunil Kumar Khatri	Amity University, Uttar Pradesh, India
Tao Zhou	North China Electric Power University, China
Vong Chi-Man	University of Macau, Macau, China
Xianzhong Yi	Yangtze University, China
Yan Pei	Kyushu University, Japan
Yuan Gao	Tsinghua University, China
Yuduo Wang	Beijing Information Science and Technology University, China
Zhongtian Jia	University of Jinan, China

Part IV
Cloud Computing

Chapter 81
Design of Mobile Electronic Payment System

Ting Huang

Abstract Multi-bank mobile electronic payment system uses mobile terminals for electronic payments, which can circulate in multiple banks and cannot limit from the bank that issues the e-cash. The paper researches electronic payment on the withdrawal agreement, the pay agreement, the deposit agreement, the update protocol of the e-cash based on elliptic curve cryptography. The design of the system is more suitable for mobile payment terminals with limit of calculation capacity, storage, network bandwidth, and power supply, which meets the needs of the day-to-day transactions.

81.1 Introduction

Nowadays the mobile electronic payment is popular in the world that the user has done dynamic payment with his mobile terminal, such as Visa DPS. But a lot of e-cash cannot be in circulation in the number of financial institutions because the amount of e-cash issue is fixed and the transaction can only be done in the system [1]. Based on elliptic curve cryptosystem this paper proposes multi-bank mobile electronic payment system. The micro-payment transacts off line by the k-ary tree and the macro-payment transacts on line, which meet the needs of the actual transaction in the e-cash circulation of the banks. Divisibility of the mobile e-cash [2] is realized by the K-tree ($K > 1$). The solutions of multi-bank electronic payment [3] are that multiple banks issue the e-cash.

T. Huang (✉)
College of Computer and Information Technology, China Three Gorges University,
Yichang, Hubei 443002, China
e-mail: 14863403@qq.com

81.2 Design of a Project

The symbols of the whole text involve: x,y are the point of abscissa and vertical axis on the elliptic curve, respectively; S_U, S_C are the user and payee's private key; P_U, P_C are the user and payee's public key. Each bank is classified by the number, such as ICBC is the bank$_1$, etc.; the other bank is the bank$_n$, of which the corresponding private key and the signature private key are defined as S_{Bn}, S_{1n}, S_{2n} and the corresponding public key is P_{Bn}, P_{1n}, P_{2n}, ($n \in (1,\infty)$). The model of the mobile electronic payment system is shown in Fig. 81.1.

The use cycle of the e-cash is defined as T (T is the time cycle T_1 to T_3). The time T_1 generates when the user fetches the e-cash from the bank. The time cycle T_1 to T_2 is an effective use cycle of the e-cash T' (T' is the given value). When the time T_2 reaches, the e-cash is unavailable. Only before the time T_3 (The time cycle T_2 to T_3 is the updating cycle of the e-cash T" that is the given value) the user goes to the bank for updating the e-cash.

81.2.1 Design of Off-Line Payment(Micro-Payment)

1. Withdrawal agreement

    Step 1: The user fetches M yuan from the bank$_n$ and gives the segmentation parameters K of the e-cash. Then he generates $A \in {}_R Z_n$ and $\rho = H(A + S_U)$. At the same time, he gets the K-ary tree by M yuan, saves $\alpha_{(n)}$, ρ, K, T_1, B to the database, and sends to the bank$_n$ after encrypting ID_{nU}, M, $\alpha_{(n)}$, $\beta_{(n)}$, $\gamma_{(n)}$, T_1 by the shared key.

    Step 2: The bank$_n$ gets the root node of the divisible e-cash: $\alpha_{(n)} = H[(MS_{1n} + \rho S_{2n}) \| S_{Bn} \| B \| T_1]$, and calculates $\beta_{(n)} = H[\alpha_{(n)} \| (CG)_x \| M \| T_1]$, $\gamma_{(n)} = \beta_{(n)} S_{Bn} + C$, B, $C \in {}_R Z_n$.

    Then it saves $\alpha_{(n)}$, ρ, K, T_1, B to the database, encrypts ID_{nU}, M, $\alpha_{(n)}$, $\beta_{(n)}$, $\gamma_{(n)}$, T_1, and distributes to the user.

    Step 3: The user decrypts ID_{nU}, M, $\alpha_{(n)}$, $\beta_{(n)}$, $\gamma_{(n)}$, T_1 and stores.

2. Payment agreement
    Payments in two ways are described in Tables 81.1 and 81.2.

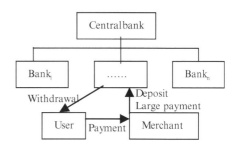

Fig. 81.1 The model of the mobile electronic payment system

81 Research and Design of Mobile Electronic Payment System

Table 81.1 Payment agreement (M′ = M yuan)

User	Payee
M, $\alpha_{(n)}$, $\beta_{(n)}$, $\gamma_{(n)}$, T_1 is sent	The payee tests $\beta_{(n)} \stackrel{?}{=} H\left[\alpha_{(n)} \| (\gamma_{(n)}G\text{-}\beta_{(n)}P_{Bn})_x \| M \| T_1\right]$, saves M, $\alpha_{(n)}$, $\beta_{(n)}$, $\gamma_{(n)}$, computers $\eta = H[(S_C P_U)_x]$, and sends η, the user's name, M, $\alpha_{(n)}$
The user tests $\eta \stackrel{?}{=} H\left[(S_U P_C)_x\right]$ and deletes entire e-cash records of $\alpha_{(n)}$, M	

Table 81.2 Payment agreement (M′ < M yuan)

User	Payee
The K-tree is generated by M′: For the branches $\alpha_{(n)l1,\ldots lS}$, the user takes $B_1, \ldots, B_{k,\ldots,k}, C_1, \ldots, C_{k,\ldots,k} \in_R Z_n$, $p = \{\rho, l1\}$, $\alpha_{(n)l1} = H(B_{l1} \| S_U \| \alpha_{(n)} \| T_1)$, $\beta_{(n)l1} = H[\alpha_{(n)l1} \|(C_{l1}G)_x \| \rho \| p[l1] \| K \| M' \| T_1]$, $\gamma_{(n)l1} = \beta_{(n)l1} S_U + C_{l1}$, $\ldots$, $\alpha_{(n)l1,\ldots,lS} = H(B_{l1\ldots,lS} \| S_U \| \alpha_{(n)l1,\ldots,l(S-1)} \| T_1)$, $p = \{\rho, l1, \ldots lS\}$, $\beta_{(n)l1,\ldots,lS} = H[\alpha_{(n)l1,\ldots,lS} \|(C_{l1,\ldots,lS}G)_x \| \rho \| p[l1] \| \ldots \| p[lS] \| K \| M' \| T_1]$, $\gamma_{(n)l1,\ldots,lS} = \beta_{(n)l1,\ldots,lS} S_U + C_{l1,\ldots,lS}$, saves $B_1, \ldots, B_{k,\ldots,k}, C_1, \ldots C_{k,\ldots,k}$ to the database. M′, p, $\alpha_{(n)l1,\ldots lS}$, $\beta_{(n)l1,\ldots,lS}$, $\gamma_{(n)l1,\ldots,lS}$, $\alpha_{(n)}$, T_1 are sent. The branches that are not used can be paid until all the K-tree branches are run out.	
	The payee tests $\beta_{(n)l1\ldots lS} \stackrel{?}{=} H\left[\alpha_{(n)l1\ldots lS} \|(\gamma_{(n)l1\ldots lS}G\text{-}\beta_{(n)l1\ldots lS}P_U)_x \| \rho \| p[l1] \| \ldots \| p[lS] \| K \| M' \| T_1\right]$, saves M′, p, $\alpha_{(n)l1,\ldots lS}$, $\beta_{(n)l1,\ldots,lS}$, $\gamma_{(n)l1,\ldots,lS}$, K, computers $\eta = H[(S_C P_U)_x]$, and sends η, the user's name, M′, $\alpha_{(n)l1,\ldots lS}$
After the user tests $\eta \stackrel{?}{=} H\left[(S_U P_C)_x\right]$, the amount of e-cash will subtract M′ from the database and $\alpha_{(n)l1,\ldots lS}$ will be deleted.	

3. Deposit agreement

After the businesses send the relevant transaction information stored in the payment protocol to the bank, the bank verifies the correctness of this information to detect double-spending. When the merchant deposits to the bank$_n$, the e-cash deposits to the bank$_n$; merchant deposits to the bank$_m$ (not the bank$_n$), the bank$_n$ sends depositing issued to the bank$_m$. After the e-cash in the bank$_m$ is dealt the merchant finishes the depositing process. The central bank that has higher level than ordinary bank verifies the authenticity of the transfer information on a regular time, detects the untrue transfers, and increases the penalty.

In this section the payment node $\alpha_{(n)l1,\ldots lS}$ of an e-cash sent by the business as an example: The e-cash is checked whether it is out of date in deposit database, and whether it has the same path p. If it has the same path, you can conclude the reuse of the e-cash. Then the bank can find the dishonest user's identity who takes e-cash ρ through the withdrawal database. Otherwise, the bank accepts the deposits.

4. Update of the e-cash

When the usage time T' of the e-cash reaches, the user must update through the following protocol, while the e-cash will set aside more than the time T_3.

Step 1: The user regenerates $A \in {}_R Z_n$, $\rho = H(A + S_U)$, regenerates the K-tree by the residual e-cash M, saves ρ, K to the database. After they are encrypted by the bank and the user's shared key, ρ, K are sent to the bank.

Step 2: The bank regenerates the root node of the divisible e-cash according to the user's update time T_1: $\alpha_{(n)} = H[(MS_{1n} + \rho S_{2n}) \| S_{Bn} \| B \| T_1]$ and computes B, C $\in {}_R Z_n$, $\gamma_{(n)} = \beta_{(n)} S_{Bn} + C$, $\beta_{(n)} = H[\alpha_{(n)} \| (CG)_x \| M \| T_1]$, saves $\alpha_{(n)}$, ρ, K, T_1, B to the database. After they are encrypted, ID_{nU}, M, $\alpha_{(n)}$, $\beta_{(n)}$, $\gamma_{(n)}$, T_1 are sent to the user.

After it has been updated, the e-cash can be transacted by the previous protocol.

81.2.2 Design of On-Line Payment (Large Payment)

1. Withdrawal agreement

Step 1: The user withdraws M yuan from the bank$_n$. After he encrypts M with the shared key of the user and the bank$_n$, the user sends it to the bank$_n$.

Step 2: The bank$_n$ gets the root node of the divisible e-cash: $\alpha_{(n)} = H[(MS_{1n} + S_{2n}) \| S_{Bn} \| B \| T_1]$, calculates $\beta_{(n)} = H[\alpha_{(n)} \| (CG)_x \| M \| T_1]$, $\gamma_{(n)} = \beta_{(n)} S_{Bn} + C$, B, C $\in {}_R Z_n$, and saves $\alpha_{(n)}$, T_1, B to the database. After he encrypts ID_{nU}, M, $\alpha_{(n)}$, $\beta_{(n)}$, $\gamma_{(n)}$, T_1, the bank$_n$ will distribute them to the user.

Step 3: After he decrypts ID_{nU}, M, $\alpha_{(n)}$, $\beta_{(n)}$, $\gamma_{(n)}$, T_1, the user will save them.

2. Payment agreement

   Step 1: The user pays the e-cash to the payee, M, $\alpha_{(n)}$, $\beta_{(n)}$, $\gamma_{(n)}$, T_1 are sent to the payee.
   Step 2: The payee inspects the authenticity of the e-cash: The payee tests $\beta_{(n)} \stackrel{?}{=} H\left[\alpha_{(n)} || (\gamma_{(n)}G - \beta_{(n)}P_{Bn})_x || M || T_1\right]$, and sends the user name, $\alpha_{(n)}$, M, T_1 to the bank.
   Step 3: The bank checks the authenticity of $\alpha_{(n)}$. After it gets the validation of $\alpha_{(n)}$, the bank will inform the payee.
   Step 4: The payee saves M, $\alpha_{(n)}$, $\beta_{(n)}$, $\gamma_{(n)}$, calculates $\eta = H[(S_C P_U)_x]$, sends η, the user name, $\alpha_{(n)}$, M to the user.
   Step 5: The user confirms that the payment is successful, tests $\eta \stackrel{?}{=} H\left[(S_U P_C)_x\right]$, deletes the entire e-cash records of $\alpha_{(n)}$, M.

3. Deposit agreement
   The course is the same as the deposit agreement in off-line payment. For example, the business sends the e-cash $\alpha_{(n)}$, which issame as the payment node $\alpha_{(n)11,...}$ is of an e-cash in off-line payment.

4. Update of e-cash
   The e-cash will set aside more than the time T_3. The user must update the e-cash through the following protocol.

   Step 1: After he encrypts M with the shared key of the user and the bank, the user sends it to the bank.
   Step 2: The bank regenerates the divisible e-cash according to the time T_1 of the user updating the e-cash $\alpha_{(n)} = H[(MS_{1n} + S_{2n}) || S_{Bn} || B || T_1]$, calculates B, $C \in_R Z_n$, $\beta_{(n)} = H[\alpha_{(n)} || (CG)_x || M || T_1]$, $\gamma_{(n)} = \beta_{(n)} S_{Bn} + C$, saves $\alpha_{(n)}$, T_1, B to the database. After he encrypts ID_{nU}, M, $\alpha_{(n)}$, $\beta_{(n)}$, $\gamma_{(n)}$, T_1, the bank will distribute them to the user.

   After it has been updated, the e-cash can be the same with the previous transaction protocol.

81.3 Security and Efficiency Analysis

1. Blind signature: $\alpha_{(n)}$, $\beta_{(n)}$, $\gamma_{(n)}$ are the right signature of the e-cash M based on blind signature protocol.
   Proof: If the validation of the e-cash $\beta_{(n)} \stackrel{?}{=} H\left[\alpha_{(n)} || (\gamma_{(n)}G - \beta_{(n)}P_{Bn})_x || M || T_1\right]$ is established, that proofs $\gamma_{(n)}G - \beta_{(n)}P_{Bn} = CG$. $\gamma_{(n)} = \beta_{(n)}S_{Bn} + C$, so $\gamma_{(n)}G - \beta_{(n)}P_{Bn} = (\beta_{(n)}S_{Bn} + C)G - \beta_{(n)}P_{Bn} = CG$. Similarly we can prove that $\alpha_{(n)11}, \ldots, \alpha_{(n)l1,\ldots,1S}$, $\beta_{(n)l1}, \ldots, \beta_{(n)l1,\ldots,1S}$, $\gamma_{(n)l1}, \ldots, \gamma_{(n)l1,\ldots,1S}$ are correct signature of the e-cash M'.

2. The signatures and the e-cash are unforgeable

    The signatures and the e-cash embed the private key S. Any attacker must get the private key S so that he can forge the signatures and the e-cash, which must solve the ECDLP problem. This problem remains unsolvable, so the signatures and the e-cash are unforgeable.
3. When the e-cash is spent repeatedly, the bank will reveal the user's anonymity

    When he consumes the e-cash, the user will send the e-cash and p to the merchant. If the user spends repeatedly, the bank will find the reuse of the e-cash or p when the business deposits. The bank inquires about the withdrawal database, thus it can identify the user's identity.

    So it is not possible to forge a correct e-cash. When the e-cash is spent repeatedly, merchant can detect the user's identity through the bank. Therefore the protocol is safe and fair (The bank must be safe and reliable). If the user trades normally, the user's anonymity is guaranteed. The bank will reveal the user's identity only the illegal transaction. Thus the anonymity of the user will be ensured, the interests of the banks and the merchants will be ensured too.

    The time of protocol implementation and storage consumption in multi-bank mobile electronic payment system are the key to efficiency. 160 bit length of the key in ECC has the same powerful functions as 1,024 bits length of the key in RSA [4]. The divisible e-cash protocol [5] bases on the zero-knowledge proof, strong RSA problem. The achieve agreement of this system bases on ECC. Compared to RSA, ECC need not exponentiation compute. The computation of ECC can be negligible. Therefore the efficiency of this protocol runs faster. Paying a divisible e-cash [5] needs to save (S, π), $S = g^{\frac{1}{s+j+1}}$, π = (R, info, pk_M, time, T, C_r, C_{pk_u}, C_S, C_t, C_d, C_j, Φ). The storage space needs at least $1024 + 128 + 1024*7 + 1024*5 = 13440$ bit $= 1680$ byte. While paying a divisible e-cash in this system simply saves M', $α_{(n)l1,l2,l3}$, $β_{(n)l1,l2,l3}$, $γ_{(n)l1,l2,l3}$, $p = \{ρ,l1,l2,l3\}$, K (Assuming the K-tree assigns to the third branch). The storage space is approximately $32 + 128 + 128 + 192 + (128 + 32*3) + 32$ bit $= 736$ bit $= 92$ byte, which reduces the 94.5 % of the storage amount and the network bandwidth. The 2-tree is used [5, 6]. Because the divided branches in splitting the e-cash are too much, the amount of computation in generating the e-cash is increased.

    The e-cash in this system contains the use cycle. When the end of the use cycle is coming, the bank can delete the e-cash. Efficient divisible e-cash scheme [7] needs storage (S, T, $\prod$ s, I, ξ), $S_{i,0}, \ldots, S_{i,2^l-1}$. Thus a lot of storage space is saved in this paper. The efficiency of the bank retrieving the database and the deposit agreement are improved. The reliability and maintainability of the data are improved too. When the end of the use cycle of the e-cash is coming, the user can update the e-cash in the bank through implementing update protocol of the e-cash, which guarantees the legitimate rights and interests of the user.

    The implementation of the multi-bank e-cash [8] requires the central bank to issue the e-cash. A financial institution is added, which makes the protocol more complicatedly.

81.4 Conclusion

To sum up, the protocol of multi-bank mobile electronic payment system is safe, simple, efficient, and suitable for the mobile payment terminals of which calculation capacity, storage, network bandwidth, and power supply are very limited.

References

1. Ziba Eslami, Mehdi Talebi. (2011). A new untraceable off-line electronic cash system. *Electronic Commerce Research and Applications, 10(1)*, 59–66.
2. Ting Huang, Shou-zhi Xu. (2010). Study on mobile divisible e-cash based on elliptic curve. *Journal of Wuhan University of Technology, 32(23)*, 150–153 (In Chinese).
3. Ting Huang. (2012). Study on multi-bank mobile electronic payment. *Theoretical and Mathematical Foundations of Computer Science* (Vol. 38, pp. 507–511). USA: Information Engineering Research Institute.
4. Menezes, A., Okamoto, T., & Vanstone, S. (1993). Reducing elliptic curve logarithms to logarithms in a finite field. *IEEE Transactions on Information Theory, 39(5)*, 1639–1646.
5. Xinyu He. (2010). *Research on e-cash protocol based on multi-level proxy blind signature and each node can be paid.* Master dissertation for Yanshan University, Qin Huang Dao (In Chinese).
6. Jiuhong Wang. (2010). *The research on efficient divisible e-cash based on ECC.* Master dissertation for Yanshan University, Qin Huang Dao (In Chinese).
7. Yong-bo Yu, Xiao-zhu Jia, Feng Qing-feng. (2010). Efficient divisible e-cash scheme based on one-way accumulator. *Computer Engineering and Applications, 46(10)*, 206–208 (In Chinese).
8. Xiangwen Meng, Baohua Zhao. (2011). Fairness-based multi-bank e-cash. *Computer Applications and Software.28, 163(10)*, 195–197 (In Chinese).

Chapter 82
Power Saving-Based Radio Resource Scheduling in Long-Term Evolution Advanced Network

Yen-Yin Chu, I-Hsuan Peng, Yen-Wen Chen, Chi-Fu Yi, and Addison Y.S. Su

Abstract It is well known that power saving is one of the most important issues for mobile device in accessing network services. The efficient conservation of energy for longer operation times of a mobile station is vital to the success of various mobile applications. This paper proposes a systematic approach to allocate radio resources in Long-Term Evolution Advanced (LTE-A) network by considering the channel condition and QoS requirements, while the power saving is a centric issue during the scheduling process. The proposed scheme includes the selection of component carriers (CC) and the allocation of radio resource to satisfy the QoS demands while minimize the power consumption of user equipment (UE). Additionally, exhaustive simulations were performed to examine the performance of the proposed scheme. Both http and video streaming traffic models were applied during the simulations. The experimental results show that the proposed scheme achieves better performance when compared to the other scheme.

82.1 Introduction

The mobile multimedia applications, such as video streaming and video telephony, enables the needs of broadband access in wireless mobile networks. Although Long-Term Evolution (LTE) network provides higher bandwidth than that of current 3G network, its channel bandwidth is limited to 20 MHz [1] without flexibility.

Y.-Y. Chu • Y.-W. Chen (✉) • C.-F. Yi
Department of Communication Engineering, National Central University, Taoyuan, Taiwan
e-mail: ywchen@ce.ncu.edu.tw

I.-H. Peng
Department of Computer Science and Information Engineering, Minghsin University of Science and Technology, Hsinchu, Taiwan

A.Y.S. Su
Research Center for Advanced Science and Technology, National Central University, Taoyuan, Taiwan

The emerging LTE-Advanced (LTE-A) provides four additional technologies, which are carrier aggregation (CA), advance multi-input multi-output (MIMO), coordinated multiple points (CoMP) transmission and reception, and relaying, to improve transmission performance. Among them, carrier aggregation and advance MIMO are designed to overcome the issue of limited channel bandwidth [2]. The user equipment (UE) with LTE-A capability could aggregate more than one component carrier (CC) for more channel bandwidth. However, the LTE-A capable UE requires more antennas to receive the radio signal from different CC, which results in more power consumption. For the mobile devices, such as UE of LTE, it is critical to save power so as to lengthen the usage after charge.

The Orthogonal Frequency Division Multiple Access (OFDMA) technology is used for downlink transmission in LTE/LTE-A network. The eNode B (eNB) is responsible for the allocation of radio resource. The radio resource allocation is considerably more correlated and critical to the diversity channel conditions among UEs and the effective usage of the adaptive modulation and coding (AMC) scheme for transmission in OFDMA systems. A higher level of modulation and coding scheme (MCS) achieves higher spectrum efficiency and, therefore, higher system throughput if channel condition is acceptable. As the channel conditions of UEs may be different and changeable from time to time, if it is not well allocated, the service quality will be downgraded due to insufficient bandwidth. In LTE-A network, eNB can allocate either resource blocks (RB) of the same CC or the RBs that belong to different CC to LTE-A capable UE to satisfy its required bandwidth. However, the LTE UE can only be allocated with the RBs of the same CC. Then, for a specific LTE-A UE, if its channel conditions of all RBs are the same then it will save more power to be allocated with RBs of the same CC. Furthermore eNB may allocate RBs of the same CC to an UE even this allocation will sacrifice system throughput from the power saving point of view. Thus it is a tradeoff when the issues of system throughput and power saving are encountered. And the satisfaction of quality of services (QoS) is the compromise to balance these two issues. The objective of this paper is to propose a systematic scheme to properly allocate radio resource for LTE UE and LTE-A UE coexistence environment with different QoS requirement and under changeable channel condition.

The rest of this paper is organized as follows. Section 82.2 overviews some background and related works. The proposed radio resource allocation scheme is described in Sect. 82.3. Section 82.4 provides the comparison of simulation results with discussion. We conclude our study in the last section.

82.2 Related Works

The use of CA in LTE-A provides UE to aggregate the radio signals from different component carriers. And the component carriers can come from different spectrum bands. Thus UE can aggregate carriers from either the contiguous CCs or the

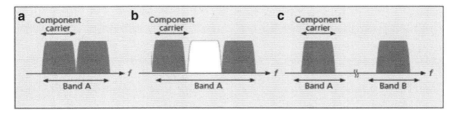

Fig. 82.1 Types of carrier aggregation

noncontiguous CCs of the same spectrum band. UE can also aggregate carriers from the CCs that belong to different spectrum bands as shown in Fig. 82.1 [3].

Each UE can use activation/deactivation mechanism to manage the CA function. And eNB can inform UEs about the allocation information, including CC and RB, through the radio resource control (RRC) message. UE can turn off the antenna without receiving the signals of physical uplink shared channel (PUSCH) and physical downlink control channel (PDCCH) that is temporally unused. As the carrier aggregation provides the flexibility in allocating radio resource, several schemes were proposed to study this issue [4–8]. The LTE-A UE and LTE UE are different in using CC, therefore, it is better to use different policies for both kinds of UE [4]. The authors also designed the measurement functions as the reference to be adopted in both independent packet scheduling per CC and cross-CC packet scheduling [4]. In addition to introducing the random selection and round robin (RR) selection for CC, the reference signal received power (RSRP) value was defined for the use of CC selection [5]. And the authors also presented the proportional fair (PF) and modified largest weighted delay first (M-LWDF) scheduling algorithms for fair resource allocation among UEs. The separated burst-level scheduling (SBLS) was proposed to improve the separated random user scheduling (SRUS), which may introduce unbalance load of CC [6]. The possible structures of LTE-A UE were proposed in with detail analysis of their power consumption [7]. The discontinuous reception (DRX) mechanism provides sleep mode for UE to save power. The authors proposed that LTE-A UE can arrange separated DRX parameters for each CC by considering the QoS requirement [8]. The authors gave their observations on the relationship between power saving efficiency and packet loss ratio by adjusting the inactivity timer and on duration in LTE-A environment. Thus the resource allocation can be designed from different view point such as QoS, fairness, throughput, power saving, etc. In this paper, we propose the radio resource scheduling scheme by trading off several criteria. Although the power saving is the centric in the proposed scheme, QoS satisfaction of each UE and the resource utilization shall also properly considered.

82.3 The Proposed Power Saving Centric Radio Resource Scheduling

The conceptual architecture of LTE UE and LTE-A UE coexistence environment is illustrated in Fig. 82.2. It shows that LTE-A UE can access more than CC, however, the LTE UE can only access one CC. The subcarriers of each CC can be grouped into physical resource block (PRB) and each CC consists of several PRBs. It is noted that LTE/LTE-A users have different channel condition, represented as signal to interference plus noise ratio (SINR), on each PRB. The objective of radio resource allocation is to properly arrange PRB(s) to UEs in accordance with their bandwidth requirement and channel condition. Each UE can either periodically or aperiodically report its channel condition, i.e. channel quality indication (CQI), to eNB through the Physical Uplink Control Channel (PUCCH) and PUSCH. The channel condition can be reported by either wideband CQI or subband CQI. The wideband CQI is the average channel condition over the entire channel and cannot precisely describe the quality difference among channels. On the contrarily, channels are grouped in to resource block groups (RBG) in subband CQI and each RBG reports its CQI.

The proposed scheme can be subdivided into two parts. The part 1 deals with the selection of CC, and part 2 allocates radio resource to UE as shown in Fig. 82.3. In practical application, UE may have different applications. In this paper, we categorize the UEs into two types of services, which are guaranteed bit rate (GBR) and non-GBR. As the QoS requirements of GBR UE and non-GBR UE

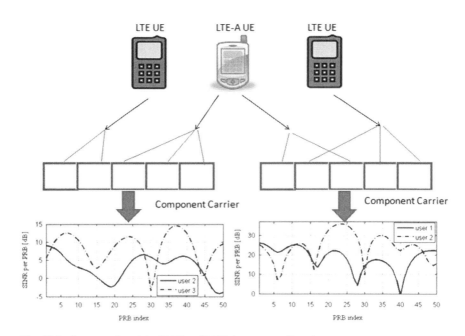

Fig. 82.2 Conceptual model of LTE and LTE-A resource allocation

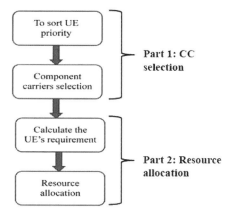

Fig. 82.3 The proposed radio resource allocation framework

are different, two weighting functions are designed to differentiate their scheduling priority. Generally, the delay budget is the major concern of GBR traffic, while the packet loss is critical for the non-GBR traffic. And in order to avoid wasting network resource, the GBR packet tends to be dropped if its queuing time exceeds the delay budget. The weighting functions of GBR UE and non-GBR UE are illustrated in (82.1) and (82.2), respectively. It is noted that, although the weighting functions of GBR and non-GBR are obtained in different way, their weightings are compared to each other to decide the priority.

$$W_{G(i)} = \left(\frac{\gamma^* M_{ave(i)}}{M_{max}}\right)\left(\frac{d_i}{D_i}\right) \quad (82.1)$$

$$W_{N(i)} = \left(\frac{M_{ave(i)}}{M_{max}}\right)\left(\frac{q_i}{Q_i}\right) \quad (82.2)$$

The $M_{ave(i)}$ is the average MCS level of the UE i received for all un-allocated resource block groups (RBG) and M_{max} is the maximum MCS level. In (82.1), d_i and D_i denote the longest waiting time of the packets that are in the queue and the maximum delay budget of UE i, respectively. The values of q_i and Q_i indicate current queuing length and the queue size of UE i, respectively. The parameter γ is applied to adjust the balance of both weightings because, as the scales of delay time and queue length may be incompatible, it may introduce bias between the ratios of the second terms in both equations. The UE with larger weighting value has higher priority to choose CC. Each UE calculates its preferences of all CCs and chooses the CC with the highest preferences for radio resource allocations. The preference of UE i on CC n is calculated by the summation of the available bandwidth of all k RBGs of CC n by referring to the MCS level of UE i as (82.3).

$$P_{i,n} = \sum_{j=1}^{k} (R_{n,j}|m_{i,n,j}) \quad (82.3)$$

```
Algorithm: Power-saving Resource Allocation
1.   Select a high-priority UE according to the priority queue
2.   IF high-priority UE is LTE
3.     use high priority CC for UE on the system
4.     WHILE any RBG is not allocated or the UE doesn't get enough resource
5.       IF the bandwidth of MCS level < UE_Requirement && MCS level != 1 THEN
6.         Reduce the MCS level
7.       ELSE
8.         allocate resource to the high-priority UE
9.       END IF
10.    ENDWHILE
11.  ELSE IF high-priority UE is LTE-Advanced
12.    FOR priority CC on the system < LTE-A UE the number of antenna
13.      WHILE any RBG is not allocated or the UE doesn't get enough resource
14.        IF the bandwidth of MCS level < UE_Requirement && MCS level != 1 THEN
15.          Reduce the MCS level
16.        ELSE IF MCS level == 1 && use CC < LTE-A UE the number of antenna
17.          Change CC
18.        ELSE allocate resource to the high-priority UE
19.        END IF
20.      ENDWHILE
21.    END FOR
22.  END IF
```

Fig. 82.4 The proposed power saving-based resource allocation algorithm

where $m_{i,n,j}$ denotes the MCS level of UE i in RBG j of CC n and $(R_{n,j}|m_{i,n,j})$ represents the residual bandwidth of RBG j of CC n for UE i by referring to its MCS level. It is noted that each LTE UE can only choose one CC and the number of CC that the LTE-A UE can choose depends on the number of its antenna. The bandwidth to be allocated for each GBR UE follows its guaranteed bit rate. However, the bandwidth to be allocated for each non-GBR UE can be calculated in a statistic manner for fairness consideration. In this paper, we calculate the bandwidth to be allocated for the non-GBR UE i in each frame, B_i as the following equation

$$B_i = B_{ave(i)}^* \left[\beta + (1-\beta)^* \frac{q_i}{Q_{ave(i)}} \right] \quad (82.4)$$

where $B_{ave(i)}$ and $Q_{ave(i)}$ are the average allocated bandwidth and average queue length of UE i till now, respectively, and q_i is its current queue length. The parameter β is designed to adjust whether B_i tends to follow the previous allocated bandwidth $B_{ave(i)}$ or to reflect current queuing condition. For example, if β is equal to 1, then the UE will always be allocated with the average allocated bandwidth.

The proposed power saving-based radio resource allocation algorithm is shown in Fig. 82.4. It is known that UE shall adopt the same MCS level for all received RBGs. If eNB allocates two RBGs to a specific UE and the channel conditions of these two RGBs are different, then the RBG with better channel quality shall yield to the RBG with worse channel quality and downgrade its MCS level. As LTE-A

UE has more than one antenna, it can choose RBG with the same MCS level from different CC without downgrading its MCS level to maximize its spectrum efficiency. However, from power saving point of view, eNB can allocate RBGs of the same CC to LTE-A UE and the UE only needs to open one antenna. It may introduce the downgrade of MCS level and then sacrifice some system throughput. It is noted that the more RGBs UE gets does not mean the more acquired bandwidth because of the downgrade of MCS level when getting more RGBs. Therefore eNB needs to check whether it is worthwhile or not to allocate additional RBG.

82.4 Experimental Simulations

In order to examine the efficiency of the proposed scheme, exhaustive simulations were conducted to compare the performance of the proposed scheme with the compared scheme [6]. The video streaming service and http service were applied as GBR and non-GBR traffic, respectively [9]. The channel bandwidth was assumed to be 10 MHz and the ITU Veh-A channel model was adopted during the simulations. The number of CC, each with 17 RBGs (3 RBs/RBG), and the number of antenna of each LTE-A UE were assumed to be 4 and 2, respectively. We fixed the number of non-GBR UE to be 20 and varied the number of GBR-UE in the simulations. The simulation results of mean throughput and the average numbers of antenna used per GBR UE and non-GBR UE are shown in Figs. 82.5 and 82.6, respectively.

It is noted that both of the proposed scheme and the compared scheme use the same priority decision and resource allocation scheme during the simulations. The only difference is the selection of CC. The proposed scheme selects CC according

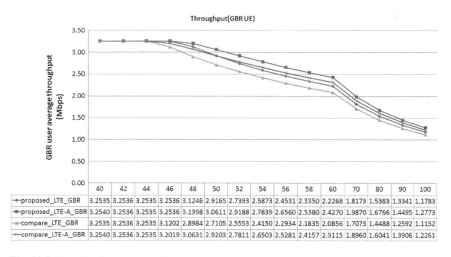

Fig. 82.5 Results of mean throughput

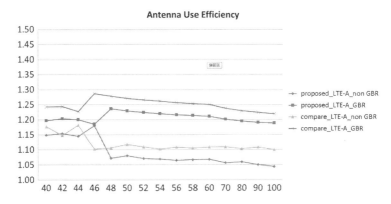

Fig. 82.6 Results of numbers of antenna used

to (82.3) while the least load first scheme is applied in the compared scheme. The simulation results show that the proposed scheme has higher mean throughput than the compared scheme. The proposed scheme uses less number of antennas than that of the compared scheme as shown in Fig. 82.6. It also shows that the number antenna decreases when the number of UE exceeds 48. This phenomenon is mainly owing to that eNB actively discards GBR packets, which exceeds the delay budget, and, as the load is a little decreased, LTE-A UE may use less number of antennas.

82.5 Conclusion

The power saving-based radio resource allocation scheme in LTE-A network is proposed in this paper. The proposed scheme considers the coexistence of LTE and LTE-A UE with GBR and non-GBR quality requirements. Exhaustive simulations were performed to examine the efficiency of the proposed scheme. The simulation results demonstrate that the proposed scheme can achieve the desired performance in throughput and number of antenna used when compared to the other scheme. As the power consumption model highly depends on the design of hardware architecture, the study of power saving shall be flexible enough to be adopted for new hardware design. And it may be a valuable research direction in the future.

Acknowledgements This research work was supported in part by the grants from the National Science Council (NSC) (grant numbers: NSC 98-2221-E-008-063, NSC 99-2218-E-159-001, NSC 100-2221-E-008-097, and NSC 101-2221-E-159-026) and Research Center for Advanced Science and Technology, National Central University, Taiwan, ROC.

References

1. 3GPP TS 36.300 V10.3.0 Evolved Universal Terrestrial Radio Access (E-UTRA) and Evolved Universal Terrestrial Radio Access Network (E-UTRAN); overall description; Stage 2.
2. Iwamura, M., Etemad, K., Fong, M. H., Nory, R., & Love, R. (2010). Carrier aggregation framework in 3GPP LTE-advanced. *IEEE Communications Magazine, 48*(8), 60–67.
3. Akyildiz, F. I., Gutierrez-Estevez, M. D., & Reyes, E. C. (2010). The evolution to 4G cellular systems: LTE-advanced. *Physical Communication, 3*(4), 217–244.
4. Wang, Y., et al. (2010). Carrier load balancing and packet scheduling for multi-carrier systems. *IEEE Transaction on Wireless Communication, 9*(5), 1780–1789.
5. Tian, H., Gao, S., Zhu, J., Chen, L. (2011). Improved component carrier selection method for non-continuous carrier aggregation in LTE-advanced systems. *IEEE Vehicular Technology Conference*, San Francisco, CA, 2011 (pp. 1–5).
6. Zhang, L., Zheng, K., Wang, W., & Huang, L. (2011). Performance analysis on carrier scheduling schemes in the long-term evolution-advanced system with carrier aggregation. *IET Communications, 5*, 612–619.
7. Wang, Y. J., Xiao, D. K., & Wang, W. J. (2010). A research on power consumption of receiver in CA scenarios. *ICIE, 1*, 247–250.
8. Yin, F. (2012). An application aware discontinuous reception mechanism in LTE-advanced with carrier aggregation consideration. *Annals of Telecommunications, 67*, 147–159.
9. 3GPP TR 25.892 V6.0.0, Technical Specification Group Radio Access Network, Feasibility Study for Orthogonal Frequency Division Multiplexing (OFDM) for UTRAN enhancement.

Chapter 83
Dispatching and Management Model Based on Safe Performance Interface for Improving Cloud Efficiency

Bin Chen, Zhijian Wang, and Yu Wang

Abstract In order to solve the performance problem of the cloud computing environment, a dispatching and management model (JDRMSP), which is based on safe performance interface is proposed in this chapter. By using the performance interface integrated into the safe DPI as the original basic data capture, agent-based job scheduling algorithm as the job dispatching method, and ant colony algorithm resource scheduling strategy as the resource management method, the integrated cloud performance is enhanced. For illustration, a simulation experimental example is utilized to show the effect of the model. From the experimental results we can get the conclusion that The JDRMSP model can analyze the cloud environment performance of various cloud components distribution pattern more accurately, and this is the basis of configuration control of the performance of the entire cloud environment, ultimately achieving the purpose of enhancement of the performance of the cloud. The JDRMSP model can effectively solve the performance data capture accuracy problem and take advantage of the dispatching and management algorithm to optimize the cloud environment.

83.1 Introduction

With the rapid development of cloud computing, more and more industries took cloud computing uses in business models into account [1]. As the cloud performance is deeply impacted by the system dispatching and resource management strategy, providing valid job dispatching and resource management model is essential, and the model must build on the premise of safety and efficiency. In this chapter, the performance data interface is mainly performance agent and server interface method (PASI), and it combines with the deep packets inspection (DPI).

B. Chen (✉) • Z. Wang • Y. Wang
Computer & Information Engineering College of HoHai University, Nanjing 210098, China
e-mail: robininblue@hotmail.com

Moreover, continuous monitoring and large system scale lead to the overwhelming volume of data collected by health monitoring tools. So, collecting exactly sufficient and valid data for performance analysis is necessary. It means, appropriate selection method and collection frequency of working data are very important for the effect and exactitude of the performance analysis. The collected information is used to take part in the analysis of enhancement and optimization of the dispatching and management model (JDRMSP).

83.2 Related Work

Job dispatching has become an important issue in cloud computing [2]. The job combination and dispatching strategy algorithm with dynamic programming (JCDS-D) was proposed to focus on the job allocation and the communication overheads minimizing in cloud system [3]. In contrast to the traditional job managers based on scheduler with push, light-weight job coordinator is designed to process data request by pull mode in the proposed job management system [4].

Meanwhile, the resource management is also a key factor for the performance of the cloud environment. There are various significant issues in resource allocation, such as maximum computing performance and green computing [5]. Green computing has become more and more important in recent times[6]. Energy saving is achieved by continuous consolidation of VMs according to the current utilization of resource, virtual network established between VMs, and thermal state of computing nodes [7].

The job dispatching and resource management strategy needs essential performance data as proof. The performance agent and server interface method (PASI) which consists of PMC, PMA, and PMS effectively evaluates the performance of the cloud center [8]. The PASI model can collect performance information automatically and give the analysis report according to the different granularities. However, the cloud architectures also increase dynamic data communications which inherently increase security risks. Deep Packer Inspections (DPI) is essential in protecting the cloud against malicious threats such as Web exploits, zero-day attacks, and mal-ware. [9].

In this chapter, we focus on job dispatching combined with resource management (JDRMSP) based on the VMs which improve cloud efficiency.

83.3 JDRMSP Model for Cloud Computing

Job dispatching and resource management based on safe PASI architecture is divided into three layers as shown in Fig. 83.1. On the top is the clients layer; it is the portal of the cloud environment. Request messages from different terminals are gathered here, synchronized, and sent to the job dispatching mechanism.

Fig. 83.1 Job dispatching and resource management based on safe PASI architecture

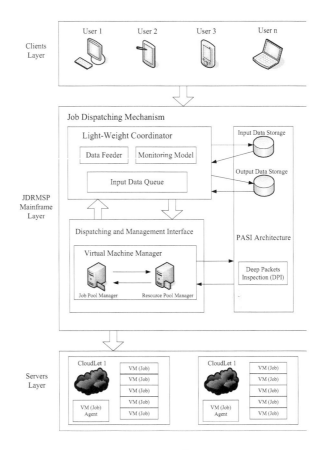

The JDRMSP model uses the agent-based job scheduling algorithm as the job dispatching method, the ant colony algorithm resource scheduling strategy as the resource management method, and performance interface integrated into the safe DPI as the original basic data capture.

The job dispatching mechanism, which plays the central role in the JDRMSP Mainframe Layer, is consisted by Light-Weight coordinator, Data Storage, Dispatching, and Management Interface. The proposed job dispatching system is composed of a data feeder, queue of input data, and monitoring module. The coordinator communicates with the data storage system, which belongs to the PASI architecture, for necessary performance information exchange. The light-weight coordinator performs the essential resource analysis based on the performance statistic data and response to the dispatching and management interface for the core data source request. The essential performance analysis data is supplied by the PASI which is embedded in DPI [10]. The architecture is given in Fig. 83.2.

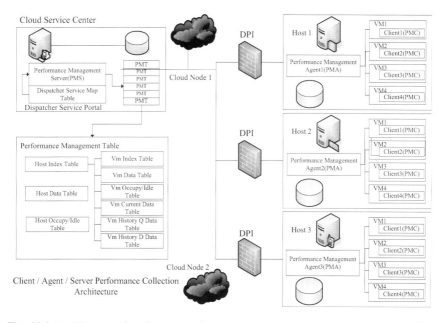

Fig. 83.2 Architecture of performance client, agent, and server collection model embedded with DPI

83.4 Modeling and Analysis

In reality, the JDRMSP architecture plays as a powerful coordinate system for the cloud computing environment. We assumed that the jobs that come to the queue abide by the Poisson distribution process with an arrival rate of λ, while the process time to each job by the queue has a genera distribution. So, we can build an M/G/1 queuing model with a non-preemptive system. It means that M jobs dispatching is done by the coordinator by 1 server at G time. To measure the characteristics of jobs and computing resources, we assume that jobs are in the same priority and are submitted to the cloud according to Poisson distribution with rate λ_i, and job dispatching system in the cloud will arrange some resources in the cloud to process each job with a general service time distribution of "T." First, we can get the total rate λ_i of the arriving of the jobs:

$$\lambda = \sum_{i=1}^{n} \lambda_i$$

Then, we can get the relationship between the rate and time and the length of the queue in the system which is named L_S:

$$L_S = \lambda E\{T\} + \frac{\lambda^2(E^2\{T\} + VAR\{T\})}{2(1 - \lambda E\{T\})}$$

The average number of jobs in each cloudlet waiting queue is K_q^i; E[T] means service time probability, μ means the service rate, and VAR[T] means the traffic intensity:

$$R_i = Lim_{\alpha->\infty}\left[\frac{M(T)E[X^2] + L(T)E[V^2]}{2T}\right] \qquad W_i = R_i + \sum_{i-N_i}^{i-1} X_j$$

From the above equation, we can derive the K_q^i and we can get the result of the cost function of each cloudlet, and finally we can get the job dispatching cost function of the cloud environment $\phi_i(\mu)$:

$$K_q^i = \lambda_i * W_i \qquad \phi_i(\mu) = \eta_i \mu_i + \psi_i K_q^i \qquad \phi(\mu) = \sum_{i=1}^{n} \phi_i(\mu_i)$$

We assume that the maximum storage amount of resource is S. At the initial time T = 0, the resource pool is filled with S units of resource. In the resource management, only four kinds resource costs of cloud computing should be considered. The storage cost per resource unit in resource pool (E_1), the maintenance cost per resource unit in resource pool (E_2), the shortage cost of per resource unit within shortage time unit (E_3), and the shortage cost of per resource unit (E_4). F(S) means the total expectation cost in per unit time, P_n is the probability of the number of resources ordered but not delivered. To find the optimal resource pool capacity means to find S* which makes a minimum worth of F(S). We can derive the optimal resource in job dispatching function.

$$F(S) = E_1 \sum_{n=0}^{S}(S-n)P_n + E_2 S$$

$$+ E_3 \sum_{n=S}^{\infty}(S-n)P_n + \lambda E_4 \sum_{K=1}^{N} Kq_K \sum_{n=S}^{\infty} P_n \qquad JR = \phi(\mu) * F(S)$$

As the JDRMSP architecture is based on PASI which is embedded by DPI, the proportional allocation of cores for request G_i can be calculated by:

$$\theta_i(t) = size_i(t) \qquad G_i = \frac{\theta_i(t)}{\sum_{j=1}^{M} \theta_i(t)} * R = \frac{size_i(t)}{\sum_{j=1}^{M} size_i(t)} * R$$

We let Q_i be the probability that core I has the largest multiplied hash value, and if we set $\prod_{k=1}^{N} x_k = 1$, Then we can derive:

$$Q_i = \sum_{j=1}^{i} \frac{\left(\prod_{k=1}^{j-1} x_k\right)\left(x_j^{N-j+1} - x_{j-1}^{N-j+1}\right)}{N-j+1}$$

$$Q_i = Q_{i-1} + \frac{\left(\prod_{j=1}^{i-1} x_j\right)\left(x_i^{N-i+1} - x_{i-1}^{N-i+1}\right)}{N-i+1}$$

The D_{cloud} is the most important index for the performance of the cloud [8].

$$D_{cloud_{Q/D}} = \frac{T_m * S_E(i)}{P_{Q/D}} \qquad S_E(i) = \sum_{j=i}^{y} S(j)$$

$$S(y) = \text{Prob}\{M_{y-1} \le x \le K\} \qquad S(r) = \text{Prob}\{M_{i-1} \le x \le M_i - 1\}$$

$$= \prod_{j=1}^{y-1} \eta^{M_j - M_{j-1}} \eta_y \left(\frac{1 - \eta_y^{K - M_{y-1}+1}}{1 - \eta_y}\right) S_0 \qquad = \prod_{j=1}^{i-1} \eta^{M_j - M_{j-1}} \eta_i \left(\frac{1 - \eta_i^{M_i - M_{i-1}}}{1 - \eta_i}\right) S_0$$

The advantage of the model is the combined use of job dispatching and resource management strategy and making use of the PASI architecture as the performance data source and DPI as the safe guarantee [11]. We can reach the conclusion that the JDRMSP model can effectively solve the performance data capture accuracy problem.

83.5 Numerical Illustration

In this section, we will demonstrate the mathematical model of comprehensive effect of JDRMSP index $Comp_{cloud}$, which consists job dispatching factor $\phi(\mu)$, resource management factor $F(S)$, DPI factor Q_i, and PASI factor $D_{cloud_{Q/D}}$. From Tables 83.1 and 83.2, we can get the relationship between the change of JDRMSP factor and unidirectional and bidirectional compound cloud performance. Meanwhile, we can summarize the result from Fig. 83.3. It illustrates the influence brightly.

We can reach the following conclusion: The adjustment of PASI is the most effective method for the compound performance of the cloud environment. The job dispatching method influence is inferior to PASI. The control of resource management engenders less effect compared to dispatching of job and PASI. DPI adjustment produces a negative effect in the whole cloud environment, while the cloudlets' safe index is promoted.

Table 83.1 Unidirectional compound cloud performance coordinate result

$\phi(\mu)$	$F(S)$	Q_i	$D_{cloud_{Q/D}}$	$Comp_{cloud}$
↑ Δφ	↑ ΔF	–	–	↑ 2.08ΔC
↑ Δφ	–	↑ ΔQ	–	↓ 0.11ΔC
↑ Δφ	–	–	↑ ΔD	↑ 4.15ΔC
–	↑ ΔF	↑ ΔQ	–	↓ 0.32ΔC
–	↑ ΔF	–	↑ ΔD	↑ 3.36ΔC
–	–	↑ ΔQ	↑ ΔD	↑ 1.73ΔC

Table 83.2 Bidirectional compound cloud performance coordinate result

$\phi(\mu)$	$F(S)$	Q_i	$D_{cloud_{Q/D}}$	$Comp_{cloud}$
↑ Δφ	↑ ΔF	↓ ΔQ	↓ ΔD	↑ 0.33ΔC
↑ Δφ	↓ ΔF	↑ ΔQ	↓ ΔD	↓ 0.94ΔC
↑ Δφ	↓ ΔF	↓ ΔQ	↑ ΔD	↑ 1.47ΔC
↓ Δφ	↑ ΔF	↑ ΔQ	↓ ΔD	↓ 1.28ΔC
↓ Δφ	↑ ΔF	↓ ΔQ	↑ ΔD	↑ 1.12ΔC
↓ Δφ	↓ ΔF	↑ ΔQ	↑ ΔD	↑ 0.28ΔC

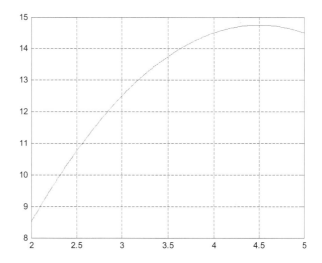

Fig. 83.3 Compound cloud performance chart

83.6 Conclusion

In this chapter, we have described a job dispatching and resource management model based on safe performance interface-PASI for improving cloudlets' efficiency. In this model, we can adjust the performance of the cloud environment by relevant dispatching and management method dynamically. The research result is useful for the analysis of the average performance of the current running cloud environment, affords some relevant revision, and adjusts the strategy for the cloud architecture.

In our future works, we will concentrate on the performance suppression strategy in the cloud environment, as the performance statistic methods cause some problem.

Acknowledgements [Foundation] The natural science foundation of Jiangsu Province in 2012 "The data in the cloud environment safe recovery of applied research."

References

1. Qinlong Jiang, Weibing Feng, Junjie Peng, Fangfang Han, Qing Li, Wu Zhang, et al. (2011). Inventory-based resource management in cloud computing, 2011. In *Tenth International Symposium on Distributed Computing and Applications to Business, Engineering and Science* (pp. 242–243).
2. Tai-Lung Chen, Ching-Hsien Hsu, & Shih-Chang Chen. (2010). Scheduling of job combination and dispatching strategy for grid and cloud system. *5th International Conference on Advances in Grid and Pervasive Computing, GPC 2010, 23*(2), 109.
3. Kushal Dutta. (2012). A smart job scheduling system for cloud computing service providers and users modeling and simulation. In *First International Conference on Advances in Information Technology* (pp. 192).
4. Haehyun Kim. (2011). Light-weight cloud job management system for data intensive science, 2011. In *Fourth IEEE International Conference on Utility and Cloud Computing* (pp. 625–630).
5. Huang, C.-J., Guan, C.-T., Heng-Ming Chen, Y.-W., Wanga, S.-C. C., Li, C.-Y., & Wenga, C.-H. (2013). An adaptive resource management scheme in cloud computing. *Engineering Applications of Artificial Intelligence, 26*(2), 141.
6. Younge, A. J., von Laszewski, G., Wang, L., Lopez-Alarcon, S., & Carithers, W. (2010). *Efficient resource management for cloud computing environments* (pp. 534). IEEE.
7. Beloglazov, A. (2010). Energy efficient resource management in virtualized cloud data centers, 2010. In *Tenth IEEE/ACM International Conference on Cluster, Cloud and Grid Computing* (pp. 303–305).
8. Chen Bin, Wang Zhijian, & Wang Yu. (2012). Performance collection model with agent and server interface for cloud computing, ICCCT2012. In *Seventh International Conference on Computing and Convergence Technology (ICCIT, ICEI and ICACT)* (pp. 118–119).
9. Smallwood, D., & Vance, A. (2011) Intrusion analysis with deep packet inspection, 2011. In *International Conference on Cloud and Service Computing* (pp. 281–284).
10. Huang, C.-J., Guan, C.-T., Heng-Ming Chen, Y.-W., Wang, S.-C. C., Li, C.-Y., et al. (2013). An adaptive resource management scheme in cloud computing. *Engineering Applications of Artificial Intelligence, 26*(3), 45–48.
11. Matsumoto, H., & Ezaki, Y. (2012) Dynamic resource management in cloud environment. *Computer Programming—723 Computer Software, Data Handling and Applications* (pp. 367–371.

Chapter 84
A Proposed Methodology for an E-Health Monitoring System Based on a Fault-Tolerant Smart Mobile

Ahmed Alahmadi and Ben Soh

Abstract In the development of general system design approaches, the main concern is whether the approach meets the proposed system's specifications and the ability of the system to operate for a specified period within those specifications. However, with the expansion in the field of sensitive and complex systems such as e-health monitoring systems, a greater emphasis is placed on the behaviour of the system with the presence of fault (i.e., fault-tolerance). Consequently, when the system is being built, tasks such as fault-tolerance requirements are essential to ensure the quality of the resulting reliable e-health monitoring system. By considering the fault-tolerance requirements as functional requirements in the requirement phase, the completeness of reliability requirements for an e-health system can be developed. This paper proposes a methodology that conceptually studies fault-tolerance in relation to a smart mobile e-health monitoring system. The methodology aims to contribute towards standardising the fault-tolerant requirements of a reliable e-health monitoring system.

84.1 Introduction

The distractions of the operation of different major complex systems have underlined the need to develop and propose novel mechanisms that reduce the effect of disruptions and enhance the reliability level of these systems. An e-health monitoring system is one such complex system that needs to be studied and analysed in order to improve and develop its reliability. Fault-tolerance plays a significant role in achieving this aim. It may be defined as the operation of avoiding system failures in the presence of faults. However, fault-tolerant design is one of the most significant dependability attributes that are required to achieve reliability. As a

---

A. Alahmadi (✉) • B. Soh
Department of Computer Science and Computer Engineering, La Trobe University, Melbourne, VIC 3083, Australia
e-mail: ahalahmadi@students.latrobe.edu.au

result, the importance of fault-tolerance introduces the need to study and analyse the concepts of fault-tolerance and its mechanisms in such complex systems.

The clear expansion in the field of e-health systems, and the aim to initiate a highly reliable application relating to human life, has created the importance of such a performance requirement for both stakeholders and developers. To date, very little research has addressed fault-tolerance requirements for e-health monitoring systems. Accordingly, it appears that no clear conclusions can be drawn from previous work to clarify the mechanism of integrating the fault-tolerance concept and requirements into a smart mobile e-health monitoring system.

Thus, this paper proposes a methodology for an e-health monitoring system based on a fault-tolerant smart mobile. Our methodology aims to contribute towards standardising the fault-tolerant requirements for a reliable e-health monitoring system.

84.2 Proposed Methodology

With a view to studying and analysing the correlations between the attributes indicating the fault-tolerant design and the complex system QoS level required by users, we propose a methodology that is categorised into three phases, as shown in Fig. 84.1. First, phase 1 defines the concept of fault-tolerance in terms of definitions, attributes and requirements that need to be clearly understood and then developed to identify the QoS level expected by the patient, which is the second phase. The third phase describes a particular complex system—in our case, an e-health monitoring system—and the performance attributes that need to be recognised. One way to develop the system design methodology is to consider the characteristics of e-health monitoring systems and the theory associated with the concept of fault-tolerance. For instance, a design approach based on an understanding of the concept of fault-tolerance leads to the advantage of the capability of developing the design utilising the main essence of that concept. Qualitative studies and approaches are significant and required to afford a better understanding and design of a fault-tolerant system such as an e-health monitoring system [1].

Fig. 84.1 Proposed methodology

84.3 Context of the Proposed Methodology

In this section, we discuss the context of our proposed methodology, which includes smart mobile e-health monitoring systems' perspectives, fault-tolerance concept clarification, and investigation and analysis of the fault-tolerance concept in an e-health monitoring system.

84.3.1 Smart Mobile E-Health Monitoring System Perspectives

Towards the aim of investigating and analysing the correlation between the concept of fault-tolerance and the proposed e-health monitoring system, it is necessary to first identify and clarify the perspectives of the system. The network architecture of the system is proposed based on a high level of scalability. The design of three levels of interconnected networks is flexible enough to guarantee the availability of the system in all expected situations. Figure 84.2 shows the overall configuration of the system's network, which consists of body sensors connected via Bluetooth to a smart phone. The exact location of the patient in the home environment is recognised using radio-frequency identification (RFID) tags distributed in each part of the house. These tags also help to determine the current activity of the patient. Through the Internet, the smart phone can contact healthcare providers or paramedics directly if needed.

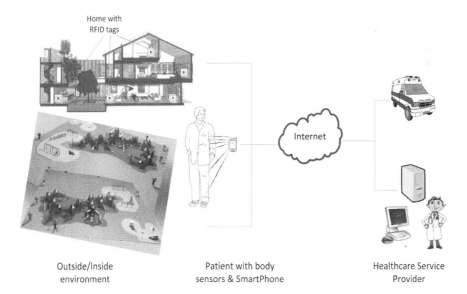

Fig. 84.2 Overall configuration of the system's network

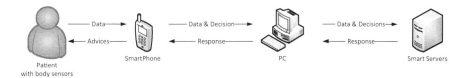

Fig. 84.3 The physical components and the system framework [2]

To study and identify the appropriate way of integrating fault-tolerance concept perspectives into a complex e-health monitoring system, the three main components of our system—physical, network and software—should be identified. The network component was introduced and discussed earlier. Both the physical components and the roles of each component (i.e., software component) must be clarified based on the specifications of the system and users' requirements. In our system, the physical component and the system framework are shown in Fig. 84.3. More details about the proposed system can be found in our previous paper [2].

84.4 Investigation and Analysis of the Fault-Tolerance Concept in an E-Health Monitoring System

In this section, we contribute to identifying a clear understanding and pre-analytical study of the concept of fault-tolerance in a specific system—smart mobile e-health monitoring. Based on the proposed system objectives and requirements mentioned in Sect. 84.3 and the fault-tolerance conceptual analysis (Fig. 84.4), this section aims to clarify and analyze our system's fault-tolerance definitions, threats, attributes and requirements in order to clarify the efficient methods of integrating the results into the design and implementation phases of our smart mobile e-health monitoring system.

84.4.1 Definition and Measures Analysis

Fault-tolerance is the ability to avoid the service failure; indeed, it is the capability of the system to keep executing the correct implementation of its programme and all other functions despite the occurrence of a fault [3, 4]. In some works, it is defined as the aim of delivering a specified service in the existence of an active fault [5, 6]. In addition, a fault-tolerant system has the ability to continue performing its operations with the appearance of faults. As mentioned in many works fault-tolerance is considered one of the significant requirements in order to obtain a dependable and reliable system [6–8].

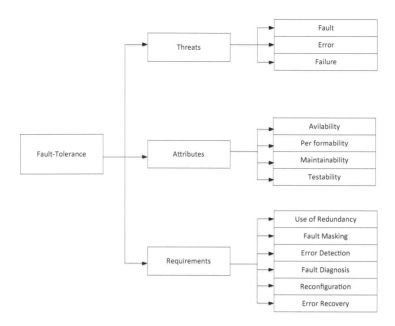

Fig. 84.4 The fault-tolerance concept taxonomy

The system fault-tolerance can be analysed and measured quantitatively and qualitatively [9]. Some design characteristics that qualitatively evaluate a fault-tolerant design include a safe-shutdown mode, no single point of failure and consistency specifications. In contrast, due to fault-tolerance being one of the main requirements of reliability, the quantitative measures are typically derived from the quantitative reliability parameters, including Mean Time Between Failure (MTBF), Mean Time To Failure (MTTF), Mean Time To Repair (MTTR), Failure Rate (λ) and Repair Rate or Maintainability Parameter (μ). All of these parameters can be calculated and measured for the physical, network and software components of our system in order to demonstrate the level of fault-tolerance.

In our e-health monitoring system, the fault-tolerance concept is the ability to achieve the requirements of the system, patient and medical staff with the appearance of faults in any of the system's three components (physical, network or software). As mentioned above, our system is considered a sensitive system due to the relations between the system objectives and patient's health, consequently, life. As a result, denying its services completely (i.e., system failure) is not an acceptable scenario, especially for essential/critical functions such as emergency situations when a patient needs the paramedics' attendance. The above-mentioned facts show the clear need for a fault-tolerant design in our e-health monitoring system to ensure the delivery of its services.

84.4.2 Threats (Faults, Errors and Failure Concept) Analysis

The various definitions of the fault-tolerance clearly show that the main threat to fault-tolerant design is fault, followed by error and failure. These three terms, as well as the cause and effect correlations between them, are the basic and key terms in fault-tolerant design [7]. In our observation, accepting that a fault will occur in the system at some point is the secret behind the need for fault-tolerant design.

The concept of faults, errors and failures has been understood differently by different people. A system failure occurs when the achieved service deviates from the specified service, where the service specification is an agreed description of the expected service. A failure occurs because the system is erroneous: the error is the part of the system state that is liable to lead to failure (i.e., the delivery of a service that does not comply with the specified service). In its phenomenological sense, the cause of an error is a fault. Another observation is that since the relationship between faults, errors and failures is an inheritance (i.e., cause–effect) relationship, both of the concepts—fault-tolerance and error-tolerance—are usually used interchangeably.

84.4.3 Attributes Analysis

Although many authors demonstrate different features of fault-tolerant design, some of the most common attributes are availability, performability, maintainability and testability [7, 9]. In some works, "graceful degradation" is defined as another attribute of fault-tolerance; however, the majority mention it as the system performability due to the very close relationship between them. Each attribute has its own definition, threats, features and means that introduce the need to study and analyse each one. Some of these attributes are discussed below.

Availability: Unlike reliability, availability is defined as the capability of the system to perform its requirements correctly at a given time t [3]. To remove ambiguity between the reliability and availability concepts, it is important to understand that availability measures the correctness of the required performance at a given instant of time, whereas reliability measures the correctness of the required performance during a specific period (interval) [7].

Performability (Graceful Degradation): The performability of a system is defined as the probability that it will either perform correctly or it will be allowed to continue operating at a diminished performance level.

While reliability is a measure of likelihood that all of the system functions are performed correctly, performability is the likelihood that a subset of the critical functions is performed correctly. However, the assumption in the analysis of availability and reliability concepts in relation to our system is that the system's possible states are binary—either up or down. This unsophisticated view holds true for our system if tolerating faults is not an option, but in our proposed fault-tolerant smart mobile e-health monitoring system design, many additional system states become significant—one for each possible fault prototype.

84.4.4 Requirements Analysis

Many studies show a variety of requirements for obtaining a fault-tolerant design [6, 7, 9]. In general, the use of redundancy is one of the logical and common techniques that play a significant role in achieving the goal of a fault-tolerant concept. In addition, fault-masking, error detection, fault-diagnosing and reconfiguration followed by error recovery are other techniques used in fault-tolerant design. Some of these attributes are discussed below.

Use of Redundancy: In fault-tolerance, redundancy is the use of ancillary elements in a system to achieve the same or similar functions as the main elements for the purpose of tolerating possible faults [3]. In a more detailed definition, it is the use of extra resources beyond what is required for a normal system operation. However, this involves some inherent costs, including hardware, software, effects on system performance and the penalties of space and power. This observation forces designers of fault-tolerant systems to cautiously and comprehensively answer the questions of "where, why and how" they use redundancy techniques in relation to their system objectives. Usually, redundancy techniques can be classified into hardware, software, information and time redundancy.

Fault Masking: Fault masking is the process of preventing faults from constructing an error [3]. From the above-mentioned concepts of faults, errors and failures, faults are initially the cause of errors in a system, and errors are the first effect of a fault. The aim of fault-masking techniques is to cancel the effects of faults (i.e., prevent the system from being erroneous). For example, for a negative value that will cause an error, the system should prevent its input rather than handle it.

Error Detection: This is the identification of errors in a system. The detection of errors ensures the system's continual correct operation. Therefore, from a fault-tolerance design standpoint, the failure to detect an error will lead to a system failure. In general, error-detection mechanisms may be classified into two main categories: a Self-Checking (SC) mechanism, where a module can detect internal errors concurrently with normal operation before propagating results to the upper levels of the system; and a Non-Self-Checking (NSC) mechanism, where a module does not have internal error-detection capability and the error-detection mechanism is performed by the upper levels of the system.

Reconfiguration: Once an error has been detected and/or fault diagnosis has been carried out, system reconfiguration must be initiated, in that the faulty unit must be isolated and replaced (if spares are available). If spares are not available the mentioned performability/graceful-degradation attribute is needed. This graceful-degradation permits the system to continue operating with a reduced performance level.

84.5 Conclusion

The important factor that influences the quality of an e-health monitoring system is the level of dependability. Towards this aim, and respecting that faults will occur in the system at some point, we first need to obtain a system with an ability to tolerate faults. The significance of fault-tolerant design introduces the need to study and analyse the concepts of fault-tolerance and its mechanisms in a complex smart mobile e-health monitoring system. This paper presents a pre-analytical framework of the concept of fault-tolerance in relation to an e-health monitoring system. One of the best ways to develop our system design methodology is to consider the characteristics of e-health monitoring systems and the theory associated with the concept of fault-tolerance. This paper first clarifies our proposed e-health monitoring system by showing its three components with the system specifications and user requirements (physical, network and software). We then identify the concept of fault-tolerance, including definitions, threats, attributes and means. Based on these two steps, we propose a pre-analytical conceptual study of a fault-tolerant smart mobile e-health monitoring system as a third step towards implementing the results of our study in the execution phase of our e-health monitoring system in future work.

References

1. Orlandi, E. (1990). Computer security: A consequence of information technology quality, *Proc. 1990 I.E. Int. Carnahan Conference on Crime Countermeasures, Security Technology*, 1990 (pp. 109–112).
2. Alahmadi, A., Soh, B. (2011). A smart approach towards a mobile e-health monitoring system architecture. *2011 International Conference on Research and Innovation in Information Systems (ICRIIS)*, 2011 (pp. 1–5).
3. IEEE Std 610. (1990). *IEEE Standard Computer Dictionary. A Compilation of IEEE Standard Computer Glossaries*.
4. von Neumann, J. (1956). Probabilistic logics and the synthesis of reliable organisms from unreliable components. In C. E. Shannon & J. McCarthy (Eds.), *Automata studies, annals of math studies* (Vol. 34, pp. 43–98). Princeton, NJ: Princeton University Press.
5. Edwards, N. (1994). *Building dependable distributed systems. ANSA*. Cambridge, UK: APM Ltd.
6. Avizienis, A., Laprie, J. C., Randell, B. (2001). *Fundamental concepts of dependability* (Research Report No. 1145). Toulouse, France: LAAS-CNRS.
7. Pradhan, D. (1996). *Fault-tolerant computer system design* (1st ed., pp. 5–14). Upper Saddle River, NJ: Prentice Hall Inc.
8. Oliveto, F. E. (1997). The four steps to achieve a reliable design, *Proc. 1997 National Aerospace and Electronics Conference, (NAECON)*, 1997 (Vol. 1, pp. 446–453).
9. Heimerdinger, W., Weinstock, C. (1992). A conceptual framework for system fault tolerance (Technical Report CMU/SEI-92-TR33. ESC-TR-92-033). SEI.

Chapter 85
Design and Application of Indoor Geographical Information System

Yongfeng Suo, Tianhe Chi, and Tianyue Liu

Abstract For the present situation of shortage in GIS indoor theories and insufficiency in indoor GIS applications, a set of indoor GIS research theories, indoor map cartography specifications, and related technologies closely integrated with fire-fighting industry were proposed in this paper. The indoor map cartography specifications included technological processes of the map cartography and matched data updating mechanism, convenient for fast, accurately, timely producing professional indoor map. The key technologies, such as symbol dynamic drawing, indoor outdoor seamless integration, map updating and path analysis, were preliminary applied in fire-fighting emergency rescue platform, so as to realize functionalities such as indoor and outdoor seamless expression, POI updating periodically, and the best rescue path analysis, and improve the transparent command level of the fire rescue site, and it also may have certain reference value to other emergency rescues.

85.1 Introduction

As indoor activities are increasingly frequent, the requirements of interior space service have become more and more urgent. When people get into the large complicated building like marketplace or airport, it is very easy to get lost or can't find the right destination because of distinguishing ambiguous direction. Therefore, the market needs indoor geographic information "indoor map" as the carrier of public space information personalized. At abroad, the Google takes the lead in

---

Y. Suo (✉)
Institute of Remote Sensing Applications, Beijing 100101, China

College of Navigation Jimei University, Xiamen 361021, China
e-mail: yfsuo@qq.com

T. Chi • T. Liu
Institute of Remote Sensing Applications, Beijing 100101, China

releasing indoor maps of Android version based on Google Map, when the user get into the interior building, the detailed building floor plans will appear automatically. Indoor map production mainly adopted the traditional data processing methods at present, relying on data collection network accumulated by outdoor map. But as a result of field data acquisition channel and low accuracy, slow update cycle, it is difficult to ensure professional application requirements.

From the application-driven angle, this paper put forward indoor geographic information processing chain combined with the public security fire control industry with depth, and find out the indoor map data production specifications which are from data preparation to final work output. With the purpose of solving the difficult indoor data acquisition, updating large-scale batch production, long update cycle, and other key issues efficiency, the indoor GIS platform was designed, and also indoor map-related technologies were applied in fire emergency rescue, enhanced the complex building fire rescue efficiency and relief effect from the overall ascension.

85.2 Indoor GIS Theory Analysis

Just like exploration in outdoor geographical space, people in research of interior space also need to develop suitable methods and technologies for interior space to ensure that the Geographic information system into the interior space with the scientific theoretical basis and the effective implementation method [1].

85.2.1 Conception

Indoor GIS (short for "IDGIS"), relative for Outdoor Geographic Information System, it is a kind of computer system to collect, storage, manage, analysis, show the internal space information of the huge constructions, such as hotel, university, school, museum, trains, subway station, shopping center parking, and any other type of building. It is the general technical expression in a microcosmic and large-scale angle to express high precision space and attribute information of interior space entities. Indoor map is the direct expression of Indoor Geographic Information System. Indoor map included different floor POI (Points of Interest) information, the mapping of doors, windows and walls, different attributes of the polygon information, extended data itself, such as the height of the 3d properties and captured additional semantic information [2].

85.2.2 Indoor GIS Coding

Indoor GIS coding is composed of five levels, namely, "city area code," "building," "floor," "room," "entity," the "entity" of room is minimum management unit, and each level corresponding to the only coding of indoor map, namely the division of indoor GIS coding code is from the "building" to "floor," to "room" (or "corridor"), and finally to entity according to the hierarchical coding [3].

85.3 Design of Fire-Fighting Indoor GIS

85.3.1 Indoor Map Cartography

To meet the large-scale indoor GIS needs of firemen's rescue work, the internal fire electronic map of building must embody POI point layer, partition wall line layer and room corridor surface layer. On that basis, setting the fire rescue plan of Key Unit of Fire Safety, provide data support for fire-fighting indoor GIS platform, and effectively carry out fire rescue work. Compared with the traditional field collection, indoor map field collection used the first acquisition with a combination of fire key position and fire control facilities Enterprise fill tool (short for "Enterprise fill tool") which updated regularly. First acquisition information included important fire control facilities, fire unit-related information, etc.; "Enterprise fill tool" is the updated source of indoor map. The fire key units provided the fire regularly indoor change information, and Modified part of enterprise terminal will be real-time updated to fire terminal database, assuring indoor map data acquisition real-time and accuracy.

- Overall drawing efficiency
  In order to ensure overall working efficiency in drawing indoor map, the overall production process should include data preparation, field acquisition, office mapping, and quality inspection to achievements management, as shown in Fig. 85.1.
- Fire indoor two-dimensional map
  Fire indoor two-dimensional map element types concluded POI layer, partition layer, and surface layer. POI layer showed important fire-fighting facilities point information in indoor map, which was aimed to show different types of important fire control facilities with different point symbols, including fire-fighting facilities, alarm facilities, smoke facilities, evacuation indicator facilities, dangerous goods. Most of the fire POI symbol referred to the fire equipment graphic symbol national standard [4]; Partition layer showed the linear information such as partition way and wall material, which distinguished area effectively; surface layer distinguished reflect regional function division, through the filling different colors to show [5, 6].

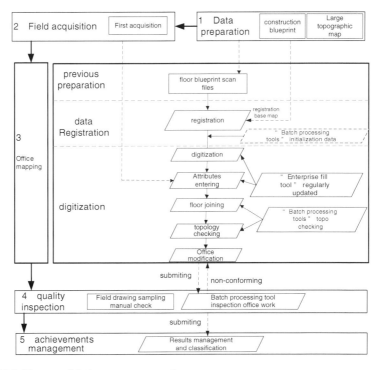

Fig. 85.1 The overall indoor map cartography

85.3.2 Overall Architecture

- Overall architecture design
  To meet the demands in fire fighting and rescue work of the high-rise buildings, we need to build "fire-fighter indoor GIS platform" on the basis of indoor GIS, which contained five-layered architecture, fire-fighter GIS platform Running deck layer, and provided indoor and outdoor map-based data services and function services. The platform was based on WebService-based SOA architecture, and service was provided for all kinds of application model all the way. By using the unified description language, this platform can support application model integrated in cross-platform, professional field, heterogeneous environment [7–9].
- "Enterprise fill tool"
  As the simply main source of indoor map data, "Enterprise fill tool" included regularly updating maintenance records such as the important fire control facilities and the latest attribute information from the fire safety key units, and realized online graphics editing function. Enterprise users created the task form, filled in or modify the returning data, and submitted it to fire department,

whose user submitted data and audited if agreed, they directly filed; if disagreed, then back to the submitter. The data modified by enterprise was finally real-time updated to fire database.

85.4 Key Technologies

85.4.1 Symbol Dynamic Drawing

The map symbols were important part of the indoor map, which mainly were drawn in the following two ways. Styled Layer Descriptor (referred to as "SLD") sign extension can be divided into three forms like point, line, and area. Different Layer corresponded to different SLD files, which were associated with the service of releasing map by GeoServer. Symbol dynamic render referred to those elements of point, line, and surface type can be redraw on the basis of extending Openlayers original symbols. For example, drawing the factor of every fault line was a dynamic way by transforming line to area, then mapping those areas, so as to produce different line symbols.

85.4.2 Data Exchanging and Indoor and Outdoor Map Seamless Displaying

Multi-source heterogeneous data exchanging was achieved through the specific trigger. The moment the trigger was met the requirements that it began to exchange, and the relevant settings in switching strategy description file documents used standard XML format, describing the data with the corresponding relations, the trigger strategy, rules script exchanging, processing plug-ins exchanging, and other information. At the same time, the platform also uses open-source tools for a variety of format conversion, such as the usage of the Ogr2Ogr exchanging ESRI Shapefile format geographic data into GeoJson format and usage of the shp2pgsql in PostgreSQL exchanging shapefile data into PostGIS database and so on. Indoor and outdoor map seamless displayed through the setting zoom level value of the indoor map, namely amplification to a certain proportion, ensuring in the premise of the same outdoor and indoor map projection, indoor map automatically overlay displaying outside the outdoor map.

85.4.3 Indoor Map Updating

Indoor map data updating was proceeded by "enterprise fill tool," the regular feedback mechanism, and the regular updated data was managed through the release mechanism. It can be divided into two kinds that one type was the "default version," namely local area network indoor map version, also known as the father version; the other type was enterprise networks updated version the enterprise regularly updated Version, also known as subversion. Enterprise fill tool update adopted the Web Feature Service Transactions (short for "WFS-T"), this way of which also followed OGC standards, the client adding, modifying, deleting, saving operation through the WFS map Service elements.

85.4.4 Path Optimization Algorithm

Based on the traditional outdoor network data structure and path analysis algorithm, path optimization algorithm comprehensively considered the open-source database software (PostGIS) and commercial GIS software (ESRI ArcGIS) team work abilities, increased stair, fire channel, fire door, fire shutter, and the fire special dynamic topology characteristic, formed the indoor and outdoor unified network data structure, and used node segment network to model more floor space between floors and buildings path analysis. Across the floors path analysis generated indoor channel network by the key nodes like stair, elevator, and then according to the network automatically generate starting points and end points and added the weight value to channel, which adopted indoor and outdoor integration variable weight path analysis algorithm, considering the fire shutter extreme situation that Fire shutter down cause weight changing. It was suitable for special emergency use. Across the buildings path analysis automatic generated indoor and outdoor unified network based on building entrances and outdoor road nodes.

In a building, optimal path analysis was firstly to call corresponding network data services; then we can call field analysis algorithm, each get the nearest point number in the network; according to entering the position of starting point and end point; at last, we can calculate the optimal path between start point to end point and any critical distance between nodes by querying statement and calling optimal path analysis algorithm. At the same time, we can also set end point and obstacle points through the network data, then recreate a new optimal path. The generating path results can control the generation of language description by establishing rich lexicon and rationalization of the rule base, specifically including generating direction and distance and so on along the route.

85.5 Implementation and Application Example

"Indoor GIS platform" mainly was realized by the Openlayers client technology, following OGC standard completely. This platform was simple, lightweight, and easy for coding and transmission which was as practical application in for fire fighting in Tianjin Binhai Hi-tech Zone.

The platform provided the fire commanders with the standardization and normalization of spatial information, meeting the demands of fire rescue in large buildings. Among them, "Enterprise fill tool" provided a method for rapid firefighting indoor rapid mapping, and increased overall efficiency of the mapping after verifications; on the other hand, "enterprises updating mechanism" was put forward for the first time, which was important to indoor facilities maintenance, guaranteeing to form a "live" fire indoor geographic information database every three months.

85.6 Conclusion

In view of that current indoor GIS theories were not very mature, cartography specification and application system have not formed unified standards. Combined with public security fire control industry, this paper formed a set of indoor map production standards, extended fire indoor GIS application to traditional map processing production. It will play a reference role in large-scale constructing other industry indoor map. The technology of indoor map visualization, map update, and optimal path analysis in fire indoor GIS platform can meet the preliminary application requirements in interior space. The indoor map production standards and key technologies can be copied and expanded from public security fire department to other fields, such as business, emergency, flood, quakeproof, disaster prevention, and earthquake terror. In addition, indoor geographic information system is an important technology to promote the construction of digital city or wisdom city. Geographic information system from outdoor to indoor is the inevitable trend in the future.

References

1. Sen Xiao, Xiang Li. (2010). The research and application of geographic information system in indoor space. *Geomatics & Spatial information technology, 33*(5), 38–40 (in Chinese).
2. Summary of Indoor OSM [EB/OL]. (2012-07-15) http://wiki.openstreetmap.org/wiki/IndoorOSM
3. The 2000 technical guidance of Existing surveying and mapping results into the national geodetic coordinate system. (2010-04-25) http://wenku.baidu.com/view/39d3067102768e9951e738e5.html
4. GB/T 4327-2008. (2008). *Fire protection technical documents graphic symbol with fire-fighting equipment*. Beijing: China Standards Press (in Chinese).

5. Schafer, M. (2011). Automatic generation of topological indoor maps for real-time map-based localization and tracking, Indoor Positioning and Indoor Navigation. *Indoor Positioning and Indoor Navigation* (*IPIN*), *2011 International Conference on*, *IEEE*, 2011, Guimaraes (pp. 1–8).
6. Bernhard Hohmann. (2010). A GML shape grammar for semantically enriched 3D building models. *Computers & Graphics*, *34*(*4*), 322–334.
7. Jianjie Chen, et al., (2006). Implementation of spatial information web services based on ontology. *Journal of Zhejiang University* (*Engineering Science*), *40*(*3*), 376–380 (in Chinese).
8. Hailong Yu, et al., (2006). A study of integration between GIS and GIS-based model based on web services. *Science of Surveying and Mapping*, *35*(*2*), 153–161 (in Chinese).
9. Lina Yang, et al., (2011). Design and implementation of digital city share-and-exchange platform based on SOA. *Science of Surveying and Mapping*, *36*(*6*), 230–232 (in Chinese).

Chapter 86
Constructing Cloud Computing Infrastructure Platform of the Digital Library Base on Virtualization Technology

Tingbo Fu, Jinsheng Yang, Yu Gao, and Guang Yu

Abstract In order to improve hardware resource utilization, reduce maintenance and management costs, to build a new IT infrastructure platform for the user to provide a stable, efficient access to services. Taking the library of Harbin Institute of Technology, using VMware cloud computing solutions to build private clouds as an example, through the introduction of VMware vSphere to build a virtual architecture data center, integrate various application services, the introduction of VMware View software system provides "cloud + terminal" desktop cloud of office desktop solution. It is illustrated application of virtual technology by example. It can realize the unified management and deployment of hardware resources and the application of the data center and provide applications with high reliability, high availability, and service of mobile office environment. The IT platform can effectively gather or carrying spare computing capacity, corresponding the IT resource and service priority, improve IT management level.

86.1 Introduction

Modern digital library, we strengthen the construction of electronic resources in order to realize the resources to perfect, and use all kinds of high and new technology, through the powerful data statistical analysis to provide all kinds of personalized service to the readers [1]. With informationization of library constantly deepening, the traditional IT architecture can not gradually adapt to the rapid development of business needs. Virtualization technology is the most basic and core technology of cloud computing, construction of cloud computing infrastructure platform becomes the new development direction of Library Infrastructure Based on virtualization technology.

T. Fu (✉) · J. Yang · Y. Gao · G. Yu
Library of Harbin Institute of Technology, Harbin 150001, China
e-mail: futb@hit.edu.cn

86.2 Current Situation of Platform Infrastructure of Digital Library and Analysis of Virtualization Requirements

Network, server, storage, and PC together constitute the infrastructure platform of Digital Library. This platform can run stably and reliably and meet the growing needs of applications and services. This is a guarantee to the normal operation of digital Library. Servers of Harbin Institute of Technology library mainly are based in X86, they are running more than 30 kinds of database and application system. 100 TB: local storage in IBM DS5020. PC clients mainly include more than 120 sets of office computer, nearly 500 sets for public are distributed in electronic reading room, multimedia classroom, the zone of terminal retrieval. With the development of digital resources increase and expand the application service demand of readers, equipments need to buy have gradually increased, this will give more investment and maintenance costs.

How to build the sharing system in the existing platform? Realize the unified management and deployment of hardware resource and the application of the data center are problems which Harbin Institute of Technology library faced. Based on the sufficient research, for economic and management security considerations, the VMware virtualization solutions are choice to optimization of digital library services of Harbin Institute of technology.

86.3 The Construction of Cloud Computing Infrastructure of Digital Library

Cloud computing is representative of the IT technology. However, the deployment of cloud computing technology should be a gradual process; the way of overturning is not desirable. At present, according to the system application in our library and the physical equipment rate of existing and future business development needs to build a private cloud is the most realistic choice. In the construction of library virtual private cloud, virtualization is played a key role, but also can be said to be a part of cloud computing.

The overall design scheme is taken server, storage device, PC, and network seamless aggregated to "computing resources platform of distribution according to need" cloud by the VMware virtualization technology, as shown in Fig. 86.1. The first stage: it uses VMware vSphere to build a virtual architecture data center, integrates existing 30 kinds of application service, and through resource scheduling technology dynamically allocates and balances calculation resources, provides readers with the efficient 7×24 h access service. The two stages: library office desktop will be provided "cloud + terminal" desktop cloud solutions through VMware View software system it will realized all desktop unified management and deployment. Users install the client software in the PC, IPAD, MAC, and

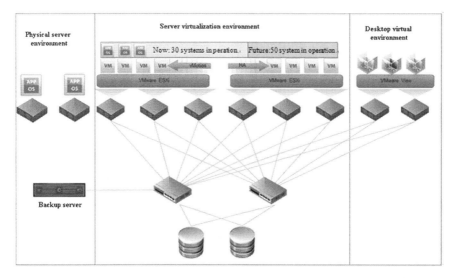

Fig. 86.1 Cloud computing infrastructure platform in library

ANDROID system. It is not restricted by time and place, user login to the server to obtain its own desktop environment, and realize application virtualization by the ThinApp component of VMware View. The application will be isolated and encapsulated to the executable file, it is separated from the underlying operating system, running in the data center server, in order to improve the compatibility and simplify application management. It is accessed through the shortcut virtual desktop.

This platform will be innovated and adjusted from the Desktop Deployment model of the original based on physical server and PC server to structure of physical and virtual combination, including physical server, server virtualization, and desktop virtualization environment. The new platform with unified computing resource pool, unified storage network, and uniform desktop environment can meet the library service on the increasing hardware infrastructure needs.

86.3.1 Server Virtualization Solutions

Server virtualization is taking the virtual hardware, operating system and application program "packaging" into a file, it is called a virtual machine (VM) [2].

The main part of the scheme is the six sets of installed the VMware vSphere software dual blade server. The internal connection of Virtualization architecture is as shown in Fig. 86.2 (Two sets as an example).

VMware vSphere is used for virtualization technology to construct the cloud computing infrastructure, and provides virtualization infrastructure, high

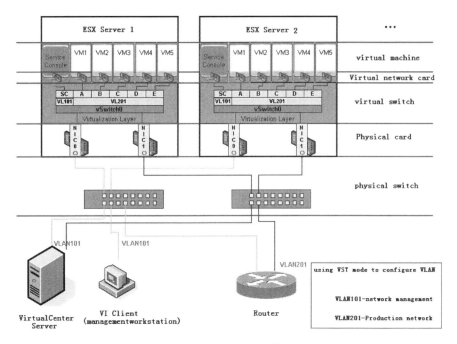

Fig. 86.2 Virtualization architecture internal connection diagram

availability, centralized management, monitoring, and a set of solutions. Including the key components of VMware vCenter server and ESX/ESXi, DRS, VMotion, HA, FT, Data Recovery, etc. ESXi Server is based on VMware virtual architecture suite vSphere components, which are mounted directly on the bare machine of physical server. The processors, memory, storage, and cyber source in the physical server are abstracted to multiple virtual machines and across a large number of virtual machines sharing of hardware resources [3]. Operating system windows, Linux, and other kinds of required application and database is installed on each virtual machine on the server, and then install a variety of application software, so as to make full use of the existing resources in the server, realizing the integration of business system in server level.

86.3.1.1 Hardware Platform

According to the analysis of performance, prior to the existing server application system and storage usage, identified by six IBM HS22 blade servers and IBM DS5020 storage array to build virtual platform, expansion after the six servers, each capable of generating 5–10 virtual server, configuration is shown in Table 86.1.

Table 86.1 The table of the server virtualization platform configuration and extension

Cycle	Name	Unit	New number	The basic configuration or processing capacity requirements
Stage 1	IBM HS22 Blade Server	Set	Memory/CPU 6	Dual quad-core Xeon CPU, 96 GB Memory
	IBM DS5020 Storage Array	Block	10	Increase of 600 GB/SAS hard disk

86.3.1.2 Software Platform

The main configuration is in method of "12 + 1," it is 12 sets of virtual enterprise architecture enhanced version of VMware vSphere (calculation of license according to the CPU number) and used authorization for one sets of VMware vCenter Management Server standard edition. VMware vCenter server is the core of virtual cluster management, for the IT environment provides centralized management, automation, resource optimization, and high availability. We should try to improve the stability and reliability of operation, so approaching the virtual machine (VM), there is a unified management by introducing the VMware vCenter virtual device.

86.3.1.3 The Server of Application Integration

The system integration included the library website, the papers submitted, integrated retrieval, collections search, mobile library, social networks, CALIS interlibrary loan, electronic reading room billing system, reference system, etc. In some not suitable for transfer to the virtual server platform, such as two IBM minicomputer which is loading of library automation system, for the larger demand of storage and system resource, retain the original physical server load mode.

86.3.2 Desktop Virtualization Solutions

Desktop virtualization can provide an independent virtual machine for each user to desktop computing, all desktop are running in the server platform, realize the unified management, when the user uses any terminal at any time and any place to login to server will call your own desktop environment, we can obtain the full PC experience at the same time.

Table 86.2 Desktop virtualization platform configuration expansion table

Cycle	Name	Unit	Number	The basic configuration
Stage 2	Four IBM HS22 and two IBM 3650M3 server expansion		Memory CPU	Each set Up to 96 GB memory Dual CPU
	IBM DS5020 hard disk storage expansion	Block	10	2 TB/SATA

86.3.2.1 Hardware Platform

Desktop virtualization platform is used the four blade servers available and two IBM X3650M3 server, configuration is shown in Table 86.2. In desktop virtual environment, just take the terminal as access equipment, not for data processing, because of cost considerations, the original PC as terminal.

86.3.2.2 Software Platform

Software configuration is the 120 set of View 5 platinum edition software package, including vSphere for desktop user, vCenter x1, View Manager x1, Composer, ThinAPP, View Agent, View Client. VMware View is based on VMware vSphere virtualization platform built in the virtual machine, to build a complete desktop environment—operating system, application program, and configuration. The desktop is changed from the delivery of cloud service, key applications, and data on terminal equipment are safely packaged in cloud computing infrastructure with virtual environment of high number encryption [4]. Staff enjoys the hitherto unknown desktop access free, and improves the work efficiency greatly. Later it will consider providing cloud desktop service for the public and teachers who have special needs.

86.4 Application Effect of Vmware Virtualization Technology in Digital Library

The introduction of virtual technology do not subvert the essence of the existing network architecture, it is based on the existing physical infrastructure to build virtual cluster and unified management. The original server device can be operated normally and with the virtual server together, building VLAN from network, data sharing, service isolation, etc. In the user access patterns, for the virtual server exchange data operation, it is equivalent to the traditional physical server access pattern will not cause any adverse impact on the business system.

86.4.1 Reduce the Cost, Improve the Overall Resource Utilization Rate

The number of physical devices was reduced through server consolidation. In the new IT system can satisfy 50 systems. The migration of more than 30 of the original application can be virtualized to virtualization platform. The average utilization server rate is increased from 5–15 % to 60–80 %.

86.4.2 Improve Operational Efficiency

All physical host, virtual machine, and virtual desktop were centralized management and maintenance, unified planning applications, and security configuration through the VMware vCenter Server, it can shorten the new application of on-line time, improve flexibility. The desktop is run on the server platform, which can ensure the data security, large maintenance work and can reduce the desktop, realize mobile office.

86.4.3 Improve the Security of the System

Platform management module provides VMotion function, it can immediately in the operation of virtual machine migration to another server, perform without disruption of the IT environment maintenance. With the use of HA and FT, maintain and upgrade zero downtime hardware, guarantee business system efficient and stable and uninterrupted operation [5]. The DataRecovery function can be backed up the management of all server and desktop by backup and management, further providing server and desktop data security.

86.4.4 Improve the Service Level, so That the Resources and the Priority of Business Correspondence

The VMware DRS function can balance the resource automatically and intelligently between the virtual machine, the deployment in a virtual machine running process play a role online, so that any one application can ensure the full resources to stable operation, at the same time, the application is not to use the resources at this time can also be other more resources application of temporary borrowing, it is a very good solution to some business system resource occupied a large quantity of problems in application of the peak.

86.5 Conclusion

Through the implementation of server virtualization and desktop virtualization solution, realizes the IT infrastructure platform in the management efficiency and resource efficiency increase, provides mobile office environment safe and convenient for the staff, while achieving system high availability and business continuity. Application of virtual technology in the library can effectively gather or carrying spare computing capacity, with the constant expansion of old, or buy high configuration server, expand the scale of virtual cluster gradually, it formed IT framework like a "cloud computing" architecture finally, the purpose was realized for raising the level of IT management.

References

1. Ping Liang. (2012). Application of server virtualization technology in the library in Colleges and Universities. *China Computer & Communication, 24*(6), 116–117.
2. Qiusheng Dong, Wen Huang. (2009). Application of server virtualization technology in the digital library server consolidation. *Information Studies: Theory & Application, 32*(1), 119–121.
3. VMware. (2011). Getting started with ESX. http://www.vmware.com
4. Hua Chen. (2011). VMware View desktop virtualization solutions. http://wenku.baidu.com/view/6226bd34ee06eff9aef807e6.html
5. Haitao Zhu. (2012). Construction and application of virtual VMware system in the library in Colleges and Universities. *New Technology of Library and Information Service, 16*(1), 68–72.

Chapter 87
A New Single Sign-on Solution in Cloud

Guangxuan Chen, Yanhui Du, Panke Qin, Lei Zhang, and Jin Du

Abstract In order to deliver centralized visibility for login activity, reduce identity proliferation and confusion, increase user adoption and security, reduce administrative costs, and support for entire identity management lifecycle in cloud, a new single sign-on solution is proposed in this paper. By introducing OAuth protocol combined with identity federation mechanism and identity mapping, the new single sign-on model can give the cloud user that has succeed through an identity authentication the permission to access other cloud services in a reasonable time period without entering the username and password repeatedly. Empirical results show that the solution will be used as an impactful measure in scenarios where frequent interactions among different cloud services and clouds that result significant impact across multiple security domains. The OAuth-based single sign-on solution can effectively solve the problems of complexity of identity management and cross-domain authentication in cloud environment and thus increased the security and improved the user's efficiency.

87.1 Introduction

Since its emergence, the cloud has become one of most vigorous forces in the industry. Providing greater reliability, improved flexibility, simpler deployment, and lower costs, the cloud can bring benefits to both users and businesses [1].

G. Chen (✉) • Y. Du • J. Du
People's Public Security University of China, Beijing 100876, China
e-mail: ericcgx@163.com

P. Qin
State Key Lab of Information Photonics and Optical Communications, Beijing 100876, China

Beijing University of Posts and Telecommunications, Beijing 100876, China

L. Zhang
The Logistics Academy, Beijing 100036, China

Apparently, cloud has undeniable prospect and potential. However, security remains a thorny issue and also the most convincing reason for users not moving their business to cloud.

In the cloud era, customers may use various kinds of cloud services at the same time or their data may distribute on different clouds. Thus, the customers may possess different identities in different clouds that may have different security mechanism. It is rather inconvenient to repeat the process of inputting the user name and password for each identity authentication when they need to call multiple resources that distributed on different clouds to jointly accomplish certain task. Meanwhile, it is really a nightmare for the user to manage numerous accounts and passwords. They may write down the passwords on the notepad or in the document. When the notepad is lost or the document is been accidently deleted, then the passwords will be lost too. Furthermore, the passwords in the notepad and document are easy to be peeped by these with ulterior motives, resulting password disclosure.

Single sign-on (SSO) is just a good way to solve these problems. Single sign-on is a transparent user authentication mechanism whereby a single action of user authentication and authorization can permit a user to access multiple protected services and network resources without the need to enter multiple passwords. Considering the identity problems that obsessing the cloud users and cloud providers, single sign-on is of great significance to the promotion and development of cloud.

This paper proposed a new sign-on solution based on OAuth for cloud by combining identity federation mechanism with identity mapping. The proposed solution is shown to deliver a centralized visibility for login activity and reduce identity proliferation and confusion, while increasing user adoption and security in practice.

The remainder text is organized as follows: Section 2 introduces the principle of OAuth protocol. In Sect. 3, we provide a detailed analysis of single sign-on flow in cloud and propose a new single sign-on model for cloud. And Sect. 4 gives the experimental evaluation of the proposed solution. Finally, Sect. 5 concludes with a summary of the new solution and suggests future work.

87.2 Single Sign-on Based on OAuth

There are two standards making the implement of single sign-on in cloud available. One is SAML (Security Assertion Markup Language), an open standard based on XML evolved from Security Services Technical Committee of OASIS. The other one is OpenID, an open user-centric digital identity authentication frame that allows users to be authenticated by certain cooperating sites using a third party service [2]. Due to the deficiency of these two tentative approaches, we proposed the SSO solution in cloud based on OAuth in this paper.

OAuth is an open protocol to allow secure authorization in a simple and standard method from web, mobile, and desktop. It provides a method for the users that allow the third party applications to access users' protected resources stored on certain site without the need to provide the username and password to the third party application.

OAuth permits user providing a token rather than username and password to access their data stored on special sites. Each token authorize a special third party application to access certain resources in a certain period of time (e.g., allowing the application to browse the tutorials in a photo editing site in the next 3 h). Thus, OAuth allows users authorize the third party application access their data stored on other sites without sharing their accessing permission or the whole content of the data.

The authentication and authorization involve three roles:

- Service provider: providing software and hardware resource, platform and service for the users.
- User: the owner of the protected resources stored on the site provided by the service provider.
- Client: the third party application that want to access the resource of the service provider.

Figure 87.1 shows the flow of OAuth authentication.

The basic process of the OAuth authentication can be summed up as follow:

1. Client requests Request Token
   Client (third party application) asks OAuth service provider for unauthorized Request Token and makes a request to Request Token URL. The request includes oauth_client_key, oauth_signature_method, oauth_signature, oauth_timestamp, oauth_nonce, and oauth_version (optional).
2. Service provider grants Request Token
   OAuth service provider grants client's request and issues the oauth_token and corresponding oauth_token_secret that without the user's authorization to client.
3. Client directs user to service provider
   Client requests OAuth service provider for user-authorized Request Token and makes a request to User Authorization URL. The request includes the unauthorized token and secret key obtained from the previous step.
4. Service provider directs user to client
   OAuth service provider directs the user to authorize. The user may be prompted which protected resources are authorized for the client to access. The authorized Request Token may be return or not in this step.
5. Client requests Access Token
   When Request Token is authorized, client will request Access Token URL to replace the Request Token with Access Token. The request includes oauth_client_key, oauth_token, oauth_signature_method, oauth_signature, oauth_timestamp, oauth_nonce, and oauth_version (optional).

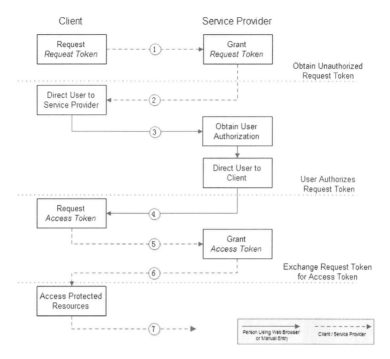

Fig. 87.1 The flow of OAuth authentication

6. Service provider grants Access Token
   OAuth service provider grants the client's request and issues the Access Token and corresponding secret key to client. The response includes oauth_token and oauth_token_secret.
7. Client accesses protected resources
   Client now can use the returned Access Token to access the resources authorized by the user.

The entire OAuth authentication and authorization can be summarized as: Obtaining unauthorized Request Token; Obtains user-authorized Request Token; Replaces the authorized Request Token with Access Token.

87.3 Single Sign-on Model for Cloud

An effective identity authentication mechanism should give the entity that has succeed through a identity authentication the permission to access other resources in a reasonable time period without entering the username and password repeatedly. Such security mechanism must take the problems like federation of authentication domains and identity mapping that brought by the request of across multiple

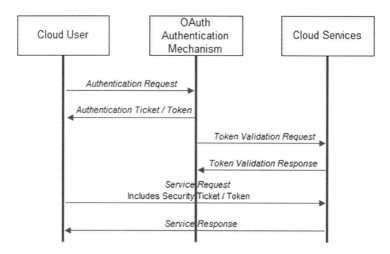

Fig. 87.2 OAuth-based single sign-on flow in cloud

security domains into consideration. So, we proposed the single sign-on solution in cloud [3]. The solution gives the user in a special logical security domain the permission to access the authorized resources through one identity authentication.

The principle of the solution of SSO in cloud is to share an SSO Token_ID object. When the cloud user request for accessing an application system on the cloud, he will be directed to authentication system for identity authentication [4]. If the user is certified, he will receive a ticket, i.e., Token_ID. The Token_ID will be labeled as a user session mark and will be sent to the authentication system for checkout when user wants to access other cloud resources. As for Web application in the cloud, the Token_ID will exist as cookie. According to this principle, we designed the SSO flow in cloud, showed in Fig. 87.2.

As can be seen from Fig. 87.2, the SSO flow in cloud can be concluded as: Cloud user requests and obtains unauthorized Request_Token_ID; Cloud user obtains authorized Request_Token_ID; Request_Token_ID validation.

In order to realize mutual authentication and multi-party authentication, this authentication mechanism has token full consideration of cross domains, agent, and identity federation. Through SSO, cloud user can access cloud services more conveniently.

According to the principle of SSO flow and OAuth protocol, we designed the SSO model of cloud, shown in Fig. 87.3. When a cloud user accesses cloud service A, cloud service A will direct him to a special OAuth-based authentication system. After authorized, he will receive a ticket that generated by the authentication system. In a reasonable time period which can be customized according to the security level, when the cloud user want to access cloud service B, he will send the ticket to cloud service B for checking up. Then the cloud service B will check the ticket with the authentication system. As the ticket can be recognized by the authentication system through identity federation mechanism, the cloud user now

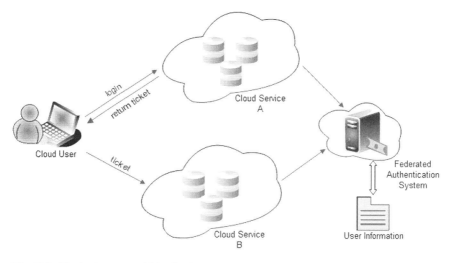

Fig. 87.3 Single sign-on model in cloud

Fig. 87.4 GetFederationTokenRequest

can access cloud service B freely. In this process, the federated identity authentication will interact with the user information database so as to accomplish effective authentication.

Due to the characteristic of cloud, trust is essential to the realization of SSO among different cloud services which distribute in different corners [5]. So, we intruded federated authentication system that providing federated identity authentication service. The federated authentication system which contains one or more federated servers that sharing public trust policy can meet the requests of cloud users in the same domain or other clouds [6]. Here, we show the two important functions in identity federation example: GetFederationTokenRequest and GetFederationTokenResponse.

The mechanism retrieves the Access Policy for the authenticated user and requests temporary security credentials by calling GetFederationTokenRequest with a valid name, an expiration time set to 2 h and the policy, shows in Fig. 87.4.

The temporary security credential is returned as a response that contains: AccessKeyID (the access key identifier for the temporary credentials), SecretAccessKey (the key used to sign requests), and SessionToken (the security token).

87.4 Experimental Study

Here, we designed two applications (named App1 and App2) to simulate two separated cloud services. Actually, the number of application can be extended on demand. The single sign-on process that covers these two applications can be realized as follows:

When we want to access App1, we can enter the URL http://localhost:8080/App1 and input the username and password for authentication. The OAuth authentication mechanism then will verify the correctness of the account. If verified, the Token_ID for the App1 will be generated and the user is redirected to the URL of APP1. And thus, user can now access the resources of App1 freely. When the user wants to access App2 at the same time, he just needs to click the link of App2 (providing the URL of App2 is http://localhost:8080/App2) without reenter username and password. In the valid time period (we set is as 2 h), client can access these two service distributed in different "cloud" freely through a single sign-on mechanism.

In this process, the login activity presenting a centralized visibility and identity proliferation and confusion are weakened.

87.5 Conclusion

Summary, this paper analyzes the plight of the cloud user when they managing numerous accounts and password in their work. Then a new single sign-on solution combined with identity federation and identity mapping in cloud is proposed. This OAuth-based SSO solution delivers a centralized visibility for login activity and can reduce identity proliferation and confusion, increase user adoption and security in practice. Future work will focus on cross-cloud identity federation.

References

1. Ravich, Y. I. (1995). "Selective carrier scattering in thermoelectric materials", Chap 7. In D. M. Rowe (Ed.), *CRC handbook of thermoelectrics* (pp. 67–81). Boca Raton, FL: CRC Press.
2. Ravich, Y. I. (1995). "Selective carrier scattering in thermoelectric materials", Chap 7. In D. M. Rowe (Ed.), *CRC handbook of thermoelectrics* (pp. 67–81). Boca Raton, FL: CRC Press.
3. Ravich, Y. I. (1995). "Selective carrier scattering in thermoelectric materials", Chap 7. In D. M. Rowe (Ed.), *CRC handbook of thermoelectrics* (pp. 67–81). Boca Raton, FL: CRC Press.
4. Ravich, Y. I. (1995). "Selective carrier scattering in thermoelectric materials", Chap 7. In D. M. Rowe (Ed.), *CRC handbook of thermoelectrics* (pp. 67–81). Boca Raton, FL: CRC Press.
5. Ravich, Y. I. (1995). "Selective carrier scattering in thermoelectric materials", Chap 7. In D. M. Rowe (Ed.), *CRC handbook of thermoelectrics* (pp. 67–81). Boca Raton, FL: CRC Press.
6. Celesti, A., Tusa, F., Villari, M., Puliafito A. (2012). Evaluating a distributed identity provider trusted network with delegated authentications for cloud federation[C]. *Proceedings of the 2nd International Conference on Cloud Computing, GRIDs, and Virtualization* (pp. 80–85). UK: Curran Associates Inc.

Chapter 88
A Collaborative Load Control Scheme for Hierarchical Mobile IPv6 Network

Yi Yang, QingShan Man, and PingLiang Rui

Abstract With consideration of the invalid registration flows and load balance problems in hierarchical mobile IPv6 (HMIPv6) networks, a collaborative load control scheme (COLC) for HMIPv6 networks is proposed to reduce registration flows and balance load. In COLC, mobile anchor point (MAP) is allowed to transfer part of its packet delivery load to its neighboring MAPs with lower load, by which the invalid registration flows decrease, and more mobile nodes (MNs) register with their favorite MAPs without capacity expansion. The validities of the scheme in reducing registration flows of HMIPv6 and performing better load balance are examined in the simulations.

88.1 Introduction

The registration traffics supporting mobility lead to magnitude pressure in wireless/mobile networks. In order to reduce the registration traffics, the work group of Internet engineering task force proposed an improved mobile IPv6 protocol named hierarchical mobile IPv6 (HMIPv6) [1], which introduced the mobile anchor point (MAP) to separate micro-mobility from macro-mobility with the objective of reducing the registration traffics.

As important intermediate nodes in HMIPv6 networks, MAPs are often deployed hierarchically to avoid overload problem. And each mobile node (MN) is allowed to register with one of those MAPs whose domains cover MN's current location. Most MNs would like to register with the MAP covering larger domains in order to stay at a relatively longer time and avoid home registration

Y. Yang (✉) • Q. Man • P. Rui
The 28th Research Institute of China Electronics Technology Group Corporation,
Nanjing 210007, China
e-mail: yiyang0803@yahoo.cn

frequently, but it's hard to keep the load balance between MAPs. Moreover, the popular MAPs can't afford all the MNs' registration especially when the number of MNs is large. In this case, most of the registration will be refused, leading a large amount of invalid registration flows. Many threshold-based load control schemes [2–4] are proposed to decrease the invalid registrations and force MN to register with other MAPs when the popular MAP is busy. But in this way, most MNs can't register with its favorite MAP, which leads to frequent home registrations and brings heavy registration flows into the network. Some studies [5, 6] have proposed the point forwarding scheme to decrease the home registration frequency, but as the point forwarding scheme lengthens the routing path, the schemes are proper for the network with small flow of data packets only.

In this paper, we introduce a collaborative load control (COLC) scheme to solve the problem described above. The proposed load control scheme allows MAP to transfer part of its packet delivery load to its neighboring MAPs with lower load and helps MAP to accept more registration from MNs without capacity expansion. This scheme helps to decrease the invalid registration flows, and allow more MNs to register with their favorite MAP, and plays good effect on the registration flow. We evaluate the performance of the proposed scheme through simulations, which shows that the collaborative load control scheme has a significant improvement on decreasing registration flows especially for scale HMIPv6 network and has a better load balance performance.

88.2 Collaborative Load Control Scheme

In this section, the COLC scheme running by MAP is introduced, which helps to decrease the invalid registration flows and allow more MNs to register with their favorite MAP.

88.2.1 Load Control Scheme

In our scheme, each MAP saves the topology map locally, which records the deployment of local network and is updated with the periodic RA from the surrounding MAPs including their coverage and load. Using topology map, MAP can search for a capable MAP list (CML) according to MN's location dynamically, which contains information (address, distance, etc.) about MAPs whose domain covers the AR.

Figure 88.1 shows the key mechanism of COLC and describes the decision-making procedure of MAP when receiving a registration message from MN.

Receiving a regional registration message, MAP searches the MN in its binding cache (BC) with the RCoA encapsulated in the registration message. Binding cache holds all the MNs primarily served by the MAP with their LCoA and RCoA recorded.

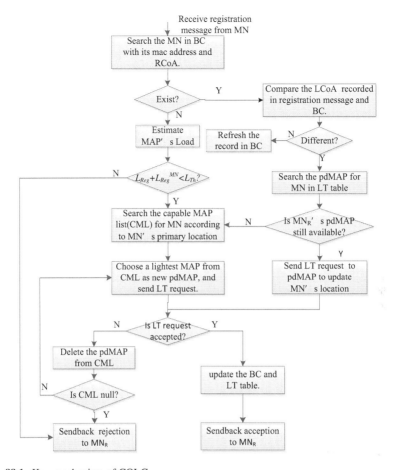

Fig. 88.1 Key mechanism of COLC

When finding an entry for MN in BC, MAP compares the LCoA recorded in entry and encapsulated in the registration message. If the two LCoA are the same, MAP just refreshes the entry. Otherwise, MAP will search the LT table to find the previous packet delivery MAP (pdMAP) for MN, which will direct packets to MN's new location. If the MN still stays in its previous pdMAP's coverage, MAP will start the load transition procedure to update the new LCoA to the previous pdMAP. If the MN has moved out of its previous pdMAP's coverage, MAP needs to choose a new pdMAP with the lightest load from MN's CML and then sends load transition message to the new pdMAP.

If there is no entry for MN in BC, the MN must be a newly accessed node. MAP should check out its load condition to determine whether to accept or reject the newly coming registration. In our scheme, MAP's load consists of packet delivery load (LPd) and registration processing load (LReg). Packet delivery load describes the occupied system capability for packet delivery, which can be computed according to the packet delivery ratio of MAP. Registration processing load describes the occupied system capability for processing registration message, which can be computed according to the registration processing ratio of MAP. In order to ensure MAP's load does not exceed its capability, the total load threshold (LTh) of MAP is set in our scheme. The registration from newly accessed MN will be accepted with the following conditions met:

$$L_{Reg} + L_{Pd} + L_{Reg}^{MN} < L_{Th} \tag{88.1}$$

$$L_{Reg}^{MN} = \text{Ratio}_{moving} \times \varnothing \tag{88.2}$$

In the above condition, L_{Reg}^{MN} is an estimation of the increasing registration processing load when MAP accepts MN's registration, which can be computed with the packet receiving ratio of MN as (88.2). $\varnothing$ is the load estimation parameter. Then MAP needs to validate the RCoA with DAD detection and continue the process only when the detection is passed. After the DAD detection, MAP needs to choose a new pdMAP with the lightest load from MN's CML and then sends load transition (LT) request to the new pdMAP. Otherwise, the registration will be rejected.

When a load transition message is received, which is extended from ICMPv6 [7] with MN's LCoA and RCoA encapsulated, MAP will check whether its load meets the conditions as follows:

$$L_{Reg} + L_{Pd} + L_{Pd}^{MN} < L_{Th} \tag{88.3}$$

$$L_{Pd}^{MN} = \text{Ratio}_{PktRec} \times \delta \tag{88.4}$$

L_{Pd}^{MN} is the estimation of increased load when MAP accepts the load transition request, which can be computed with the packet receiving ratio of MN as (88.4). δ is the load estimation parameter. If MAP's load meets the condition show in (88.3), the MAP will accept the load transition request and record the MN in LT table with the source address of load transition request as RMAP. Otherwise, the load transition request will be denied.

When the load transition response is received, MAP will continue its registration processing procedure. If the LT request is accepted, MAP updates binding cache with MN's new location and pdMAP. Otherwise, the pdMAP should be deleted from CML, and MAP continues the load transition procedure with the new pdMAP chosen from CML until CML is null.

88.2.2 Registration Procedure

The regional registration procedure in the COLC scheme is shown in Fig. 88.2.

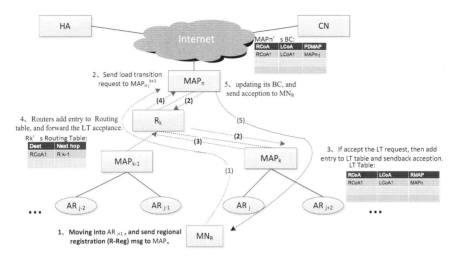

Fig. 88.2 Regional registration procedure

When an MN moves into the area covered by a new access router, AR$_j$, it firstly sends a regional registration message (binding update message) to the MAP covering AR$_j$ (Step 1 in Fig. 88.2), which might be its previous MAP or a new one depending specially on MN's MAP selection scheme. The MAP-received registration message, MAPn, decides whether to accept this registration following the decision-making procedure shown in Fig. 88.1 and sends load transition request to the selected pdMAP (Step 2 in Fig. 88.2), MAPk. When the load transition request is accepted, MAPk adds an entry to its LT table which records MN's RCoA–LCoA address pair and MAPn and then sends the load transition response to MAPn (Step 3 in Fig. 88.2). As load transition message is extended from ICMPv6 [7], each router on the way from MAPk to MAPn will extract the message and add a routing entry which will direct packets destined to MN's RCoA to MAPk (Step 4 in Fig. 88.2). When MAPn received the load transition response from MAPk, MAPn updates its binding cache with MN's RCoA–LCoA address pair and pdMAP's address and then sends back the registration response to MN (Step 5 in Fig. 88.2). If MAPn is not the previous MAP of MN, then MN needs to register with its HA to announce its new RCoA after the regional registration.

In our scheme, binding cache is extended with pdMAP field, which records the address of the MAP sharing the responsibility of packet delivery. If the pdMAP field is set null, then the MAP has to take both the responsibility of registration processing and packet delivery.

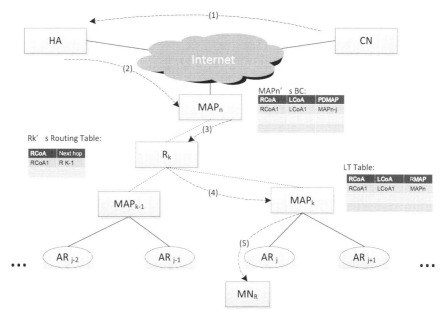

Fig. 88.3 Packet delivery procedure

88.2.3 Packet Delivery Procedure

After the regional registration procedure, MN continues its previous communication. The flow chart of the packet delivery procedure is shown in Fig. 88.3.

When a CN has packets to be sent to MN, the CN firstly sends the packets to the MN's home address (Step 1 in Fig. 88.3), which will be directed to MN's home agent (HA). Then, the HA intercepts the packets (Step 2 in Fig. 88.3) and tunnels them to the registered MAP of the MN (Step 3 in Fig. 88.3), MAPk. Since MAPn and all the routers on the way to MN's pdMAP maintain the routing information about RCoA, the received packets will be transferred directly to MN's pdMAP, MAPk (Step 4 in Fig. 88.3). MAPk maintains the mapping information between the RCoA and the LCoA in LT table; the MAP will re-tunnel the received packets to MN's primary location (Step 5 in Fig. 88.4).

In the above procedure, MN's pdMAP is responsible for packets re-tunneling, which will largely reduce the load of its registered MAP. In some cases, MN's pdMAP can also be its registered MAP.

88.3 Simulation

In this section, the performance of the proposed scheme is evaluated through the network simulator of OMNET++ [8] with extension of xMIPv6 [9]. By rewriting xMIPv6 with the load control scheme proposed in our paper, we simulate the HMIPv6 and the threshold-based load control scheme proposed in RFC3775, respectively, as to compare the performances of each scheme.

88.3.1 Simulation Environment

Figure 88.4 shows the simulation topology consisting of 64 ARs and 21 MAPs deployed hierarchically.

We deploy 64 ARs uniformly with wireless radius of 50 m in the rectangular area of 720 m × 720 m, and three layers of MAPs are deployed upon the ARs, i.e., 16 first-layer MAPs, 4 second-layer MAPs, and 1 third-layer MAP. The first-layer MAPs connect directly to the ARs. Each first-layer MAP covers four neighboring ARs, respectively. The second-layer MAPs are two hops away from the nearest AR, which directly covers four neighboring first-layer MAPs, respectively. The third-layer MAPs are three hops away from the nearest AR, which directly covers four neighboring second-layer MAPs, respectively. Not all the MAPs from the same category do crossover.

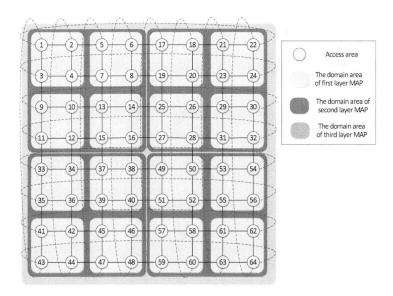

Fig. 88.4 Simulation topology

Table 88.1 Parameter values for simulation

Φ	δ	L_{Th}	$Ratio_{PktRec}$	Init value of $Ratio_{moving}$
1	1	100	1	0.1
				0.2
Number of MNs deployed				Speed range
50	200	600	1,000	5–15 m/s
				15–25 m/s

The MN' mobility follows the random-walk mobility model [10], in which the routing probability for each direction is identical. In addition, the wrap around model [11] is adopted to eliminate the boundary effects. That is, the possible directions from cell 1 are 2, 3, 22, and 43. The velocity of the MN follows the uniform distribution. In order to examine the performance under deferent load condition, we simulate the mobile environment with 50 MNs, 200 MNs, 600 MNs, and 1,000 MNs with speed ranges of [5, 15] or [15, 25], respectively. Each MN follows the distance-based MAP selection scheme proposed in RFC 3775 [1], in which MNs will choose the farthest MAP from MN.

Table 88.1 shows the parameter values used in the simulation.

88.3.2 Simulation Result

Simulation evaluated the load control scheme with three performance metrics: registration admission ratio, registration cost, and load condition of each level MAP.

Figure 88.5 illustrates the average registration admission ratio (RAR) under different load conditions with speed ranges of [5, 15]. RAR is the rate of successful regional registration. The larger RAR there is, the less regional registrations are rejected by the MAP, which helps to reduce the registration flows in the network and decrease the delay spent for registrations. Comparing the performance of the two load control schemes, we can find that the performance of the two schemes is closely related with the load conditions. When there is 50 MNs simulated, the average RAR of two schemes is practically the same. With the increasing of MNs, COLC performs better and better than the threshold-based load control scheme (ThrLC) in RAR. When the number of MNs is increased to 1,000, the average RAR of COLC is significantly lower than ThrLC. It is because MAPs can transfer part of their packet delivery load to neighboring MAPs to accept more registrations in collaborative load control scheme. Comparing with the threshold-based load control scheme in RAR, we can conclude that our scheme can significantly improve the RAR of HMIPv6 network especially for the high-density networks.

Figure 88.6 illustrates the total registration costs under different load conditions with speed ranges of [5, 15] and [15, 25], respectively. Registration cost is a metric for the registration traffic in network, which can be computed based on the number and transmission range of registration messages as follows:

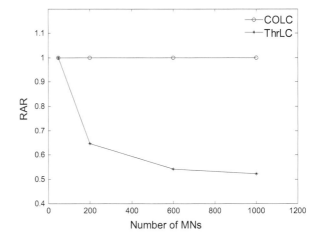

Fig. 88.5 Average registration admission ratio (RAR)

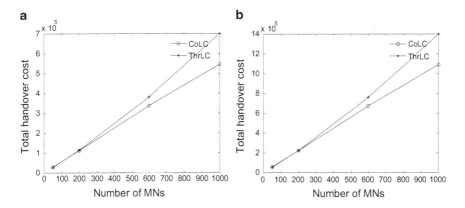

Fig. 88.6 Registration cost. (**a**) Speed range [5, 15]. (**b**) Speed range [15, 25]

$$C_{Reg} = \sum Hops_{ri} \times Weight_{ri} \qquad (88.5)$$

Where **Hops**$_{ri}$ is the transmission hops of registration message ri and **Weight**$_{ri}$ is the one-hop transmission cost of ri. In our simulation, **Weight**$_{ri}$ is set to 1.

From Fig. 88.6a, we can see that there exist two conditions in registration cost comparison. When the number of mobile nodes is small (50 or 200 in our simulation), the total registration cost of COLC scheme is a little higher than threshold-based load control scheme. It is because that the registration procedure applied with COLC scheme is a little more complex than the threshold-based load control scheme, which leads to higher registration cost. However, the total registration cost of COLC scheme becomes much smaller than threshold-based load control scheme when the number of mobile nodes increases. It is because COLC scheme performs better and better in registration admission ratio with mobile nodes

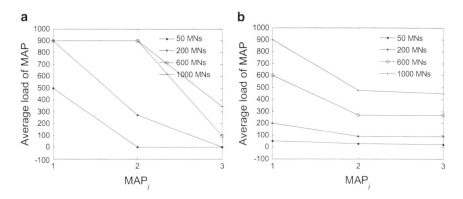

Fig. 88.7 Load balance performance. (**a**) Threshold-based load control scheme. (**b**) Collaborative load control scheme

increasing and leads to less invalid registrations and lower registration traffic. Comparing Fig. 88.6a, b, we can find that the velocity of mobile nodes has a significant effect on the registration cost. It is obvious that COLC scheme can perform much better than the threshold-based load control scheme in decreasing registration traffics for HMIPv6 network, which affords larger number of mobile nodes with high velocity.

Figure 88.7 illustrates the load balance performance of the two load control schemes under different load conditions. In our simulation, MAPs are sorted into three classes, which are identified as MAP_1, MAP_2, and MAP_3, according to the distance from the MAP to its closest AR. MAP_1 is the farthest MAP from AR (three hops away) and manages the largest domain area. MAP_3 is the closest MAP from AR (one hop away) and manages the smallest domain area. As MAP's domain has an important effect on mobile node's regional registration, mobile nodes will firstly register with the MAP_1 to acquire lowest regional registration ratio, which always leads load imbalances. Comparing Fig. 88.7a, b, we can see that COLC scheme performs better on load balance of MAPs than threshold-based load control scheme.

From the simulation results above, we can get a conclusion that the collaborative load control scheme has a significant improvement on the registration admission ratio for the large-scale HMIPv6 network, which helps to reduce the registration traffic in the network. In addition, our scheme performs better on load balance.

88.4 Conclusion

In this paper, we proposed a collaborative load control (COLC) scheme for hierarchical mobile IPv6 to improve the registration admission ratio for mobile nodes, which helps to reduce registration traffics in the network. With COLC scheme, MAPs are allowed to transfer part of their packet delivery load to neighboring

MAPs with lower load, and the MAPs accept more registration from MNs without capacity expansion. In this way, the registration traffic is reduced. The simulation results show that our scheme markedly reduced the registration flows of HMIPv6 and has a better load balance performance.

References

1. Soliman, H., Castelluccia, C., & EL Malki, K. (2005). *Hierarchical mobile IPv6 mobility management*. IETF RFC 4140.
2. Kim, Y., Kim, M., & Mun, Y. (2006). Performance analysis of the mobility anchor point in hierarchical mobile IPv6. *IEICE Transactions on Communications, E89-B*(10), 2715–2721.
3. Wang, Y. H., Huang, K. F., & Kuo, C. S. (2008). Dynamic MAP selection mechanism for HMIPv6. In *International Conference on Advanced Information Networking and Applications* (pp. 691–696). Okinawa: Institute of Electrical and Electronics Engineers.
4. Zhou, W., & Hong, P. L. (2008). A MAP-controlled load balance scheme for hierarchical mobile IPv6. In *International Conference on Wireless Communications* (pp. 691–696). Dalian: Institute of Electrical and Electronics Engineers.
5. Yi, M., & Hwang, C. (2004). A pointer forwarding scheme for minimizing signing cost in hierarchical mobile IPv6 networks. *Lecture Notes in Computer Science, 32*(7), 333–345.
6. Yi, M., Choi, J., & Yang, Y. (2007). A pointer-forwarding scheme for minimizing signaling costs in nested mobile networks. In *International Conference on Networks* (pp. 230–234). Adelaide: Institute of Electrical and Electronics Engineers.
7. Conta, A., Deering, S., & Gupta, M. (2007). *Internet control message protocol for the internet protocol version 6 specification*. IETF RFC 4443.
8. http://www.omnetpp.org/omnetpp
9. http://github.com/zarrar/xMIPv6
10. Camp, T., Bolen, J., & Davis, V. (2002). A survey of mobility models for ad-hoc network research. *Wireless Communication and Mobile Computing, 2*(5), 483–502.
11. Zeng, H., Fang, Y., & Chlamtac, I. (2002). Call blocking performance study for PCS networks under more realistic mobility assumptions. *Telecommunication Systems, 19*(2), 125–146.

Chapter 89
A High Efficient Selective Content Encryption Method Suitable for Satellite Communication System

Yanyan Xu, Bo Yang, Zhengquan Xu, and Tengyue Mao

Abstract Data transmitted by satellite communication system should be encrypted in order to provide confidentiality. A selective content encryption method suitable for satellite communication system is presented in this chapter, the key content information in the compressed stream is extracted and encrypted, and the variable modulus encryption method is proposed to solve the problem of variable length code encryption; thereby, the encrypted stream can be format compliant. This method can improve the efficiency of encryption and achieve fast, secure, and high efficient encryption of satellite communication system. The experimental results prove the effectiveness of our method.

89.1 Introduction

With the rapid development of broadband satellite communication system, its security is becoming an important issue. The data stream transmitted by satellite links is often very important and cannot be accessed by unauthorized users. Therefore, the data confidentiality is the most important security requirement of satellite network [1, 2].

At present, the most common encryption method in satellite communication is to use secure transmission protocols to do channel encryption, such as IPSec and TLS/SSL [2, 3]. Although IPSec has been successfully used in the Internet, it

Y. Xu (✉) • Z. Xu
State Key Lab of Information Engineering in Surveying, Mapping and Remote Sensing, Wuhan University, Wuhan 430079, China
e-mail: xuyy@lmars.whu.edu.cn

B. Yang
The Academy of Satellite Application, Beijing 100086, China

T. Mao
College of Computer Science, South-Central University for Nationalities, Hubei 430074, China

will cause some special problems when it is used in satellite networks, such as incompatibility with TCP performance enhancement technologies [4, 5]. SSL and TLS are only suitable for TCP connections and cannot be used in UDP connections. Moreover, the cost of secure transmission method using channel encryption is relatively high, the delay caused by encryption is high [6], and it is unacceptable to the multimedia service which has high requirement for real-time performance.

An encryption method based on DVB_RCS satellite communication network is presented in this chapter. A selective encryption method is used to extract key content information, which has most important effects on data reconstruction, and a variable modulus encryption method is proposed to solve the difficult problem of variable length coding codeword encryption; therefore, ciphertext can be format compliant. The method reduces the amount of data needed to be encrypted and improves the encryption efficiency. The experimental results prove the effectiveness of the method.

89.2 The Secure Requirement of DVB-RCS Satellite Communication System

The DVB-RCS standard can provide real-time multimedia services and it has been widely used in commercial broadband satellite systems. The structure of traditional DVB-RCS satellite communication system is presented in Fig. 89.1. The media streams are encapsulated in IP packet to transmit, while the forward link and the return link are asymmetric and the transmission delay is high. Data stream between center station and VSAT needs one-hop link to transmit and the delay is about 600 ms, while data stream between two VSATs needs two-hop links to transmit and the delay is about 1,200 ms. Thus, the DVB-RCS system has high requirements to encryption efficiency in order to ensure that data stream can be processed in high efficiency and satisfies the real-time requirement of multimedia service. At present, most satellite systems encrypt all data stream to get data confidentiality [7]; however, multimedia information, especially video information transmitted by satellite communication systems, is massive; if the data is encrypted without differential, then it will result in time and computing resource consumption and lead to high delay.

Image and video information are often compressed first before they are transmitted in order to save bandwidth. The compressed media stream is composed of several kinds of information, such as flag information, padding information, coding control information, channel encoding information, and source content information. There exist some problems if all information is encrypted: (1) Encrypting fixed format fields will cause the known-plaintext attacks. (2) Most format information does not consist of real source content information. The encryption of format information will result in computing resource consumption and lead to unnecessary overhead. (3) The encryption of some channel information such as synchronization

89 A High Efficient Selective Content Encryption Method...

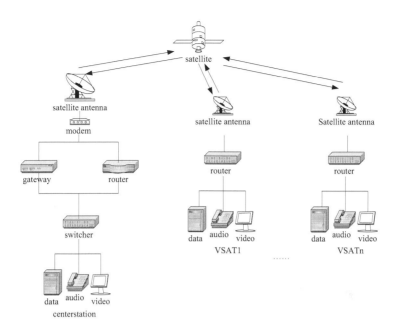

Fig. 89.1 Model of DVB-RCS satellite communication system

fields and fault-tolerant fields will cause the influence of the network adaptability and fault tolerance. According to these reasons, the encryption of DVB-RCS satellite communication systems should be format compliant, that is, the structure and syntax of ciphertext should be compliant with the standard.

Image and video information have the source feature, that is, the proportion of key data impacting on image reconstruction most is lower than 5 % [8]. A selective content encryption method is proposed in this chapter, and the high-intensity encryption method is used to encrypt key data, which has most important influence on data reconstruction, while the lightweight encryption method is used to encrypt large amount of the remaining redundancy data. By this method, the data need to be encrypted is reduced and the computing complexity is low, and the delay caused by encryption is low, too. At the same time, the difficult problem of VLC codeword encryption is solved in the chapter. The syntax information in the compressed streams such as encoding format and channel coding information will not be changed by encryption so that the encryption will not affect data compress ration and the ciphertext can be format compliant. The key data is encrypted first, then returned to original bit streams, and encapsulated to IP packet to transmit. This method will not have any influences to DVB-RCS satellite communication system data encapsulation and can be integrated to the whole system seamlessly.

89.3 A Format-Compliant Selective Content Encryption Method

A format-compliant selective content encryption method is proposed in this chapter; only key content information is encrypted; therefore, the encryption efficiency is high. A variable modulus encryption method is also proposed to solve the ciphertext format-compliant problem.

89.3.1 The Exaction of Key Information

The exaction of key information is very important in our method. Main content information is DCT coefficients and motive vector (MV) codeword.

AC coefficients are related to image contour, and DC coefficients are related to average luminance and chrominance of each MB. Therefore DC and AC coefficients can be chosen as key information.

MV includes image's motion information and is used to do motion compensation. If the MV cannot be decoded correctly, the motion information cannot be recovered and the image quality will be reduced remarkably. Therefore, MV can be used as key information.

Except for DCT coefficients and MV, different coding standard of visual media has different content information. For example, the predict mode codeword and quantization information in H.263 standard can be treated as key information.

89.3.2 The Encryption of VLC

Keeping encrypted stream format compliant is the difficult issue of VLC encryption. In image and video compression standards, transformation and quantization process are followed by the variable length coding process to get higher compression ratio. The length of VLC codeword is variable and cannot occupy the whole codeword space, and direct encryption of VLC codeword will result in invalid VLC codeword and the altering of encrypted stream structure so that the ciphertext will not be compliant to the syntax of compression standard. Take 3-bit codeword for example; there are totally 8 different codeword values (000–111) corresponding to 3-bit codeword; however, only 2 VLC codewords are valid: 100 and 101. Because of the randomness of encryption operation $E()$, two valid plaintext codeword will map to the random position in ciphertext space, which includes the grey region and translucent grey region of Fig. 89.2a. If the valid plaintext codeword is mapped to the translucent grey region, the ciphertext will not be format compliant. Only mapping plaintext to the grey region can get valid ciphertext codeword, as shown in Fig. 89.2b.

89 A High Efficient Selective Content Encryption Method...

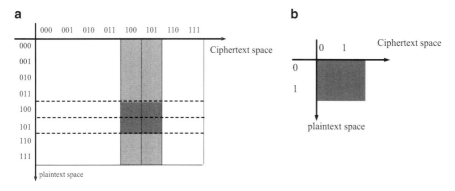

Fig. 89.2 The valid field and invalid field of encryption mapping. (**a**) The mapping from plaintext space to ciphertext space. (**b**) *Grey* region-valid VLC region

Table 89.1 MPEG4 VLC codeword

row	VLC	length(bit)	Last	Run	Level	modulus	Sequence number
1	10 s	3	0	0	1	2	1, 2
2	110 s	4	0	0	2	2	1, 2
3	1111 s	5	0	0	3	4	1, 2
4	1110 s		0	1	1		3, 4
...	......	...	...	...	...	...	...

A variable modulus encryption method is proposed in this chapter in order to solve the problem of VLC codeword encryption. Firstly, valid equal-length VLC codeword table is constructed. Take AC coefficients for example; after run-length coding, the data is represented as an event of a combination of (Last, Run, Level). VLC codeword is found in the table according to (Last, Run, Level). In our method, VLC codewords having the same length are classified to the same group, and each group corresponds to an alphabet. Codewords in the same alphabet are sequenced, each codeword is allocated to a sequence number, and the number of codewords in the group is the modulus of the alphabet. By this way a new table is constructed, where two columns including modulus and sequence number are added.

For example, Table 89.1 shows the MPEG4 VLC for intra-luminance and -chrominance TCOEF, the "s" in the VLC codeword is the sign bit (1 means negative and 0 means positive), and the first row expressing event (Last $= 0$; Run $= 0$; Level $= \pm 1$) has 2 3-bit codeword: 100 and 101, and the modulus is 2; the sequence number of 101 is 1 and 102 is 2. The modulus of the second row is also 2, where the sequence number of 1101 is 1 and 1100 is 2. After the table is constructed, the plaintext can be encrypted. The theory is shown in Fig. 89.3.

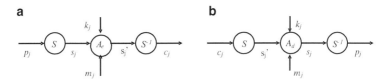

Fig. 89.3 The theory of variable modulus encryption. (**a**) Encryption. (**b**) Decryption

Assuming plaintext sequence $P = p_1 p_2 \ldots p_j \ldots$; the modulus sequence is $M = m_1 m_2 \ldots m_j \ldots$; the plain text P can be mapped to $S = s_1 s_2 \ldots s_j \ldots$, ($s_j \in [0, M_j - 1]$, $j = 1, 2, \ldots$) according to Eq. (89.1):

$$s_j = S(p_j) \tag{89.1}$$

For each s_j and its modulus m_j, k_j is generated in $[0, m_j-1]$, $j = 1, 2, \ldots$, and by Eq. (89.2) we get $s_j{'}$:

$$s'_j = (s_j + k_j) \bmod m_j = (s_j + k_j) \% m_j \tag{89.2}$$

According to Eq. (89.3) we get ciphertext $C = \{c_j : j = 1, 2, \ldots\}$:

$$c_j = S^{-1}(s'_j) \tag{89.3}$$

In the decryption process, with the same mapping rule $S()$, c_j can be mapped to $s_j{'}$, as shown in Eq. 89.4:

$$s'_j = S(c_j) \tag{89.4}$$

With the same key sequence k_j using in the encryption, and because $0 \leq s_j$, $k_j < m_j$, and $s_j \bmod m_j = s_j$, $k_j \bmod m_j = k_j$, we get s_j by Eq. (89.5):

$$(s' - k_j) \bmod m_j = (s_j + k_j - k_j) \bmod m_j = s_j \bmod m_j = s_j \tag{89.5}$$

Remapping s_j to p_j we get plaintext sequence P, as shown in Eq. (89.6):

$$p_j = S^{-1}(s_j) \tag{89.6}$$

For the discrete random variable k_j, when j is different, the sample space is different, and the modulus is variable; thus, the $K = \{k_j : j = 1, 2, \ldots\}$ is called as variable random sequence. For each mod m_j add operation, the value of m_j is variable, and the mod add operation is called as variable mod add operation.

For example, the first row in Table 89.1 is the alphabet of 3-bit codeword, and the valid VLC codewords are 100 and 101. According to the standard VLC codeword table there are two events: Event 1 (Last = 0, Run = 0, Level = 1) and Event 2 (Last = 0, Run = 0, Level = −1). If the jth event waiting for coding is Event 1 and the modulus of this alphabet is 2, then a random number R_j in [0, 1] is generated. If $R_j = 0$, we can get$(0 + 0)\%2 = 0$, and then the encrypted VLC codeword is 100; if $R_j = 1$, then the encrypted VLC codeword is 101. In the authorized decryption side, with the same R_j, the plaintext Event 1 can be recovered. Otherwise, the unauthorized user cannot determine which one is the right result because R_j is unknown.

89.4 Experiment Results

Using our method to encrypt four MPEG4 testing video sequences, the results are shown in Fig. 89.4. The encrypted video is unrecognizable, and the unauthorized users cannot get any video information.

The proportion of key information in the whole compressed stream of the testing video sequences is shown in Table 89.2. We can see that the proportion of key information is no more than 20 % of the compressed stream, and the information

Fig. 89.4 Encryption results of MPEG4 map of experiment collocation

Table 89.2 The computing complexity of MPEG encryption

Test sequence	Key information/ compressed stream (%)	Time spent in the exaction of key information (ms/frame)	Time spent in encryption/coding (proportion)
Foreman	18.69	2.10	18.47
Mother_daughter	13.72	1.16	15.26
News	11.98	1.38	20.03
Tempete	18.24	2.60	22.27
Mobile	17.28	2.91	21.60
Hallmoniter	15.71	2.12	17.02

needed to be encrypted only accounts for a small part of the whole compressed stream. We can also see that the time spent in encryption is only equal to 20 % of encoding time. By this way the computing complexity is reduced and the high encryption efficiency is guaranteed. Therefore our method is suitable for real-time application.

89.5 Conclusion

A high efficient selective content encryption method suitable for DVB-RCS satellite communication system is presented in this chapter, the key content information in the compressed stream is extracted and encrypted, and the variable modulus encryption method is proposed to solve the problem of variable length code encryption; thereby, the encrypted stream can be format compliant. The experimental results prove the effectiveness of our method.

Acknowledgments This work is supported by National Natural Science Foundation of China (No. 41101416), National Basic Research Program of China (No. 2011CB302204), and Open Research Fund of The Academy of Satellite Application (No. 20121689).

References

1. Pillai, P., & Yim-Fun Hu. (2006). Design and analysis of secure transmission of IP over DVB-S/RCS satellite systems. *Proceedings of the International Conference on Wireless and Optical Communications Networks* (pp. 1–5). Santiago, Chile: Springer
2. H. Cruickshank, S. Iyengar, S. Combes, L. Duquerroy, G. Fairhurst, & M. Mazzella. (2007). Security requirements for IP over satellite DVB networks. *Proceedings of the Sixteenth Mobile and Wireless Communications Summit, IEEE* (pp. 1–6). Piscataway, NJ
3. Qi wang, & Shengwu Wang (2009). Securing your satellite network and its contents. *Satellite and Network, 90*(12), 42–45.
4. Peng, C. (2010). *Research on key security technologies in space networks*. Changshai: National University of Defense Technology (in Chinese).
5. Guevara Noubir, Laurent von Allmen. (1999). Security issues in internet protocols over satellite links. *Proceedings of the IEEE Vehicular Technology Conference, IEEE* (pp. 2726–2730). Piscataway, NJ
6. Jonah, P. (2007). Performance implications of instantiating IPsec over BGP enabled RFC 4364 VPNs[C]. *Proceedings of IEEE Military Communications Conference, IEEE Piscataway, NJ* (pp. 1–7)
7. Xie, D. (2010). Discussion on long-distance encrypt technology of satellite HDTV. *Radio and TV Broadcast Engineering, 37*(10), 147–150 (in Chinese).
8. Yang, Z. (2005). An overview of encryption scheme for digital video. *Geomatics and Information Science of Wuhan University, 30*(7), 570–574 (in Chinese).

Chapter 90
Network Design of a Low-Power Parking Guidance System

Ming Xia, Yabo Dong, Qingzhang Chen, Kai Wang, and Rongjie Wu

Abstract A parking guidance system can help a driver quickly find an available parking space. Most currently available parking guidance systems require wire deployment in installation, thus entailing high installation costs. In this chapter, we discuss the network design of a low-power parking guidance system. We developed a tiered communication architecture including Wireless Sensor Network (WSN), General Packet Radio Service (GPRS) network and Internet to realize wireless parking space availability data transmission, and thus installation complexity can be greatly reduced. In order to reduce the battery replacement frequency of the WSN, we designed a power-minimized Medium Access Control (MAC) protocol. The proposed MAC protocol divides one network working cycle into four dedicated intervals to realize robust network organization and energy-efficient data delivery. Experimental results showed that the proposed MAC protocol can extend the battery lifetime of the WSN to more than ten years. Based on the collected parking space availability data, we built a portable parking guidance terminal to let drivers locate available parking spaces conveniently.

M. Xia (✉)
College of Computer Science and Technology, Zhejiang University, Hangzhou 310027, China

College of Computer Science and Technology, Zhejiang University of Technology, Hangzhou 310023, China
e-mail: xiaming@zjut.edu.cn

Y. Dong
College of Computer Science and Technology, Zhejiang University, Hangzhou 310027, China

Q. Chen • K. Wang • R. Wu
College of Computer Science and Technology, Zhejiang University of Technology, Hangzhou 310023, China

90.1 Introduction

Searching for an available parking space in urban areas is becoming more and more annoying to drivers. This is partly caused by the rapid increase of the number of cars in large cities, and partly by the lack of parking space availability information. In order to alleviate the problem, parking guidance systems, which detect and guide drivers to available parking spaces, were developed in recent years. However, many currently available parking guidance systems require communication wire to transmit parking space availability information [1], thus entailing high costs in system deployment. Researchers proposed to use Wireless Sensor Networks (WSNs) to realize full wireless deployments [2]. A WSN typically consists of a large number of low-power, low-cost wireless sensor nodes to collect, process and transmit data collaboratively. As a result, sensor nodes can be battery powered, and the deployment complexity can be greatly reduced. Current researches on applying WSNs to parking guidance systems have discussed sensor design [3], vehicle detection algorithm design [4] and parking space searching policy design [5]. Nevertheless, we found that communication strategies including communication architecture and protocols are also critical issues in system design. As a result, we will focus on the design of efficient communication architecture and protocols for parking space availability information transmission in this chapter. First, we will introduce our tiered communication architecture including WSN, General Packet Radio Service (GPRS) network and Internet to realize wireless and wide-area parking space availability information transmission. Then, we will present the design of a power-minimized WSN Medium Access Control (MAC) protocol to maximize the battery lifetime of parking space sensors. We chose battery lifetime maximization as our major optimization object because parking space sensors are installed on the ground and battery replacement is relatively difficult. In order to reduce maintenance costs, long battery lifetime is a must. After that, we will discuss system implementation on customized hardware and a portable parking guidance terminal based on smartphone to enable drivers to acquire parking space availability information at any place and any time. At last, the experimental deployment of the system at a parking lot verifies the effectiveness of the proposed network design.

90.2 Network Design

We developed a tiered communication architecture which integrates WSN, GPRS network, and Internet to meet the deployment requirement of parking space availability monitoring and parking guidance, as shown in Fig. 90.1.

WSN: The communication between parking space sensors and gateway nodes employs WSN technology. In order to maximize the battery lifetime of parking space sensors, we designed a power-minimized MAC protocol.

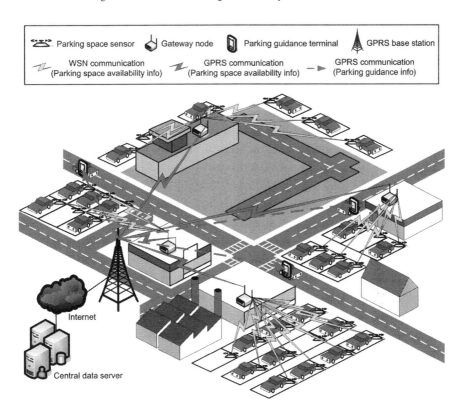

Fig. 90.1 Communication architecture

GPRS network and Internet: The communication between gateway nodes and the central data server uses GPRS network and Internet to ensure the large coverage area. At the same time, central data server will send the parking guidance information to the portable parking guidance terminal through GPRS network and Internet.

We will then elaborate on the details of the MAC protocol design. One of the major tasks of our MAC protocol for WSNs is to reduce the power consumption of sensor nodes. The design principles of WSN MAC protocols can be roughly categorized as either contention-based or Time Division Multiple Access (TDMA)-based. We chose to employ TDMA in our MAC protocol design because it does not suffer from collisions and thus can frequently achieve lower power consumption.

Generally, the gateway node is responsible for the establishment of the network, and the parking space sensors will select and join the "best" network established by one gateway node. The gateway node divides one working cycle into four intervals, including JOINING, TESTING, COLLECTING and UPLOADING, as shown in Fig. 90.2. In different intervals, the gateway node broadcasts different types of beacons periodically to notify parking space sensors.

Fig. 90.2 Four intervals of the proposed MAC protocol

The proposed MAC protocol works as follows:

(1) A gateway node scans all usable channels one by one, and chooses one silent channel to establish the network.
(2) A parking space sensor scans all usable channels one by one when powered on. It records all beacons captured in scanning and compares the signal strength and will try to select the gateway node with the strongest signal strength to transmit parking space availability information. In order to notify the selected gateway node, the parking space sensor waits for the JOINING beacon from the selected gateway node and immediately replies a *JOINING_REQUEST* after the beacon. Here we use a simple Carrier Sense Multiple Access/Collision Avoidance (CSMA/CA) protocol to avoid collision.
(3) The gateway node that receives the *JOINING_REQUEST* will reply a *JOINING_CONFIRMATION* to the parking space sensor if it is able to adopt more sensors. In the *JOINING_CONFIRMATION*, the gateway node tells the parking space sensor the time slot assigned to it in the TESTING interval. If the gateway node is unable to adopt more sensors, it will reply a *JOINING_REJECTION* to the parking space sensor and the sensor will repeat step (2) and select another gateway node.
(4) A parking space sensor that receives the *JOINING_CONFIRMATION* will go sleep and wait for the time slot assigned to it to wake up in the TESTING interval. It will start testing (i.e., transmitting several packets to the gateway node and calculating the transmission success rate) immediately once it receives the TESTING beacon broadcasted by the gateway node. We designed a dedicated TESTING interval to ensure the communication quality between the parking space sensor and the selected gateway node because the received signal strength frequently cannot accurately reflect the link quality. If the transmission success rate is high enough, the parking space sensor will send a *REGISTERING_REQUEST* to the selected gateway node. Otherwise, it will go back to step (2) and select another gateway node.
(5) A gateway node that receives the *REGISTERING_REQUEST* will reply a *REGISTERING_CONFIRMATION* to the parking space sensor telling it the time slot assigned in the COLLECTING interval.
(6) A parking space sensor that receives the *REGISTERING_CONFIRMATION* will go sleep and wait for the time slot that is assigned to it to wake up in the COLLECTING interval. Once it wakes up and receives a COLLECTING beacon, it will transmit parking space availability information to the gateway node immediately. The gateway node will cache the received parking space availability information for uploading.

(7) In the UPLOADING interval, the gateway node uploads the received parking space availability information to the central data server via GPRS network.
(8) In the next round, a parking space sensor will only wake up in its time slot in the COLLECTING interval, and the COLLECTING beacon will be the time synchronization signal to avoid clock drifting. In order to conserve energy, the sensor will transmit data only if the state of the parking space (occupied/available) changed. Otherwise, the sensor only reports its state at a low frequency. If the sensor encounters continuous data transmission error, it will go back to step (2) to re-join a network.

90.3 Hardware Design

In this section, we will discuss the design of parking space sensors, gateway nodes and parking guidance terminal.

Parking space sensor. Our parking space sensor is directly installed on the ground of a parking space, and thus we designed a special robust and water-resistant enclosure for it. The design of the enclosure and its photo are given in Fig. 90.3. We chose an STMicroelectronics STM32F103 as microcontroller and an Atmel AT86RF212 as radio chip, which works on 700/800/900 Mhz. We chose a Honeywell HMC5883L geomagnetic sensor to monitor if the parking space is occupied. If there is car on the sensor, the earth's magnetic field will be changed, and we can monitor this event through measuring the output of the sensor.

Gateway node. The gateway node also adopts an STMicroelectronics STM32F103 as its microcontroller, and it has two radio chips, one is the 700/800/900 Mhz Atmel AT86RF212 radio chip for the communication between the parking space sensors and the gateway node and the other is a GPRS modem for the communication between the gateway node and the central data server. The gateway node uses a rechargeable battery as its power supply, because the gateway node may be installed on a street lamp and directly uses the power from the street lamp. Unfortunately, the street lamp is frequently powered on only at night, and thus we designed a recharging circuit to charge the battery at night, and the battery will provide power supply to the gateway node in day time. The gateway node is

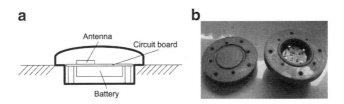

Fig. 90.3 Parking space sensor. (**a**) Sensor design concept (**b**) The photo of the sensor

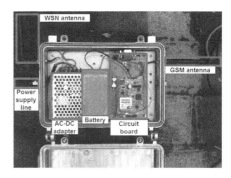

Fig. 90.4 Gateway node

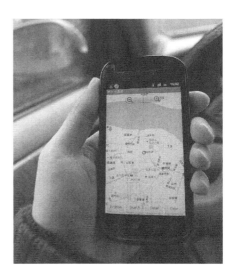

Fig. 90.5 Parking guidance terminal

encapsulated in a water-resistant enclosure. The inner structure of a gateway node is given in Fig. 90.4.

Parking guidance terminal. The parking guidance terminal is built based on an Android smartphone, as shown in Fig. 90.5, in which the small circle on the screen indicates the current position of the car, and the blue arrow on the screen indicates the position of the detected available parking space. We used a Global Positioning System (GPS) receiver to mark the positions of parking spaces in system deployment. Because the position measuring error is only several meters, we believe that user experience will not be affected. The parking guidance terminal generally works as follows: (1) when a driver arrives at the destination, he just presses the "search" button on the screen; (2) the terminal communicates with the central data server to find nearby available parking spaces and then tells the driver; (3) the driver selects his favourite available parking space and the terminal will guide the driver to the chosen available parking space.

90.4 System Deployment and Evaluation

We have deployed an evaluation system covering about 12 parking spaces at a parking lot. Figure 90.6 shows the deployment of a parking space sensor and a gateway node. In deployment, we let parking space sensors transmit 100 packets in the TESTING interval and set the transmission success rate threshold to 80 %. Under these settings, parking space sensors typically spent no more than three working cycles to establish stable connections to gateway nodes.

Because of the extremely low active time of our parking space sensors after network joining, the battery lifetime can be extended to several years in deployment. We estimate parking space sensor's battery lifetime based on its typical working parameters. Table 90.1 gives the parameters.

We can estimate the battery lifetime (T_B) of parking space sensors according to (90.1), and the result is about 10.8 years.

$$T_B = W_B T_D / (I_A T_A + I_S (T_D - T_A)) \tag{90.1}$$

Fig. 90.6 System deployment

Table 90.1 Typical working parameters of parking space sensors

Item	Value
Active current (I_A)	40 [mA]
Maximum active time (T_A)	0.05 [s]
Sleep current (I_S)	0.03 [mA]
Working cycle (T_D)	60 [s]
Battery capacity (W_B)	6000 [mAh]

90.5 Conclusion

In this chapter, we focused on the network design of a low-power parking guidance system. We developed a tiered communication architecture including WSN, GPRS network and Internet, and thus the deployment complexity can be greatly reduced. In order to maximize the battery lifetime of parking space sensors to reduce battery replacement costs in system maintenance, we designed a power-minimized TDMA-based MAC protocol for the WSN. Experimental results showed that the proposed MAC protocol can extend the battery lifetime of parking space sensors to more than 10 years.

Acknowledgments This work is supported by the Research Program of Department of Science and Technology of Zhejiang Province (2012C33073), the Key Scientific and Technology Project of Zhejiang Province (ZD2009011), and the Collaborative Industry-University Research Project of Hangzhou (20112731E54).

References

1. Yao, G. Z., Wang, J. Q., Li, Z. S., Ran, X. J. (2010) The design of parking guidance and information system based on CAN[C]. *Proceedings of the 2010 International Conference on Intelligent Control and Information Processing IEEE, Dalian, China* (pp. 171–174).
2. Bi, Y. Z., Sun, L. M., Zhu, H. S., Yan, T. X., & Luo, Z. J. (2006). A parking management system based on wireless sensor network[J]. *ACTA AUTOMATICA SINICA, 32*(6), 968–977.
3. Idris, M. Y. I., Tamil, E. M., Noor, N. M., Razak, Z., & Fong, K. W. (2009). Parking guidance system utilizing wireless sensor network and ultrasonic sensor[J]. *Information Technology Journal, 8*(2), 138–146.
4. Yoo, S., Chong, P. K., Kim, T., Kang, J., Kim, D., Shin, C., Sung, K., Jang, B. (2008). PGS: Parking guidance system based on wireless sensor network[C]. *Proceedings of the 3rd International Symposium on Wireless Pervasive Computing IEEE, Santorini, Greece* (pp. 218–222).
5. Wang, H. W., He, W. B. (2011). A Reservation-based smart parking system[C]. *Proceedings of the 2011 International Conference on Computer Communications Workshops. IEEE, Shanghai, China* (pp. 690–695).

Chapter 91
Strategy of Domain and Cross-Domain Access Control Based on Trust in Cloud Computing Environment

Bo Li, Ming Tian, Yongsheng Zhang, and Shenjuan Lv

Abstract Under the current cloud computing environment, a reasonable and practicable access control strategy is needed, which is a guarantee to protect cloud computing suppliers to provide services and many cloud users access to services. In this paper, based on analysis of many cloud computing safety features, trust management is introduced into the cloud computing service access control, within the domain of a trust-based access control strategy, in domain, presents a trust-based access control policy. Credible value will be given through the comprehensive treatment of the entity, and then AAC (authentication and authorization center) authorizes the appropriate access rights to achieve the control of the monomer in the domain. Combined with the characteristics of the existing cloud computing environment, in multiple management domains, this paper proposes a role mapping, with the role mapping relationship between the domain, which can make the inter-domain access to resources and security shared access between different domains, in order to avoid the problem of permission penetration and privilege escalation, this paper presents the mirror role based on role mapping, ultimately solves the problem.

B. Li
Academic Affairs Office Shandong Polytechnic, Jinan 250104, China

M. Tian (✉) • Y. Zhang • S. Lv
School of Information Science and Engineering, Shandong Normal University, Jinan 250014, China

Shandong Provincial Key Laboratory for Novel Distributed Computer Software Technology, Jinan 250014, China
e-mail: Tiancius@163.com

91.1 Introduction

Cloud computing era poses a huge shock to the traditional information industry, changes the traditional IT applications, and has created an enormous change to our lives. Individual or enterprise users can access to IT services only by a cloud service provider who provides a simple operation interface; users no longer need to upgrade and maintain hardware and software. Cloud computing brings great convenience to our lives, but it also presents some security risks [1]. According to the understanding of the definitions and concepts of cloud computing, the mode of cloud computing operation is to provide the user data and the corresponding computing tasks to the server. Storage of user's data, as well as operations such as handling and protection of user data, is in the "cloud" to complete. In this way, it will inevitably make the user's data in a potentially unsafe state of destruction and theft and also have more detailed personal information exposed on the network, which is a very large exposure. Judging from today's cloud computing development, the security of user's data, user privacy information protection, data stored and cloud computing to their own security and stability, and many other regulatory aspects of cloud computing issues directly relate to cloud computing user acceptance; thus, security becomes the most important factor affecting the development of cloud computing business [2]. A reasonable and practicable access control strategy is needed, which is a guarantee to protect cloud computing suppliers to provide services and many cloud users access to services.

91.2 Related Works

Cloud computing acts as a new information service mode, which brings new security risks and challenges, but has no essential difference with traditional IT security information service requirements [3,4]. It remains the core requirement of data and application of confidentiality, integrity, availability, and protection of privacy, and key technologies to meet these security requirements are access control technologies [5,6].

The document [7] analyzes dynamic demand for access control in a cloud computing environment and makes role-based access control (RBAC) model to a cloud computing environment to meet the need of complex access control management of cloud computing and dynamic management for access control and increase the maintainability. But the RBAC model is based on identification and close, whose access control mechanisms in centralized closed network environments, and does not apply to large-scale, distributed and open network, especially unable to meet the security needs of cloud computing environments [8]. In addition, the RBAC model, in assigning roles to users, only verifies the authenticity of the identity of a user, without taking into account the user's credibility.

RBAC model is used for allocating roles for access authorization, users who actually use permission do not be supervised and controlled, and insufficient against the RBAC model to extend it, Blaze's "trust management", based on trust introduced the concept of access control mechanisms, proposed a trust-based access control model TRBAC (Trust Role-Based Access Control Model) [9]. But some of the above study did not take into account the characteristics of the cloud computing security management domain, and did not explicitly give access control method for cloud computing environments.

91.3 The Trust in the Cloud

Human society is a complex system. The interactions between the entities in the system depend on trust relationships between each other. Cloud computing researchers now introduce the trust mechanism from the human society into the cloud computing environment as a basis of the exchange between entities in the cloud computing environment [10].

Trust in real life is a subjective concept, depending on the person's experience, and trust relationships would be difficult to reference to a cloud computing environment by using empirical measure.

We can use trust value to determine the degree of trust. The degree of trust allows the definition of security strategy to be more clear, and for different trust systems, we can define different security strategies [11].

91.3.1 Access Control Trust Relationship in a Domain

In accessing other entities that are in the same domain security management, trust can be directly introduced to the access control model for secure operation.

Definition 1 (Domain Trust Value): In the same domain, an entity and another entity complete an interaction, there will be an assessment for the entity to another entity. Use T to denote $-1 \leq T \leq 1$. Negative values are not satisfactory; will reduce the trust, in contrast, expressing satisfaction with the integrity; and will enhance the trust. Entity n_j after the completion of the interaction of n_i k times which gives trust values can formalize for $T(n_i, n_j)^k$.

Definition 2 (Service Satisfaction Degree): In the same domain, an entity and another entity complete an interaction; another entity to the entity's overall service satisfaction is denoted by S. After k times services entity n_j to n_i s overall satisfaction with the formula as follows:

$$S(n_i, n_j)^k = \beta \times S(n_i, n_j)^{k-1} + (1-\beta) \times T(n_i, n_j)^k \qquad (91.1)$$

Definition 3 (Direct Trust Degree): An entity's direct trust degree is related to domain trust value, the higher assessed value in a domain is, the higher entity credibility is, also the direct trust value is high, notation DTD. For two entities that never interact, DTD value is usually set to zero. In the domain entity n_i and n_j, direct trust degree after the completion of the k times interaction formula is as follows:

$$DTD(n_i, n_j)^k = \alpha \times DTD(n_i, n_j)^{k-1} + (1-\alpha) \times T(n_i, n_j)^k \qquad (91.2)$$

Definition 4 (Credit): The credibility of entities in the domain has to interact with all other entities within the domain in order to obtain satisfaction, expressed in Rp. Entity n_i in domain A can be expressed as to the credibility of $Rp(n_i,A)$; the specific formula is as follows:

$$Rp(ni) = \frac{\sum_{j=1, j \neq i}^{k} S(n_i, n_j) \times Rp(n_i, A)}{k} \qquad (91.3)$$

Definition 5 (Domain Trust Degree): In the domain an entity's trust degree is the credibility degree in the field of domain, directly made up by direct trust degree and credit, represented by symbol TD.

The i-th entity n_i trust degree in domain A can be formalized representation for TD (n_i, A); the formula is as follows:

$$TD(n_i, A) = \gamma \times \frac{\sum_{j=1, j \neq i}^{k} DTD(n_i, n_j)^j}{k} + (1-\gamma) \times Rp(n_i, A) \qquad (91.4)$$

Above all, $\alpha, \beta, \gamma = 0$, the value of weight parameter associated with the local security policy, which is stored in the local domain authentication and authorization center and, through the definition above, decides the proper policy access control to the cloud users and the service providers.

91.3.2 Access Control Strategy in the Same Domain

We introduce the trust degree which is the basic property of cloud users and cloud services or resources into role-based access control; the certification center AAC (authentication and authorization center) is responsible for access control authentication, authorization, and trust management.

Fig. 91.1 Access control strategy in the same domain

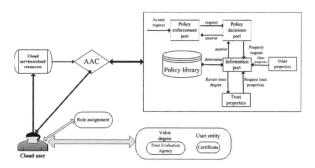

In the same domain, each time a user requests access to cloud services or cloud resources, AAC will examine cloud users' trust degree and ensure that the trust degree reaches its threshold, and then cloud users' requests for access will be permitted. The process is shown in Fig. 91.1.

Domain Access Control Process

1. Under the role-based access control, firstly the user needs to request role assignments before he/she wants to access control requests, thereby indirectly obtaining the appropriate access control permissions; in this policy, users can use access permissions or not as determined by trust management stage.
2. Users send access control requests to the AAC; the request information includes user ID, password, and the ID of access to resources or services. AAC firstly certificates user's authentication information, once passing through, trust management will authorize user's rights correspondingly. Authorization process includes the following:

   - Policy library initializes a security policy in this domain.
   - Policy implementation which sends user's access request is passed to the policy decision port.
   - Policy decision side passes the requests to the policy information port.
   - Policy information port obtains user's trust degree and other property information, returns to policy decision port.
   - In accordance to the user's information and the current security policy, policy information port makes decisions and returns the policy to enforcement port.
   - Policy enforcement port feedbacks the results to the user entity.

3. User performs the appropriate access control permissions and accesses cloud services or cloud resources.
4. Following the requests, providers of cloud services or cloud resources make assessment of the user and feedbacks to AAC.

91.4 Cross-Domain Access Control Policy

Because users often need to access different cloud services or cloud resources in different domain, safe and effective cross-domain access control policy is necessary. This paper proposes a role-based access control model and presents a new role mapping through a domain relationship, reaching to the result of resources sharing between domains.

Firstly we define two different security management domains which are Domain A and Domain B. The two roles Role A and Role B are in the separate domain.

Role A and Role B can access their cloud resources or enjoy the cloud service in their own domain. Suppose logical domains a and b are two partnership units, Domain A is an enterprise, and Domain B is a scientific research institute; two domains are using role-based access control management; now, both the enterprise and research institute will develop a research project, which needs to achieve the shared resources. At this time, the Domain A user wants to access the information of Domain B, which is cross-domain access. Through the relationship of role mapping, Domain A user by role exchanging can access resources of Domain B.

Cross-domain role mapping is shown in Fig. 91.2; the dotted line represents the mapping meaning. R_A represents role collection within the safe management Domain B. R_B represents role collection within the safe management Domain B. $R_A R_B$ stands for the role mapping relation from R_A to R_B. With mapping relationship any $R_A R_B$ ordered pair ($R_A n$, $R_B m$) from Domain A's role $R_A n$ maps to the Domain B's role $R_B m$, Domain B will assign their own domain role $R_B m$ to domain's role $R_A n$, and then the domain A's role $R_A n$ has the same rights with $R_B m$ to access to resources in the domain B, so $R_A n$ can achieve access to some resources on the domain b. This mapping can be written as $R_A n \rightarrow R_B m$.

Based on the role mapping, we enable cross-domain resource sharing to become more convenient. Meanwhile, it may also bring the problem of permission penetration and privilege escalation. For example, Domain A's role $R_A 1$ maps Domain B's role $R_B 3$; at the same time, $R_B 3$ maps Domain C's role $R_C 1$, and then it is inevitable that $R_A 1$ inherits the rights of $R_C 1$. This situation is not what we want to see.

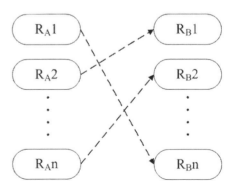

Fig. 91.2 Cross-domain role mapping relationship

Fig. 91.3 Domain A, B, and C mapping example with mirror

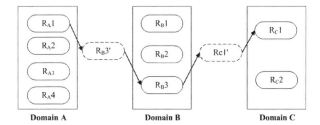

In order to avoid this situation, we introduce a concept which is called the mirror role, and the specific circumstance is shown in Fig. 91.3. From Fig. 91.3 we will know both R_B3' and R_C1' are mirror roles which come into being. In such conditions, we can avoid the problem of permission penetration and privilege escalation. Although R_A1 can obtain the same operating authority as R_B3 in Domain B and have the resource access qualifications, R_A1 cannot map R_C1 which is in Domain C through the R_B3 and cannot get the corresponding authority of R_C1 in Domain C as well. Because the identity that R_A1 gets is the mirror role R_B3' only, R_B3' cannot map R_C1' because of the interrupting of R_B3; in other words, there is no mapping relationship between R_B3' and R_C1'. Thus it can be seen that the mapping through the mirror role restricts the role of a bridge played by some roles in the process of role mapping, so executing the role mapping transfer finally makes some roles acquiring authority which they should not have. Therefore, the access control based on the mirror role mapping avoids the problem of permission penetration and privilege escalation well which appears in the role mapping.

91.5 Conclusion

In this paper, we discuss the current access control issues in the environment of cloud computing and present an in-domain and cross-domain access control policy based on trust. In the same domain through the calculation of user's trust degree, which reaches to the allocation of their according privileges, accomplishes the result of an access to resources or the corresponding service. Cross-domain we provide a role mapping to achieve resource sharing, but it may cause the problem of permission penetration and privilege escalation, in order to avoid the problem, we present the mirror role which based on role mapping, ultimately solve the problem. Through the access control strategy of in-domains and cross-domains, it can achieve the purpose of the security of cloud computing environment, users, and platform.

Acknowledgements This research was supported by the Project of Shandong Province Higher Educational Science and Technology Program under Grant No. J12LN61 and Grant No. J13LN64. In addition, the authors would like to thank the reviewers for their valuable comments and suggestions.

References

1. Feng, D. G., Zhang, M., Zhang, Y., & Xu, Z. (2011). Study on cloud computing security [J]. *Journal of Software, 22*(1), 71–83. In Chinese.
2. Li, W., Ping, L., & Pan, X. (2010). Use trust management module to achieve effective security mechanisms in cloud environment [C]. *IEEE International Conference on Electronics and Information Engineering (ICEIE2010), Kyoto, Japan (pp. 14–19)*.
3. Wang, S., Zhang, L., & Li, H. (2010). Evaluation approach of subjective trust based on cloud model [J]. *Journal of Software, 21*(6), 1341–1352. In Chinese.
4. Takabi, H., Amini, M., Jalili, R. (2007). Trust-based user-role assignment in role-based access control [C]. *IEEE Proceedings of the ACS/IEEE International Conference on Computer Systems and Applications 2007, Amman, Jordan* (pp. 807–814).
5. Jie Zhao, Nanfeng Xiao, Junrui Zhong. (2009). The behavior trust control based on Bayesian network and behavior log mining [J]. *Journal of South China University of Technology (Natural Science Edition), 37*(5), 94–100 (In Chinese).
6. Hur, J., & Noh, D. K. (2011). Attribute-based access control with efficient revocation in data outsourcing systems [J]. *IEEE Transactions on Parallel and Distributed Systems, 22*(7), 1214–1221.
7. Chuang Lin, Fujun Feng, Junshan Li. (2007). Access control technology under the new network environment [J]. *Journal of Software 18*(4), 955–966 (In Chinese).
8. Shouxin Wang, Li Zhang, Hesong Li. (2010). A subjective trust evaluation method based on cloud model [J]. *Journal of Software, 21*(6), 1341–1352 (In Chinese).
9. Wu Liu, Haixin Duan, Hong Zhang, Ping Ren, Jianping Wu. (2011). TRBAC: Trust based access control model [J]. *Journal of Computer Research and Development, 48*(8), 1414–1420 (In Chinese).
10. Guangwei Zhang, Jianchu Kang, Hesong Li. (2007). Research on subjective trust management model based on cloud model [J]. *Journal of System Simulation, 19*(14), 3310–3317 (In Chinese).
11. Chunhua Hu, Xinxing Luo, Sichun Wang, Yao Liu. (2011). Approach of service evaluation based on trust reasoning for cloud computing [J]. *Journal on Communications, 32*(12):72–81

Chapter 92
Detecting Unhealthy Cloud System Status

Zhidong Chen, Buyang Cao, and Yuanyuan Liu

Abstract In this paper, in order to detect the unhealthy status in the cloud system, a Basic Detection Strategy and a Threshold Strategy based on mathematic theory and statistical knowledge is proposed to solve this problem. By introducing unhealthy status percentage parameter α, both Basic Detection Strategy and Threshold Strategy are combined to detect and monitor the unhealthy cloud system status. For illustration, an eBay company example is utilized to show the feasibility of Basic Detection Strategy and Threshold Strategy. Empirical results show that Basic Detection Strategy with setting a suitable value to α can pinpoint most of unhealthy status in the cloud system, however, for some special unhealthy status, it must adopt the Threshold Strategy to pinpoint. The combination of Basic Detection Strategy and Threshold Strategy can effectively detect and pinpoint the unhealthy status in the cloud system and help staff to improve the performance of cloud system.

92.1 Introduction

With the development of the cloud computing, more and more enterprises are willing to adapt the strategy of deploying services on Cloud Platforms and are keen to improve the utilization of resource and reduce costs. Cloud Computing System shoulders an important mission to provide a healthy system environment for the different server applications and to ensure that each application service request can timely access required resources (CPU, memory, disk space, network and so on) [1–3]. In an ideal scenario, Cloud Computing System is able to provide sufficient resources for application requests to consume while these application

---

Z. Chen (✉) • B. Cao
School of Software Engineering, Tongji University, Shanghai 201804, China
e-mail: chenfang3376@gmail.com

Y. Liu
Site Reliability Engineering, eBay Engineering and Research Center (Shanghai), Shanghai 201210, China

Fig. 92.1 The data sample of 2 weeks

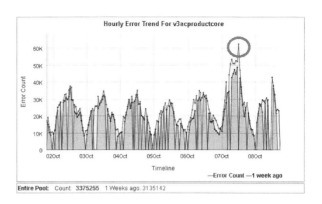

requests yield the desired results and return them to the clients (service invokers) [4, 5]. Furthermore, there should be no exception thrown while the requests are being processed. Although it is not official and not quite accurate, we define the Cloud Computing System to be in a healthy status if the scenarios described above are applicable. Unfortunately, these ideal scenarios may never exist. In order to provide customers with better and more robust services, we need to identify exceptions in time and be able to take corresponding actions effectively.

The status detection problem comes from one of the world largest e-commerce companies. The company has several huge data centers worldwide where the private cloud platforms are hosted. The company has designated technical staff members or system administrators who monitor the system status, process exceptions, and resolve system problems to ensure good services required by large amount of customers. To monitor the system status and find abnormalities effectively is crucial for this company. In order to find exceptions or abnormalities, the company had defined the rules of determining the system status (healthy or unhealthy) and conducting their manual interventions when the Cloud Computing System malfunction. For an efficient detection of system's unhealthy status we need to convert these rules established by this company into an actionable model.

A data center of the company (eBay) keeps the log files recording the number of errors occurring at different time periods. Figure 92.1 depicts the number of errors recorded in the log files of a data center over a given time period. The x-axis represents the time while y-axis indicates the number of recorded errors at different time points. Upon different scenarios, the company sets the rules below to identify the healthy status of the system:

- At some time points the curve suddenly rises or falls (we call them *spiking points*) and the changing range exceeds the predefined normal range, then these moments potentially are considered as unhealthy or in unhealthy status.
- When the spiking point in rule 1 is compared with the one point at the same moment last week and their trends are similar, and then it is considered to be healthy at the moment.

- The number of the recorded errors at a certain time point is above or below the specified threshold, and then the system could be thought being in unhealthy status at this moment.

Albeit the rules can be applied to monitor the system status manually, it is an extremely inefficient way to monitor the system status manually. Here we will propose a model-based methodology to identify the unhealthy status of the system effectively. The solution is developed by incorporating a set of algorithms based upon three rules mentioned. With the help of this approach the unhealthy status of the system can be detected in time and the corresponding message will be pushed to the staff members of the data center in time for possible actions. We first introduce the following two important definitions.

Definition 1. a recorded point $P_n(x_n, y_n)$ is called a *Peak*: if $(x_{n-1} < x_n < x_{n+1})$ && $(y_{n-1} < y_n$ & & $y_n > y_{n+1})$ where $P_{n-1}(x_{n-1}, y_{n-1})$, and $P_{n+1}(x_{n+1}, y_{n+1})$ are left and right adjacent points of $P_n(x_n, y_n)$.

Definition 2. a recorded point $P_n(x_n, y_n)$ is called a Valley if
$(x_{n-1} < x_n < x_{n+1})$ & & $(y_{n-1} > y_n$ & & $y_n < y_{n+1})$ where $P_{n-1}(x_{n-1}, y_{n-1})$, and $P_{n+1}(x_{n+1}, y_{n+1})$ are left and right adjacent points of $P_n(x_n, y_n)$.

This paper is organized as follows: The second section proposes two status detection algorithms. The computational experimental results are presented in the third section to demonstrate the effectiveness of the algorithms. The paper is concluded with the summary.

92.2 Status Detection Algorithms

The system status detection approach proposed here employs the recorded number of errors (the information is usually stored in the log files of the system) over the given time periods as shown in Fig. 92.1. As it is mentioned above, there are two curves mirroring the number of recorded errors over two time periods with the same length, e.g., this week and last week. The technical staff member usually determines the system healthy status by comparing the values for these two time periods. The criterion or rule of defining healthy or unhealthy status of the system is formed based upon the operational experience of the underlying e-commerce company. According to the rues and his own working experience, the technical staff member generally focuses on "peak" and "valley" points defined above to find unhealthy status of the system. A "peak" or "valley" point is usually caused by the exceptions when the Cloud Computing System is at an unhealthy status. In this case the number of recorded errors goes up or down sharply. The points where "peaks" or "valleys" occurs are called spiking points (for the differentiation purpose, one is called upward spiking where a peak occurs while the other is named as downward spiking where a valley appears). We might pinpoint the times when the unhealthy status of the system could occur by checking the spiking points. Nevertheless, it is tedious and ineffective to find the potential unhealthy status of the system and take actions by checking the diagram manually. We are going to propose a method that is able to detect the unhealthy status automatically.

92.2.1 The Basic Detection Strategy

Basic idea: in order to find a spike (a sudden up or down in the number of recorded errors) that may represent an unhealthy status, we first compute the slope k of two adjacent points so that we can collect a new sample including much slope k, and then for the new sample we will make statistics, meanwhile, though setting a percent that much larger slope k are filtered out of the new sample and consist another sub-sample [6–8]. With the help of the measurement described above, we are able to initially acquire some suspected spikes or potential unhealthy points according to the sub-sample. After the data is obtained, we will compare them to the ones of the same time points of last week to find out the tendency. If their tendencies are the same, then they are considered to be in healthy status otherwise they are in unhealthy status. An unhealthy status therefore may be identified.

Approach detail description: for a given set of recorded number of errors occurring in a data center, it can be plotted as shown in Fig. 92.1. The y-values of a curve may appear up and/or down abnormally and the variances are pretty bigger that generate different tendencies. The situations with sharp up or down y-values may reveal that the system encounters some exceptions at certain times.

According to the definitions discussed above it is obvious that the sharp changes in y-value form peaks and valleys of the curve or functions. For the convenient purpose these peaks or valleys will be named as local extreme values (or extreme values for short) of the function. Let one point be $P_n(x_n, y_n)$ of the function, its left adjacent point be $P_{n-1}(x_{n-1}, y_{n-1})$, and its right adjacent point be $P_{n+1}(x_{n+1}, y_{n+1})$, respectively. If the product of slope K_n for linear segment $P_{n-1}P_n$ and slope K_{n+1} for linear segment $P_n P_{n+1}$ is less than zero, namely:

$$\frac{y_{n+1} - y_n}{x_{n+1} - x_n} \times \frac{y_n - y_{n-1}}{x_n - x_{n-1}} < 0 \tag{92.1}$$

Then $(y_{n+1} - y_n) \times (y_n - y_{n-1}) < 0$, because the number of errors is recorded with the equal time interval, that is $x_{n+1} - x_n = x_n - x_{n-1}$. Based on the definitions for peaks and valleys we conclude P_n is the extreme value since its y-value is either bigger or smaller than its adjacent ones [9–11]. Using this formula we are able to define a set of extreme value points and name it as EVS (Extreme Value Set). Suspected spike points that may indicate abnormal status usually appear in the EVS.

Apparently in this way we may get a lot of extreme values for the given set of data while most of their up or down trends are more tempered that can be considered as healthy status. Therefore we need to focus on these extreme value points whose y-values vary sharply while filtering out the points whose y-values don't change sharply.

The following methodology will be applied to EVS (Extreme Value Set) sorted in non-decreasing order of y-value changes. According to the above discussions, it is conceivable that non-health spike points usually vary severely in terms of their y-values. Using the definitions presented above we conclude that the value of $|y_n - y_{n-1}|$ or $|y_{n+1} - y_n|$ for point P_n is greater than the difference between y-values of two healthy adjacent points. Furthermore if a threshold T is set properly, then $|y_n - y_{n-1}| < T$ or $|y_{n+1} - y_n| < T$ indicates that the y-value of P_n has not changed that much and the system should be in healthy status. Based upon the real applications, the probability of system exceptions occurring is relatively small and therefore unhealthy spike points should occur with low probabilities. The probability of unhealthy spike points appearing can be defined as follows:

$$P(|y_n - y_{n-1}| > T) < \alpha \qquad (92.2)$$

Where is small positive number. Then:

$$t = [\text{Total}(\text{EVS}) - \text{Total}(\text{EVS})^{*\alpha}] \qquad (92.3)$$

Total (EVS) represent the cardinality of $|y_n - y_{n-1}|$ in EVS, variable t is the time index satisfying:

$$T = |y_t - y_{t-1}|(t < n) \qquad (92.4)$$

In this paper the value of α is determined based upon the operational experiences of technical staff members of the data center.

Provided that the α value is given then the value of T can be obtained via Eqs. (92.2), (92.3), and (92.4). Furthermore, we define a point P_n for which $(|y_n - y_{n-1}|)$ is less than T to be a healthy point. Otherwise they are possibly unhealthy.

After the value of T having been given, the elements in EVS can be classified as follows.

For a given point P_n if $|y_n - y_{n-1}| >= T$ or $|y_{n+1} - y_n| >= T$, then P_n will be added into the set called EPUHPS (Extremely Possible Unhealthy Point Set).

EPUHPS helps narrow the space to be investigated to find out the real unhealthy points. The same period of the historical data is also applied to conduct the analyzing procedure of seeking unhealthy points. The basic idea behind the analysis is to identify if any point in EPUHPS possesses the same pattern as its historical records. If the patterns are similar it is considered to be healthy otherwise it is unhealthy, where the same two time points of current and last week have the similar pattern if they both either are peaks or valleys.

The following method shows how an element in EPUHPS is identified as an unhealthy one. In Fig. 92.1 we plot two curves: l_1 and l_2, where l_1 the data (number of recorded errors) is curve of current week and l_2 is the one of last week. Each P_n

that is in EPUHPS of l_1 curve is either a "peak" or a "valley". If its trend is different from the trend ("valley" or "peak") of the same period in l_2, then P_n will be added into Unhealthy Point Set (UHPS), that is, it is an unhealthy point. However if the trend is similar, we will have to compare their local variance s_p (the variance computed based upon the y-values of the three points: P_n, P_{n-1}, P_{n+1} of l_1, and the y-values of the corresponding P_n, P_{n-1}, P_{n+1} of l_2) and the global variance s_t (calculated based upon the y-values of all the points of l_1 and the corresponding ones of l_2) described below respectively. If $|s_t - s_p| < \xi$ (ξ is a very small positive number determined by the system administrator or technical staff members of the data center), then the point P_n is a healthy point; otherwise P_n will be added into the UHPS.

s_p and S_t are calculated as follows: let $V = \{P_{l_1 1}(x_{l_1 1}, y_{l_1 1}), \ldots, P_{l_2 n}(x_{l_1 n}, y_{l_1 n})\}$ be the set of the l_1 points and $V = \{P_{l_2 1}(x_{l_2 1}, y_{l_2 1}), \ldots, P_{l_2 n}(x_{l_2 n}, y_{l_2 n})\}$ be the set of the points of l_2 for the give time periods with the same length (1 week, for instance), then

$$s_t = \frac{\sum_{i=1}^{n}(y_{l_1 i} - y_{l_2 i})}{n} \qquad (92.5)$$

Let y_1, y_2, y_3 be the y-values of points P_n, P_{n-1}, P_{n+1} of l_1, and y'_1, y'_2, y'_3 be the y-values of the corresponding points of curve l_2, then

$$s_p = \frac{(y_1 - y'_1)^2 + (y_2 - y'_2)^2 + (y_3 - y'_3)^2}{3} \qquad (92.6)$$

The results of the experiments demonstrate the effective filtering ability for narrowing the searching space of possible unhealthy status.

In addition to the spike points detected by the basic approach, there are some special cases to be considered. For instance, some point's y-value is very large and way above the normal, but the trend is relatively smooth. In order to find this type of spikes, we propose the following Threshold Strategy.

In this method threshold value T is set by the system administrators or technical staff members of the data center depending on their experience or the given rules. The y-value of the observed point $P_n(x_n, y_n)$ will be compared with T. If the former is bigger, then the observed point is considered as an unhealthy one and the corresponding point is added to UHPS.

The set of unhealthy points is therefore formed via methods described above and the corresponding alerts will be disseminated to the interested subscribers (system administrators or technical staff members) for taking proper actions.

92.3 Experiment Results

The experimental data was collected from the data center of one of the largest e-commerce companies of the world. Three experiments conducted: experiments A, B, and C. The basic approach are performed in experiment A, and in experiment B, the combination of basic approach and Threshold strategy is applied. In both experiments we set up $\alpha = 0.05$ in the experiment A, B. In the experiment C, we set $\alpha = 0.1$ and apply the combination of Basic Detection Strategy and the Threshold Strategy. The purpose of these experiments is to investigate the impacts of the combination of the detection methods and the parameter in addition to the effectiveness of the proposed methods. The results can provide the guidance of establishing business rules in detecting the system status.

Based on the obtained results yielded by different methods, we will evaluate their performances and provide the guideline of applying these methods in detecting the system status.

The actual unhealthy spike points are listed in Table 92.1:

Experiment A
In this experiment, we will use the sample in Fig. 92.1 and adapt the Basic Detection Strategy, where $\alpha = 0.05$.

The result shown in the Table 92.2 demonstrates that the Basic Detection Method is able to find unhealthy spike points partially. It detects about 45 % reported unhealthy spiking points shown in the Table 92.1 under the current parameter setting.

Experiment B
In this experiment, we will use the sample in Fig. 92.1 and apply the combination of Basic Detection Strategy and the Threshold Strategy, where $\alpha = 0.05$ the threshold as 50,000 for the Threshold strategy.

The result in the Table 92.3 shows the improvement in detecting unhealthy status of the system as it is able to find 54% of reported unhealthy spiking points.

Experiment C
In this computational experiment we set $\alpha = 0.1$ and the threshold to be 50,000 for the combination of Basic Detection Strategy and the Threshold Strategy.

The outcome in the Table 92.4 demonstrates the superiority of the combined strategies with the parameter setting. It is able to obtain much better results than those in experiments A and B. More than 90 % of the unhealthy spiking points can be detected that in turn can provide necessary alerts in time to handle the exceptions.

Based on the experimental results, the combination of the Basic Detection Strategy and the Threshold Strategy is proven to be an effective procedure to detect unhealthy spiking points with a proper setting of α. The methods discussed in this paper provide a solid base to detect the unhealthy status of a Cloud Computing System.

Table 92.1 Actual unhealthy spike points in the Fig. 92.1

Date/time	Error count
10/02/2012 10:00:00 AM	0
10/02/2012 02:00:00 PM	0
10/03/2012 08:00:00 AM	31,633
10/04/2012 01:00:00 PM	0
10/04/2012 02:00:00 PM	28,601
10/06/2012 08:00:00 AM	35,103
10/06/2012 10:00:00 AM	0
10/06/2012 11:00:00 AM	33,916
10/07/2012 04:00:00 AM	0
10/07/2012 12:00:00 PM	62,455
10/08/2012 01:00:00 PM	36,509

Table 92.2 Result in experiment A

Date/time	Error count
10/02/2012 02:00:00 PM	0
10/04/2012 01:00:00 PM	0
10/06/2012 08:00:00 AM	35,103
10/07/2012 04:00:00 AM	0
10/08/2012 01:00:00 PM	36,509

Table 92.3 Result in experiment B

Date/time	Error count
10/02/2012 02:00:00 PM	0
10/04/2012 01:00:00 PM	0
10/06/2012 08:00:00 AM	35,103
10/07/2012 04:00:00 AM	0
10/07/2012 12:00:00 PM	62,455
10/08/2012 01:00:00 PM	36,509

Table 92.4 Result in experiment C

Date/time	Error count
10/02/2012 10:00:00 AM	0
10/02/2012 02:00:00 PM	0
10/04/2012 01:00:00 PM	0
10/06/2012 07:00:00 AM	0
10/06/2012 08:00:00 AM	35,103
10/06/2012 10:00:00 AM	0
10/06/2012 11:00:00 AM	33,916
10/07/2012 04:00:00 AM	0
10/07/2012 12:00:00 PM	62,455
10/08/2012 01:00:00 PM	36,509

92.4 Conclusion

In this paper we first introduce the business background of detecting unhealthy status of a Cloud Computing system. We conduct the brief analysis on the data recording the number of errors during the system operation periods to lay the foundation of detecting algorithm development. A basic detection strategy and a threshold-based method are proposed, which can help finding the spiking points where the system could be unhealthy. The computational experiments are conducted to demonstrate the effectiveness of the proposed methods and impacts of various parameter settings. A system administrator or technical staff member is able to adjust the parameter upon his experience/desire to find unhealthy spiking points. Together with the proposed methods the computational results reveal the direction of establishing business rules for detecting unhealthy status effectively.

We are planning to collect more real datasets and perform more computational experiments to seek the further improvements of the algorithms. The topic of triggering an efficient business process to handle exceptions when unhealthy status is detected is also one of our future researches.

References

1. Saripalli Prasad, Kiran, G. V. R., Shankar R. Ravi, Narware Harish, Bindal Nith. (2011) Load prediction and hot spot detection models for autonomic cloud computing. In: *4th IEEE/ACM International Conference on Cloud and Utility Computing (UCC 2011). IEEE Computer Society, Los Alamitos* (pp. 397–402).
2. Xiaojun Yu, Qiaoyan Wen. (2010). A view about cloud data security from data life circle. In: *International Conference on Computational Intelligence and Software Engineering. Peking University Press, Beijing* (pp. 203–208).
3. Donglin Chen, Mingming Ma, Qiuyun Lv. (2012). Study on transaction management system in cloud service market. In: *2012 International Conference on Technology and Management. Springer Verlag, Germany* (pp. 479–483).
4. Zhengping Wu, Nailu Chu, Peng Su. (2012). Improving cloud service reliability. In: *2012 I.E. International Conference on Services Computing. IEEE Computer Society, Los Alamitos* (pp. 90–97).
5. Jianwei Yin, Yanming Ye, Bin Wu, Zuoning Chen. (2011). Cloud computing oriented network operating system and service platform. In: *2011 I.E. International Conference on Pervasive Computing and Communications Workshops. IEEE, Piscataway* (pp. 111–116).
6. Sunahara, Y. (1982). Treatment of irregular data. I. Probability models and statistics. *Systems & Control, 26*(4), 228–236.
7. Mitchell, M. (2003). Constructing analysis of variance. *Journal of Computers in Mathematics and Science Teaching, 21*(4), 381–410.
8. Hommes, S., State, R., Engel, T. (2012). A distance-based method to detect anomalous attributes in log files. In: *2012 IEEE/IFIP Network Operations and Management Symposium. IEEE, Piscataway, NJ, USA* (pp. 498–501).
9. Frei Adrian, Rennhard Marc. (2008). Histogram matrix: Log file visualization for anomaly detection. In: *3rd International Conference on Availability, Security, and Reliability. IEEE, Piscataway, NJ, USA* (pp. 610–617).

10. Stermsek, G., Strembeck, M., Neumann, G. (2007). A user profile derivation approach based on log-file analysis. In: *2007 International Conference on Information and Knowledge Engineering. Las Vegas, NV, USA: CSREA Press* (pp. 258–264).
11. Cheng, Y.-H., & Huang, C.-H. (2008). A design and implementation of a Web server log file analyzer. *WSEAS Transactions on Information Science and Applications,* 5(1), 8–13.

Chapter 93
Scoring System of Simulation Training Platform Based on Expert System

Wei Nie, Ying Wu, and Dabin Hu

Abstract In order to reduce the cost of operation training and improve efficiency of examination, the development of simulation training platform has achieved very good results. An intelligent scoring system based on expert system plays the role of the teacher and gives the student a just assessment. It uses the professional theory and practical experience as the evaluation criteria and analyzes the operator's operation process to realize the automatic scoring through the program algorithm. The application of scoring system evaluates the operation level of students and gives students guiding opinions and error analysis.

93.1 Introduction

Simulation training [1, 2] is now used to all walks of life because it is more efficient and affordable, compared with traditional training methods. In recent years, simulation training is a necessary training tool in many companies. They make simulation training as an important part of job training and a way of identification of technical and skills contest.

Simulation training platform establishes a virtual ship operation environment to make the operator with the feel of real boat scene. It can improve the rapid response capability of ship operation and capacity of safety operation and reduce operation accident of virtual equipment [3].

How would you assess the level of actual operator? The scoring system uses the professional theory and practical experience as the evaluation criteria and analyzes the operator's operation process to realize the automatic scoring through the program algorithm. This paper is based on the practical experience of the automatic

W. Nie (✉) • Y. Wu • D. Hu
College of Naval Architecture and Power, Naval University of Engineering,
Wuhan 430033, China
e-mail: niewei213@163.com

grading system exploitation, and the platform of simulation training is ship power system. The scoring system [4, 5] uses programming techniques to assess automatically operation ability of candidate, which is based on operation experience of ship power system of technical personnel that are restored in the database as the assessment standard. The scoring system has been applied to practical ship power simulation system and has received a good effect. It not only has improved the examination efficiency and fairness of the assessment but also has saved a lot of funding. The basic module and function of composing the system have been discussed. The key technologies to achieve the system have been described.

93.2 The Principle of Expert System Theory in the Scoring System

The expert system [6] is one of the main areas of artificial intelligence research. The expert system is also known as knowledge-based systems. The expert system is designed to simulate the work of the expert thinking and ways of tackling problems in certain areas by an intelligent computer program. It takes advantage of the human expert knowledge and problem-solving approach to solve the problems in the field.

In the traditional way of training and examination, the ability of candidate operation relies on the coaches' assessment. The training and examination of simulation training platform also rely on the coaches. After the candidates finish the operation of the devices, the coach judges the level of operation of the candidates and finally gives grades. The professional knowledge and practical experience of coaches play a decisive role in this process. It is an invisible criterion in the assessment process. The scoring system is used to replace the role of coaches and implements the scoring method of coaches through a computer. The realization of scoring system is combined with expert systems theory and data analysis method. The expert system theory makes all aspects of specialist expertise and practical experience as the standard of the automatic scoring. Certain reasoning mechanism can recognize the operation steps of the candidates and make reasonable judgment through the computer program. Data analysis is the basis of the scoring system which obtained candidates operation time steps, instrumentation and indicator of changes in the situation.

The expert system mainly consists of rules knowledge base, comprehensive database, inference engine, the interpreter, knowledge acquisition, and man-machine interface. Figure 93.1 shows the expert system structure of the scoring system of simulation training platform.

Rules knowledge is the key of quality of the expert system, including the quality and quantity of knowledge in database. In general, the expert system knowledge base and expert systems program are independent to each other. The user can change the content of knowledge base to improve the performance of the expert system. The inference engine matches rules in the knowledge base to access to new

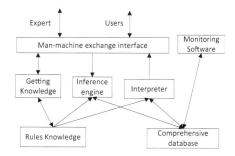

Fig. 93.1 The expert system structure of the scoring system

conclusions and get the results of problem solving. The inference engine of the scoring system realizes by programming language. The inference engine is the organization control mechanism of the expert system. Reasonable inference engine can make use of the rules of the knowledge base to solve practical problems.

93.3 The Knowledge Base of Expert System

Appropriate knowledge representation could convert the expert knowledge to the handle expression of computer system. The knowledge base of the scoring system is expert knowledge and practical experience of technical staff of simulation training platform operation, which is stored in the database.

93.3.1 Knowledge Acquisition

Knowledge acquisition [7, 8] is one of the main works of the expert system development, which is the basic technique of artificial intelligence and knowledge engineering. The most important sources are expert knowledge and practical experience. In the scoring system, field experts and technical personnel work together to complete the acquisition of knowledge. The process of development mainly divided into the following steps.

First, the developer must have the complete mastery of simulation training platform, including the method of operation, its composition, functions, and training. They could apply books and operating manual to practice and exchange with skilled operators.

Second, the obtained knowledge was systematized. The developer must have actual operating experience and repeat discussions with experts and professors to do further research on the knowledge of science and rationality.

Fig. 93.2 The basic form of knowledge representation of the scoring system

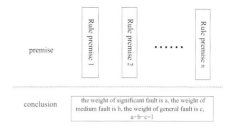

93.3.2 Knowledge Representation

Knowledge representation of knowledge is a formal and symbolic process, which uses a computer language to encode the domain knowledge. It has an important impact on the performance of the system. Knowledge representation methods relate to the design of various data structures. The purpose of knowledge representation is to be able to use this knowledge to reason and make decisions through effective representation of knowledge. Figure 93.2 shows the basic form of knowledge representation of the scoring system.

In this paper, the knowledge representation is production rules. The basic form of rule representation is as follows:

$$\text{If } P \text{ then } Q$$

Where P is the premise of the rule and Q is the conclusion. For example, if the valve of A has not turned on before you the switch B, This situation was recorded as a medium error. Where, the premise of the rule is a compound condition that constitutes by two simple conditions. In the mark-reducing method, the conclusion can be divided into several levels from low to high; the higher the level, the more the points. If the degree of operation mistake is between the medium fault and the significant fault, you can introduce the idea of fuzzy mathematics.

The knowledge representation is determined mainly by the following aspects: knowledge-use efficiency, understandability, and the degree difficulty of knowledge maintenance. The same knowledge can be expressed through a variety of methods, but the effect will be very different. Every kind of knowledge representation has its own characteristics, which also have their own advantages and disadvantages to different areas. Thus, it can combine several representations to achieve a particular field of knowledge representation.

The form of production rules knowledge is simple to understand and explain. The rules are independent to the extraction and formalization of knowledge. But constraints and interactions of production lead to the low inference efficiency. In the scoring system, each production rule is one of the scoring criteria. The knowledge base of each item consists of a certain rule.

93.4 The Scoring System of Simulation Training Platform

In the training of simulation training platform, the teacher makes the judgment to the operational level of students after examination. The teacher's specialized knowledge and practical experience have played the decisive role in this process. Now, we have built a system in which the scoring plays the role of the teacher.

Simulation training platform achieves the purpose of training students through restoring the operating environment and feel of the real equipment. The simulation training platform consists of simulation modules, monitoring software, communication interface, terminal node modules, database, and so on. Figure 93.3 shows the basic principle of the scoring system of simulation training platform.

Simulation modules can simultaneously perform multiple simulation models, including the dynamic and steady state simulation of the main equipment of simulation training platform, the operation simulation of normal and fault conditions of the power system, and the data interconnection with other systems. Monitoring software can exchange the data with other software and obtain the data recording the student's operation information from the terminal node modules. The function of database is to record the process of students' operation information.

The scoring system is mainly made up of the database and software programming. The database stored all kinds of data information, including scoring criteria of expert knowledge, operation information, and basic personal information and examination results. Those data could exchange with other software and obtain the data recording the operation information from the consoles. Through analyzing those data from the database, the system can identify the operation steps of students. The scoring rules stored in the database can give a reasonable grading of operation level by programming technology.

The reasoning implementation of the scoring system is forward reasoning, which is based on the known facts as the starting point of reasoning. The basic idea of forward reasoning is that the computer identify the current applicable knowledge from the initial facts, and then constitute a set of applicable knowledge and some conflict resolution strategies. Figure 93.4 shows the reasoning mechanism of single armature start-up operation.

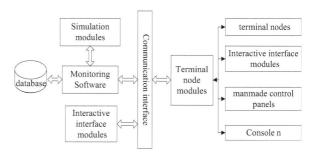

Fig. 93.3 The basic principle of the scoring system of simulation training platform

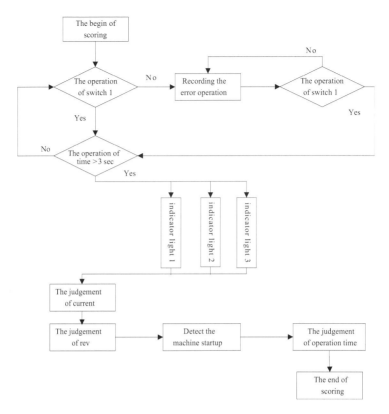

Fig. 93.4 The reasoning mechanism of single armature start-up operation

In the reasoning process of the scoring system, the key is the entire production rule. A rule expresses that if the premise is satisfied, you can launch the conclusion that the conclusions under the operation will be performed. In order to promote the operation of the scoring system of simulation training platform, the establishment of a knowledge base must follow the next regulations.

1. All rules constitute the rules knowledge base of scoring system operation judge, which contains all possible operational errors. Each rule can consist of certain sliver rules. If any of the operation steps meets this rule, the conclusion of this rule will be executed.
2. When an operation error may satisfy the two rules processing and trigger by the two rules, it should get the greater point as the coefficient of operation error.
3. The weights of the scoring system are initially drawn up according to the actual staff of long-term experience and the advice of experts, which are stored in the database as the initial default value. According to the reasonableness of the results of the scoring, those weights can be dynamic changes in the process of the scoring system debugging, making the weights more reasonable.

4. In the scoring rules, there are many compound checking rules. It considers mainly the following two aspects: First, whether is the time of the operation right which is accordance with the operation instructions. Second, whether is the steps correct which includes the operation of button and the button in the correct order. If the student gives a correct operation after drain operation, it is the operating part of the operating mistakes.

93.5 The Example of Program Algorithm of Scoring System

After the end of the examination, the operation information of the candidates would be stored in the database. Automatic scoring system identifies the operation of candidates through access database data and saves it to an array variable. The most critical of the data analysis is the algorithm of the recognition process. Each record of discrete variables in the database corresponds to an operation of the candidates. It can uniquely identify candidate operation by detecting the changes in the database, such as the operating button or switch. It is an important foundation of the recognition algorithm. Figure 93.5 is a simplified schematic of the implementation process.

Analysis of the data is that procedure of the candidates was got from the recorded data. The steps identified were mainly realized through certain procedure algorithm.

The candidate's operation identification mainly includes two aspects. First, combing with the time variable and the value of the operating variables, operating

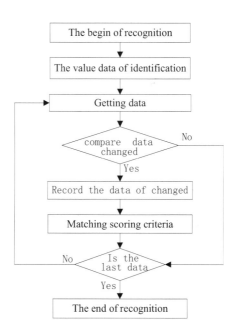

Fig. 93.5 A simplified schematic of the implementation process

time of the button or switch was achieved by similar methods of code program. Second, the problem is how to determine the change direction of the button or switch operation.

93.6 Conclusion

The development of the scoring system of simulation training platform is a great help for training and teaching. The design and implementation of the scoring system involve many fields of knowledge. It is a new way that an expert system theory was applied to the scoring system of simulation training platform. The emphasis of system development is the establishment of a knowledge base. The established methods and guidelines of the knowledge base were particularly introduced. In practical applications, the automatic scoring system also achieved good results.

References

1. Qingfu Kong, Jingyang Shong, Fanming Zeng. (2010). Technological condition and developmental trend of marine engine simulation training equipment. *Ship Science and Technology, 32* (2), 138–140.
2. Wei Nie, & Ying Wu. (2010) Study of automatic scoring system for simulation training. *2010 3rd International Conference on Advanced Computer Theory and Engineering* (pp. 403–406). Chengdu: IEEE Conference Publications.
3. Abhishek Kumar, Seung, Kyum Choi. (2011). Tolerance allocation of assemblies using fuzzy comprehensive evaluation and decision support process. *The International Journal of Advanced Manufacturing Technology, 55*(4), 379–391.
4. Xingtao Zhao. (2008) Improvement of automatic scoring of the power plant simulator training. *Automation of Electric Power Systems, 18*(3), 74–75.
5. Guojiang Bao. (2008). *Research on the ship maneuvering evaluation system based on ship handing simulator*. Dalian: Dalian Maritime University.
6. Shenghua Cai, Zhuxiao Liang. (2008). Student evaluation of large power plant simulation system. *Journal of System Simulation, 20*(21), 5989–5992.
7. Hong Zhang, Feng Xiu, Biguang Jin. (2005). Simulating test of ship navigation safety evaluation using ship handling simulator. *Ship Science and Technology, 26*(5), 567–571.
8. Yongsheng Fan, Fangzhen Cheng. (2000). Study on the scoring system of power plant training simulator. *Journal of System Simulation, 12*(3), 282–28.

Chapter 94
Analysis of Distributed File Systems on Virtualized Cloud Computing Environment

Tiezhu Zhao, Zusheng Zhang, and Huaqiang Yuan

Abstract Although various performance characteristics of distributed file system have been documented, the potential performance efficiency of distributed file system on virtualized cloud computing infrastructure is not clear. This chapter focuses on the performance of Hadoop Distributed File System (HDFS) on virtualized Hadoop. We construct a virtualized Hadoop platform and perform a series of experiments to investigate the performance of HDFS on the virtualized Hadoop cluster. Experimental results verify the efficiency of distributed file system on virtualized Hadoop to process the mass-intensive application.

94.1 Introduction

Distributed file systems can effectively solve the problems of the mass data storage and I/O bottlenecks in the mass distributed storage system and become the research hotspot of the storage industry and academia. Distributed file systems are key building blocks for cloud computing applications. Therefore, the industry is witnessing distributed file systems for large data center storage [1]. The performance of distributed file system directly affects the efficiency of the whole distributed computing environment. Therefore, the performance of distributed file system is the key research issue.

Hadoop is a highly scalable compute and storage platform for implementing the Google MapReduce algorithms in a scalable fashion on commodity hardware. The core Hadoop project solves two problems with big data: fast, reliable storage (HDFS) and batch processing (MapReduce) [2]. The HDFS cluster consists of a single NameNode, a master server that manages the file system namespace and regulates access to files by clients. In addition, there are a number of DataNodes,

T. Zhao (✉) • Z. Zhang • H. Yuan
Engineering and Technology Institute, Dongguan University of Technology, Dongguan 523808, China
e-mail: tzzhao83@163.com

usually one per node in the cluster, which manage storage attached to the nodes that they run on. HDFS exposes a file system namespace and allows user data to be stored in files.

Although the HDFS have been widely studied for several years, there are relatively few studies on HDFS with the virtualized Hadoop platform. The potential impact to application performance of distributed file system on the virtualized Hadoop platform is not clearly understood. In this chapter, we focus on the performance of HDFS on virtualized cloud computing environment. The main contribution of this chapter can be summarized as follows (1) A virtualized Hadoop platform is constructed to investigate the performance of HDFS on the virtualized Hadoop cluster; (2) We verify the performance characteristics of HDFS in the context of different application scenarios.

The remainder of this chapter is organized as follows. We begin by introducing related work in Sect. 94.2. The virtualized Hadoop platform is proposed in Sect. 94.3. We investigate the performance of HDFS on the virtualized Hadoop cluster, discuss the experiment results in Sect. 94.4, and conclude the chapter in Sect. 94.5.

94.2 Related Work

Existing research for distributed file system can be classified into four categories: (1) Performance analysis of distributed file system with the specific application scenario. The use of clustered file systems as a backend for Hadoop storage has been studied previously. The performance of distributed file systems such as Lustre, PVFS and GPFS with Hadoop has been compared to that of HDFS [3]. Most of these investigations have shown that non-HDFS file systems perform more poorly than HDFS, although with various optimizations and tuning efforts, a clustered file system can reach parity with HDFS [4]. (2) Metadata management and query optimization. Metadata management is critical in scaling the overall performance of large-scale data storage systems and a large-scale distributed file system must provide a fast and scalable metadata lookup service. Wang et al. proposed a two-level metadata management method to achieve higher availability of the parallel file system while maintaining good performance [5]. (3) Performance parameter analysis and tuning. Yu et al. indicated that excessively wide striping can cause performance. To mitigate striping overhead and benefit collective IO, authors proposed two techniques: split writing and hierarchical striping to gain better IO performance [6]. Yu et al. presented an extensive characterization, tuning, and optimization of parallel I/O on the Cray XT supercomputer (named jaguar) and characterized the performance and scalability for different levels of storage hierarchy [7]. (4) Optimizing data distribution strategy and data access strategies. Li et al. modeled the whole storage system's architecture based on closed Fork-Join queue model and proposed an approximate parameters analysis method to build performance model [8]. Yu et al. adopted a user-level perspective to empirically reveal the implications of storage organization to parallel programs running on

Jaguar and discovered that the file distribution pattern can impact the aggregated I/O bandwidth [9]. Piernas et al. adopted a novel user-space implementation of active storage for Lustre and the user-space approach has proved to be faster, more flexible, portable, and readily deployable than the kernel-space version [10].

94.3 Virtualized Hadoop Platform

94.3.1 Execution Engine of Hadoop

To get an idea of how data flows between the client interacting with HDFS, the NameNode, and the DataNode, consider Fig. 94.1, which shows the execution engine of Hadoop. The I/O flow of execution engine consists of four main modules: Application Launcher module, JobTracker module, Tasktracker module, Hadoop Distributed File System (HDFS) module. All the four modules cooperate with each other to implement MapReduce application, using HDFS for storage.

The execution engine workflow is shown as follows: (1) Application Launcher submits a job to JobTracker and saves jars of the job in the HDFS file system. (2) JobTracker monitors all jobs' execution status and makes scheduling decision. JobTracker divides the job into several map/reduce tasks and schedules tasks to TaskTracker for execution. According to the implementation situation, JobTracker reports execution status to JobTracker. (3) TaskTracker runs on every node and manages the status of all tasks which run on that node. TaskTracker fetches task jar from the HDFS file system and saves the data into the local file system. (4) HDFS is responsible for storing data, which come from Application Launcher and TaskTracker.

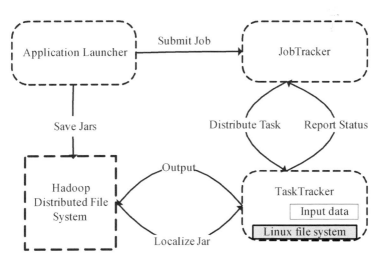

Fig. 94.1 Execution engine of Hadoop

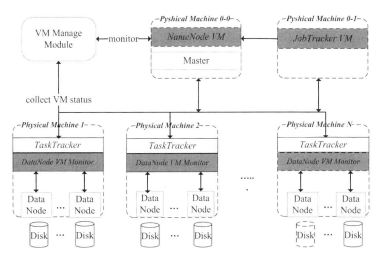

Fig. 94.2 Architecture of virtualized Hadoop platform

94.3.2 Architecture of Virtualized Hadoop Platform

Figure 94.2 illustrates the architecture of the virtualized Hadoop cluster platform. In this platform, VM manage module, which manages all virtual resources, is the key component. It is responsible for monitoring the virtual machine resource status and tuning the configuration parameters of the virtualized Hadoop cluster. DataNode VM monitor module is responsible for monitoring the utilization status of DataNode, including the utilization of CPU, memory, disk, and network status.

By using the virtualization technology, one physical machine can be shared by several virtual machines. For the performance consideration, NameNode VM and JobTractor VM are deployed in two physical machines, respectively. Each physical machine can deploy multiple DataNode VMs, which are monitored by DataNode VM monitor.

94.4 Experiment Analysis

94.4.1 Experiment Setup and Scenarios Design

The experiment was performed on the Hadoop cluster with nine physical nodes and the configuration is shown in Table 94.1. The experiment Hadoop cluster is constructed as: two nodes for severs (one for NameNode, one for JobTracer) and the remaining nodes are DataNodes.

94 Analysis of Distributed File Systems on Virtualized Cloud...

Table 94.1 Configuration of experiment environment

Component	Detailed description
Hadoop version	Hadoop 0.20.1
Physical machine OS	Red Hat Enterprise Linux 5
Virtual machine OS	CentOS 5.5
NameNode	6 core Xeon X5650 processors at 2.66 GHz and 48 GB Memory
DataNode	6 core Xeon X5650 processors at 2.66 GHz and 48 GB Memory
Network	10-Gigabit TCP/IP Ethernet
Xen version	Xen 3.4.4
Benchmark	TeraSort, TestDFSIO

Our experiment chooses two typical benchmarks to evaluate the performance of HDFS on virtualized Hadoop platform: (1) TeraSort benchmark, represents one typical use case, is probably the most well-known Hadoop benchmark that combines testing the HDFS and MapReduce layers of the Hadoop cluster. (2) TestDFSIO benchmark, an I/O-intensive Hadoop standard benchmark, is used to test the backend file system performance of HDFS and Lustre in the context of a Hadoop job. It is helpful for tasks such as stress testing HDFS and Lustre, to discover performance bottlenecks in your network, and to give you a first impression of how fast your cluster is in terms of I/O.

The experiment mainly considers three aspects: read/write, datasize scale, and DataNode scale. The experiment cases are designed as follows:

Case 1: Analyze the read/write running time corresponding to different dataset size using TeraSort benchmark. The detailed configuration is as follows: 7/28 virtual DataNodes (vDataNodes) with the configuration of 2 vCPUs and 2,048 MB vMemory; datasize = 100 MB, 400 MB,..., 2,200 MB; blocksize = 64 MB; replication = 3.

Case 2: Analyze the read/write throughput corresponding to the different number of vDataNodes using TestDFSIO benchmark. The detailed configuration is as follows: num. of vDataNodes=1, 5, 9, 13, 17, 21, 25, 29, 33; each vDataNode with 2 vCPUs and 2,048 MB vMemory; datasize = 5 GB; blocksize = 64 MB; replication = 3.

94.4.2 Experiment Result Analysis

For all experiment scenarios, the writing/reading performance is tested three times and a median value will be compared to the other to avoid outliers. In case 1, the running time is selected as the performance metric. Figure 94.3 plots the read/write running time for different datasizes, which vary from 100 to 2,200 MB.

As shown in Fig. 94.3, we note that as the number of datasize scales, the running time increases quickly. When the datasize is small, the time difference is relatively small. However, when the data size exceeds 1,300 MB, the running time difference

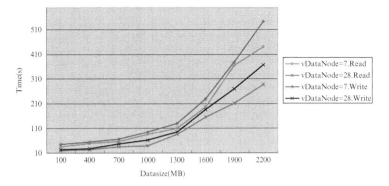

Fig. 94.3 Running time comparison of the different datasizes

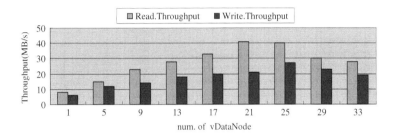

Fig. 94.4 Throughput comparison of the different number of vDataNodes

increases quickly. The read performance of HDFS on the virtualized Hadoop platform is better than the write performance. The performance of HDFS is better when the number of vDataNodes is 28.

In case 2, the throughput (MB/s) is selected as the performance metric. Figure 94.4 shows the performance (throughput) changes with the increase of the number of vDataNodes.

As illustrated in Fig. 94.4, the read throughput is better than write throughput. The read throughput continues to rise with the increase of the number of vDataNodes, and arrives at the maximum value when then vDataNode number is 21. Later, the read throughput is declining as the number of vDataNodes increases. The main reason is the competition for physical resources when the number of vDataNodes is too much (>21). The change of the write throughput is similar to the read throughput. The results clearly indicate that the distributed file system can maintain better efficiency on the virtualized cloud computing environment.

94.5 Conclusion

In this chapter, we study the performance and efficiency of distributed file system on the virtualized computing environment. We first introduce the I/O flow characteristic of HDFS and construct a virtualized Hadoop platform. Then, we perform a series of experiment to investigate the performance of HDFS on the virtualized Hadoop cluster platform. The experiment result shows that the distributed file system can maintain better efficiency on the virtualized cloud computing environment. It is necessary to study the performance of distributed file system in some specific application environments.

Acknowledgements This work is supported by the Natural Science Foundation of Guangdong Province, China (Grant No. S2012040007746),the Scientific Research Foundation for Doctors of DGUT(ZJ130604), the National Natural Science Foundation of China (Grant No. 61170216, 10805019, 61272200).

References

1. Cheng, K., & Wang, N. (2012). The feasibility research of cloud storage based on global file system. In *Proceeding of 2012 9th International Conference on Fuzzy Systems and Knowledge Discovery* (pp. 2507–2511). Piscataway, NJ: IEEE.
2. Konstantin, S., Hairong, K., Sanjay, R., et al. (2010). The Hadoop distributed file system. In *Proceedings of the 2010 I.E. 26th Symposium on Mass Storage Systems and Technologies* (pp. 1–10). Washington, DC: IEEE Computer Society.
3. Sun Microsystems Inc. (2010) *Using Lustre with Apache Hadoop*. White Paper. pp. 1–25.
4. Xyratex Inc. (2011). *Map/reduce on Lustre*. White Paper. pp. 1–16.
5. Wang, F., Yue, Y. L., Feng, D., et al. (2007). High availability storage system based on two-level metadata management. In *Proceedings of the 2007 Japan–China Joint Workshop on Frontier of Computer Science and Technology* (pp. 41–48). Washington, DC: IEEE Computer Society.
6. Yu, W., Vetter, J. S., Canon, R. S., et al. (2007). Exploiting lustre file joining for effective collective IO. In *Proceeding of the Seventh IEEE International Symposium on Cluster Computing and the Grid* (pp. 267–274). Washington, DC: IEEE Computer Society.
7. Yu, W., Vetter, J. S., & Oral, H. S. (2008). Performance characterization and optimization of parallel I/O on the Cray XT. In *Proceeding of the 2008 I.E. International Symposium on Parallel and Distributed Processing* (pp. 1–11). Piscataway, NJ: IEEE.
8. Li, H. Y., Liu, Y., & Cao, Q. (2008). Approximate parameters analysis of a closed fork-join queue model in an object-based storage system. In *Proceeding of the Eighth International Symposium on Optical Storage and 2008 International Workshop on Information Data Storage* (pp. 1–8). Bellingham, WA: SPIE.
9. Yu, W., Oral, H. S., Canon, R. S., et al. (2008). Empirical analysis of a large-scale hierarchical storage system. In *Euro-Par 2008, LNCS, 5168* (pp. 130–140). Berlin: Springer.
10. Piernas, J., Nieplocha, J., & Felix, E. J. (2007). Evaluation of active storage strategies for the lustre parallel file system. In *Proceeding of the 2007 ACM/IEEE Conference on Supercomputing* (pp. 1–8). New York, NY: ACM.

Chapter 95
A Decision Support System with Dynamic Probability Adjustment for Fault Diagnosis in Critical Systems

Qiang Chen and Yun Xue

Abstract In order to locate and remove the faults in the critical systems where the faults occur, this paper proposes a three-layer decision support system for fault diagnosis, in which both static information and dynamic information of the system are used to find out suspicious components. In the process of locating the faults, a bipartite graph is applied to describe the relation between the symptom and the components, on the basis of which a method is proposed to calculate the value of fault evidence of a component. Then, the components whose values are larger are chosen as the result. Meanwhile, the decision support system adjusts the data of the bipartite graph according to the actual situation in order to improve the effectiveness of the diagnosis. The experiment shows that the fault diagnosis process in the decision support system can locate the fault more effectively.

95.1 Introduction

The traditional small application system becomes large and critical, which is widely used in military, defense, finance, and other important areas. Due to the limitation of time, personnel, and technical conditions, the practical application of the system is impossible to be developed without defect. Once these important applications failed and cannot be fixed quickly, that may result in immeasurable loss and serious consequences.

The traditional fault diagnosis is achieved by experts, which may take a long time, and the accuracy of the results is closely related to the individual ability.

Q. Chen (✉)
School of Management, Huazhong University of Science and Technology,
Wuhan 430074, China
e-mail: kurt_cq@163.com

Y. Xue
Naval Academy of Armament, Beijing 100055, China

Meanwhile, the ascendants may change frequently, so the personal experience in this analysis is difficult to be shared with the others. Therefore, the traditional analysis process which emphasizes individual ability obviously cannot meet the actual requirements.

This paper presents a fault diagnosis decision support system, which applies the bipartite graph to build a system fault diagnosis model to analyze the possible reason to the failure. What is more, the model parameters can be adjusted dynamically according to the actual result of the fault. The result of the experiment indicates that the algorithm in this paper can get a better diagnostic effectiveness.

95.2 Related Work

Fault diagnosis is an analysis process to obtain the final cause of the failure based on the system through the fault symptoms and related information [1]. The fault diagnosis process can be taken as a black box, and its input is the information of the failed system which is organized in a particular way, and its output is the possible reason.

The existing diagnosis methods usually make use of the certain types of information of the failed system. Gunjan Khanna creates a real-time causality diagram by analyzing the information interaction between the components and analyzes the possible cause of the fault with the legal sequence of interactions defined according to the rules [2]. Wen generates a structured system model by calculating the program dependence graph and then uses the program slice to determine the reason of the fault [3].

Steinder used heuristic method to create a set of fault hypothesis which can explain all the events received and calculated the assumptions set of all the assumptions confidence. Then assumption with the largest confidence is selected as the results [4]. Zheng proposed a three-layer belief network model and analyzed the relationship between network events [5]. Zhang improves the accuracy of fault diagnosis by analyzing the relationship of the alarm information in the adjacent window of time [6].

There will be inconsistencies in the diagnosis result with the actual system operation, by reasoning based on the static information or symptoms. For example, the diagnosis result may indicate that the component A is failed, but A is not involved in the actual operation and cannot cause the generation of the fault. So the result greatly reduces the efficiency of diagnosis. If we only use the run-time information for diagnosis, the set of components which may result in the system failure can be generated, but it is difficult to analyze which is of the largest possibility, and it will also reduce the efficiency of diagnosis. Thus, we should use the two types of information for diagnosis, and the result can meet the actual needs.

According to the above analysis, we provide a support system for fault diagnosis, which makes use of the static information and run-time information for diagnosis, and can adjust the result in the diagnosis to improve the efficiency.

95.3 Decision Support System for Fault Diagnosis

95.3.1 System Framework

As shown in Fig. 95.1, the decision support system consists of three layers. From the bottom to the top, the diagnostic information is processed through the three levels to generate the result.

The capture and abstraction layer injects probes to capture the related information and extracts the static information from related documents. Based on this information, the analysis and diagnosis layer generates the result graphically displayed in the human–computer interaction layer. After each diagnosis, the decision support system should adjust the information for diagnosis according to the actual cause of the fault, in order to improve the efficiency of subsequent diagnosis.

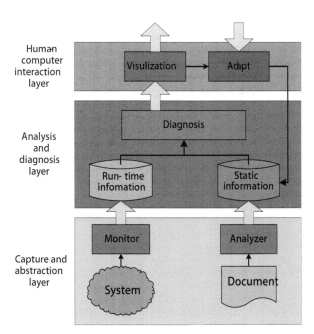

Fig. 95.1 Framework of the decision support system

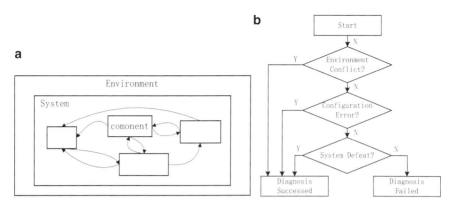

Fig. 95.2 (**a**) Schematic diagram of a failed system. (**b**) Diagnosis process

95.3.2 The Fault Diagnosis Strategy

As shown in Fig. 95.2a, the analysis of a failed system on the actual environment consists of the internal system environment, system boundary, and the system itself. Any problem in each part may result in failure. For example, the system is developed in a lab environment; there are some differences between the simulation environment and the actual environment, which are likely to cause the system to fail. If the configuration of each subsystem is incorrect, it will also result in failure. At the same time, each component of the system may also trigger fault. Therefore, the diagnosis strategy provided in this paper is to analyze the system from the outside to the inside.

Diagnosis strategy is shown in Fig. 95.2b. When a fault occurs, we first decide whether the system is incompatible with the environment. If the answer is no, then we will judge the correctness of the system configuration. Based on the exclusion of configuration problems, we will analyze each component and the interaction between them to find out the cause of the failure, and then the final diagnosis results are given.

In this paper, we focus on the third part of the diagnosis, and we provide an algorithm of fault location to achieve the diagnosis in the next section.

95.3.3 Fault Location Algorithm

Because of the spread of fault, the fault which occurs in one component may cause other component to fail which will generate different symptoms. At the same time, there are multiple mapping relationships between the symptom and the cause of the fault, in which multiple faults show the same symptoms or one fault results in some different symptoms. The bipartite graph is competent for describing multiple mapping between the symptom and the cause, so we use the bipartite graph as the analysis model.

95.3.3.1 Analysis Step

The fault diagnosis process consists of the following steps:

First, the execution path t_r and bipartite graph model G_B of the failed system are extracted. Bipartite graph model can be defined as $G_B = \{C, S, P_E, S_o\}$. C stands for the component set $C = \{(c_i, p_{ci})\}$, in which c_i is the ith component and p_{ci} is the probability of the component to be failed. S is the symptom set including all the possible symptoms in the system. P_E is the set of the relationship between component and symptom, and $P_E = \{p_{ci,sj}\} = \{p(s_j|c_i)\ |s_j \in S, c_i \in C\}$, in which $p(s_j|c_i)$ is the probability of component c_i to generate symptom s_j. If $p_{ci,sj} > 0$, component c_i can generate symptom s_j. $E(c_i) = \{s_j|\ p_{ci,sj} > 0\}$ is the set of symptoms which can be generated by c_i. $S_o \in S$ is the set of observed symptoms.

Second, defeat exists in the execution paths t_r, so we can focus on the bipartite graph G_B slice G_{Br} according to the t_r.

The next step is to find out a set of components which can explain all the observed symptoms. In this paper, a heuristic greedy algorithm is applied to find out the suspicious component. In this process, we should calculate the evidentiary value of the component to select the suspicious component; therefore, the algorithm of calculating the component failure evidentiary value is important.

95.3.3.2 Algorithm of Calculating the Component Failure Evidentiary Value

Actually, the failure evidentiary value of the component c_i is affected by the following:

1. In the observed symptom set S_o, the more the symptoms of the component c_i, the greater the evidentiary value of the component; thus, the evidentiary value can be defined as

$$W_{eo}(c_i, S_o) = \sum_{s_j \in E(c_i) \land s_j \in S_o} p(s_j, c_i) \tag{95.1}$$

2. Symptom s_k can be explained by component c_i but has not been observed. Such symptoms can appear in the following cases. First, other failure symptoms that make s_k cannot be observed, so these symptoms should enhance the evidentiary value of the c_i. $P_L(s_k)$ is defined as the probability of s_k that occurs without observation, and the evidentiary value can be defined as

$$W_{euc}(c_i, S_O) = \sum_{s_k \in E(c_i) \wedge s_k \in S - S_O} p_L(s_k) \cdot p(s_k, c_i) \quad (95.2)$$

Another case is that these symptoms without observation have not been produced actually. The more such symptoms, the smaller the probability of failure of c_i; thus the evidentiary value can be defined as

$$W_{eue}(c_i, S_O) = \frac{1}{\sum_{s_m \in E(c_i) \wedge s_m \in S - S_O} 1 - p(s_m, c_i)} \quad (95.3)$$

3. In the set of the observed symptoms which can be explained by c_i, there are some false symptoms because of the incorrect threshold and inaccurate monitoring methods and so on. The observed false symptoms may affect the evidentiary value of c_i. So, the value can be calculated as

$$W_{es}(c_i, S_O) = \sum_{s_n \in E(c_i) \wedge s_n \in S_O} p_S(s_n) \cdot p(s_n, c_i) \quad (95.4)$$

According to the analysis above, the evidentiary value $W(c_i, S_O)$ of c_i is defined as follows:

$$W(c_i, S_O) = \frac{W_{eo}(c_i, S_O) + W_{euc}(c_i, S_O)}{W_{eue}(c_i, S_O) + W_{es}(c_i, S_O)}$$

$$= \frac{\sum_{s_j \in E(c_i) \wedge s_j \in S_O} p(s_j, c_i) + \sum_{s_k \in S_U} p_L(s_k) \cdot p(s_k, c_i)}{\sum_{s_m \in E(c_i) \wedge s_m \in S - S_O} \frac{1}{1 - p(s_m, c_i)} + \sum_{s_n \in E(c_i) \wedge s_n \in S_O} p_S(s_n) \cdot p(s_n, c_i)} \quad (95.5)$$

Then, the fault diagnosis algorithm can be provided as follows:
Algorithm GFDA
Input: bipartite graph model $G_B = \{C, S, P_E\}$ and the observed symptom set S_o;
Output: a set of suspicious components H;
GFDA (G_B, S_o) {
$H = \phi$; $C_e = \phi$; $S_D = \phi$;
   while$(S_o \neq \phi)$ {
      analyze each $s_j \in S_o$, compute the $C_e = \{c_i | c_i \in C, s_j \in E(c_i)\}$; calculate the evidentiary value $W(c_i, S_O)$ of each c_i in the set of C_e; choose the component cm with the greatest value;
      $H = H \cup \{c_m\}$; $S_o == S_o - \{s_n | s_n \in E(c_m), s_n \in S_o\}$;
      $S_D = S_D \cup \{s_n | s_n \in E(c_m), s_n \in S_o\}$;
      $F_e = \phi$;
   }
   OutPut(H);
}

95.3.3.3 Algorithm Complexity Analysis

The GFDA outer cycle removes a symptom from the observed symptom set S_o at each time until all symptoms are removed. Thus, the outer cycle needed loop $|S|$ times at most. Calculating the failure evidentiary value of the component c_i executes $|C_e|$ times, of which the maximum times is $|C|$. Therefore, the complexity of GFDA is O ($|S| \times |C|$).

95.3.4 Probability Adjustment Strategy

In the diagnosis process, some probabilities are used to calculate the component failure evidentiary value. These probabilities are obtained by analyzing the system historical data and configuration information based on the expert experience, so there is uncertainty in estimating these probabilities. Thus, it is very necessary to adjust the probability according to the actual diagnosis result.

P_{ci} is the probability of c_i to be failed, and it is in the range [0, 1]. If P_{ci} is 0, it means that c_i does not fail. If P_{ci} is 1, it means that c_i will fail. For the component with a higher failure probability, its growth rate based on a successful confirmation should be less than the growth of the low probability. Therefore, P_{ci} can be adjusted as follows:

$$p_{c_i} = \alpha(1 - p_{c_i})^2 \quad (0.1 \leq \alpha \leq 0.2) \tag{95.6}$$

95.4 Experiment

In this experiment, we randomly generated a system with n components and the set S and P_E, according to the model $GB = \{C, S, P_E, S_o\}$. Then, the fault set is selected, and the set S_o is also selected from the set S randomly. In order to simulate the symptom loss, some symptoms in S_o are removed randomly. In addition, some of the symptom is not contained in S_o initially and will be added into S_o to simulate the mendacious symptoms. The number of this removed or added symptom is decided randomly. Finally, the algorithms proposed by Steinde, Zhang, and our paper will be performed ten times to compare their effectiveness, in which the number of component is increased gradually.

We use two marks to assess the fault location algorithm: detection rate and the false-positive rate, which may indicate the effectiveness of the algorithm. And Fig. 95.3 shows the result of the experiment.

Figure 95.3a shows that the detection rate of our algorithm is higher. What is more, as the number of component is increased, there are fluctuations in the other two algorithms. But our algorithm can still maintain a high level due to the adjustment strategy which can increase the detection rate as the experiment is

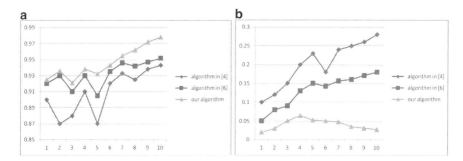

Fig. 95.3 (a) Detection rate of algorithms. (b) False-positive rate of algorithms

repeated. As shown in Fig. 95.3b, our algorithm is the best in the false-positive rate. Although incensement of the false-positive rate of the other two algorithms is not rapid in the later experiments, our algorithm can reduce the rate as the repeat of the experiment process.

In conclusion, the result of experiment indicates that the algorithm in this paper can get a better diagnostic effectiveness especially in the multiplication of the experiment.

95.5 Conclusion

This study constructed a decision support system with three layers to assist the maintainer to fix the failure. We used the bipartite graph to describe the correspondence relationship between the components and the symptoms. Then the set of suspicious components which can explain all the observed symptoms was generated by calculating the probability of component to be failed. The probability adjustment mechanism was introduced in the support system in order to improve the efficiency of fault diagnosis.

We focus on the faults caused by the components in the system. These faults caused by the environment or configuration should be studied further to improve the comprehensiveness of the fault diagnosis decision support system.

References

1. Małgorzata Steindera, & Adarshpal S. Sethi. (2004). A survey of fault localization techniques in computer networks. *Science of Computer Programming, 53*(2), 165–194.
2. Khanna, G., Cheng, M. Y., Varadharajan, P., Bagchi, S., Correia, M. P., & Verıssimo, P. J. (2007). Automated rule-based diagnosis through a distributed monitor system. *IEEE Transactions on Dependable and Secure Computing, 4*(4), 266–279.

3. Wanzhi Wen, Bixin Li, Xiaobing Sun, & Cuicui Liu. (2013). Technique of software fault localization based on hierarchical slicing spectrum. *Journal of Software, 24*(5), 977–992 (In Chinese).
4. Steinder, M., & Sethi, A. S. (2004). Probabilistic fault diagnosis in communication systems through incremental hypothesis updating. *Computer Networks, 45*(4), 537–562.
5. Zheng, Q., Hu, W., Qian, Y., Yao, M., Wang, X., & Chen, J. (2008). A novel approach for network event correlation based on set covering. *The Proceeding of Fifth International Conference on Fuzzy Systems and Knowledge Discovery, 3*(3), 122–126.
6. Zhang, C., Liao, J., Li, T., & Zhu, X. (2012). Probabilistic fault localization with sliding windows. *Science China Information Sciences, 55*(5), 1186–1200.

Chapter 96
Design and Implementation of an SD Interface to Multiple-Target Interface Bridge

Guoyong Li, Leibo Liu, Shouyi Yin, Dajiang Liu, and Shaojun Wei

Abstract The design and implementation of an SD card controller circuit architecture for multiple-target interface, suitable for communication function extension of existing electronic device for UBICOMP, are presented in this paper. The SD to multiple targets bridge includes an SD memory controller, a ping-pong FIFO, and a target selectable interface, such as UART, SPI, parallel, and NAND Flash IO. The bridge follows SD memory card v2.0 specification so that it is fully flexible in terms of portable device without any special drivers. The ping-pong FIFO increases the throughput of this system, and the availability of UART, SPI, parallel, and NAND flash interfaces provides flexibility for implementation of applications that requires the conversion of data to feed the SD bus. A tidy NAND flash is also implemented in the multiple-target interface for FTL of NAND flash. The new design has been verified and implemented in FPGA. It has also been synthesized and will be taped out through a 0.18 μm CMOS technology. Experiment reveals that the proposed architecture presents superior performance in platform-independent, interface-scalability and integrality compared with existing works.

96.1 Introduction

Ubiquitous computing (UBICOMP) [1] is the trend towards increasingly ubiquitous connected computing devices in the environment. Information processing of UBICOMP in human–computer interaction has been thoroughly integrated into everyday objects and activities. Meanwhile, peer-to-peer (P2P) communication [2]

G. Li
Research Center for Mobile Computing, Tsinghua University, Beijing 100084, China

L. Liu (✉) • S. Yin • D. Liu • S. Wei
Research Center for Mobile Computing, Tsinghua University, Beijing 100084, China

Institute of Microelectronics, Tsinghua University, Beijing 100084, China
e-mail: liulb@tsinghua.edu.cn

is needed by electronic devices in UBICOMP system. However, most existing electronic terminals are not available for P2P communication. Therefore, how to extend these terminals with function of P2P communication handily is very important. Security Digital (SD) protocol is an industry-leading memory card storage standard that simplifies the use and extends the life of consumer electronics, e.g., mobile phones, for millions of people every day. Most of modern computers and mobile terminals are equipped with SD slots, such as mobile phone and PDA. This universality makes the SD-based function extensions, especially P2P communication, become a new and hot research area, where SD interface does not only serve as a memory access protocol but also becomes a channels of function extensions of existing mobile terminals.

The SD Memory card slave controller is designed to reside within an SD memory card. It serves as a bridge between the SD bus and user logic that provides the actual function of the card. Using the standard SD bus protocol, computers and mobile terminals could access various application devices such as blue tooth, WiFi, and ZigBee. Thus, this bridge makes computing devices have the ability of P2P communication and UBICOMP of existing terminals becomes possible with this extending bridge.

Currently, most SD controllers support standard SD interface on one side and 8-bit flash IO on the other side, such as the one introduced by inCOMM [3]. So these controllers are most used in memory storage card. iWave company shows an SDIO to UART bridge controller [4], which could convert data only from UART to SDIO interface. Although it supports SDIO Interrupt feature, SDIO aware host controller is needed in the mobile terminals, which is much different from the normal SD memory host controller in hardware and software. Altera company also gives a SD slave controller IP core [5], SD/SDIO/MMC Slave Controller, whose user interfaces are mainly master parallel interfaces, and the common interfaces, such as UART and SPI, are not included. In addition, microcontrollers would be added to these SD controllers to implement the flash translation layer (FTL) for NAND flash management.

Taking all the above considerations into account, the goal of our proposed SD slave controller is a platform-independent, interface-extensible and function (FTL)-complete bridge which could convert the data from the common interfaces of existing mobile terminal's transceiver to standard SD interface. This bridge contains an SD slave controller with a ping-pong FIFO in it and a multiple-target interface. The ping-pong FIFO increases the throughput of this bridge and the multiple-target interface is designed to extend interfaces for various electronic devices. Moreover, in order to implement the function of SD memory, a tidy NAND Flash controller, including a flash translation layer (FTL), is designed in hardware in the multiple-target interface. It will present better performance in integrality compared with FTL implemented in software in microcontroller. One application on ZigBee of this bridge is shown in Fig. 96.1, where an SD-ZigBee card is implemented with this bridge.

The architecture of the remainder of this paper is as follows. The next section gives a design consideration of our bridge, including the overview of hardware

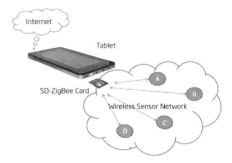

Fig. 96.1 Demonstration of SD-ZigBee card with our proposed bridge

architecture, the design of the SD slave controller and the multiple-target interface. Then the implementation of the bridge, performance analysis, and demonstration are presented in Sect. 96.3. Finally, conclusion is discussed in Sect. 96.4.

96.2 Design Consideration of Proposed Bridge

The architecture of proposed SD to multiple-target interface bridge, as shown in Fig. 96.2, consists of an SD card slave controller meeting SD memory card v2.0 specification [6] and a multiple-target interface for various electronic devices. In this section, structure and functions of main components will be described in details.

96.2.1 Platform Independent SD Card Slave Controller

The SD card slave controller is designed to comply with SD Physical Layer Specification Version 2.0. It supports the standard interface of 9-pin and is designed to operate at a maximum frequency of 50 MHz. The bus interfaces supported are SD 1-bit and 4-bit modes. The controller hands hot insertion, removal of card. As shown in the left side of Fig. 96.2, the controller consists of a command handler, a data handler, a card status state machine, and a ping-pong FIFO.

With this SD memory controller, any terminal that could access SD memory card is able to access the application cards based on the proposed bridge, such as SD-ZigBee and SD-WiFi, without any driver or hardware change. Compared with this, an SDIO device needs an SDIO aware host and a custom driver to support its operation. Therefore, this proposed bridge is platform independent and could be applied to various OS, such as Android, WinCE, and Symbian.

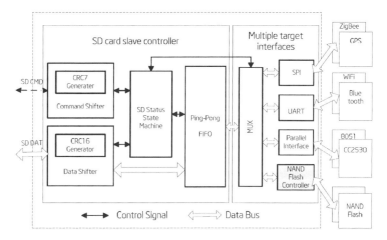

Fig. 96.2 Architecture of proposed SD to multiple targets bridge

96.2.1.1 Command Handler

The command handler receives the 48-bit command and transfers it from serial to parallel command. All commands are protected by cyclic redundancy check (CRC) bits. If the CRC check fails, the handler will not respond and the command will not be executed. Meanwhile, handler will not respond and change its status if an illegal command is received, such as a command that is not defined, or not supported by the card controller. The state machine diagram is shown in Fig. 96.3.

In the beginning, the handler is in "IDLE" state. If a command arrives, it transfers to "RAED CMD" state and starts to read and then checks the command. If the received command has a response, the handler transfers to "SEND RESPONSE" state and starts to send a response to host. Otherwise, it will return to "IDLE" state directly.

The CRC7 check is used for all commands, all response except type R3, and the CSD and CID registers. The CRC7 is a 7-bit value and is computed as follows:

$$G(x) = x^7 + x^3 + 1. \text{ (Generator polynomial)}$$
$$M(x) = (1\text{st bit}) * x^n + (2\text{nd bit}) * x^{(n-1)} + \ldots + (\text{last bit}) * x^0.$$
$$CRC[6\ldots0] = \text{Remainder}\left[(M(x) * x^7)/G(x)\right].$$

As shown in Fig. 96.4, CRC7 Generator/Checker is composed of seven shift registers and two adders. Command or response is shifted into the circuit serially. When all the bits are shifted, the CRC7 check code is presented from the seven registers.

Fig. 96.3 Command state machine diagram

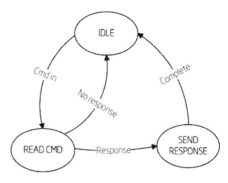

Fig. 96.4 CRC7 generator/checker

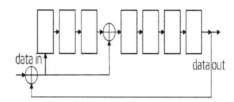

96.2.1.2 Data Handler

The data handler handles data transactions in the controller, including a transmitter and a receiver. If a data read command is received, the receiver will get data from FIFO, convert them into serial with additional CRC check and send it to SD host. If a data write command is received, the hander will receive serial data from SD interface and store them in the data buffer for further processing. If all write buffers are full, and as long as the card is in Programming State, transmitter will send busy status to host to indicate host the slave card is busy. The state diagram of data handler is shown in Fig. 96.5. Data access also needs CRC check and it uses a 16-bit CRC generator and checker for a more accurate data check. The CRC16 circuit is similar to CRC7 module in command handler.

96.2.1.3 SD Card State Machine

The card state machine handles the card state described in SD 2.0 Physical specification. It includes Inactive State, Idle State, Ready State, Identification State, Stand-By State, Transfer State, Sending-Data State, Receiving-Data State, Programming State, and Disconnect State, where the state transitions depend on the received command.

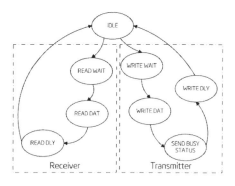

Fig. 96.5 Data state machine diagram

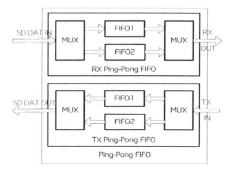

Fig. 96.6 Ping-pong FIFO operation multiple-target interface

96.2.1.4 Ping-Pong FIFOs

The card slave controller has two different clock domains, one is SD clock whose frequency range is 0 to 50 MHz and the other is programming clock with maximal frequency 50 MHz. To handle the data stream of different clock domains, we use ping-pong FIFOs [7]. The treatment scheme of ping-pong operation is as below: The input data stream is divided into two FIFOs via input data multiplexer (MUX). In the first buffer cycle, the input data fills in FIFO1; In the second buffer cycle, the input data stream fills in FIFO2, meanwhile, the data stored in FIFO1 in the first buffer cycle outputs to processing unit via an output data multiplexer. In the third buffer cycle, the input FIFO and output FIFO are exchanged. The cycle changes like this flow and goes round and round. With the rhythmed switching of input data MUX and output data MUX, the data via buffer could be send to processing unit without a pause. The diagram of ping-pong FIFO is shown in Fig. 96.6.

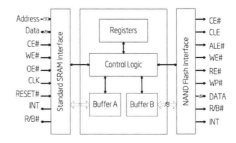

Fig. 96.7 NAND flash controller functional block diagram

96.2.2 Multiple-Target Interface

96.2.2.1 Target Selectable Interface

In order to provide flexibility for implementation of applications that require the conversion of data to feed SD bus, a multiple-target interface is proposed, including SPI, UART, parallel interface, and NAND Flash interface. As shown in the right part of Fig. 96.2, this multiple-target interface module is composed of an SPI host controller, a UART controller, a standard Parallel interface, a tidy NAND Flash interface, and a multiplexer. With the MUX, one of the target interfaces is selected at a time. We allocate a different address for each target interface. Generally, we allocate one sector (512 bytes) for each non-memory target such as ZigBee device [8], WiFi device, and blue tooth device. Through analyzing the address from SD memory host, this module decides which target to access.

96.2.2.2 Tidy Flash Controller Including FTL

The NAND Flash controller comprises a control logic module with two buffers and two different interfaces. One side of the tidy NAND Flash controller is a standard SRAM interface so that SD controller could access NAND flash as a SRAM conveniently, the other side is a standardized NAND Flash device interface which is compliant to the open NAND Flash interface specification (ONFi). The function block diagram is shown in Fig. 96.7.

Since there is not an MCU in our bridge, the bad block management and address mapping is performed in hardware in the control logic module. Using a RAM, logic address to physic mapping is implemented [9], which is shown in Fig. 96.8. A block-level mapping method is adopted for the sake of memory size. The selected NAND flash has totally 8,192 physical blocks, where 7,680 blocks are mapped to user data area and the remaining 512 blocks are mapped to free pool for bad block management. The bad block mark and logic block number (LBN) are written in the

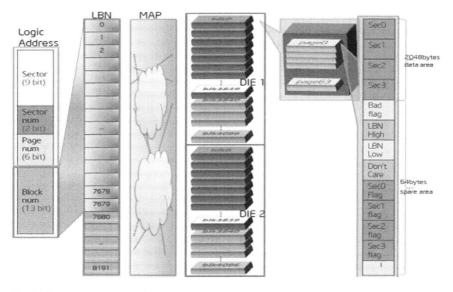

Fig. 96.8 The address map of NAND flash

Fig. 96.9 Circuit implementation of verification system

extra space of pages of every physical block. Thus, this proposed bridge could be adopted to implement the storage function without extra MCU and presents good performance in integrality.

96.3 Implementation and Analysis

96.3.1 Implementation

This bridge has been prototyped on an FPGA-based verification system [10], which consists of a Xilinx Spartan-3E FPGA, a Hynix 8 Gb NAND Flash, and a CC2530 ZigBee SoC. The circuit implementation of the verification system is shown below (Fig. 96.9).

Table 96.1 Comparison between the proposed bridge and other controllers

Interface	IN371AC [3]	iW-SDIO slave controller [4]	EP560 [5]	SDZ537 [11]	Proposed bridge
SD	√	√	√	√	√
SDIO			√	√	
UART		√		√	√
SPI					√
Parallel master interface			√		√
Flash IO	√				√

The entire bridge architecture has been FPGA verified and implemented. It also walked through a 0.18 μm CMOS technology and would be taped out. Synthesis experiments show our bridge reaches 50 MHz with the implementation size of 10 mm^2. The equivalent gate count is about 80,000 and the total memory is 47 Kb.

96.3.2 Performance Analysis

Verification on this subsystem is based on simulation except that the wireless communication interface is verified on FPGA. The programming frequency is set at 50 MHz and SD clock frequency is set at 50 MHz. Interface-scalability compared with that of other controllers is shown in Table 96.1.

The results in Table 96.1 give comparisons between our proposed bridge and other SD controllers. The proposed bridge could support UART, SPI, and parallel and NAND flash IO for connection to various target devices, while others can just support part interface. Therefore, it shows superior performance in interface-scalability.

96.3.3 Demonstration

As Fig. 96.1 shows, with an FPGA implemented bridge, a ZigBee SoC CC2530 and other basic components, an SD-ZigBee card is made. Communication interface between our bridge and CC2530 is parallel interface. With other four sensor nodes (A, B, C, D), we could establish a wireless sensor network. Via the general SD memory interface, a portable tablet with an upper machine application software could read environment information instantaneously, which could be recorded in magnanimous NAND Flash storage at the same time. Some comparisons are shown in Table 96.2 below.

Table 96.2 Comparisons between SD-ZigBee card and SDZ 537

Parameter	SD-ZigBee card (proposed bridge)	SDZ 537 [11]
Embedded MCU	×	√
SDIO aware host	×	√
Custom drivers	×	√
Power voltage	2.7–3.6 V	2.8–3.6 V
Operation current	40–50 mA (max)	24–29mA (average)
Communication distance	40 m	30–100 m
Data rate	250 kbps	250 kbps

From the results shown in the Table 96.2, for the approximate power (Voltage and Current) and performance (Distance and Data Rate), an application card equipped with this bridge could implement the NAND flash storage function without extra MCU and an SD memory host controller could access it without any hardware(SDIO aware) or driver change. The integrality and platform independent features would reduce the difficulty and period of system design. Other SD-based communication devices could be implemented easily with our proposed bridge such as SD-WiFi card and SD-Blue Tooth card.

96.4 Conclusion

This paper presented an SD interface to multiple-target interface bridge. The bridge was designed to adapt to different electronic devices and could extend the function of existing ubiquitous computing devices. A tidy NAND Flash controller was also implemented in hardware so that an 8 Gb NAND Flash could be connected directly to build up an SD memory card. The implementation and demonstration showed that it is very convenient to design various SD-based communication devices using our proposed bridge, and thus many existing terminals could be extended with P2P communication functions, which will make the UBICOMP of existing electronic device possible. Experiment reveals that the proposed bridge presents good performance in platform-independent, interface-scalability and integrality.

References

1. Lyytinen, K., & Yoo, Y. (2002). Ubiquitous computing. *Communications of the ACM, 45*(12), 63.
2. Bruda, S. D., et al. (2012). A peer-to-peer architecture for remote service discovery. *Procedia Computer Science, 10*, 976–983.
3. inCOMM Technologies Co. Ltd. (2009). Flash card controller.
4. iWave Company. (2012). *iW-SDIO slave controller*. SDIO slave controller datasheet, Rel. 1.4.
5. Eureka Technology Inc. (2009). Ep560 datasheet.
6. S. C. Association. (2006). SD specification, version 2.0.

7. Huang, P., He, H., & Xu, D. (2008). Asymmetric asynchronous FIFO design in navigation receiver. *Journal of Projectiles, Rockets, Missiles and Guidance, 1*, 77.
8. Z. Alliance. (2005). *Zigbee specification*. ZigBee document 053474r06, version 1.
9. Kim, J., Kim, J., Noh, S., Min, S., & Cho, Y. (2002). A space-efficient flash translation layer for compact flash systems. *IEEE Transactions on Consumer Electronics, 48*(2), 366–375.
10. Dajiang Liu, Shouyi Yin, Jianfeng Chen, Hui Gao, Shaojun Wei. (2011). A portable environmental monitoring system based on WSN for off-the-shelf sensors. *International Conference on Computational Problem-Solving (ICCP), 2011* (pp. 569–572).
11. L. Spectec Computer Co. (2009). SDZ-537 microSD ZigBee card datasheet.

Chapter 97
Cloud Storage Management Technology for Small File Based on Two-Dimensional Packing Algorithm

Zhiyun Zheng, Shaofeng Zhao, Xingjin Zhang, Zhenfei Wang, and Liping Lu

Abstract In order to improve storage efficiency of small files in the cloud storage systems based on HDFS (Hadoop Distributed File System), this paper proposed a merging process approach based on a two-dimensional packing algorithm, called TDPHDFS (two-dimensional packing for HDFS). In it the correlations between file size and arrival time are comprehensively considered to assist the small files to be merged into large ones. The simulation results demonstrate that the storage efficiency of small files is improved, while the stability remains the same, yet less resource is consumed. The TDPHDFS algorithm can effectively reduce the performance penalty in both storage space and memory consuming while managing massive small files.

97.1 Introduction

HDFS (Hadoop Distributed File System) is a distributed file system model with highly fault-tolerant performance [1], which can be deployed on ordinary machine or virtual machine. It supports Java runtime environment and provides high-throughput data access. The metadata of the file system is placed in memory by NameNode [2]; if there are a large number of small files, the system will undoubtedly reduce the storage efficiency and storage capacity of the entire storage system. Therefore, how to solve the problem of small file storage efficiency has been researched as a popular topic of cloud storage.

Z. Zheng • S. Zhao • X. Zhang • Z. Wang (✉)
School of Information Engineering, ZhengZhou University, ZhengZhou 450001, China
e-mail: iezfwang@zzu.edu.cn

L. Lu
Department of Information and Engineering, Henan College of Finance and Taxation, ZhengZhou 450001, China

This paper is focused on merging small files before they were uploaded to HDFS, and it proposes a method of merging small files by draw packing algorithm. The paper is organized as follows. In Sect. 2 we describe the relevant aspects of HDFS. In Sect. 3 we give an algorithm to merge small files. In Sect. 4 we describe our experiments and evaluation, and in Sect. 5 we present conclusions and future works.

97.2 HDFS Basic Framework and Related Research

A typical HDFS cluster is composed of a single NameNode and multiple DataNodes. The NameNode main function is maintaining file system namespace and managing all files and directories in the file system tree and the whole tree [3].

As the file system work node, the DataNode stores the actual data block, retrieves data blocks according to the need, and responds to the command which includes create, delete, and copy the data block.

The NameNode is the single point and stores the metadata information [4]. When the number of small files increases to a certain extent, the NameNode resource consumption will become a bottleneck of the system performance.

Currently, the mainstream idea of research is consolidating or combining small files into large files; there are usually two methods to be used: one is using Hadoop Archive (HAR) technology to merge small files (file size is less than 10 MB) [5]; the other is approaching a certain method of the small files' combination for specific applications.

The research results by using archiving technique to process small files are as follows: Mackey takes advantage of HAR technology to achieve the merging of small files, thereby improving the efficiency of HDFS metadata storage [6]. Yu Si and Gui Xiaolin uses the SequenceFile to combine small files, integrate the multi-attribute decision theory and experiment, and then obtain an optimal file merging scheme [7].

The research results by approaching a certain method for specific application are as follows: Liu combines small files into large files by using the WebGIS access mode features and establishes a global index for them; the storage efficiency of small files is improved [8]. Liu L optimizes the concurrent access of small files in distributed storage system [9].

97.3 Small Files Merging Algorithm

In order to solve the problem of low efficiency for store small files in HDFS, this paper proposes TDPHDFS (two-dimensional packing for HDFS) algorithm during pre-merging small files with two stages. In the first stage, sort these files and generate the optimal solution file sequence. In the second stage, use file stream to merge small files to ensure the efficiency of the cost of the merger, and then submit the merged file to the DataNode by client.

97.3.1 Two-Dimensional Packing Problem

Bin-packing (BP) problem [10] is a combinatorial optimization problem having very strong application background, and its solution is extremely difficult. In it, objects of different volumes must be packed into a finite number of bins or containers each of volume in a way that minimizes the number of bins used [11]. Treat each block of HDFS as a box, and small file waiting to be merged as the object; drawing on the idea of bin-packing algorithm can maximize the utilization of the HDFS storage space.

One representative part of the two-dimensional packing problem is 2 BP (two-dimensional bin-packing problem) [12], described as follows: given n rectangular items, $F = \{1, 2, \ldots, n\}$ and the infinite plurality of rectangular boxes with same size. The width of the ith item is wi, and the height is hi ($i = 1, 2, \ldots, n$). The rectangular box's width is W, and the height is H. All items are put into the box with the least number.

97.3.2 TDPHDFS Algorithm

With the idea of dynamic delay, TDPHDFS does not require users to upload small files immediately, only packing when there are enough (total file size is greater than a threshold value such as 64 MB) small files in the buffer. Considering the waiting time of the file in the buffer, TDPHDFS increases the file's priority level as the waiting time increases at speed A (set 2).

TDPHDFS proposes the following definition: there is only one box and the size of box is set to 63 MB; 1 MB space is reserved as the compressed file index information.

For case of description, Table 97.1 describes the parameters used in TDPHDFS. The TDPHDFS algorithm's description is as follows:

1. Initialization process defines a two-dimensional array named FileArray[MD5][key]; the key is initialized as file's size.
2. Read the size of all files in buffer; update the FileArray[MD5][key]; if the array already exists, the same MD5 value file, the corresponding weight is multiplied by 2.
3. Sort the file by weights from the largest to smallest. If total weights are greater than 63MB, $\sum_{i=1}^{n} \text{FileArray}[i][2] \geq 63M$, start packing or else exit the program, and wait for the next packing scheduling.
4. Begin packing from the maximum weight of the file. If file f_i is boxed and the total size of the files which are boxed already is less than 63 MB, then store f_{i+1} and delete the f_i records from the array. Loop step 4.

Table 97.1 The parameters used in TDPHDFS

Symbol	Description
fi	The ith file to packing
F	A collection on files in buffer F = {f1,f2,....,fn}
n	The number of files in buffer
a	The rate change of file priority, set 2
FileArray[MD5][key]	File information; the first column indicates the MD5 value of the file, and the second column indicates the weight of the file
Key	The weights of the file
Flevel	The priority of the file
y	The file number waiting for merge

5. Calculate the actual size of the files in the box; if it is less than 62 MB (packing threshold), go to step 2.
6. Calculate the number of files to be merged, and set y. Generate the index of ranked files, and attach the index to the post-merger file; combine them.
7. Upload merged file, and go to step 2.

In the fifth step, the reason for TDPHDFS finding and packing files again is because in step 2, there will be some small files that are not loaded while weights are increased. After priority loading these small files, there will be some virtual space occupied. So it is necessary to confirm the actual size after packing. The best time complexity of the proposed algorithms is $T(n) = O(n \times \log n)$, and the worst time complexity is $T(n) = O(n^2)$.

97.4 Simulation and Result Analysis

97.4.1 The Configuration of Simulation

In order to verify the algorithm performance on improving the storage utilization and reducing resource consumption, the following experiments are designed, detailed experimental environment configuration is shown in Table 97.2.

97.4.2 Simulation Analysis

In the experiment, replication is set to 1; when all the files are uploaded to HDFS, use "du" command to show the size of all files in HDFS. The results are shown in Table 97.3. When we use DS1, the total size of files in HDFS is 488 G (upload directly), which is over 200 times the space cost than when using TDPHDFS. The DS2 files' space cost is almost a differential of 60–1.

97 Cloud Storage Management Technology for Small File...

Table 97.2 Experimental environment

Components	Description
Hadoop version	0.2.3
NameNode number	1
DataNode number	7
OS	Ubuntu 10.04
CPU	NameNode: Intel I3-2130 3.4 G
	DataNode: Intel Core Duo T2410 2.0 G
Memory	NameNode, 4 G; DataNode, 2 G
Net situation	LAN

Table 97.3 The total size of files after upload

Dataset	Description	Upload way	Total size
DS1	8571 files, total size 1.49 G	Use TDPHDFS	2.3 G
		Directly	488 G
DS2	1000 1MB files, total size 1 GB	Use TDPHDFS	1.1 G
		Directly	62.5 G

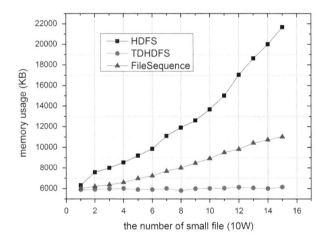

Fig. 97.1 Memory usage in NameNode with three methods

As it is shown in Fig. 97.1, when the number of small files increased, the memory usage in NameNode increases, and compared with the FileSequence, the TDPHDFS can reduce this consumption more effectively. As the number of small files increased, TDPHDFS can control the memory usage cost under a very low level.

A comparison of the memory usage in 4 DataNodes is shown in Figs. 97.2 and 97.3 with or without the use of TDPHDFS algorithm. The average memory

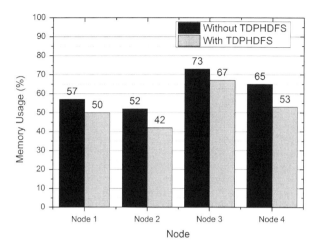

Fig. 97.2 Memory usage comparisons during upload files

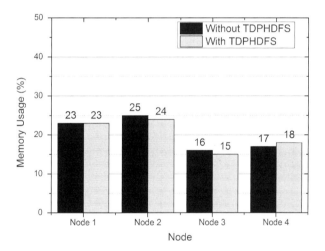

Fig. 97.3 Memory usage comparisons while HDFS is idle

occupancy rate is 56 % when uploading files to HDFS, and if the server is idle, the average memory occupancy rate could be 21 %. By using TDPHDFS, the average memory occupancy rate could be 42 and 19.5 % separately. This test shows that with TDPHDFS, the pretreatment could reduce the consumption of system resources.

97.5 Conclusion

In this paper, we have analyzed the HDFS basic framework and proposed a TDPHDFS algorithm to solve low storage performance when storing small files in HDFS. This algorithm uses improved packing algorithm to merge small files to achieve the maximum usage of HDFS' default block. There is a greater improvement in time and space to store small files with TDPHDFS compared with the origin HDFS.

References

1. Borthakur. *HDFS architecture guide*. Hadoop Apache Project. Retrieved from http://hadoop.apache.org/common/docs/current/hdfs_design.pdf.
2. Shengjun Xue, Wu-Bin Pan, & Wei Fang. (2012). A novel approach in improving I/O performance of small meteorological files on HDFS. *Applied Mechanics and Materials 17* (11), 1759–1765.
3. Ming Chen, Wei Chen, Likun Liu, & Zheng Zhang. (2007). An analytical framework and its applications for studying brick storage reliability. *IEEE, 7*(26), 242–252.
4. Zhanjie Wang, & Lijun Zhang. (2012). Mix-P2P architecture of distributed storage system based on HDFS. *Advanced Materials Research 382*(7), 92–95.
5. Shvachko, K., Huang, H., Radia, S., & Chansler, R. (2010). The hadoop distributed file system. *MSST, IEEE 26th Symposium, 4*(1), 1–10.
6. Sehrish, M. S., & Wang, J. (2009). Improving metadata management for small files in HDFS. In *Cluster computing and workshops, 2009* (pp. 1–4). CLUSTER'09. IEEE International Conference, IEEE.
7. Yu, S., Gui, X., Huang, R., & Zhuang, W. (2011). Improving the storage efficiency of small files in cloud storage. *Journal of Xi'an Jiaotong University., 10*(1109), 65–72.
8. Yang, C. T., Huang, K. L., Liu, J. C., Chen, W. S., Hsu, W. H. (2012). On construction of cloud IaaS using KVM and open nebula for video services. In *Parallel processing workshops* (ICPPW), 2012 41st International Conference, IEEE.
9. Thusoo, A., Sarma, J. S., Jain, N., Shao, Z., Chakka, P., Anthony, S., et al. (2009). Hive: A warehousing solution over a map-reduce framework. *Proceedings of the VLDB Endowment, 2* (2), 1626–1629.
10. Lodi, A., Martello, S., & Monaci, M. (2002). Two-dimensional packing problems: A survey. *European Journal of Operational Research, 141*(2), 241–252.
11. Côté, J. F., Gendreau, M., & Potvin, J. Y. (2013). An exact algorithm for the two-dimensional orthogonal packing problem with unloading constraints. *Interuniversitaire Research Centre on Enterprise Networks, 6*(1), 1–32.
12. Andrea Lodi, Silvano Martello, Michele Monaci, Claudio Cicconetti, Luciano Lenzini, Enzo Mingozzi et al. (2011). Efficient two-dimensional packing algorithms for mobile WiMAX. *Management Science, 57*(12), 2130–2144

Chapter 98
Advertising Media Selection and Delivery Decision-Making Using Influence Diagram

Xiaoxuan Hu and Fan Jiang

Abstract The influence diagram (ID) is introduced into advertising media selection and delivery strategy making by reducing uncertainty in the process of decision-making. This paper conducts a survey and selects relevant variables including product category, advertising budget, target audience, media selection, authority, and coverage. The topology layer of the ID model is constructed by distinguishing the causal relationship among variables, and the parameter layer is defined through the judgment of conditional probability. Empirical results show that scientific assessments of the various expected utility values in the decision-making program are put by probabilistic reasoning. Based on it, the larger profit can be obtained under a smaller cost with the principle of expected maximization. Therefore, the model does an effective job and provides reference for decision-makers.

98.1 Introduction

The huge role of advertising is more and more valued by many enterprises as an important part of corporate marketing activities. At the same time, the effect of advertising is declining seriously because many audiences respond to it with intensive indifference and resistance. Thus, how to select the suitable media and make a right decision are the primary decision-making problems in the corporate advertising works. However, due to the presence of many uncertainties, it greatly increases the risk and difficulty of decision-making. Therefore, the use of appropriate tools and methods, the quantitative analysis for uncertainties, and having scientific assessment of the various decision-making programs are all important issues.

X. Hu • F. Jiang (✉)
School of Management, Hefei University of Technology, Hefei 230009, China
e-mail: huxiaoxuan@vip.sina.com; caocaoyatou66@163.com

By now, a large number of researchers do qualitative work to have the advertising media selection and decision-making from the point of view of expert knowledge and experience. Very small part does quantitative analyzing to get optimal strategy, such as, mathematical programming, multi-attribute evaluation method. For example, we can use the dual simplex method of linear programming problem to select the best combination of advertising media. Buratto made use of the linear programming model to obtain the maximum profit media solutions and did research about the advertising channel selection in the market segments [1]. Viscolani applied nonlinear programming method to solve the general advertising decisions assuming that each market segment demand function was linear and the function of media advertising cost was quadratic [2]. Jha adopted the same objective programming method to maximize the total profit in all markets, taking into account the different characteristics of different markets [3]. Ngai got the best website for online advertising by making use of AHP [4]. Other researchers applied constrained goal programming model, multiple criteria decision-making model, or double goal programming model to solve the problem of advertising decision-making [5–7]. In a word, these methods are certain specific solutions to advertising decision-making, however, when facing other similar scenarios, it's necessary to re-plan all processes, thus, resulting in large amount of calculation and failure to do the universal advertising decision-making.

In this paper, we will introduce influence diagram (ID) to construct the model of advertising media selection and delivery strategy, analyze each factor which will effect decision-making qualitatively and quantitatively using powerful ability to describe problems and inference probability, and ultimately, put a sequence for various decision-making programs by the calculation of ID, thus providing a reference for policy makers.

98.2 Decision-Making Modeling

Professor Howard proposed the graph model based on uncertain information expression and solved complex decision problems in 1984 [8], consisting of the dependencies between variables, conditional independence relationships, and decision-makers' preference information. The theory and application are developing more and more rapidly so that it could build related model for any statistical analysis problem [9–11]. The generation process of ID is a hierarchical process and the specific process is shown in Fig. 98.1.

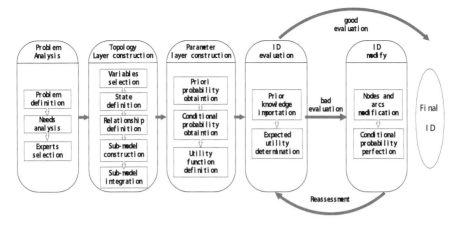

Fig. 98.1 The generation process of ID

98.2.1 Selection of Variables and the Corresponding State Space

To build the graph of ID, determining all variables related is key, usually finished by the knowledge and expertise of experts. There are a lot of variables during the process of corporate advertising media selection and delivery; however, not every variable has an important influence on the final decision-making. So we will conduct a questionnaire survey, discriminate the importance among variables, and delete some irrelevant variables. Then, we distribute these questionnaires to experts in the field and ask them to give their points ranging 1–5, "1" representing unimportance of variables, while "5" representing a large extent. Ultimately, the associated variables can be shown in Table 98.1.

98.2.2 Construction of Topology Layer for Advertising ID

According to the variables previously obtained, we begin to build the topology layer based on consultancies of the experts. In order to reflect accurately the conditional relationship among these variables, we repeatedly ask the experts, such as persons of experience in the advertising industry, the media people, and business owners. The specific information of questionnaire can be seen in Fig. 98.2.

Based on the above questionnaire, we could draw influence relationships among variables properly, resulting in two network diagrams: media selection and delivery strategy, as shown in Figs. 98.3 and 98.4, respectively.

Table 98.1 The state spaces of each variable

Variables	State space	Remark
Product category	Hotel, food, clothing, tobacco and alcohol, computer	For many categories of all kinds of products on the market, it's complex to be covered in the "Product Category" variable, just to pick a few representative kinds
Advertising budget	Low	Deterministic variable with the lowest cost, hoping to get a larger income by the lowest budget
Target audience	Old man, middle-aged, youth	It's a simple division in order to obtain the conditional probability, assuming that they have the purchasing power
Product life cycle	Introduction, growth, maturity	The recession is not the introduction of state space because of a sharp decline in product sales in this period, and the profit will fall significantly so that companies no longer put into advertising the products, but consider what strategy out of the market to adopt as soon as possible
Media selection	TV, radio, Internet, outdoor, newspaper	It's a decision node
Delivery strategy	Centralized, continuous, intermittent	It's a decision node
Authority	High, middle, low	It's a random variable, indicating that the type of media has a different degree of authority
Coverage	High, middle, low	It's a random variable, representing that the type of media has a different degree of coverage
Audience awareness	High, middle, low	It's a random variable. The probability value would change with the difference among delivery strategies in a certain stage for different products
Audience preference	High, middle, low	Different delivery strategies will lead to different audiences' preference
Expected sub-income 1	High, middle, low	It's influenced by the target audience, audience preference, and the corresponding coverage
Expected sub-income 2	High, middle, low	Both audience awareness and audience preferences will affect its conditional probability value
Total revenue	High, middle, low	The final state will be decided by two sub-incomes
Media costs	High, low	The media costs are different due to the diversity type of media
Delivery costs	High, low	Choosing different delivery strategies will result in different delivery costs
Total costs	High, low	Both the media costs and delivery costs determine its value

Q1: What are the key factors (ie, a priori node) in the choice of media clearly?
A1: Only clearing out product categories, budget of advertising, target audience and product life cycle, we could be able to select the media category.
Q2: After selecting the established media, which factors will impact the decision-making program?
A2: Based on historical experience and data, we can mainly divide into the following several aspects—media authority, media coverage and media costs. At the same time, the type of audience will affect the final income.
Q3: What criteria should be determined when evaluating the result of decision-making?
A3: In order to facilitate our research, the economic benefits of advertising could be considered, such as, the expected revenue and cost.

Fig. 98.2 A simple questionnaire

Fig. 98.3 Media selection module

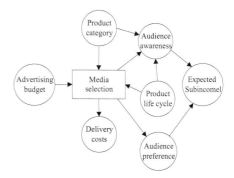

Fig. 98.4 Delivery strategy module

The overall model should be built by the joint of some intermediate nodes, connecting the main and common nodes in two sub-modules, such as *Total income* and *Total cost*. Thus, the integrated ID could be built on the basis of sorting out the logical relationship among variables, as shown in Fig. 98.5.

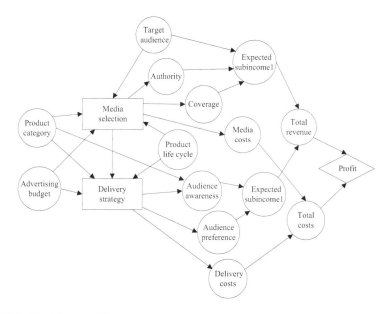

Fig. 98.5 The integrated ID

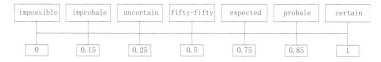

Fig. 98.6 The probability assessment scale

98.2.3 Construction of Parameter Layer for Advertising ID

The next step is to quantify conditional relationship among all variables, using the knowledge and experience of experts and applying historical statistic data usually. Since there exists some difference between different enterprises, and different types of data are not homogeneous, so it is difficult to obtain accurate a priori probability and conditional probability in modeling. Therefore, the conditional probability is solved by using the first method, consulting experts in the field. We will employ the easiest but most effective way for description of the probability—the probability assessment scale [12]—as shown in Fig. 98.6.

The key of building ID properly is to establish an appropriate function. Accordingly, we will also invite experts in the field and some experienced staffs to describe functions. In order to be able to get the opinions of experts, we will make the conditional probabilities in the ID into questionnaire, completed by all participating

Table 98.2 The conditional probabilities of "Authority"

Media	Authority		
	High	Middle	Low
TV	0.7	0.2	0.1
Radio	0.3	0.4	0.3
Internet	0.3	0.4	0.3
Outdoor	0.5	0.3	0.2
Newspaper	0.6	0.3	0.1

experts; finally, all assignments will be updated and stated averagely so that the conditional probability tables are obtained, as seen in Table 98.2, and the others are similar to Table 98.2.

For whole ID, since the date of some input node is difficult to get, we temporarily simplify the processing to have the same probability for each state of all input nodes, when determining some a priori probabilities.

98.3 Evaluation of Advertising ID

Nowadays, the research of ID has been quite mature and the algorithms are also very complete. A number of software have been developed to solve the problem of the calculation and optimization of ID in foreign countries, such as Analytica, Netica, Pulcinella, Hugin, BayesialLab, Smile, SmileX, and GeNIe. In this paper, the GeNIe2.0 was applied to help us solve some calculation, which is powerful and has a wealth of graphic elements and a variety of decision-making tools and user-friendly interface.

The integrated ID is derived with some a priori probabilities and conditional probabilities based on the above questionnaire, applying GeNIe2.0 software. The next step is to verify the ID. For example, there is a kind of wine, which is a mature brand and whose target audience is middle-aged person, then we input some priori information based on it—target audience, product category and product cycle and update the ID, thus, the expected utility values are as shown in Table 98.3.

From Table 98.3, the largest expected utility EU equals 36.697281, meaning that the enterprise should select the program of TV and intermittent delivery strategy when there is the smallest advertising budget. By entering different data, the corresponding maximum expected utility values are obtained by virtue of the ID, whose classification results are consistent with the actual situation, reflecting the excellent decision-making function. In the process of choosing kinds of advertising media by actual corporate, we usually try to combine with other media when there is sufficient funds in order to bring greater profits. Thus, it is possible to sort the expected utility values through the ID, and further work may also be needed to select the programs with larger utility values and to study the specific advertising media composition problem.

Table 98.3 Expected utility values

	Centralized	Continuous	Intermittent
TV	26.593488	31.972454	36.697281
Radio	21.706331	27.103728	24.704282
Internet	25.752049	31.811325	30.957273
Outdoor	23.511136	28.949377	29.166374
Newspaper	27.636540	34.238384	32.237819

98.4 Conclusion

Advertising media selection and delivery strategy making are important ways to promote products in a corporation. This paper proposes the model of ID to analyze and calculate the process of decision-making, reducing complexity and uncertainty to a large extent. Applying this model has the following advantages: (1) representing intuitively the relationships among corresponding factors in the process of choosing advertising media and increasing data correlation and logic, overcoming some uncertainty in advertising program; (2) estimating quantitatively the relationships by virtue of the form of conditional probability; (3) reducing some subjectivity and macro considerations comparing to some enterprise experience decision-making and avoiding prejudices made by some leaders; and (4) making the decision more convenient and more realistic with collection of two-stage decision-making process in the ID model. Therefore, our results indicate that the approach is a promising one.

Acknowledgements This research is supported by the National Natural Science Foundation of China (No. 71071045, 71131002, 71001032) and Humanities and Social Science Projects of Ministry of Education of China (13YJC630051).

References

1. Buratto, A., Grosset, L., et al. (2006). Advertising channel selection in a segmented market [J]. *Automatica, 42*(8), 1343–1347.
2. Viscolani, B. (2009). Advertising decisions for a segmented market [J]. *Optimization, 58*(4), 469–477.
3. Jha, P. C., et al. (2011). Optimal media planning for multi-products in segmented market [J]. *Applied Mathematics and Computation, 217*(16), 6802–6818.
4. Ngai, E. W. T. (2003). Selection of web sites for online advertising using the AHP [J]. *Information & Management, 40*(4), 233–242.
5. Bhattacharya, U. K. (2009). A chance constraints goal programming model for the advertising planning problem [J]. *European Journal of Operational Research, 192*(2), 382–395.
6. Kwak, N. K., et al. (2005). An MCDM model for media selection in the dual consumer/industrial market [J]. *European Journal of Operational Research, 166*(1), 255–265.

7. Perez Gladish, B., et al. (2010). Planning a TV advertising campaign: A crisp multiobjective programming model from fuzzy basic data [J]. *Omega, 38*(1–2), 84–94.
8. Howard, R. A., & Matheson, J. E. (2005). Influence diagrams [J]. *Decision Analysis, 2*(3), 127–143.
9. Smith, J. Q. (1989). Influence diagrams for statistical modeling [J]. *The Annals of Statistics, 17*(2), 654–672.
10. Cobb, B. R., & Shenoy, P. P. (2008). Decision making with hybrid influence diagrams using mixtures of truncated exponentials [J]. *European Journal of Operational Research, 186*(1), 261–275.
11. Kjaerulff UB, Madsen AL (2013) Bayesian Networks and influence diagrams: A guide to construction and analysis [M]. *Springer* (pp. 70–95).
12. Witteman, C., & Renooij, S. (2003). Evaluation of a verbal–numerical probability scale [J]. *International Journal of Approximate Reasoning, 33*(2), 117–131.

Chapter 99
The Application of Trusted Computing Technology in the Cloud Security

Bo Li, Shenjuan Lv, Yongsheng Zhang, and Ming Tian

Abstract For the lack of safety and reliability of the information in the cloud computing environment, in order to create a more flexible and adaptable security mechanism, the combination of cloud computing and credible concept is a major research direction in today's security. Based on the view mentioned above, this paper strengthens the research of trust computing technology to solve the security issues in cloud and cloud-based trust transfer, on the basis of the practical work of the experts and scholars on the trust transfer technology, and expands the theoretical model of the trust chain. This paper uses the stochastic process algebra and Petri nets as a modeling tool to build two trust chain models, demonstrates the credibility of certain behavioral characteristics of the chain, analyzes several constraints of credible chain, and provides a valuable reference for engineering practice of the credible chain.

99.1 Introduction

Cloud computing obtains much of the industry's attention since its birth, and with the increasing globalization of the world economy and the development of information technology, cloud-based applications can be promoted in all areas of society [1]. At the same time, the advantages of cloud computing can reflect in all walks of life in society, from business to science to many key sectors of national security,

---

B. Li (✉)
Academic Affairs Office Shandong Polytechnic, Jinan 250104, China
e-mail: lvshenjuan2011@163.com

S. Lv • Y. Zhang • M. Tian
School of Information Science and Engineering, Shandong Normal University, Jinan 250014, China

Shandong Provincial Key Laboratory for Novel Distributed Computer Software Technology, Jinan 250014, China

and can see the trail of cloud computing. However, with the development of information technology, people give more importance on the information security and cloud data security, and privacy protection is generally considered to be an important aspect of cloud computing. So far, however, the customer does not have any means to prove the integrity and confidentiality of the applications and the upload of their own data, resulting in the issue of trust between the client and the cloud service provider, which will become the biggest obstacle to hinder the development of the cloud computing world. Currently, cloud computing technology is facing many security technology crises [2, 3], such as counterfeiting electronic signature, electronic signature repudiation, Trojan attacks, and viruses' damage, which seriously threatens the Internet cloud computing trust and cloud security. For a range of issues, the combination of trust computing technology and cloud computing and of the application of trust computing technology in the cloud security is the direction to solve this problem; therefore, it's important to strengthen the trust technology [4].

99.2 Related Works

In recent years, many experts and scholars use different theories and methods in cloud security issues and achieve certain results.

Today, the most active organization on cloud security is Cloud Security Alliance (CSA). CSA as an industry-recognized research on Cloud Security Alliance do more in-depth cloud security research. According to the CSA on cloud security research, other organizations of international numbers, such as CAM (Common Assurance Metric beyond the cloud) [5], Microsoft and Green League, and other institutions also do some research on the field of cloud security and put forward the framework of computing security and cloud security technology solutions research. But overall, the international research on cloud security issues is not deep enough. In connection with cloud security issues, this paper will begin with trust computing technology and strengthen the research of trust computing technology to solve the security issues in cloud and cloud-based trust transfer as the research object, at the conclusion on the basis of the practical work of the experts and scholars in the trust transfer technology; expand the theoretical model of the trust chain; analyze several constraints of credible chain; and provide a valuable reference for engineering practice of the credible chain [6].

99.3 Trusted Computing Technology

Nowadays, implementation of the trusted computing platform is trusted PC; the module TPM (Trusted Platform Module) is an SOC chip, which is the root of trust of the trusted computing platform. TSS (TCG Software Stack) is the TPM support

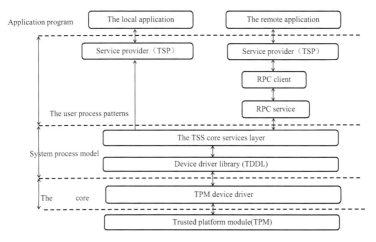

Fig. 99.1 Trusted computing system. TPM: Trusted Platform Module, it is a security chip. OS: Operating system, it is the system software collection which manages hardware resources of computer, controls the other program's running and provides interactive operation interface for users

software of a trusted computing platform [7]; its main function is to provide the environment of application software development compatible with heterogeneous trusted platform module. TNC (Trusted Network Connect) ensures the integrity of the website visitor through the network to access, collect, and verify the requester's integrity information. We evaluate this information according to certain security policy and then decide whether to allow the requester and the network connection [8, 9]. Trusted computing system is shown in Fig. 99.1.

99.4 The Research of Trusted Computing Technology in Cloud Security

99.4.1 The Data Model

Definition 1: The activity of atomic component is a four-tuple as shown in Eq. 99.1:

$$\text{Activity} = (\alpha, I_p, r, t) \tag{99.1}$$

Among them, $\alpha \in Act, p \in P$, when component p executes behavior α, I_p expresses the integrity property; $I_p = true$ represents component p is complete,

$I_p = false$ represents that component p is incomplete, r represents the integrity metric size of component p, and t represents the duration of the activity.

Definition 2: The operator refers to the relationship between atomic components, such as synchronization, concurrency, and competition, and structure composite components by operator. Operators are expressed as in formula 99.2:

$$P ::= (\alpha, I_p, r, t) \cdot P | P + Q | P_L Q | Nil | A \stackrel{def}{=} P \tag{99.2}$$

Among them, $(\alpha, I_p, r, t) \cdot P$ expresses performing (α, I_p, r, t) before operation P; $P + Q$ expresses performing activities of P and Q; $P_L Q$ expresses components p and Q performing together; L expresses communication between P and Q; Nil expresses computing platform deep in a deadlock situation and stop the transfer of trust; at present, the platform is in an incredible state; $A \stackrel{def}{=} P$ expresses performing activity P in a circle.

Definition 3: Define two trust transfer rules in trust transfer process as follows:

Rule 1 The prefix operator transfer rules

$(\alpha, I_p, r, t) \cdot P$ occurred prefix operation and migrated to P after (α, I_p, r, t) as shown in 3:

$$(\alpha, I_p, r, t) \cdot P \xrightarrow{(\alpha, I_p, r, t)} P \tag{99.3}$$

The state transition of trusted chain is shown in formula 99.4:

$$I_{TC} \rightarrow I_{TC} \oplus (I_p, r) \tag{99.4}$$

Among them, I_{TC} represents the state of trusted chain, $I_{TC} = true$ expresses platform is in a credible state, and $I_{TC} = false$ expresses the transmission of trust is stopped and platform is in an incredible state. The initial value of I_{TC} is true. $\oplus$ expresses extending trusted chain with integrity metric size r. The calculation method is shown in formula 99.5:

$$I_{TC} = \begin{cases} I_{TC} \quad and \quad I_p & (under \quad r) \\ I_{TC} & (otherwise) \end{cases} \tag{99.5}$$

The trust transfer rule is shown in formula 99.6:

$$\frac{I_{TC} = true}{(\alpha, I_p, r, t) \cdot P \xrightarrow{(\alpha, I_p, r, t)} P} I_{TC} \rightarrow I_{TC} \oplus (I_p, r) \tag{99.6}$$

Rule 2 Select operator transfer rules

$P + Q$ migrate to P' under the premise of $P \xrightarrow{(\alpha, T_p, r, t)} P'$. Its trust transfer rule is shown in 7:

$$\frac{I_{TC} = true \& P \xrightarrow{(\alpha, I_p, r, t)} P'}{P + Q \xrightarrow{(\alpha, I_p, r, t)}} P' I_{TC} \to I_{TC} \oplus (I_p, r) \quad (99.7)$$

Under the premise of $Q \xrightarrow{(\alpha, I_Q, r, t)} Q'$, we measure the integrity of Q with integrity metric size r; its trust transfer rule is shown in formula 99.8:

$$\frac{I_{TC} = true \& Q \xrightarrow{(\alpha, I_Q, r, t)} Q'}{P + Q \xrightarrow{(\alpha, I_Q, r, t)}} Q' I_{TC} \to I_{TC} \oplus (I_Q, r) \quad (99.8)$$

99.4.2 Semantic Model Design

In accordance with the transmission process of trust in TPM, OS(Operating System) and the user mode process, this paper establishes semantic model of TPM, OS and the user mode process, after which, we link each model, making them become trusted chain model.

99.4.2.1 The Semantic Model of TPM

In the process of trust transfer, activities which need to be completed by TPM are extending and extended; if the two atomic behaviors are completed, then platform is in an idle state. The semantic model of TPM is expressed in formula 99.9:

$$TPM \stackrel{def}{=} (extending, I_{TPM}, r, t) \cdot (extended, I_{TPM}, r, t) \cdot (TPM\_idle, I_{TPM}, r, t) \cdot TPM \quad (99.9)$$

99.4.2.2 The Semantic Model of OS

OS is an important part in the process of trust transfer. First we should complete integrity measuring of OS and then extend the results to trusted chain by extending. Its semantic model is expressed in formula 99.10:

$$OS \stackrel{def}{=} (run, I_{OS}, r, t) \cdot (mean, I_{OS}, r, t) \cdot (extending, I_{OS}, r, t) \cdot (OS\_idle, I_{OS}, r, t) \cdot OS \quad (99.10)$$

99.4.2.3 The Semantic Model of the User Mode Process

The user mode process starts to run after it gains scheduling and finally gets into the waiting queue. The semantic model of the user mode process is expressed in formula 99.11:

$$U \stackrel{def}{=} (schedule, I_U, r, t) \cdot (run, I_U, r, t) \cdot (U\_idle, I_U, r, t) \cdot U \qquad (99.11)$$

99.4.2.4 The Semantic Model of Trusted Chain

Based on the above semantic model, we can build the semantic model of trusted chain. It is expressed in formula 99.12 as follows:

$$TC \stackrel{def}{=} (U_1 U_2 \cdots U_n)_{\{schedule\}} OS_{\{extending\}} TPM \qquad (99.12)$$

In this model, the interaction between user and OS is realized by schedule; the interaction between OS and TPM is realized by extending.

99.5 The Simulation Results and Analysis

In this paper, we compare the performance of the algorithm in two aspects through the simulation experiments:

1. The analysis of credibility's accuracy.
2. Efficiency analysis of computing reliability.

In the same situation, if the value of untrust is bigger, the confidence probability is smaller and the accuracy is lower; if the value of pratio is bigger, the shorter the trust transfer time and the higher the efficiency.

The experimental environment is Pentium(R) Dual-Core 2.1 GHZ CPU, 2 G memory, 500 G hard disk and Windows 7 ultimate operating system; experimental data analysis software environment is MATLAB 7.11.

If the value of untrust is not equal to zero, set a different value of pratio and untrust, comparing transmission efficiency and accuracy probability of credibility under different conditions. The experimental results are shown in Fig. 99.2.

From Fig. 99.2, we can find when values of pratio are equal; if the value of untrust is bigger, the less the number of completed activities are and the shorter trust transfer time is; when values of untrust are equal, if the value of pratio is bigger, the less the number of completed activities are, and the shorter trust transfer time is.

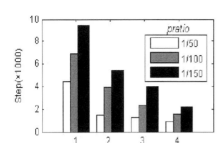 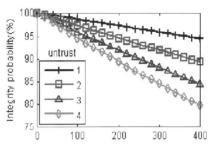

Fig. 99.2 The results of the experiment credibility efficiency and reliability accuracy experiment

99.6 Conclusion

This paper defined some transfer rules, and on the basis of trust, we designed a new semantic model for TPM, OS, the user mode processes and trusted chain; overcame the shortcomings of the original transfer rules and semantic model; and strengthened the research and application of trusted computing in the cloud security, solving some problems related to trust decision in the cloud security effectively. Simulation results showed that the model has certain advantages in the accuracy and efficiency of trust degree. In addition, the research work of cloud security based on trusted technology is still in its primary stage. For the direction, we also need to do a lot of research work.

Acknowledgements This research was supported by the Project of Shandong Province Higher Educational Science and Technology Program under Grant No. J12LN61. It was also supported by the Project of Shandong Province Higher Educational Science and Technology Program under Grant No. J13LN64. In addition, the authors would like to thank the reviewers for their valuable comments and suggestions.

References

1. Xiao-yong Li, Xiao-lin Gui, & Qian Mao. (2009). The monitoring dynamic trust model based on the adaptive behavior. *Journal of Computer, 32*(04), 664–674 (In Chinese).
2. Deng-guo Feng, & Yu Qin. (2008). Trusted computing environment method research. *Journal of computer, 31*(09), 1640–1652(2008) (In Chinese).
3. Deng-guo Feng, & Yu Qin. (2010). A property proof protocol based on TCM. *China Science: Information Science, 40*(02), 189–199 (In Chinese).
4. Yu Qin, & Deng-guo Feng. (2009). Remote attestation based on the component properties. *Journal of Software 20*(06), 1625–1641 (In Chinese).
5. Zi-wen Liu, & Deng-guo Feng. (2010). Dynamic integrity measurement architecture based on trusted computing. *Journal of Electronics and Information, 32*(04), 875–879 (In Chinese).
6. Huan-guo Zhang, Lu Chen, & Li-qiang Zhang. (2010). Research on trusted network connection. *Journal of computer, 33*(4),706–717 (In Chinese).

7. Xing Zhang, Qiang Huang, & Chang-xiang Shen. (2010). An analysis method of transmitting trust chain based on non-interference model. *Journal of Computer, 11*(1), 74–81 (In Chinese).
8. Run-lian Zhang, Xiao-nian Wu, Sheng-yuan Zhou. (2009). A Trust Model Based on Entity Behavior Risk Assessment. *Journal of Computer. 32*(04), 688-698 (In Chinese)
9. Jie Zhao, Nan-feng Xiao, & Jun-rui Zhong. (2009). The behavior trust control based on Bayesian network and behavior log mining. *Journal of South China University of Technology (Natural Science Edition), 37*(05), 94–100 (In Chinese).

Chapter 100
The Application Level of E-commerce in Enterprises in China

Yinghan H. Tang

Abstract Based on the process of corporate value formation and performance system, this chapter extracts key factors indicating E-commerce application level in enterprises and has established a set of E-commerce measurement indicator system. In addition, this chapter uses Delphi method and Analytic hierarchy process to identify the coefficients of various factors. By applying this model to measure the E-commerce application level in 23 heterogeneous enterprises in Chinese domestic market, this study proves that the proposed model can yield a relative accurate measurement of E-commerce application level in enterprises. The results also indicate that there are strong individual differences among different enterprises in China. The E-commerce application level in individual enterprises is affected by corporate strategy, informatization level, E-commerce application performance. and human resources. The nature and the size of enterprises have significant correlation with E-commerce application level. The study also finds that the big-sized enterprises will become stagnant when they develop to a certain level, which is known as a "trap".

100.1 Introduction

In the era of network economy, E-commerce has gradually penetrated into every economic and social aspect. It is becoming obvious that E-commerce is useful in saving costs, improving efficiency, increasing market share and enhancing corporate competitiveness advantages. But there exist great differences in E-commerce application among different enterprises due to some factors of the individual enterprises themselves. Then how to compare the E-commerce application level among enterprises? So far, there is no universally accepted measure method. Many

Y.H. Tang (✉)
Management School of Shenzhen Polytechnic, Shenzhen, Guangdong 518055, China
e-mail: tangyh@szpt.edu.cn

enterprises think they are E-commercialized when part of their business is done online. The answer to this question lies in some indicators to measure the E-commerce level in enterprises so as to compare the differences among them. Is the next questions are how to measure the E-commerce application level in enterprises?

100.2 The Definition of Enterprise E-commerce Application Level

E-commerce application level of enterprises reflects the situation when the enterprise applies E-commerce at a certain time point. Different definitions are given from different perspectives of study in this field. From the viewpoint of corporate value creating process, E-commerce application level is a process to manipulate modern network technology, information technology, and computer technology to manage and control people, possessions, materials, and information so as to create value for the enterprise [1]. It is a dynamic process. But from the perspective of corporate production, it involves taking advantage of E-commerce technology to efficiently exploit and utilize various resources, to improve operation, management, and production, to lower cost and enhance quality as well as to strengthen the innovation ability and competitiveness. No matter from which perspective to define E-commerce application level in enterprises, one point is clear: it is the degree that E-commerce application has on the enterprise final performance and the actual outcome that drives companies to adopt E-commerce.

From the above discussion, it can be concluded that the E-commerce application level in enterprises is a measurement tool to estimate E-commerce level in the operation and development process of enterprises as well as a contribution degree that E-commerce has to the development of enterprises. In the following paragraphs, E-commerce indicator is used to replace the phrase "the E-commerce application level."

The studies on how to measure E-commerce started from researches on the measurements of information economics. Mark Lupe is said to be the first in this field. In 1960s, Mark Lupe first used The Final Demand Method (also called the Expenditure Approach, the Final Product Approach) to measure American knowledge industry; the second scholar in this field is Borat who studied the measurement of American information economy. Many international organizations and research institutions, such as OECD, APED, and IDC, have made studies on E-commerce indicator system. Starting from the 1990s [2], EU began researching on how to analyze and compare E-commerce level between countries and to predict its future trend by means of indicator system and methodologies. UNCTD have released many editions of E-commerce and its development report since 2001. UK, USA, and Japan also published their respective E-commerce study reports successively [3]. Among these studies, the one made by a Japanese scholar who proposed the

Social Informatization Index is widely adopted because this measurement approach is easy to use.

Apart from official organizations, some consultancies and investigation firms as well as the academic circle also get involved in this field. For instance, the two biggest Internet investigation firms, Forrester Research and Jupiter Communications, Stanford University and University of Minnesota have all established their own E-commerce centers or projects to do relative researches. In 2002, PWC Consulting cooperated with Carnegie Mellon University to study measurement methods of E-commerce level and have proposed an E-commerce maturity model [4].

The network economic research center in Peking University has also made a series of studies on network economy and released a report involving IT development, E-commerce, E-government, etc. In a word, the measurement studies made in China are about informatization or network economy. Few studies are done exclusively on E-commerce application level in enterprises.

In conclusion, in order to develop E-commerce and drive it forward, a lot of countries have already made studies on E-commerce application and development as to how to set data collection, to make quantitative analysis, to evaluate, and to make predictions. But none of these studies have adopted a universally accepted measurement approach. Some studies already done both home and aboard are about how to measure information, rather than E-commerce; others do measure E-commerce level, but they are merely at a national or regional level. There are no studies in a strict sense that have measured it at a micro and medium level. However, these studies have great significances in providing references for further researches about the measurement of E-commerce application level in enterprises.

100.3 Constructing the Index System and Measurement Approaches of E-commerce Application Level in Enterprises

100.3.1 Selection and Determination of Major Indicators

There are various factors that have an impact on E-commerce application level. Many of them are interconnected and highly correlated, making it difficult to choose. According to the operation performance flow in E-commerce enterprises, E-commerce can be divided to six fundamental parts, e.g., [5] Nature and scale of the enterprise, business strategy, network informatization level, human resources, operation performance, E-business, and customer management. Then the key indicators are chosen from the six parts just discussed. (see Fig. 100.1) [6].

From the process shown in Fig. 100.1, the measurement indicators of E-commerce application level in organizations can be identified from Table 100.1.

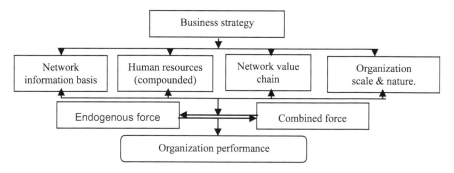

Fig. 100.1 The process showing impact of E-commerce application on organization performance

100.3.2 Determining the Indicator Weight of E-commerce Application Level in Organizations

In the comprehensive evaluation of many indicators, the determination of weight is the most fundamental but important task because it affects the outcome of comprehensive evaluation. Currently, there are many ways to determine indicator weight [7]. Generally speaking, there are two categories. One is called subjective method of weighting. The second is called objective method of weighting. Subjective method of weighting determines, from a qualitative perspective, the corresponding weight value of indicators according to their meaning and effects they show. It involves Delphi Method (also called Experts Grading Method) and the Analytic Hierarchy Process (AHP). Objective method of weighting belongs to a quantitative analysis, including Entropy method, Principal Component Analysis, and Factor analysis method. Generally speaking, Delphi Method is commonly used. Though it is somewhat subjective, it can usually reflect the real situation.

The studies discussed above on the evaluation index system of E-commerce application level have explained from various perspectives definitions and features of E-commerce application in organizations. In order to test the rationality of evaluation index systems and to explore the rules of E-commerce application level, the author of this chapter tries to construct an evaluating model of E-commerce application level in organizations.

100.3.3 Evaluating Model Constructing

This chapter uses the Simple linear weighting method (that is part of the comprehensive scoring method) to construct the index model of E-commerce application level in enterprises.

100 The Application Level of E-commerce in Enterprises in China

Table 100.1 Weight value and standard value of indicators

Target layer	Criterion layer	Evaluating indicator	Unit	Standard value	Index property	Index weight
E-commerce application level in enterprises	E-commerce application strategy	Business strategy	–	1	+	0.0549
		Goals	–	1	+	0.0395
		Network project teamwork	–	1	+	0.0424
		System ready status	%	70	+	0.0204
		System elasticity	%	60	+	0.0385
	E-commerce application performance	Ratio of annual expenses to train employee e-business technology and skills to total education training expense	%	8	+	0.0268
		Ratio of information infrastructure input expense to fixed assets	%	12	+	0.0313
		Ratio of value added by E-commerce to total value created	%	50	+	0.0328
		Rate of profit	%	24	–	0.0125
		Contribution rate of network assets	%	25	+	0.0281
	E-commerce business situation	Rate of network customer	%	35	+	0.0312
		Rate of network procurement	%	75	+	0.0291
		Rate of network booking	%	80	+	0.0391
		Sales rate online	%	50	+	0.0442
		Sales rate offline	%	30	–	0.0338
		Customer relationship management	%	100	+	0.0301
		Ratio of business online to offline	%	100	+	0.0321
	Network informatization status	Ratio of E-commerce investment to infrastructure, upgrading and reconstruction	%	25	+	0.0220
		Number of computers per 100 people	%	75	+	0.0231
		Number of telephones per 100 people	%	90	+	0.0126
		Rate of Internet access to computers	%	80	+	0.0313
		Rate of online information release	%	80	+	0.0304

(continued)

Table 100.1 (continued)

Target layer	Criterion layer	Evaluating indicator	Unit	Standard value	Index property	Index weight
		Ratio of E-commerce personnel to total employees	%	50	+	0.0314
		Average band width of Internet access per capita	*1	20	+	0.0231
		Modes of Internet access	*2	1	+	0.019
		Total capacity of Web server	–	2	+	0.0202
		Average of daily online time/work time of front line staff	%	60	+	0.0391
		Degree of getting information through Internet	%	65	+	0.0218
		Degree of exchanging information through Internet	%	90	+	0.0314
	HR condition	Number of professional technicians per 100 person	%	6	+	0.0307
		Number of students per 100 people	%	13	+	0.0261
		Ratio of the amount of labor done online to that of offline	%	100	+	0.0270
	Enterprise condition	Industry/sector	–	1	+	0.0218
		Enterprise size	–	1	+	0.0222

Note 1: Business strategy, business goals, network project teamwork, and ways of Internet connection are given coefficient by levels
Note 2: The reference standard value comes from conclusions drawn from statistics and analysis of organizations [1]
Note 3: *1 = M/100 people, *2 = ADSL/fiber(0.6/1)

$$I = \sum p \cdot \omega i \qquad (100.1)$$

In formula (100.1), I refers to E-commerce index; p means the value of the ith evaluation index after dimensionless treatment. ω means the ith weight of evaluation index. In order to make it easy to calculate p, after comparing the real value and standard value in enterprises, this model choose the value after dimensionless treatment. That is, $p = pi/\tau i$ where τi is a standard value. Then formula (100.1) can be changed into:

$$I = \sum p \cdot \omega i = \sum \left(\frac{pi}{\tau i}\right) \cdot \omega i \qquad (100.2)$$

This is the E-commerce index of enterprises. It can be used to measure the E-commerce application level in enterprises and also can be made to compare the E-commerce application level in one enterprise with that of others.

100.4 A Case Study of Applying E-commerce Application Level and Its Analysis

To apply the above index system model to practice, we randomly choose 23 enterprises (mainly located in Shenzhen) for interview to get the data we want (Table 100.2).

From Table 100.2, it is obvious to see that there are striking differences in E-commerce application level of enterprises. In some enterprises, the E-commerce application level is relatively quite high. In those enterprises, E-commerce is adopted in every business transactions and has significant impact on organization performance; while in some other enterprises, E-commerce application level is quite low. Mean 50.15 indicates that the overall E-commerce application level of Chinese enterprises is rather low.

Table 100.3 indicates the differences in E-commerce application level among enterprises of different industries. The manufacture industry shows generally low E-commerce application level while the information software industry has an E-commerce application level that is obviously above the average. From the viewpoint of enterprise size, those super large and medium-sized enterprises are reported to have high level of E-commerce application. Nevertheless, the small-sized ones have relatively low levels. This investigation finds out that there is a "trap" in E-commerce application level of enterprises, that is, when the application level

Table 100.2 E-commerce application index of Chinese enterprises

	Mean	Maximum	Minimum	Std deviation	No. of samples
E-commerce application index	50.15	110.15	8.00	46.52	23

Source: Data comes from the result of questionnaire with managers in 23 enterprises

Table 100.3 Distribution of the E-commerce application index of enterprises

		Mean	Maximum	Minimum	No. of samples
Nature of the enterprise	Manufacture	31.19	76.43	8.00	6
	Information software service	72.56	110.15	59.67	5
	Business, trade	55.36	90.21	50.16	6
	Service	41.49	95.13	17.32	6
Scale of the enterprise	Super large	74.06	82.14	65.98	2
	Large	51.11	65.22	35.26	3
	Medium	61.04	75.2	38.43	6
	Small	19.38	88.86	8.00	12

Note: there are 23 enterprises in this survey. Scale of enterprises is divided according to the number of people in an enterprise, e.g., 500 people or more is super large; 100–499 is large; 30–99 is medium; 30 or fewer is small

in the super large and large-sized enterprises develops to a certain degree, it will become stagnant, while once the small-sized enterprises achieve a breakthrough, their E-commerce will develop into its maximum extreme.

100.5 Conclusions and Suggestions

Firstly, there exist sharp differences in E-commerce application among Chinese enterprises. Large organizations generally have a higher level of E-commerce application than small and medium-sized enterprises. A main reason to this lies in the fact that large enterprises tend to invest more to infrastructure facilities, such as network information. Moreover, large enterprises have relatively more advantages over human resources. However, when the application level comes to a certain stage, it will fall into stagnation. Therefore, to look for a new breakthrough in E-commerce application is an urgent issue that most Chinese E-commerce enterprises face to tackle.

Secondly, the development strategy of E-commerce specialization has substantial effect on E-commerce application in enterprises. No matter what industry the enterprise belongs to, the degree of importance it attaches to E-commerce decides the level of E-commerce application. Without a long-term development strategy, some small-sized enterprises get lost in the wave of the current E-commerce tide. Though the quality of human resources occupies only a small proportion of weight, it is an inherent factor that is critical to E-commerce application level index of enterprises because it drives their endogenous growth of E-commerce. Besides, some enterprises are difficult to survive just because they have no sufficient compounded talents in the field of E-commerce.

Finally, the nature of industry decides the E-commerce application level of enterprises. For manufacturing industry, their need for E-commerce application is not as strong as other industries because it is still a traditional industry and its modes of production are also traditional. However, in the long run, it should lay some emphasis on E-commerce as E-commerce will help it for transformation and upgrading, and also for better competitiveness. Many indicators of the E-commerce application level index have close relationship with the online and offline businesses, but the final enterprise performance directly decides whether the enterprise will continue with E-commerce and reinvest in E-commerce or not. However, this kind of "circle" needs adjusting based on the development strategy of the organization. After all, E-commerce application in enterprises is still in a process of exploration. Our investigation shows that the current combination of online and offline business mode is a main approach to apply E-commerce in enterprises. This mixed management model is a product of traditional industry combined with new technology and new services. The new development of E-commerce application in enterprises is an innovation of E-commerce application model for organizations.

References

1. Huang, J. Q., Shui, M., & Cai, W. J. (2012). E-commerce ready level model and its case study. *Statistics and Decision., 2012*(18), 100–103. In Chinese.
2. OECD (1998) Measuring the ICT sector. http://www.oecd.org/, retrieved on May 21, 2011
3. OECD (2002) Measuring the information economy [R/OL]. http://wwww.oecd.org, retrieved on March 20, 2012.
4. Committee for Information, Computer and Communications Policy, OECD. (1997) Measuring electronic commerce. http://www.oecd.org/, retrieved on Oct. 30, 2012.
5. Yang, H. H., & Li, P. (2011). Analysis on new industrialization evaluation index and measurement. *Economic Management, 10*, 121–125. In Chinese.
6. Zhao, J., Zhu, Z., & Wang, F. (2010). E-commerce performance evaluation model of enterprises based on process. *Manage Eng J, 2010*(1), 17–24. In Chinese.
7. Chen, X. D., & Fu, L. S. (1999). Study on the measurement of Chinese industry informalization level. *Social Science Management, 6*, 21–31. In Chinese.

Chapter 101
Toward a Trinity Model of Digital Education Resources Construction and Management

Yong Huang and Qingchun Hu

Abstract This chapter aims to solve the problem of how to construct and manage digital educational resources effectively. It puts forward a trinity mode based on system architecture, workflow, and technology system. The trinity model consists of "Pre-Stage, Mid-Stage, Post-stage," "Theory, Practice, Regulation" and "Approach, Tool, Rule." By combining the trinity mode with case studies, the issues concerning construction and management of digital educational resources are to be analyzed, including topic selection, relationship between quantity and quality, implementation. Over the past year, results have showed that the trinity model could shorten more than 50 % of the development cycle of the project. The model could greatly help improve the construction and management of digital educational resources.

101.1 Introduction

With popularity of the mobile storage technology and equipment, the digital teaching and learning would play a prominent role in the future. A number of e-learning materials have been made. The traditional teaching and e-learning materials will coexist in current classrooms [1, 2]. There has been researches focusing on the effect of digital-integrated education on national innovation systems [3, 4]. Our study is based on the project of "To Carry Out the Construction of Digital Curriculum and Learning Environment Change" supported by the Shanghai Municipal Education Commission [5]. In the past 10 years, we have done a lot to

Y. Huang (✉)
Shanghai Audio-Video Education Center, Shanghai Distance Education Group,
Shanghai 200086, China
e-mail: hyong@shtvu.edu.cn

Q. Hu
School of Information Science and Engineering, East China University of Science and Technology, Shanghai 200237, China

build Shanghai Education Resource Center (http://www.sherc.net). Now, we need to summarize and reflect on our past 10 years' work. We find that the following three aspects should be improved.

The first aspect goes with topics selection. Most resources in Shanghai Education Resource Center are not fully used. The contents in the textbooks are updated frequently, such as the learning contents in Chinese and English for instance have been updated by more than 50 %. However, some digital materials in the Shanghai Education Resource Center are not updated subsequently [5].

The second aspect is the relationship between quantity and quality. The huge number of resource in Shanghai Education Resource Center makes users difficult to search for their favorite resources. And also, the poor quality of some resources reduces the usability of resources.

Thirdly, there is a point at implementation and publishing. With the popularity of electronic equipment, digital teaching and learning will become a mainstream in the future [6, 7]. The importance of efficient construction and management of high-quality digital education resources grows with increasing government investment on educational resources [8, 9]. It is worthy to have a study on how to improve the quality and the usability of the digital education resources. It is quite complex, for it involves many factors, such as storages [10], materials, applications [11], teachers' views [12], marketing reports, copyright [13], and so on [14].

For the past 2 years, our focus has gradually shifted from the construction of the repository to the development of digital educational resources. Till now, we have issued 10 digital textbooks and more than 200 APPs (Nada Online http://www.ndapk.com). Our work of digital education resources construction and management usually start from the perspective of education or engineering, but we are often at a loss in this way.

Currently, we had a lot of successful cases, but there still exist a lot of problems and cases of failures. It is time to summarize and study on more effective methods to construct and manage the resources. On basis of the past years' research and development activities, this chapter discusses a digital education resources construction management model—a Trinity Model.

101.2 The Components and Meanings of the Trinity Model

The word "trinity" is commonly used to describe three individuals, three things, or aspects in a tight inseparable whole. The concept of trinity is to be used, and more appropriate in our practices of constructing and managing digital education resources.

In our practice, digital education resources construction and management consist of three parts: workflow, system architecture, technology.

Furthermore, the construction and management of digital education resources can be seen as a big project. Workflow in the project could be divided into "Pre-stage, Mid-stage, and Post-stage," these three stages are whole inseparable.

Fig. 101.1 A trinity model

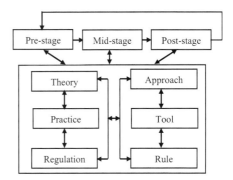

The system architecture of construction and management can be divided into a threesome "Theory, Practice, Regulation." The technology used in the construction and management can be divided into an inseparable "Approach, Tool, Rule."

The workflow, system architecture, and technology combine to form a unity, named as a trinity development model together (Fig. 101.1).

101.2.1 Pre-stage

Pre-stage is the first stage of the project. It includes mainly literature review of the relative theory, preparation of topics, the top-level designs, data collection, feasibility analysis, copyright issues, regulations, methods, and tools. This stage should start from the users' demand to consider which topics to be designed. Only those topics what can really improve the users' learning interest, should be worthy to be done. The key points during the first stage include the following:

Pre-stage is a continual and iterative process, focusing on data accumulation, analysis and research. The work during the pre-stage builds library that benefits the following work, including cases library, theoretical library, methods library, rules library, tools library, Apps library. All these six libraries provide decision-making managers with effective help.

Pre-stage is a decision-making process. The aim of decision-making is to reach a certain goal. It is an analysis and judgment process, which adopts scientific methods, selecting a feasible method among two or more. The decision-making process chooses the best in the process of the optimal solution through analysis and comparison.

101.2.2 Post-stage

There is no little thing after finishing resource construction. In fact, work during post-stage is also very important and rich. It includes collecting students' or teachers' feedback; revision and updating; products issued online; customer service; research activities; information dissemination; developing new topics; and project evaluation. The work in post-stage is to improve the education resource, such as three-round assessment activities, discussion, performance, regular meetings of assessment by experts, and feedback mechanisms.

101.2.3 Mid-stage

Mid-stage is in the mid-term in workflow, including content, technology, evaluation, application, feedback, releases, and upgrade. The development process involves a lot of methods, tools, and rules, including technical level, education level, government level, and resources on ecology. If the members in a project team do not communicate and collaborate well, it could lower the development efficiency and the quality.

We need to systematically deal with pre-stage, mid-stage, and post-stage of the project, and deal them with theory, practice, regulations, techniques (methods, tools, rules) in order to improve usability of the digital resources.

101.2.4 The Relationships

Communications among the three stages is very important.

From Pre-stage to Mid-stage, there should be documents of requirements specification, bidding, experts' assessments, project contracts, etc.

The documents from Mid-stage to Post-stage involve experts' position paper, usability testing reports, project inspection reports, products backup, etc.

Documents from Post-stage to Pre-stage should be evaluation reports, new project proposals, seminars, etc.

The advantage of this model lies in two aspects.

1. It helps to find the reason of inefficiency. In order to develop digital resources better, beside a focus on Mid-stage of a project, we need to have an in-depth research and attention to the relationship among these three stages. Sometimes, one project is inefficient because of ignoring the key role of the Post-stage to Pre-stage. And, there is no good technology system and relevant theories at hand.
2. It helps to find the right way to solve the problem. The practice of optimization methods organized into libraries is very valuable and reused. A comparative study of different tools for decision-making and different methods for developing efficiently has a significant impact. With establishment and update of rules library and methods library, it is easy to form a coordination mechanism.

101.3 Implementation of Trinity Model

We have developed a lot of digital resources. And some are still in development. The implementation of "Theory, Practice, Regulation" and "Approach, Tool, Rule" reflects past practices, as shown in Tables 101.1 and 101.2.

We cannot reach the essence of the technology if we view technology only as a tool or means. The technology is composed of a variety of methods; tools and rules of the system work together for a purpose. In the past 2 years' practice, we frequently encounter three typical problems: topics, quality and quantity, implementation.

101.3.1 Selecting Topics

Some APP topics, such as learning English APPs, are expected to be popular; however, the download record is poor. On the other hand, sports APPs, which seem to be less-popular topics, came out with many downloads. This phenomenon can be analyzed with Long Tail theory, which is a new theory appearing in the Internet Age proposed by Americans Chris Anderson [15]. APPs are virtual products, as Google Adwords, iTunes. They are consistent with the Long Tail theory. All of their payment and delivery cost is close to zero. It can be said that the virtual products' sales are inherently suitable for the Long Tail theory. The products of online retail giant Amazon are all-inclusive, rather than just a few commodities that can create high profits. Result shows that the Amazon model is successful, and the

Table 101.1 Implementation of "Theory, Practice, Regulation"

	Theory	Practice	Regulation
Pre-stage	T1: the long tail theory	P1: topics	R1: topics regulation
Mid-stage	T2: game theory	P2: quality and quantity	R2: elevate regulation
Post-stage	T3: computational thinking	P3: implementation	R3: publish regulation

Table 101.2 Implementation of "Approach, Tools, Rules"

	Practice	Approach	Tool	Rule
Pre-stage	P1: selecting the topics	A1: brainstorm	T1: converter	R1: if average down per day >5 then upgrade APP
Mid-stage	P2: improving the quality and quantity	A2: project	T2: iBook author	R2: if the score of efficient <3 then redesign
Post-stage	P3: implementation	A3: group	T3: feedback tool	R3: if the score of user <3 then redesign

profit is not ideal if we ignore the Long Tail and focus only on a few bestsellers products.

In market analysis, we should concentrate on specific target markets, to create a new product and service. Our suggestions are listed below:

(a) Checking the products we have developed and analyzing the available resources. We need to know which of them is a success or failure? What is the reason? Is there any possibility of "redevelopment" for those products?
(b) Analyzing market demand, to clarify the nature of the issues. What are the specific issues that need to be addressed? What are the key problems to be resolved? Why is it necessary? What are the difficulties? What are the existing resources? Are there any especially good methods to solve the problem? What is the significance to solve the problem? Whether there exists a similar product?
(c) Spending more time on the topics' selection. It is the basis of the publishing business that decides whether the follow-up work is a success or failure. The successful topics can help the digital publishing institutions form and establish a good follow-up development cycle.
(d) Establishing rules for developing activities. For instance, if downloads are over five items per day in a whole month, then these products should be worthy to be updated and improved.

101.3.2 Improving Quality and Quantity

In practice, quality and quantity are often on contrary. In general, quality is difficult to guarantee if focus is on pursuit of the quantity. In this case, most products are with a small number of download. The Game Theory helps to solve this contradiction. One solution is pursuit of quality with the rise of the number; pursuit of quantity with the improvement of quality.

We have a wealth of APPs resources online. To our disappointment, we have only a small number of download with poor usability. The phenomena make us to refine our current products and publications; establish a three-round assessment system for digital publishing.

101.3.3 Implementations

It is better to understand the problem and seek prompt solution to the problem based on the concept of "Computational Thinking" [16, 17]. Computational thinking means the method to solve problems using the computer technology. We do not study digital terminal when we face to the digital terminal implementation of digital teaching and learning, just like the astronomy is not a research of the telescope. We need to solve the problem like computer scientists do, to understand the transformations of digitized teaching and learning.

Our suggestions include the following: improving digital educational resources evaluation criteria; collecting data continually for future development after issue of a product. After an assessment activity, a summary report is needed for next-stage revision and improvement.

Based on our actual situation and through the trinity model, we argue: on road of digital publishing, teacher training, and supplementary teaching resource pack.

101.3.3.1 On Road of Digital Publishing

From professional views in the industry, the most important factor in "digital transformation of teaching and learning" is teaching and learning materials. Currently, at least 70 % of domestic publishing companies in China rely on textbooks and tutorial resources to survive. There would be crises in the current digital age. In addition, the textbooks are varied in different provinces, and their update speed is not faster. There are many issues to be addressed if the "digital transformation of teaching and learning" project plan is launched all around our country by the government. With regard to e-textbooks, how much does it cost and how to resolve copyright issues involve the three departments: education, publishing company and the market. The form and content of the "digital transformation of teaching and learning" would pose a challenge to education and publishing industry, but there exists great opportunities.

101.3.3.2 Teacher Training

In the teaching system, for the intervention of the new media, there is an urgent need to train teachers. There are many good platforms in teacher training and research activities. We can collect a large number of practice cases, such as the digitized teaching and learning application in different schools.

The reform of information technology in education is irresistible. Children would be left behind if he/she is less trained by "transformation of digital teaching and learning resource." The research towards transformation includes physiological and psychological impact of its students, the applications model in education and teaching and education management.

In 2013, Chinese government plans to launch a number of e-learning materials projects for primary and secondary schools in the city of Shanghai. The massive digital resources accumulated over the past decade in Shanghai Distance Education Group. In the project, all these resource are grouped by the "cloud platform" for sharing among all schools in Shanghai.

Using the trinity model, we propose to speed up the research on digitized transformation of teaching and learning office institutional settings, such as the establishment of data centers, research centers, training centers, technical centers, and service center.

101.3.3.3 Supplementary Teaching Resource Pack

The key to the effective implementation of digital teaching and learning should have two sides, including developing resources and teaching mode. We need to consider how to provide supplementary teaching resource pack. First of all, we should make full use of educational resources and our publishing resources rather than let teachers search on internet. These resources are valuable and useful. It is very important to consider how to conduct secondary development, and how to deliver them to the hands of teachers and students.

101.4 Conclusion

In early 2011, in the project "Reform of Digital Teaching and Learning," we had developed digital resources for three courses involving mathematics, information literacy, and the English language. It took 6 months to discuss and organize. The result, however, is not satisfactory.

It fails in five parts. Requirement analysis is not enough in the Pre-stage, such as selecting tools, designing ideas, and copyright issues. There is less collaboration among developers. There is less education theory to guide the process. And there are less evaluation and feedback mechanisms.

In late 2011, we analyzed our work from the three levels: workflow, system architecture and technology. And the trinity model was proposed and used at that time. In 2011, in our projects, it generally cost 6 months to finish an e-textbook or an APP project without the trinity model. Now, we can develop it within 2 months under the model. The number of download and the amount of our APPs products in 2012 rise ten times more than in 2011. The result shows that the model could shorten the development cycle of a project by more than 50 %. The model could greatly help improve the construction and management of digital educational resources.

The trinity model makes digital education resources construction and management more effective. The Pre-stage and Post-stage work of digital education resources construction in practice is most likely to be ignored, although they are very important. We usually deal with the digital education resources construction and management with conventional thinking and theoretical support in practice. But lack of system construction and support results in a lot of disorder and inefficient activities. It will be more economic, scientific, and logical to make a decision in construction and management of the digital educational resources, based on the construction of theoretical library, method library, rule library, tool library, and case library.

Acknowledgements This research is supported by Digital Educational Resources Ecological Construction and Sharing Mode under Shanghai Municipal Education Commission by National Educational Programs Grant No. NOESP dca110194.

References

1. Nussbaum, M., & Diaz, A (2013). Classroom logistics: Integrating digital and non-digital resources. *Computers & Education, 69*(0), 493–495.
2. Chen, N.-S., et al. (2011). Augmenting paper-based reading activity with direct access to digital materials and scaffolded questioning. *Computers & Education, 57*(2), 1705–1715.
3. Wiseman, A. W., & Anderson, E. (2012). ICT-integrated education & national innovation systems in the Gulf Cooperation Council (GCC) countries. *Computers & Education, 59*(2), 607–618.
4. Allegra, E., et al. (2011). Cross-border co-operation and education in digital investigations: A European perspective. *Digital Investigation, 8*(2), 106–113.
5. Shanghai's long-term Education Reform and Development Plan (2010~2020). Available from: http://www.360doc.com/content/11/0313/21/5344705_100838514.shtml (2010) (In Chinese)
6. Loveless, A., & Underwood, J. (2010). Learning in digital worlds: A view from CAL09. *Computers & Education, 54*(3), 611–612.
7. Thompson, P. (2013). The digital natives as learners: technology use patterns and approaches to learning. *Computers & Education, 65*, 12–33.
8. He, K. (2009). The status and strategies on the construction of digital learning resources. *E-education Research, 10*, 5–9 (in Chinese).
9. Jing, Y. J., & Li, X. (2011). The study and practice of a pattern on the community of regional education information resources infrastructure. *China Educational Technology, 1*, 83–86 (in Chinese).
10. Fu, X., et al. (2011). On data integration, warehousing and software reuse in the construction of digital campus: A review on performance. *Procedia Engineering, 15*, 3109–3113.
11. Tohidi, H. (2011). Human resources management main role in information technology project management. *Procedia Computer Science, 3*, 925–929.
12. Petko, D. (2012). Teachers' pedagogical beliefs and their use of digital media in classrooms: sharpening the focus of the 'will, skill, tool' model and integrating teachers' constructivist orientations. *Computers & Education, 58*(4), 1351–1359.
13. Hunter, B. (2013). The effect of digital publishing on technical services in university libraries. *The Journal of Academic Librarianship, 39*(1), 84–93.
14. Kreijns, K., et al. (2013). What stimulates teachers to integrate ICT in their pedagogical practices? The use of digital learning materials in education. *Computers in Human Behavior, 29*(1), 217–225.
15. Anderson, C. (2009). *The long tail: Why the future of business is selling less of more* (pp. 67–80). New York: Hyperion Books.
16. Wing, J. M. (2007). Computational thinking. *Communications of ACM, 49*(3), 33–35.
17. Karp, R. M. (2011). Understanding science through the computational lens. *Journal of Computer Science and Technology, 26*(4), 569–577.

Chapter 102
Geographic Information System in the Cloud Computing Environment

Yichun Peng and Yunpeng Wang

Abstract Cloud computing has became a very popular vocabulary in recent years. The combination of cloud computing and GIS (geographic information system) can improve the performance of GIS. By analyzing the technology of cloud computing, this paper introduces the concept of GIS based on cloud computing; based on the current major GIS application development trends, key technologies of cloud GIS are proposed; finally four application modes of cloud GIS are presented. Cloud GIS can improve stability and efficiency services to end users by optimized network resource allocation of underlying data and services.

102.1 Introduction

At present, in the field of GIS, there exist two questions: firstly, in data aspect, because of widely data sources, which could cause the following question, coordinates and formats are not interchangeable, the data is not compatible, semantics are not uniform, and difficult to share, difficult to interoperate and so on, moreover, how to store, manage, update and analyze these massive data is also difficult to achieve. Secondly, in application, with the development of society, people's demand for geographical information service also continues to grow; however,

---

Y. Peng (✉)
Guangzhou Institute of Geochemistry Chinese Academy of Science,
Guangzhou 510640, China

City College of Dongguan University of Technology, Dongguan 523106, China

University of Chinese Academy of Sciences, Beijing 100049, China
e-mail: yichunpeng678@hotmail.com

Y. Wang
Guangzhou Institute of Geochemistry Chinese Academy of Science,
Guangzhou 510640, China
e-mail: wangyp@gig.ac.cn

data production unit is relatively less and a single system is also hard to own all the resources and the ability of processing, therefore, users unable to get their required data from a single source too. In addition, large-scale concurrent access and high cost of upgrading are the main factors that hinder the development of GIS. Service-based "cloud computing" has the advantages of massive data storage, large-scale computing, and in-depth data mining, which is very suitable for GIS development; in addition, GIS service provider can deploy flexibly GIS applications on cloud computing platforms and can dynamically adjust the system's software and hardware requirements. GIS end user may gain the service on demand. Therefore, GIS based on cloud computing can not only simplify system deployment and management and reduce the cost of investment, operation, and maintenance but also improve the flexibility of GIS applications and infrastructures. In this paper, the concepts of GIS based on cloud computing are introduced; based on the current major GIS application development trends, key technologies of cloud GIS are proposed; finally four application modes of cloud GIS are presented.

102.2 Cloud Computing Technologies

Since Google puts forward to the concept of cloud computing in 2006, Amazon's "Elastic Computer Cloud" service, IBM's "Blue Cloud" plan, Microsoft's Internet operating system "Midori," Sun's "Black Box" plan, SAP, Yahoo, and some other large companies have developed their own "Cloud" plans or have launched their "Cloud" products. In early 2008, IBM cooperated with the Wuxi municipal government and established Wuxi Software Park Cloud Computing Center, which began commercial applications of cloud computing in China. A number of industries and localities also have launched some cloud computing plans, such as Rising's "Cloud Security" plan, Beijing's "Auspicious Cloud Computing," Shanghai's "Yunhai Plan," Suzhou's "Fengyun Online," Guangzhou's" Tianyun Plan," China Mobile's "Tianyun Plan," China Unicom's "Woyun Plan," and China Telecommunication's "Nebula Plan."

What is cloud computing? Nowadays, cloud computing is still an evolving paradigm. Its definitions, use cases, underlying technologies, issues, risks, and benefits will be refined in a spirited debate by the public and private sectors. These definitions, attributes, and characteristics will evolve and change over time. The NIST definition of cloud computing is as follows [1]: "Cloud computing is a model for enabling convenient, on-demand network access to a shared pool of configurable computing resources (e.g., networks, servers, storage, applications, and services) that can be rapidly provisioned and released with minimal management effort or service provider interaction. This cloud model promotes availability and is composed of five essential characteristics, three service models, and four deployment models." The cloud computing industry represents a large ecosystem of many models, vendors, and market niches. This definition attempts to encompass all of the various cloud approaches.

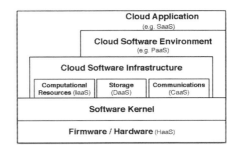

Fig. 102.1 Five-layer model of cloud computing

In 2008, Lamia Youseff, a doctoral student of the University of California Santa Barbara, Maria Butrico, and Dilma Da Silva, researchers of New York IBM T.J. Watson Research Center, published a research report entitled "Toward a Unified Ontology of Cloud Computing." This report established a five-layer model [2] as shown in Fig. 102.1.

At present, there are mainly three kinds of cloud computing service model, IaaS (Infrastructure as a Service), PaaS (Platform as a Service), and SaaS (Software as a Service) [3].

IaaS: one service of the closest underplayed in cloud computing, its output products are some resources of computing, storage, and network, such as VM, Storage, CDN and DNS. SaaS and PaaS will be created on the IaaS. IaaS provides cloud computing and cloud storage services for high availability, flexible expansion, on-demand billing, easy to use, low cost, and other advantages favored by many domestic and foreign companies. At abroad, such as Amazon AWS, IBM Smart Cloud, Microsoft Windows Azure, Rackspace, and NASA open source products: Open Stack, in addition, VMware, BlueLock, CSC, GoGrid, Savvis also launched its own IaaS technology or products; domestic such as Aliyun, Grand Cloud, and HUAWEI Single Cloud.

PaaS: in addition to providing computing, storage, and network infrastructure hardware resources, and PaaS also provides the basic framework for software development. Application developers must develop and host applications according to language and specification of platform, independent software vendors, or other third parties for vertical industries to create new solutions, but do not have to purchase the development, control the quality, or build the server. At present, there are some cloud computing platforms such as Google Application Engine, Sina Application Engine, Salesforce.com's Force.com, and Microsoft's Azure.

SaaS: is a software layout model, is a completely innovative software application model, its application is designed for network delivery, and is convenient for the user to host, develop, and access through the Internet; its output is information system, such as OA, CRM, ERP, and CMS. SaaS providers put up the network infrastructure and software hardware operating platform of which requirement in realizing enterprises informatization and will be responsible for implementation and maintenance of the system; enterprises do not need to purchase software and hardware, build a computer room, or recruit IT staff, but can use information system via the Internet. The main products are alesforce.com, NetSuite, Google Gmail, Zimbra, Zoho, the IBM Lotus Live and SPSCommerce.net, Ali software, etc.

With the deepening of cloud computing, database technology has been changing from the traditional relational database memory data grid to NoSQL, Database technology transformation has produced the fourth kind of cloud computing mode: DaaS, Data as a Service.

DaaS is a strong complement to SaaS model, a service model of web-based virtual storage; it can be for business users and business intelligence users to simplify the process of information retrieval, the user according to the actual storage capacity to pay. The benefit of which database migrate into cloud is data integration, usually in large enterprises; database needs to be shared across different departments; cloud services can be integrated into a single custody DBMS, so DaaS can reduce the problem of interior database expansion. The main products are Amazon's SimpleDB, VMware's vFabric the Data Director, Google's AppEngine and China Telecom Shanghai branch in collaboration with EMC "e cloud," etc.

IaaS is the foundation of cloud computing; DaaS is based on IaaS; SaaS can be deployed on PaaS or deployed directly on IaaS; and PaaS can be built on IaaS or be directly built on the physical resources. SaaS, PaaS, and IaaS combined with DaaS can build a complete cloud computing environment.

102.3 The Key Technology of Cloud GIS

Cloud GIS is the result of the combination of cloud computing and GIS; it is no longer a single GIS software platform, but provides storage, software, and content which can be virtually flexible, deployed, or rented, as long as through the PC desktop, mobile phone, and a web browser; the user can access on-demand data, map, spatial analysis, and Geoprocessing services which are provided by cloud GIS. At present, there have some GIS products based on cloud computing at home and abroad. At abroad, there are, such as, Google's Google Earth, Google Moon and Google Mars, ESRI's ArcGIS Online and ArcGIS10.1 etc. At home, there are, such as, SuperMap's SuperMap GIS 6R, MapGIS's MapGIS K9 SP3, GeoStar's GeoCloud etc.

Corresponding to the four-service models of cloud computing, cloud GIS also has some related concepts and technologies: cloud computing infrastructure belongs to IaaS, GIS application, cloud GIS platform, GIS cloud service platform, and GIS cloud application platform; all that belongs to PaaS; cloud GIS software belongs to SaaS; data services belong to DaaS; the composition system of cloud GIS is shown in Fig. 102.2 [3, 4].

Cloud GIS must run through data, software, and development; the user can really get the GIS resources at any time, so cloud GIS has its own unique technology in cloud computing platform and data interoperability, GIS spatial data storage, management, analysis and processing, and terminal access.

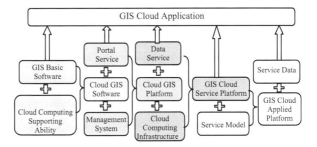

Fig. 102.2 The composition system of cloud GIS

102.3.1 Technology of Cloud Computing Platforms and Data Interoperability

Cloud GIS to realize cross operating system (Linux/Unix/AIX/Windows), cross GIS platform, support and synchronize with a variety of hardware architectures, can meet the private cloud and public cloud environments as a set of unified architecture and realize integrated connectivity and interaction of private cloud and public cloud, and its application fields including desktop, LAN, and Internet. It can support single release, automatic synchronization, frequency statistics and automatic optimization, support cloud internal data interoperability, private and public cloud interoperability, and interoperation between cloud center. In order to achieve the above operation, we must establish interoperation and integration standards of cloud computing, cloud computing service interface standards and application development standards, cloud computing interface standards between the different levels, cloud computing service catalog management, seamless migration between different cloud portability standards, cloud computing standards of business indicators, cloud computing architecture management standards, and cloud computing security and privacy standards, etc. Cloud computing technology use the multilevel structure framework, from top to bottom: business logic layer, application layer, distributed file and operating system layer, virtualization layer, hardware layer and data center infrastructure layer.

102.3.2 Technology of GIS Spatial Data Storage, Management, Processing, and Analysis

Cloud GIS uses virtual storage technology to establish a highly efficient, seam-less, multi-source, multi-scale, multi-spatio-temporal data model, which realize massive spatial information storage, management [5, 6].

Storage technology: the development of spatial data storage from the file system to a distributed file system and cloud storage system based on Internet technology entirely; spatial database has developed from enterprise database to distributed spatial database and will support for BigTable, HBase, and NoSQL to store and manage data in the future and support uniform access to it through the spatial database interface standard and REST interface. There are two main technologies at present: Google's non-open source GFS (Google File System) and Hadoop's open source HDFS (Hadoop Distributed File System). Most of the IT companies, including Yahoo's and Intel's "cloud" plans, adopt the data storage technology of HDFS. The future development will focus on large-scale data storage, data encryption, and security guarantee and continue to improve the rate of I/O.

Data management: uses virtualization technology to realize the unified management of spatial database; support for rapid migration and automatic synchronization of data between systems, departments, levels, etc.; has off-line application and online update technology; and achieves distributed, multilevel, and supporting multi-terminal spatial data security process. This method that the table divided by column and then stores it to storage is usually used for the data management. There are two main technologies at present: Google's BigTable data management technology and Hadoop's open source data management module that is similar to BigTable.

Data processing: a task-oriented asynchronous spatial data processing architecture, supports concurrent processing and process control in large clusters, supports long-time running and long transaction processing, supports for mobile terminal handling large spatial databases, has those functions of visualized design for processing flow and monitoring real-time running status of system, can cross-platform, cross-regional integrate spatial data processing, and immediately release the processing result.

Spatial analysis: has unified spatial analysis framework and rich spatial analysis model, establishes a standardized analysis model library, and supports for rapid construction and automatic operation of spatial analysis process and the immediate release of the analysis results.

102.3.3 Technology of Terminal Access

The final purpose of GIS platform based on cloud computing is to let the user access all the GIS functions through a browser. To meet smooth transmission of the massive data in a different network, it is requested that the system has unified kernel and interface, and its services can be accessed by various types of desktop, web, and mobile client, ultimately achieving those effects such as data synchronization access, consistent processing results, and elegant user experience.

102.4 The Application Mode of Cloud GIS

According to the four models of cloud computing, cloud GIS also put forward four kinds of GIS application services, namely [7, 8], GIS Software as a Service (namely, SaaS), Geographical Information Platform as a Service (namely, PaaS), Geographic Information Infrastructure as a Service (namely, IaaS), and Geographic Information Content as a Service (namely, DaaS). Figure 102.3 shows the cloud application structure of ArcGIS [9].

102.4.1 GIS Software as a Service

Also known as the "cloud of geographic information services," this service refers to the use of the Internet to provide online geographic information processing services, including map publishing, data format conversion, spatial analysis, and other services. Generally use the development of multi-tier architecture: the upper adopts SOA architecture pattern; geographic information services are packaged into standard web services and are incorporated into the management and use of the SOA system, its contents, including service interface, service registration, service search, and service access; the middle layer realizes the billing of the user using geographic information, takes charge of the load balance and map tile service; data layer is the underlying data service provided by the GIS server. This service mode, the GIS system developers, can build their own solutions, such that they can develop their own GIS solutions by using GIS platform firm's PaaS services; use cloud

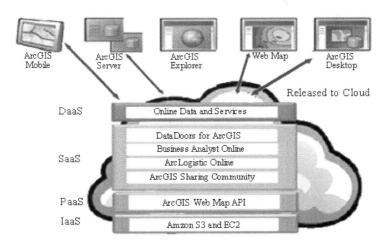

Fig. 102.3 The cloud application structure of ArcGIS

computing model to provide services for their customers, such as ESRI Business Analyst Online; it also allows user to combine GIS technology with thematic data, consumer data, and business data to achieve on-demand analysis and report and map web services, but Business Analyst Online is maintained by ESRI; users do not have to worry about data management and technical updates.

102.4.2 Geographical Information Platform as a Service

This service provides the whole GIS development environment to the user as a service. From AreGIS9.2, ESRI began to promote its ArcGIS Online, and to provide GIS services by a series of API, GIS developers can develop GIS software in the Google App Engine platform and run on Google's cloud computing infrastructure. At home, GIS platform providers are basically only using the cloud computing model to provide services for their customers and partners, for example, GIS platform provider can create PaaS service from their own GIS platform, then, when its partners develop their various GIS, they do not need to buy the GIS platform license but only need to rent the platform even; don't install or deploy the GIS platform in local server but be able to directly carry out the development of GIS on the Internet; all that will bring great convenience and save a lot of cost for GIS partners. The cloud services platform of SuperMap GIS consist of geographic information cloud services, navigation product and spatial data processing. This platform is open to the third-party, they can add some applications and provide SaaS services to those end users.

102.4.3 Geographic Information Infrastructure as a Service

Geographic information infrastructure as a service is the basis of the "cloud" model, which is the foundation of geographical information software as a service and geographic information content as a service; therefore, this infrastructure environment and service mode are indispensable parts of cloud GIS geographic information services deployed in the cloud; then cloud GIS users can pay a monthly fee to rent the commercial cloud computing platform software and hardware resources. At present, these are the main companies: Amazon, IBM, and some of telecommunications, which provide the cloud infrastructure. Amazon provides two rent modes: Elastic Compute Cloud (namely, EC2) and Simple Storage Service (namely, S3). ESRI's ArcGIS cloud applications are built in Amazon's EC2 and S3; it provides cloud map slice services which can be uploaded to the cloud and establishes the data center in the clouds; users can put the map cache in the data center of Amazon cloud.

102.4.4 Geographic Information Content as a Service

Geographic information content as a service provides data, map information, and a simple query service for GIS end-user online; the user can access the content on demand, without the need to establish and maintain data, which is the lowest level of cloud GIS application. ESRI's ArcGIS Online Map and GIS Server is a typical SaaS model, users can configure service on demand, and be able to quickly produce thematic maps, access seamless based map, etc. In addition, Baidu Map, Google Maps, Bing Maps, Yahoo Maps, etc., generally provide API for developers to use their cloud services; this API is a set of JavaScript or Flash language application programming interface, which can help users to build some function-rich, interactive map applications in their website.

102.5 Applications

Figure 102.4 shows the Geographic Information Society Service Platform of Jilin province based on SuperMap GIS cloud services platform.

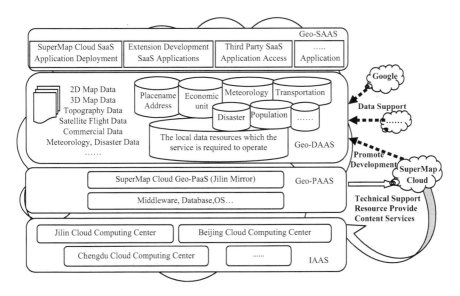

Fig. 102.4 The geographic information society service platform of Jilin province

102.6 Conclusion

With the development of cloud computing technology and the gradual and further application of GIS, cloud computing and GIS will be further fused, which can improve stability and efficiency services to end users by optimized network resource allocation of underlying data and services. But the cloud computing-based GIS in a real sense is required to a deep study; cloud GIS still has a long way to go and GIS rushes up high in the clouds also still to face great challenges. However, the arrival of the cloud era was an irresistible trend of development; cloud computing-based GIS is bound to be one of the GIS main development trends in the future.

Acknowledgements This work was supported by the Scientific and Technological Projects of Guangdong (No. 2009B010800042) and in part by Projects of Science and Technology of Dongguan (No. 201110825100119).

References

1. Peter Mell, & Timothy Grance. (2011). *The NIST Definition of Cloud Computing.* http://csrc.nist.gov/groups/SNS/cloud-computing/index.html.
2. Michael Armbrust, Armando Fox, Rean Griffith, Anthony D. Joseph, Randy Katz, Andy Konwinski et al. (2009). *Above the clouds: A Berkeley view of cloud computing.* Berkley: University of California.
3. Liqian Dai, & Na Chen. (2009). The development of GIS in the times of cloud computing. *Journal of Anhui Agricultural Science, 37*(31), 15556–15557, 15572 (in Chinese).
4. Liu Yang. (2011). Research on GIS application model based on cloud computing. Henan University (in Chinese).
5. Alexander Lenk, Markus Klems, Jens Nimis, Stefan Tai, Thomas Sandholm. (2009). What is Inside the Cloud? An Architectural Map of the Cloud Landscape, Software Engineering Challenges of Cloud Computing, CLOUD '09. ICSE Workshop on, 23–31 (2009).
6. Yonggang Wang, Sheng Wang, & Daliang Zhou. (2009). Retrieving and indexing spatial data in the cloud computing environment. *Cloud Computing, 12*(4), 322–331 (in Chinese).
7. Er qi Wang. (2011). Cloud computing and GIS technology innovation. *New Economy Weekly, 10*(2), 83–87 (in Chinese).
8. Esri China Information Technology Co., Ltd. ArcGIS and Cloud Computing Technology. Esri technical white paper (2010) (in Chinese).
9. Fang Lei, Yao Shenjun, Liu Ting, & Liu Renyi. (2010). A cloud computing application in land resources information management. *IEEE,* 388–393

Part V
Embedded Systems

Chapter 103
Memory Controller Design Based on Quadruple Modular Redundant Architecture

Yuanyuan Cui, Wei Li, and Xunying Zhang

Abstract For space application to improve the reliability of the memory operation, quadruple modular redundant (QMR) architecture is used in all registers of the memory controller. The QMR architecture in this paper can correct one-bit faults, detect two-bit faults, and also tolerate single event transient (SET). By modifying finite state machine (FSM) of the memory controller, when one uncorrectable fault is checked, the memory operation can be terminated in time and return the error information. Compared with triple modular redundancy (TMR), although the area overhead is increased by 47,530.59297 μm^2, the single event upset (SEU) failure rate is lower by 6 orders of magnitude. Experimental results show that when 1 bit-flip or 2 bit-flips are injected in QMR registers, they can be corrected or detected in time, respectively. Memory controller using QMR architecture increases the area overhead, but the advantage is the higher reliability valuable for safety system.

103.1 Introduction

Dependability issues are the most important for space application. Due to the continuous increase in the integration level of electronic systems, an acceptable degree of reliability is increasingly difficult to be guaranteed. It is necessary to design fault-tolerant memory controller which is the important interface between the processor and the memory. To protect against SEU errors, the registers of the key modules in the system can often be implemented using TMR, for example, the LEON-FT processor [1], the SCS750 single board [2], and the Virtex FPGAs [3].

Y. Cui (✉) • W. Li
Graduate Department, Xi'an Microelectronics Technology Institute, Xi'an 710054, China
e-mail: hebutcyy@126.com

X. Zhang
Research and Development Department, Xi'an Microelectronics Technology Institute, Xi'an 710054, China

For the TMR architecture, any 1 bit-flip can be corrected, but the faults with multiple bit-flips can generate false results. The probability of multiple bit-flips is increasing as integrated circuits continue to scale into the deep micron regime. The system using TMR will not meet the requirement of reliability.

To improve the dependability of access memory, the QMR architecture is adopted in this paper. In recent decades, QMR has been successfully used in a wide variety of fields such as aviation and railroad [4–6]. Compared with TMR, although the area overhead is increased by 47,530.59297 μm^2, the SEU failure rate of memory controller using QMR is lower by 6 orders of magnitude. By using separate clock trees to QMR registers, memory controller has the ability to tolerate SET. For QMR, any correctable error will be removed automatically within a clock cycle. When a group of QMR have 2 bit-flips, the detected error cannot be corrected. So in every clock cycle, the error detection signals of all QMR need to be examined. Using SMIC 130 nm standard CMOS process to synthesize, the error detection circuit's delay is 1.10 ns, meeting the design timing. By modifying FSM, any memory operation having a non-corrected error can be terminated in time and return the error information. Experimental results show that when 1 bit-flip or 2 bit-flips are injected in QMR registers, they can be corrected or detected in time, respectively. Memory controller using QMR increases the area overhead, but the advantage is the higher reliability valuable for safety system.

103.2 QMR Fault-Tolerant Technique

103.2.1 QMR Architecture

Figure 103.1 shows the QMR architecture. Every harden module needs to add three same ones. Every two modules constitute a group, namely, A group and B group. At the same time, one is a host group, another is a standby group. At the beginning, we assume the A group is the host group. If outputs of two modules from A group are the same, comparison circuit A12 will send a control signal to the output multiplexer and the output is from A group. If A12 always shows the results of A group are the same, the output is still from A group. When the results are different, B group will switch to the host group. In the same cycle, if the results from B group are the same, the output will be from B group; otherwise, the error detection signal will be set. Thus, QMR can correct and detect errors. The QMR is usually used in a subsystem which can self-repair. Because registers may be updated every cycle, they have the ability of restoring themselves naturally. So in Fig. 103.1, A1, A2, B1, and B2 can be registers.

Figure 103.2 shows the proposed circuit structure of the QMR. To simplify the design, we assume A group is still the host group, only if the results of the A group are different and that of the B group are the same; the output will be from the B group. When there are 2 bit-flips, the error cannot be corrected, but the error

Fig. 103.1 QMR block diagram

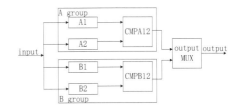

Fig. 103.2 QMR circuit structure

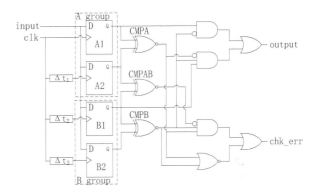

detection signal chk_errr will be set. Compared with TMR, the delay of one "xor" logic is increased.

Due to SET faults, a signal from combinational circuit can have a glitch. When the glitch is clocked in a register, the register will have an error. To avoid it being captured into all four registers, each of the four lanes can have separate clock trees. Compared with the clock tree of register A1, phase differences are Δt_1, Δt_2, and Δt_3, respectively. In the space environment, the width of the glitch is often 0.35~1.3 ns [7]; thus, the phase difference Δt is usually 1~1.5 ns. If $\Delta t_1 = \Delta t$, $\Delta t_2 = 2\Delta t$, and $\Delta t_3 = 3\Delta t$, the glitch whose width is less than Δt can be captured only by one register; the error can be corrected. When the glitch can be captured only by two registers, the error will be detected.

Compared with TMR, $3\Delta t$ is too large; system performance is reduced. A tradeoff scheme can be used; let $\Delta t_1 = \Delta t$, $\Delta t_2 = 0$, and $\Delta t_3 = 2\Delta t$; parts of glitches whose widths are less than Δt can be captured by A1 and B1 at the same time. This kind of errors cannot be corrected but can be detected. QMR with separate clock trees can protect against SEU errors and SET errors.

103.2.2 QMR Reliability

When the QMR registers have no error, or one bit-flip error, the output is correct; when two bits are flipped, the error information can be output, so this case is safe too. Let the reliability of single modular be R, so that QMR will be $R_{QMR} = 6R^2 - 8R^3 + 3R^4$. To make sure $R_{QMR} > R$, $R > 0.23$ must be satisfied. That means when R > 0.23, the QMR architecture will just be valid. Reliability is a function of time. Commonly used reliability functions have exponent distribution function, normal distribution function, and Weibull distribution function. Exponent distribution function is often adopted, because exponent distribution is a single parameter distribution type and has broad applicability. So we suppose that reliability function obeys exponent distribution, $R(t) = e^{-\lambda t}$; λ is SEU failure rate, and here, let $\lambda = 10^{-6}$ error/(bit·day). Therefore, $R_{QMR}(t) = 6e^{-2\lambda t} - 8e^{-3\lambda t} + 3e^{-4\lambda t}$; the mean time between failures (MTBF) of QMR is

$$\text{MTBF} = \int_0^\infty R_{QMR}(t)dt = \int_0^\infty 6e^{-2\lambda t} - 8e^{-3\lambda t} + 3e^{-4\lambda t} dt$$
$$= 13/(12\lambda) > 1/\lambda = \int_0^\infty R(t)dt \qquad (103.1)$$

Formula (103.1) shows the integral result of R_{QMR} is greater than that of $R(t)$ in $0 \le t < \infty$. Because $R(t)$ is a descending function, if $t > -\ln 0.23/\lambda$, then $R(t)$ is less than 0.23, and $R_{QMR}(t)$ is less than $R(t)$. In fact, due to limited battery life, undated life-span for the satellite system is impossible. The life of low Earth orbit satellites is about 3~10 years; that of geostationary Earth orbit and middle Earth orbit satellites is about 12~15 years. When $t < -\ln 0.23/\lambda = 4{,}027$ years, $R_{QMR}(t)$ is still greater than $R(t)$, so it is feasible to use the QMR architecture. For the TMR, when 3 registers have no error or one bit-flip, the output can be the correct value. Thus, TMR reliability is $R_{TMR} = 3R^2 - 2R^3$. To guarantee $R_{TMR} > R$, R must be greater than 0.5. Let $R(t) = e^{-\lambda t}$ and $R_{TMR}(t) = 3e^{-2\lambda t} - 2e^{-3\lambda t}$; the MTBF of TMR is

$$\text{MTBF} = \int_0^\infty R_{TMR}(t)dt = \int_0^\infty 3e^{-2\lambda t} - 2e^{-3\lambda t} dt = 5/6\lambda < 1/\lambda$$
$$= \int_0^\infty R(t)dt \qquad (103.2)$$

Formula (103.2) shows the integral result of RTMR is less than that of $R(t)$ in $0 \le t < \infty$. But when $t < \ln 2/\lambda = 1{,}899$ years, using TMR is also feasible. Figure 103.3 shows the relation of $R(t)$, $R_{TMR}(t)$, and $R_{QMR}(t)$. The horizontal axis represents time, the vertical axis represents reliability. With the time growing, $R(t)$, $R_{QMR}(t)$, and $R_{TMR}(t)$ are all decreasing. When $0 < t < \ln 2/\lambda$, $R(t) > 0.5$, so R_{QMR} and R_{TMR} are all greater than $R(t)$, namely, using TMR and QMR can all improve reliability. But when $\ln 2/\lambda < t < -\ln 0.23/\lambda$, $0.23 < R(t) < 0.5$, $R_{QMR}(t) > R(t)$, but $R_{TMR} < R(t)$, that is to say, using QMR is still feasible, but using TMR, the reliability is even lower than that of single modular. As Fig. 103.3 shows, the reliability of QMR is always greater than that of TMR.

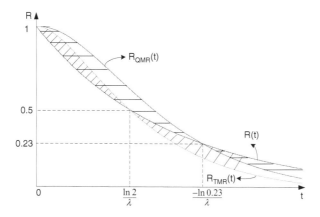

Fig. 103.3 Comparison of reliability of QMR and TMR

103.2.3 QMR SEU Failure Rate

The failure rate function subjecting to exponential distributions is λ(t) = λ. To simplify the analysis, we assume SEU failure rate of every memory element is the same. Let SEU failure rate of single modular be λ; thus, that of TMR is $\lambda_{TMR} = 3\lambda^2 - 2\lambda^3$, and that of QMR is $\lambda_{QMR} = 4\lambda^3 - 3\lambda^4$. The memory controller contains 464 registers; SEU failure rate of memory controller is $\lambda_{memctrl} = 1 - (1 - \lambda)^{464} = 4.64e - 4$ error/(bit·day), which cannot meet the requirement of reliability. When using TMR, $\lambda_{memctrl\_TMR} = 1 - (1 - \lambda_{TMR})^{464} = 1.39e - 9$ error/(bit·day); when using QMR, $\lambda_{memctrl\_QMR} = 1 - (1 - \lambda_{QMR})^{464} = 1.86e - 15$ error/(bit·day). Compared with TMR, using QMR to all registers, using SMIC 130 nm standard CMOS process to synthesize, the area overhead of the memory controller is increased by 44,892.833344 μm², but the SEU failure rate is lower by 6 orders of magnitude.

103.3 Design Implementation

Using QMR, there are two implementation modes. One is the memory controller as a subsystem; in Fig. 103.1, A1, A2, B1, and B2 are all memory controller modules. Another mode is all registers of the memory controller can be implemented using QMR. About area consumption, the front mode adds the area of combinational circuit, and the after mode adds the area of the comparison circuit and multiplexer of QMR for all registers. But SEU failure rate of the front mode is $\lambda_{memctrl\_QMR'} = 4\lambda_{memctrl}^3 - 3\lambda_{memctrl}^4 = 3.99e - 10$ error/(bit·day) and is greater by 5 orders of magnitude. So we select the after mode.

Figure 103.4 shows timing control of accessing memory operation. In an idle state, the request from the processor is still detected. When there is a read operation,

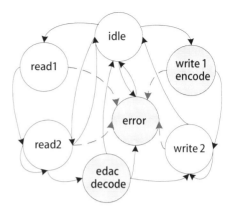

Fig. 103.4 FSM with fault-tolerance design

switch to read1; the address will be given, chip select signal and read enable signal will be valid. In read2, the data from the external memory will be locked. To protect external memory using edac (error detection and correction) circuit, an edac state is added. In the edac state, if the decoder result is no error, the data will be propagated to the processor and will switch to idle. If there is a correctable error, switch to write2; the corrected data will be written back memory and will be sent to the processor. If there is a non-corrected error, switch to error state; memory error information will be propagated to processor. When there is a write operation, switch to write1; the address will be given, chip select signal will be valid, and the data and corresponding checksum will be prepared. In write2, write enable signal is still valid, and when write operation is completed, switch to idle state.

The QMR has error detection function; all error detection signals need to be checked every cycle. The memory controller contains 464 flip-flops, namely, 464 error detection signals do "or" operation and become a final error detection signal. This combinational circuit's delay is 1.10 ns; for the system of clock frequency no more than 909 MHz, the delay meets design timing and the area overhead is 2,637.759626 μm^2. By modifying FSM, when the final error detection signal is set, in whichever state the FSM is, it will switch to error state. In Fig. 103.4, broken lines indicate timing control modification. When some faults are checked, the operation of a memory can be terminated in time and can return the error information, improving the reliability of the accessing memory.

103.4 Experimental Results

To evaluate the reliability of our approach, it is necessary to inject faults in QMR registers and to code relevant test cases. Figure 103.5 shows simulated structure. Access memory simulator simulates operation timing from AHB bus. When a test case (an access memory operation) is running, the correlative fault can be injected, test case list and fault injection (FI) list are corresponding, and every group of QMR

Fig. 103.5 Simulated structure

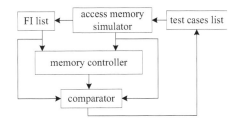

Table 103.1 The simulation results

Category	Description	Bits	Case type Read	Case type Write	FI time	1 bit-flip cycles	2 bit-flips cycles
Rw_ctrl	Read and write control	79	9	70	read1/write1	167	158
addr	Read/write address	32	0	32	write2	64	96
data	Read/write data	64	32	32	edac/write1	160	192
fsm_state	FSM state	4	0	4	write2	8	12
ahb_ctrl	Ahb control signals	14	0	14	write1	28	28
config	Config registers	102	0	102	write2	204	306
en_dec	Encode and decode info	169	160	9	edac/write1	498	658

registers has 4 bits FI signals RegN_err[3:0] corresponding to A1, A2, B1, and B2 respectively. These signals are input of memory controller. When one FI signal is valid, the relevant register will have 1 bit-flip. Comparator receives the FI information and the number of cycles of the current operation; if the fault is 1 bit-flip, the ending timing of the corresponding test case is checked whether it matches the "okay" ending timing of AHB bus. If the fault is 2 bit-flips, the ending timing is checked whether it matches the "error" ending timing.

In this experiment, read access and write access use 3 cycles and 2 cycles, respectively, by configuration. Table 103.1 lists the various categories of registers and provides a description for each, as well as the number of bits of registers. One test case is coded for every group of QMR registers of every category. When the faults injected to the registers are only correlative with read access, test cases use read operation; otherwise, to save simulation time, write operations are used. The number of different types of test cases is given in the fourth and fifth columns. Each test case is executed twice; 1 bit-flip is injected for the first time, 2 bit-flips is injected second time, and FI time is shown in the sixth column. The 1 bit-flip can be self-repaired, the timing of access memory is not affected, and the cycles of running test cases are indicated in the seventh column. The 2 bit-flips are injected in the QMR registers at FI time; after one cycle, the test case would be finished; the simulated cycles are shown in the last column. The earlier the FI time is, the less the cycles of fault detection are. To correct the 2 bit-flips fault, except for the simulated cycles in the last column, accessing memory operation must be executed again, so more cycles are needed, but the compensation is higher reliability desired for the high-safety system.

103.5 Conclusion

Instead of TMR, all inner registers of memory controller are implemented using the QMR architecture; the area overhead is increased by 47,530.59297 μm^2, but SEU failure rate of the memory controller is lower by 6 orders of magnitude. Although the error detection signal of every QMR needs to be examined every cycle, using SIMC 130 nm standard CMOS process to synthesize, this error detection circuit's critical path delay is 1.10 ns meeting design timing that is at most 909 MHz. By modifying FSM, when some faults are checked, the operation of a memory can be terminated in time, error information can be returned, and the reliability of the system can be improved. For the hardware system with high safety, if the overhead of area and power meet the design requirements, the QMR architecture can be used to reduce the failure rate of the key components.

References

1. Gaisler, J. (2002). A portable and fault-tolerant microprocessor based on the SPARC v8 architecture. In *Proceedings International Conference on IEEE, USA* (pp. 409–415).
2. Longden, L., Thibodeau, C., Hillman, R., Layton, P., Williamson, G., & Dowd, M. (2002). Designing a single board computers for space using the most advanced processor and mitigation technologies. *European Space Components Conference*, Toulouse (pp. 313–316).
3. Carmichael, C. (2001). Triple module redundancy design techniques for virtex FPGAs. *Xilinx Application Note XAPP, 197*, 1–37.
4. Chen, G., Fan, D., & Wei, Z. (2010). All electronic computer interlocking system based on double 2-vote-2. *China Railway Science, 31*(4), 138–144 (In Chinese).
5. Zhang, B., Lu, Y., Han, J., & Wei, Z. (2009). Reliability and security analysis of double 2-vote-2 redundancy system. *Journal of System Simulation, 21*(1), 256–261 (In Chinese).
6. Zhang, J., Wang, H., & Jiang, D. (2006). Analysis of double 2-vote-2 fault-tolerant architecture used in computer-based interlocking system. *Computer and Telecommunication, 15*(11), 46–49 (In Chinese).
7. Eaton, P., Benedetto, J., Mavis, D., Avery, K., Sibley, M., Gadlage, M., & Turflinger, T. (2004). Single event transient pulsewidth measurements using a variable temporal latch technique. *IEEE Transactions on Nuclear Science, 51*(6), 3365–3368.

Chapter 104
Computer Power Management System Based on the Face Detection

Li Xie, Yong He, Yanfang Tian, and Tinghong Yang

Abstract In order to reduce the unnecessary power waste of computer system, the working principle of earlier Windows power management program and the new face recognition function of Windows 8 are analyzed in this paper. And the conflict between the convenience of use and the effects of energy conservation and environmental protection is given attention to. We put forward a new method based on the detection of frontal face in front of the monitor instead of the events of keyboard or mouse. Experimental results show that the method is a fast and effective one. Particularly, when user is leaving for a moment, this method is better than the work of Windows power management program. The results tell us that this method can save electrical energy about 4.28 % than windows power management program.

104.1 Introduction

With the development of society, there is a growing demand for energy. So the energy conservation and emission reduction will undoubtedly become a very noteworthy subject nowadays. However, due to the huge increase in the number of computer users, the lack of existing power management program, and many other factors, a great waste of energy is made on the use of the PC.

According to the data released in 2008 by the market research firm Gartner Inc., the global PC had been more than ten million, and the number still kept a steady

---

L. Xie (✉)
Chongqing Electric Power College, Chongqing 400053, China
e-mail: 43329588@qq.com

Y. He
Chongqing Experimental High School, Chongqing 401320, China
e-mail: tumblerman@126.com

Y. Tian • T. Yang
Logistic Engineering University, Chongqing 401311, China

improvement in the growth rate of 12 % each year. If this improving speed continues, the number of global PC will reach 20 billion in 2014. For a so large number, a little waste on each computer will be unforgivable.

When using a computer, we often encounter emergencies which need a temporary leave such as answering phones, receiving express, and taking printed statements. For convenience, we generally do not turn off the computer. Occasionally once cannot result in much waste of energy, but the cumulative number of all the world's computers should not be ignored.

But the earlier system comes with power management program usually requires a relatively long reaction time (usually at least a few minutes) to enter a power-saving mode [1]. The traditional criterion to determine whether the system enters power saving mode is: "Is there keyboard click or mouse click?" It brings us some trouble to determine the response time. If the threshold of response time is set too long, the power saving effect is not obvious. But if it is too short when we just use the computer to read some documents and information, and do not use the keyboard and mouse for a long while, the system will in turn affect our work.

Recently, Microsoft's latest release of Windows 8 system provides a face recognition function. But this function is mainly focused on the management and login of user's accounts, and this way is found not safer than the early password security. Only needing a certain account user's photo, anyone can illegally log on to the system. In addition, the Windows 8 system's functions of detecting the user's arrival with the distance inductive sensor and automatically booting must obtain the support of sensor hardware. But for PC in current extensive use, this kind of sensor does not yet exist.

Therefore, according to the technical level of existing computer hardware, to develop a highly intelligent computer power-saving system will be of great significance for energy conservation.

104.2 Working Principle and System Structure

104.2.1 Working Principle

"Is there a keyboard or mouse action" obviously has its limitations as a decision standard to enter the power-saving state, since it is difficult to obtain the compromise between the energy-saving effect and the normal using of users. In order to resolve this contradiction, we must find a way that gives a more accurate and clear description of these two states.

We find that the face is often positive facing on the monitor when someone is normally using a computer. Based on this premise, we can determine whether someone is using the computer according to if some positive face is in front of the monitor or not.

We use camera, a kind of commonly used external equipment, to collect the image in positive front of the monitor. Then, we detect faces through a face detection algorithm and control the computer's power-saving state according to the test results. In certain detection period (off-screen time, in seconds), the system will immediately turn off the monitor into the primary power-saving state when the computer does not sense the face because of the user's temporary leave. And once the user is back to the computer, it will immediately turn on the monitor and return to normal. Further, if no human face is detected for a very long time (standby time, in minutes), the system will go into a better state of power-saving standby. In this way, if there is a temporary or a long-term leave of user, the computer will give a quicker and more accurate judgment and turn off the monitor or enter system standby in time. All those come to our purpose of power saving.

104.2.2 Feasibility Analysis

The face detection is a complex pattern recognition process. Its main difficulties are the inner face changing and the environmental impact on face to be detected. More details:

1. Very complex details in changing of faces, different physical characteristics such as face shape and color, and different expressions such as opening and closing of eyes and mouth
2. The blocking of other objects on the face, such as glasses, hair, head ornaments, and other external objects
3. The influence of light, such as image brightness, contrast variation, and shadow

All those set up obstacles for people to solve the problem of face detection [2].

In order to remove those difficulties, in this paper we take an existing approach with better robustness in the field of face detection, i.e., approximate Haar characteristics method, to classify the targets, better dealing with adverse factors such as light and color [3].

In addition, we find that the monitor power is about 30–100 W, the total power of a desktop computer is about 350 W, and the notebook is about 100 W. Compared with all those, the camera's working power is around 90 mW, which is negligible and does not increase the load on system with any operation. So our method is of a better feasibility.

104.2.3 System Structure

This system mainly consists four parts: "system tray," "parameter setting module," "video processing module," and "hardware control module." The working principle is shown in Fig. 104.1.

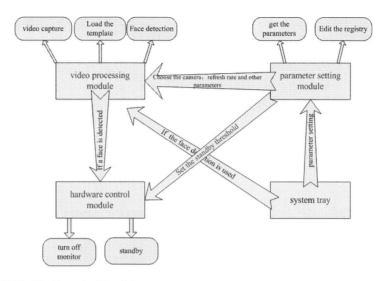

Fig. 104.1 Working principle of the system

The "system tray" is the input interface of all commands and parameters. In special cases, such as watching videos or listening to music, you can suspend or cancel the running of the system through this module. In most cases, the module is mainly used to set and modify the system parameters.

"Parameter setting module" is primarily responsible for the gaining and saving of system parameters, including the settings of power-saving parameters, video capture devices, and acquisition parameters. Through setting power-saving parameters, the system can control when the monitor should be turned off after the face detection fails and when the system enters standby mode. Through the setting of video capture devices and parameters, one can control the frequency of system detection and then control the sensitivity of face detection.

"Video processing module" is the core of this system, which is mainly responsible for detecting image in front of the monitor and examining whether there is user's frontal face in image detected by face detection algorithm. If the system cannot detect the front face, and cannot reach the turning-off time threshold set by "parameter setting module" to tell "hardware control module" to turn off monitor, the "video processing module" will keep working before the system gets into deeper power-saving state. If the face is detected sometime later, the system will tell "hardware control module" to turn on the monitor. But if the face data is still not detected until it reaches the standby threshold, the "hardware control module" should be told to enter deeper power-saving state. The flow chart of video processing is shown in Fig. 104.2.

"Hardware control module" is for controlling monitor's turning on and off, according to the system state informed by parameter setting module and the relevant parameters of video processing module. It also examines whether the system enters the standby and is responsible for the interaction with computer hardware.

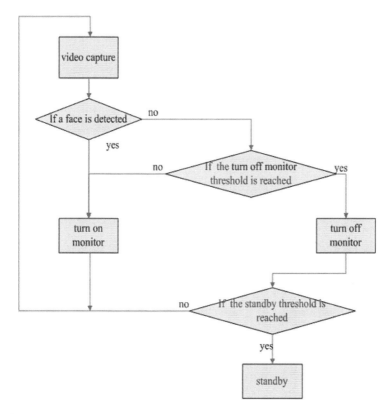

Fig. 104.2 Flow chart of video processing

104.2.4 System Implementation

In actual using, we develop the system with cross-platform programming language QT. Its core part is the face detection using a face detection algorithm supported by OpenCV (Open Source Computer Vision Library). Since it only works on examining whether there is a frontal face in all video images, but does not have to deal with face recognition, the algorithm is simple and is easy to handle with [4].

Face detection program completes three-part work, i.e., loading classifier, loading the image to be detected, as well as detecting and marking.

The target detecting classification of "haarcascade_frontalface_alt2.xml" file storage supported by OpenCV is used in this system [5]. The test results show that it basically meets the actual requirement, without the tedious steps of their own training classifiers.

The loading classifier program is as follows:

```
QString cascade = QCoreApplication::applicationDirPath().
  replace("/","\\")+
  "\\haarcascade_frontalface_alt2.xml";
```

```
if( !cascade.load( cascadeName.toStdString()) )
  {
cerr << "ERROR: Could not load classifier cascade" << endl;
  return -1;
  }
```

"Loading the image to be detected" and "detecting and marking" are two periodically performed steps. They are mainly based on the refresh time of system parameters setting to periodically collect image in front of the screen through the camera and to give the appropriate backup for calling of system detection part.

```
IplImage* iplImg = cvQueryFrame( capture );
frame = iplImg;
if( frame.empty() ) return -1;
if( iplImg->origin == IPL_ORIGIN_TL )
frame.copyTo( frameCopy );
else
flip( frame, frameCopy, 0 );
int r=detectAndDraw( frameCopy, cascade, scale ,disp);
where the"detectAndDraw" function codes are:
int Facedetect::detectAndDraw( Mat& img, CascadeClass ifier&
  cascade, double scale, bool disp)
{
int i = 0;
vector<Rect> faces;
const static Scalar colors[] =
{
CV_RGB(0,0,255),
  CV_RGB(0,128,255),
    CV_RGB(0,255,255),
    CV_RGB(0,255,0),
    CV_RGB(255,128,0),
    CV_RGB(255,255,0),
    CV_RGB(255,0,0),
    CV_RGB(255,0,255)

};
Mat gray, smallImg( cvRound (img.rows/scale),
cvRound(img.cols/scale), CV_8UC1 );
cvtColor( img, gray, CV_BGR2GRAY );
resize( gray, smallImg, smallImg.size(), 0, 0, INTER_LINEAR );
equalizeHist( smallImg, smallImg);
cascade.detectMultiScale( smallImg, faces, 1.1, 2, 0, Size
  (20, 20));
int radius=0;
```

```
for ( vector<Rect>::const_iterator r = faces.begin(); r !=
    faces.end(); r++, i++ )
{
    Point center;
    Scalar color = colors[i%8];
center.x = cvRound((r->x + r->width*0.5)*scale);
center.y = cvRound((r->y + r->height*0.5)*scale);
radius = cvRound((r->width + r->height)*0.25*scale);
if(disp==true)
{
    circle (img, center, radius, color, 3, 8, 0);
}
}
if (disp==true)
{
cv::imshow(" Face detection results ", img );
}
else
{
cvDestroyWindow("Face detection results ");
}
return radius;
}
```

104.3 System Testing

In order to test the performance of the system, we use a laptop to do a power discharging test, respectively, under normal operating conditions (programming and Internet) to use the software and not to use until the system naturally turns off. The test data is shown in Fig. 104.3. Obviously, the power-saving effect is significantly superior to the effect of not using the software. This advantage comes mainly from power saving of closing the screen system standby while there is temporary leave.

The system chooses the power-saving strategy of turning off monitor to deal with temporary leave. In actual test, there are temporary leaves about six times in former 1,540 s, each time about 5 s to 5 min (due to the density of coordinate, the display is not obvious after screen-off, but the cumulative power discharging trend after multiple screen-off is more gentle). And at the time 1,540 s, the measured remaining power is, respectively, 64.39 and 60.11 %, i.e., there is a saving approximately 4.28 % of electricity.

Fig. 104.3 Power-saving testing data

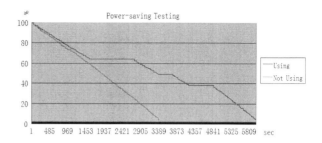

For user's longtime leaving, the system chooses standby. Shown at the 1,541–2,742, 3,371–3,771, and 4,189–4,861 s in Fig. 104.3, power consumption is almost negligible.

104.4 Conclusion

The "computer power management system based on face detection" works on the condition of existing computer hardware level. And in addition to the long system standby strategy of conventional power management system, when the user is temporarily away, this face detection system can more quickly give the accurate judgment, shut down the monitor, and save power. Therefore, using of face detection and recognition technology can solve the problems of slower response time and inaccurate judgment. It also can save unnecessary energy consumption of a computer system, and is of high practical value and promoting significance.

References

1. Zhao, X., Chen, X., Guo, Y., & Yang, F. (2008). The research progress of operating system's power management. *Computer Research and Development, 45*(5), 817–824.
2. Chen, Z., & Jiang, M. (2012). Designing of face detection system based on the OpenCV. *Electronic Design Engineering, 20*(10), 82–185.
3. He, Y., & Li, G. (2012). Face detecting, tracking and feature points positioning system. *Electronic Design Engineering, 20*(8), 189–192.
4. Zhang, Y., & Li, Y. (2011). General face detection module designing based on the OpenCV. *Computer Engineering and Science, 33*(1), 97–101.
5. Tao, Y. (2012). Face recognition applications based on the OpenCV. *Computer Systems and Applications, 21*(3), 220–223.

Chapter 105
Twist Rotation Deformation of Titanium Sheet Metal in Laser Curve Bending Based on Finite Element Analysis

Peng Zhang, Qian Su, and Dong Luan

Abstract Laser sheet bending is a new metal forming process realized by thermal stresses resulted from the irradiation of laser beam scanning. Laser forming is a new type of sheet metal forming process. The sheet metal is formed by asymmetrical thermal stresses. The three-dimensional elastoplastic thermomechanical coupled finite element model of laser bending for Ti-6Al-4V plates was established with nonlinear finite element analysis software ANSYS. The bending properties of sheet metal with different processing parameters were simulated. The results show that the twist rotation deformation of sheet metal can be influenced by laser power, spot diameter, scanning velocity, scanning path curvature, and the distance between scanning path and free end.

105.1 Introduction

The laser is a kind of tool, and common light source cannot be compared with it because of high purity, high brightness, high coherence, and high directivity, so it is used widely. Laser processing technology is greatly used in the area of cutting, welding, and surface treatment with the features of high energy injection rate, low hot influence area, easy guidance, being not affected by electromagnetism, high speed machining, no tool wear, and noise pollution [1–3]. Based on the characteristic of thermal expansion and contraction, laser sheet bending is a new metal forming process realized by thermal stresses resulted from the irradiation of laser beam scanning. Compared with other conventional machine forming methods, laser sheet bending has many advantages such as on die molding, noncontact molding, no external force molding, hard-to-deformation thermal normal cumulative forming, and laser beam mode without specific requirements [4, 5].

P. Zhang (✉) • Q. Su • D. Luan
School of Materials Science and Engineering, Harbin Institute of Technology at Weihai, Weihai, Shandong 264209, China
e-mail: pzhang@yeah.net

When trajectory of the laser beam is linear compared with sheet, V-shape parts are obtained; when trajectory is non-repeated or curve, composite curved special-shaped parts are obtained, such as cylindrical parts, disc-shaped parts, spherical part, and a variety of complex shapes three-dimensional shaped pieces [6]. The deformation process of sheet metal is very complicated, especially the residual stress of preorder deformation and the influence of geometry for deformation after unloading; with the curve scan path and because of the thermal effect of laser beam, the sheet also produces torsional deformation in addition to the bending deformation in author's previous studies [7]. Therefore, studying the torsional deformation of the sheet in the process of laser radiation, analyzing influence law of the sheet torsional deformation's laser processing parameters can lay a foundation for the implementation of the sheet laser thermal stress precision forming.

105.2 Finite Element Model of Laser Bending

Laser bending is a complex process of interaction of many factors, such as sheet performance parameters, geometric parameters, and laser processing parameters, and designing the processing technology by experimental method will spend much time and human and material resources. With the rapid development of computational mathematics and computer technology as well as the improvement and perfect of finite element algorithm, using the numerical simulation method to simulate practical production has been proved that it is an effective way and has a huge potential [8–12].

105.2.1 Physical Model

In view of the fact that laser bending process is complex and influenced by many factors, it is difficult to establish the model to accurately reflect the actual situation, so this paper makes hypothesis to simplify calculation model as follows: (1) The freedom of all nodes of one end of the sheet should be restricted to avoid the sheet occurring rigid body displacement, and this is consist with the actual situation; (2) the laser beam moves at equal speed and irradiates the surface of the sheet vertically; (3) the material is isotropic, and the thermal physics and mechanical properties change with temperature; (4) the thermal absorption coefficient of material is constant; and (5) the process of scanning is carried out under the melting point of the material.

105.2.2 Movable Heat Source Based on Scanning Path

The three-dimensional elastoplastic thermomechanical coupled finite element model of laser bending is established with nonlinear finite element analysis software ANSYS, and the model uses three-dimensional eight-node hexahedral elements. Through ANSYS secondary development, subroutine of Gauss heat source based on scanning path is established, and laser thermal load is applied to the respective units in the form of heat flux. Scanning path uses parametric curve designing, and cubic B-spline curve can describe the scanning path easily and accurately.

Figure 105.1 shows the scheme of laser curve scanning. Select the displacement of the free end of the sheet on the sideline three-point z to calibrate the influence of technical parameters for laser bending. The material of the sheet is Ti-6Al-4V, its performance parameter comes from literature [13], and geometry size is $50 \times 40 \times 0.8$ mm.

105.3 The Twist of Sheets During Laser Curve Bending

Figure 105.2 shows how the displacement field of sheets laser curve scanning distributes. Figure 105.3 shows how the movement of three points demarcated on sheets varies. It can be seen that the movement of Point 1 on x-direction always belongs to stretch deformation, and the movement of Point 3 on x-direction always belongs to compression deformation. And Point 2 hardly deforms on x-direction, only when laser beam goes by the region near Point 2 does it generate tiny deformation. There are different degree y-direction movements on the three points demarcated. Because the beginning of the laser beam scanning is asymmetric, the y-direction movement of Point 1 is minor than Point 2 and Point 3.

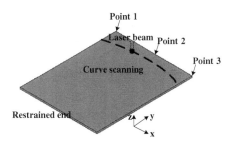

Fig. 105.1 Sketch of laser scanning scheme

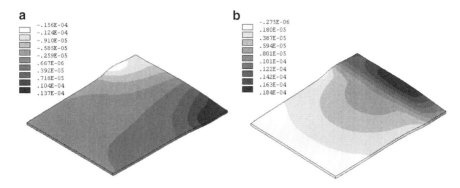

Fig. 105.2 Displacement of laser curve scanning at $t = 0.6$ s. (**a**) Displacement of x (**b**) Displacement of y

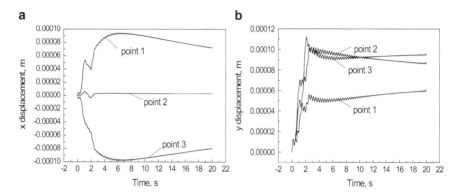

Fig. 105.3 Variation of displacement with time of laser curve scanning

105.3.1 Laser Power

Figure 105.4 shows how the laser power influences the twist of sheets. It can be seen that with the increasing of the laser power, the x-direction movement and y-direction movement of sheets increase. The main reason is that the thermal expansion increases when the input energy increases.

105.3.2 Spot Diameter

Figure 105.5 shows how the spot diameter influences the twist of sheets. It can be seen that with the increasing of the spot diameter, the x-direction movement of

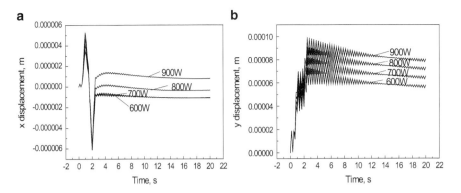

Fig. 105.4 Effect of laser power on twist rotation deformation

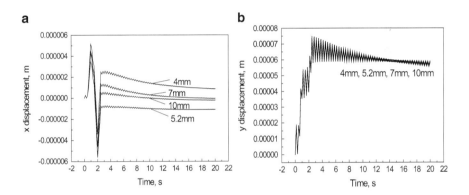

Fig. 105.5 Effect of spot diameter on twist rotation deformation

sheets decreases, while there is no significant change about y-direction movement of sheets. It can explain that y-direction movement of sheets is not sensitive to the spot diameter changes.

105.3.3 Scanning Speed

Figure 105.6 shows how the scanning speed influences the twist of sheets. It can be seen that with the increasing of the scanning speed, both the x-direction movement and the y-direction movement of sheets increase at first then decrease. When other progress parameters are certain, there is some scanning speed, which makes the x-direction movement and y-direction movement of sheets maximum.

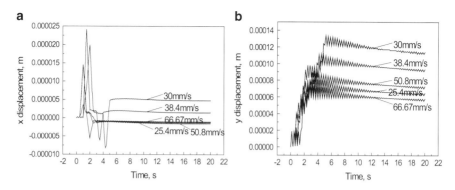

Fig. 105.6 Effect of scanning velocity on twist rotation deformation

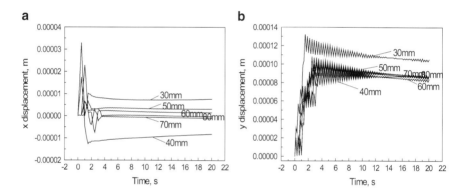

Fig. 105.7 Effect of scanning path curvature on twist rotation deformation

105.3.4 The Curvature of Scanning Paths

Figure 105.7 shows how the curvature of scanning paths influences the twist of sheets. It can be seen that with the increasing of the curvature of scanning paths, the x-direction movement of sheets decreases at first then increase, while the y-direction movement of sheets decreases at first then does not change significantly.

105.3.5 The Distance Between Scanning Paths and Free End

Figure 105.8 shows how the distance between scanning paths and free end influences the twist of sheets. It can be seen that with the increasing of the distance

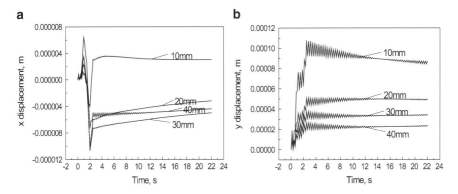

Fig. 105.8 Effect of distance between scanning path and free end on twist rotation deformation

between scanning paths and free end, the x-direction movement of sheets decreases at first then increase, while the y-direction movement of sheets decreases.

The laser power, the spot diameter, and the scanning speed are energy factors, and the increase of input energy leads to increasing in thermal expansion of sheets. The curvature of scanning paths has influence on the input energy and the size of heat-affected zone. The distance between scanning paths and free end has influence on the effect of the material between scanning line and free end on stiff constraint of heating zone.

105.4 Conclusion

1. Based upon the three-dimensional elastoplastic thermomechanical coupled finite element model of laser bending for Ti-6Al-4V plates, the bending properties of sheet metal with different processing parameters were simulated.
2. This study analyzed how the different processing parameters influence the twist of sheets. The increase of laser input energy would lead to an increase in thermal expansion of sheets, and the curvature of scanning paths can influence both the input of energy and the size of heat-affected zone, while the distance between scanning paths and free end can influence the effect of the material between scanning line and free end on stiff constraint of heating zone.

References

1. Raya, T., & Umino, T. (1991). Present status of CO_2-laser processing. *Journal of Materials Engineering, 13*(4), 299–360.
2. Geiger, M. (1993). Synergy of laser material processing and metal forming. *CIRP Annals, 43*(2), 563–570.

3. Washio, K., Ttakenaka, H., Okino, K., Aruga, S., Matsui, E., & Kyusho, Y. (1992). Welding and cutting car-body metal sheets with fiber delivered output from high-power ND-YAG lasers. *NEC Research and Development, 33*(1), 102–109.
4. Thomson, G. (1998). Improvements to laser forming through process control refinements. *Optics and Laser Technology, 2*(30), 141–146.
5. Gisario, A., Barletta, M., Conti, C., & Guarino, S. (2011). Springback control in sheet metal bending by laser-assisted bending: Experimental analysis, empirical and neural network modelling. *Optics and Lasers in Engineering, 49*(12), 1372–1383.
6. Hennige, T. (2000). Development of irradiation strategies for 3D-laser forming. *Journal of Materials Processing Technology, 103*(1), 102–108.
7. Zhang, P., Guo, B., Shan, D. B., & Ji, Z. (2007). FE simulation of laser curve bending of sheet metals. *Journal of Materials Processing Technology, 184*(1–3), 157–162.
8. Maji, K., Pratihar, D. K., & Nath, A. K. (2013). Analysis and synthesis of laser forming process using neural networks and neuro-fuzzy inference system. *Soft Computing, 17*(5), 849–865.
9. Ji, Z., & Wu, S. C. (1998). FEM simulations of the temperature field during the laser forming of sheet metal. *Journal of Materials Processing Technology, 74*(2–3), 89–95.
10. Wu, S. C., & Ji, Z. (2002). FEM simulation of the deformation field during the laser forming of sheet metal. *Journal of Materials Processing Technology, 121*(2–3), 269–272.
11. Kyrsanidi, A. K., Kermanidis, T. B., & Pantelakis, S. G. (1999). Numerical and experimental investigation of the laser forming process. *Journal of Materials Processing Technology, 87*(1–3), 281–290.
12. Chen, D. J., Wu, S. C., & Li, M. Q. (2004). Studies on laser forming of Ti-6Al-4V alloy sheet. *Journal of Materials Processing Technology, 152*(1), 62–65.
13. Ren, X. D., Zhang, R. K., Zhou, J. Z., Zhang, X. Q., & Lu, X. Z. (2006). Influence of laser parameters on laser-shock forming of Ti-6Al-4V alloy. *Chinese Journal of Nonferrous Metals, 16*(11), 1850–1854 (In Chinese).

Chapter 106
Voltage Transient Stability Analysis by Changing the Control Modes of the Wind Generator

Yu Shao, Feng Shi, and Xiang Li

Abstract The chapter studies voltage transient stability when the wind generator changes its control modes. The chapter studies the influence caused by connection with wind farms based on simulation and makes comparison between different control modes, then gives the conclusion. The chapter takes the real grid model and the result of the study has some means to the relative study.

106.1 Introduction

Problems on voltage stability have the direct bearing on the safe operation of the grid, and it hopes that the wind turbine itself could regulate the reactive power. Most studies are always using ideal models. The chapter used the real model of a regional grid to simulate transient stability under two control modes, which had great persuasion to the study of the modern grid.

The structure of the chapter is as follows: The classification of the control modes is described in Sect. 106.2. Section 106.3 presents the problem of voltage stability with the connection of wind farms. Section 106.4 presents the simulation. Section 106.5 concludes the whole chapter.

106.2 Classification of the Control Modes

The chapter compared the two control modes, such as the power factor constant mode and voltage constant mode. The characteristic of the power factor constant mode is that wind turbine follows the change of the power in transient process, regulating the

Y. Shao (✉) • F. Shi • X. Li
Zhengzhou Electric Power Supply Company, Zhengzhou 450000, China
e-mail: 672851649@qq.com

relative power to keep the power factor constant. The characteristic of voltage constant mode is that wind turbine follows the change of the voltage in transient process, regulating the relative power to keep the power factor constant [1].

106.3 Voltage Stability with the Connection of Wind Farms

106.3.1 Capacity Effect of Wind Farms

The combination of the large and small wind farms takes different problems. There are two circumstances. One is that a large wind farm connects with EHV grid and the other is that several small wind farms connect with grid intensively. Small wind farm would affect the grid in several parts. Large wind farm would induce voltage vibration in the grid sometimes [2].

106.3.2 Influence Caused by Characteristics of Wind Farm

1. Type of wind turbine. The types of wind turbine widely used in the grid are constant speed turbine based on asynchronous motor and turbine based on DFIG (Double-Fed Induction Generator).
2. Pneumatic power control technology. The technology mainly includes fixed pitch control, non-fixed pitch control, and active stall control. Pitch angle control technology could strengthen voltage stability in a fault.
3. Level penetration. It means the proportion of the whole wind capacity connected to the grid with the load. High level means large capacity of wind was connected to the grid and the proportion of conventional generators was low [3].

106.3.3 The Influence Caused by Reactive Power Compensator

The asynchronous motor needs more reactive power when it is on operation and reactive power compensator is needed to be installed. The control of condenser bank is discrete and the need of reactive power raised by the change of active power on every compensator stage still be provided by the grid. Switches of condenser bank would cause jumps of voltage. Besides, the characteristics of condenser bank cut lots of reactive power when the voltage depressed, which made the need of reactive power raise in wind farm and worsen the voltage to collapse.

106.4 Simulation

106.4.1 Models of DFIG

Dynamic models of DFIG are composed of generator model, rotation control model, and reactive power model. The voltage equation based on d, q axes is shown in formula Eq. (106.1) [4]:

$$\begin{cases} U_{ds} = p\Psi_{ds} - \Psi_{qs} + r_s I_{ds} \\ U_{qs} = p\Psi_{qs} + \Psi_{ds} + r_s I_{qs} \\ U_{dr} = p\Psi_{dr} - s\Psi_{qr} + r_r I_{dr} \\ U_{qr} = p\Psi_{qr} + s\Psi_{dr} + r_r I_{qr} \end{cases} \quad (106.1)$$

In equations, U_{ds}, U_{qs} are voltage on stator and U_{dr}, U_{qr} are voltage on rotor. I_{ds}, I_{qs} are current on stator and I_{dr}, I_{qr} are current on rotor. Ψ_{ds}, Ψ_{qs} are magnetic linkage on stator and Ψ_{dr}, Ψ_{qr} are magnetic linkage on rotor. $p = d/dt$ is differential operator. Let $p\Psi_{ds} = p\Psi_{qs} = 0$, $r_s = 0$; the dynamic equation is

$$\begin{bmatrix} U_{ds} \\ U_{qs} \\ U_{dr} \\ U_{qr} \end{bmatrix} = \begin{bmatrix} 0 & -X_{ss} & 0 & -x_m \\ X_{ss} & 0 & x_m & 0 \\ px_m & -sx_m & r_r + pX_{rr} & -sX_{rr} \\ sx_m & px_m & sX_{rr} & r_r + pX_{rr} \end{bmatrix} \begin{bmatrix} I_{ds} \\ I_{qs} \\ I_{dr} \\ I_{qr} \end{bmatrix} \quad (106.2)$$

Represent the dynamic equation on axes d and q:

$$\begin{cases} \dfrac{dE'_d}{dt} = -\dfrac{x_m}{X_{rr}} U_{qr} + sE'_q - \dfrac{1}{T'_{d0}}\left[E'_d + (X_{ss} - x')I_{qs}\right] \\ \dfrac{dE'_q}{dt} = -\dfrac{x_m}{X_{rr}} U_{dr} + sE'_d - \dfrac{1}{T'_{d0}}\left[E'_q + (X_{ss} - x')I_{ds}\right] \end{cases} \quad (106.3)$$

$$\begin{cases} E'_d = x' I_{qs} \\ |\dot{U}_s| - E'_q = x' I_{ds} \end{cases} \quad (106.4)$$

Under the influence of magnetic field on stator, $U_{ds} = 0$ and M_e are decided by voltage and current on stator: $M_e = |\dot{U}_s| I_{qs}$. And $|\dot{U}_s|$ represents amplitude of voltage.

The regulations of reactive power output are realized by current of axes d and q on rotor, called vector control. Current on rotor is regulated by U_{dr} and U_{qr}. In order to get the relation between voltage and magnetic linkage on rotor, equations of them are shown [5]:

$$\begin{cases} U_{dr} = p\Psi_{dr} - s\Psi_{qr} + r_r I_{dr} \\ U_{qr} = p\Psi_{qr} + s\Psi_{dr} + r_r I_{qr} \end{cases} \quad (106.5)$$

$$\begin{cases} \Psi_{dr} = X_{rr} I_{dr} + x_m I_{ds} \\ \Psi_{qr} = X_{rr} I_{qr} + x_m I_{qs} \end{cases} \quad (106.6)$$

When $X^{\cdot} = x_{rr} - \frac{x_m^2}{X_{ss}}$:

$$\begin{cases} U_{dr} = (r_r + X^{\cdot} p) I_{dr} - sX^{\cdot} I_{qr} \\ U_{qr} = (r_r + X^{\cdot} p) I_{qr} + sX^{\cdot} I_{dr} + r_r I_{qr} + \frac{sx_m}{X_{ss}} |\dot{U}_s| \end{cases} \quad (106.7)$$

Equation (7) is an electromagnetic transient equation. And the dynamic equation of rotor is $\frac{ds}{dt} = \frac{1}{T_J}(M_m - M_e)$. s is slip of generator. T_J is inertia time constant of rotor. M_m is machine torque. M_e is electromagnetic torque.

Electromagnetic transient equation showed that rotate speed could be changed by regulating electromagnetic torque. The control system could regulate the reactive power according to the aforementioned. The article used BPA software, and typical parameters of GE1.5MW were used in the simulation.

106.4.2 Simulation Under Different Control Modes

A small wind farm was connected to a 220 kV grid. The capacity of wind farm was 90 MW. The capacity per turbine was 1.5 MW. The wind farm was connected by a single line. A one-phase short-circuit fault happened on a 220 kV line in the grid. The fault happened at the 10th cycle; breakers at both ends of the line acted at the 16th cycle and reclosed at the 56th cycle successfully. Voltage of connection point and turbine are shown in Figs. 106.1 and 106.2. The reactive power of turbine is shown in Fig. 106.3. The red line represents for constant power factor control and the purple one represents for constant voltage control.

According to Figs. 106.1, 106.2, and 106.3, voltage of the grid could recover under two kinds of control. Differences are as follows: (1) For voltage of connection point and turbine, constant voltage control could make the level of voltage higher. (2) For reactive power of turbine, constant voltage control distributed more to the voltage by letting turbine made more reactive power output. In sum, constant voltage control is more positive to the voltage stability.

Reactive power made by turbines is far less than the shortage of the grid, and the reactive power made by automatic field forcing is 0.6 Mvar without other compensators, which distributes less to the grid voltage. So in process of a fault, automatic

Fig. 106.1 Voltage of connection point

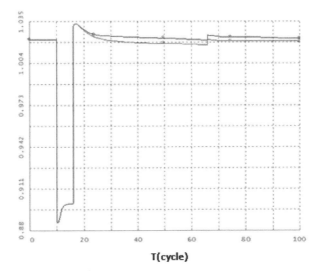

Fig. 106.2 Voltage of the turbine

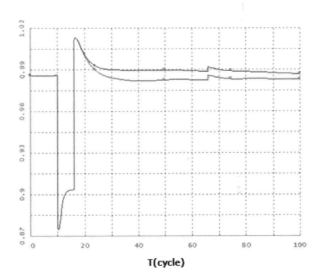

field forcing could not improve voltage of connection point and turbine obviously, and low voltage protection would cut turbines, which could cause vibration of the grid.

Installation of SVC could improve voltage transient stability. There would be a same simulation with the wind-farm-installed SVC, checking the effects of SVC. With the installation of SVC, the simulation waves are shown in Figs. 106.4, 106.5, 106.6, and 106.7. The red line represents for constant power factor control and the purple one represents for constant voltage control.

Fig. 106.3 Reactive power of turbine

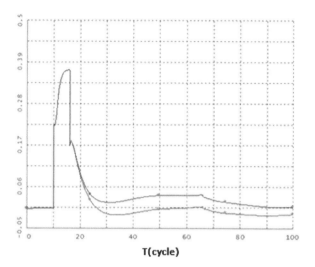

Fig. 106.4 Voltage of connection point

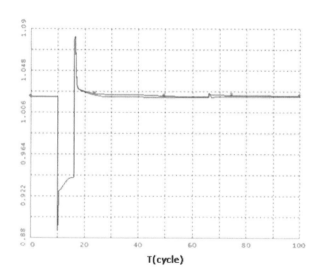

According to Figs. 106.4, 106.5, 106.6, and 106.7 and Tables 106.1 and 106.2, voltage of connection point and turbine is same under two types of control with SVC. The reason is that the reactive power generated by turbines is 0.3 Mvar, which is far less than 23 Mvar generated by SVC, and the recovery of voltage is mainly determined by SVC. When constant voltage control is used, the level of voltage is higher after recovering from the fault. Besides, time of voltage under 0.9 p.u is short and reduces the probability of cutting turbines.

Fig. 106.5 Voltage of the turbine

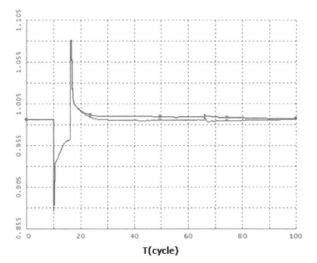

Fig. 106.6 Reactive power of turbine

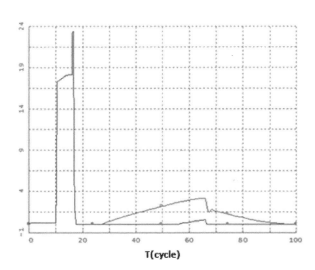

106.5 Conclusion

With the study on voltage transient stability, by changing the control modes of the wind generator, it gives the following conclusions: (1) In comparison with constant power factor control, wind farm with constant voltage control is positive to the voltage stability of the grid. (2) The installation of SVC could improve voltage transient stability effectively and the simulation of two control types is the same.

Fig. 106.7 Reactive power of SVC

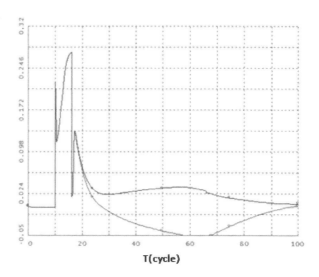

Table 106.1 Voltage of the turbine

Control mode	The original Without SVC	The original With SVC	The minimum Without SVC	The minimum With SVC	The stable Without SVC	The stable With SVC
Constant voltage control	0.985	0.985	0.868	0.868	0.986	0.985
Constant power factor control	0.985	0.985	0.868	0.868	0.985	0.984

Table 106.2 Voltage of connection point

Control mode	The original Without SVC	The original With SVC	The minimum Without SVC	The minimum With SVC	The stable Without SVC	The stable With SVC
Constant voltage control	1.022	1.022	0.89	0.89	1.023	1.022
Constant power factor control	1.022	1.022	0.89	0.89	1.019	1.02

References

1. Hansena, A. D., & Michalke, G. (2007). Fault ride-through capability of DFIG wind turbines. *Renewable Energy, 23*(32), 1594–1610.
2. Xiang, D. W., & Li, R. (2006). Control of a doubly fed induction generator in a wind turbine during grid fault ride-through. *IEEE Transactions on Energy Conversion, 21*(3), 652–662.
3. Lopez, J., Sanchis, P., Roboam, X., et al. (2007). Dynamic behavior of the doubly fed induction generator during three-phase voltage dips. *IEEE Transactions on Energy Conversion, 22*(3), 709–717.

4. Lopez, J., Gubia, E., Olea, E., et al. (2009). Ride through of wind turbines with doubly fed induction generator under symmetrical voltage dips. *IEEE Transactions on Industrial Electronics, 56*(10), 4246–4253.
5. Bueno, C., & Carta, J. A. (2006). Wind powered pumped hydro storage systems, a means of increasing the penetration of renewable energy in the Canary Islands. *Renewable and Sustainable Energy Reviews, 10*(3), 312–340.

Chapter 107
The Generator Stator Fault Analysis Based on the Multi-loop Theory

Yu Shao, Feng Shi, and Xiang Li

Abstract Interturn short circuit is a common kind of fault in generator. The chapter takes multi-loop theory to analyze the theory of fault on generator stator and puts the math model of the generator. Changes of main parameters are analyzed separately when fault happens. According to the result, the chapter analyzes the influence on the main parameters caused by the fault of generator stator and summarizes the factors of parameter changes.

107.1 Introduction

Synchronous generators are the main force in thermal power generation, which are the most important in power system. The condition of synchronous generators decides the stability of the power system directly. The monitoring system of generators comprises local thermal insulation monitoring apparatus, partial discharge monitoring apparatus, palladium barrier leak detectors [1], etc.

In order to ensure the accuracy of operation parameters, the math models are used. For the reason that the magnetic field of generator is whirling in space and the structure of stator and rotor is very huge and the convert between magnetic energy and electric energy is abstract, it is possible to use the math model to study, which makes online monitoring meaningful [1].

The common inner faults are unsymmetrical faults, such as single-phase earth fault and phase-to-phase fault. When a fault happens, the harmonic rate is high in three phases, which can generate a different rotate speed and rotate orientation of the magnetic field. The winding EMF caused by unsymmetrical current is also confused. These are all marked features of inner faults of synchronous generators, which need an exact math model to deal with.

Y. Shao (✉) • F. Shi • X. Li
Zhengzhou Electric Power Supply Company, Zhengzhou 450000, China
e-mail: 672851649@qq.com

The structure of the chapter is as follows. The math models of inner faults in synchronous generators are described in Sect. 107.2. Section 107.3 presents the multi-loop theory. Section 107.4 presents equations of voltage and magnetic linkage. Section 107.5 presents simulation and analysis. Section 107.6 concludes the whole chapter.

107.2 Math Models of Inner Faults

Interturn short circuit, interphase short circuit, and open-phase short circuit are the main faults of synchronous generators. When faults are analyzed, current and voltage are set to be status-variable.

Math models are set to describe inner faults of stator windings based on multi-loop theory, analyzing the connections between stators and rotors. Multi-loop theory helps make voltage equations and magnetic linkage equations. Models of transformer and grid are made according to reality [2].

107.3 The Multi-loop Theory

The multi-loop theory makes winding to be a study unit, which breaks the ideal law that the motor seems to be the ideal one when analyzing inner faults of stator winding. Stator windings are seen as one-phase winding of three-phase winding, and rotor windings are dealt with in non-salient pole machine or salient pole machine. Windings are seen as an independent unit to be used in making magnetic equations.

In the procession of analysis, the motor is seen as many mutual sportive circuits. When the theory is used to study winding faults, influences caused by harmonic magnetic fields should be taken into account, so does the space structure of stator winding, connection shapes, etc. [3].

107.4 Equations of Voltage and Magnetic Linkage

107.4.1 Branch Equations

Any branch equations of voltage and magnetic linkage can be made according to multi-loop theory, such as branch Q:

$$\begin{cases} u_Q = pY_Q - r_Q i_Q \\ Y_Q = -\sum_{S=1}^{N_1} M_{QS} i_S + \sum_{i=1}^{d} \sum_{g=1}^{2P} M_{Qgi} i_{gi} + M_{Qf} i_f \end{cases} \quad (107.1)$$

In formula (107.1), u_Q, i_Q, Y_Q, r_Q represent for voltage, current, magnetic linkage, and resistance of branch Q in the stator, respectively.

i_S, i_{gi}, i_f represent for current of branch S in the stator, current of damping branch "i" in pole "g," and field current of generator. M_{QS}, $M_{Q\ gi}$, M_{Qf} represent for the mutual inductors. N_1 represents for the number of branches in stator. d represents for the number of resistance branches, and P the number of poles. p is differential operator. $Q = 1, 2, \ldots\ldots N_1$.

If there are no interturn faults of stator branches and voltage of three phases is u_A, u_B, u_C, u_Q would meet the relation that [4]

$$u_Q = \begin{cases} u_A & Q = 1, 2, \cdots, a \\ u_B & Q = a+1, a+2, \cdots, 2a \\ u_C & Q = 2a+1, 2a+2, \cdots, 3a \end{cases} \quad (107.2)$$

When there is an interturn fault in the "i" branch of phase A in stator, the expression of u_Q is shown next:

$$u_Q = \begin{cases} u_A & Q = 1, 2, \cdots, i-1, i+1, \cdots, a \\ u_B & Q = a+1, a+2, \cdots, 2a \\ u_C & Q = 2a+1, 2a+2, \cdots, 3a \\ u_A - u_k & Q = i \\ u_j & Q = 3a+1 \\ u_k - u_j & Q = 3a+2 \end{cases} \quad (107.3)$$

107.4.2 Loop Equations

According to the above, the "i" loop equation of "g" pole is shown in formula (107.4):

$$\begin{cases} 0 = p\Psi_{gi} + r_{gi} i_{gi} - r_C \left(i_{g,i-1} + i_{g,i+1} \right) \\ \Psi_{gi} = -\sum_{S=1}^{N_1} M_{gi,S}\, i_S + \sum_{j=1}^{d} \sum_{k=1}^{2P} M_{gi,jk}\, i_{jk} + M_{gif}\, i_f \end{cases} \quad (107.4)$$

In formula (107.4), i_{gi}, Ψ_{gi}, r_{gi} represent for voltage, current, magnetic linkage, and resistance of loop "i" in pole "i," respectively.

$i_{g,i-1}$, $i_{g,i+1}$, r_C represent for voltage, current of loop "i-1," and loop "i+1" in pole "g," respectively. $M_{gi,S}$, $M_{gi,jk}$, $M_{gi\ f}$ represent for mutual inductors between the resistance loop and other branches. And $i=1$, $i_{g,i-1} = i_{g-1,d}$, $i=d$, $i_{g,i+1} = i_{g+1,1}$.

Fig. 107.1 Forward direction of current in stator and rotor

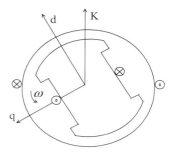

107.4.3 Equations of Field Winding

According to the forward direction in Fig. 107.1, equations of voltage and magnetic linkage in field winding are shown in formula (107.5):

$$\begin{cases} u_f = p\Psi_f + r_f i_f \\ \Psi_f = -\sum_{S=1}^{N_1} M_{fS} i_S + \sum_{i=1}^{d} \sum_{g=1}^{2P} M_{f\,gi} i_{gi} + L_f i_f \end{cases} \quad (107.5)$$

In formula (107.5), Ψ_f, r_f, u_f, i_f represent for magnetic linkage, resistance, voltage, and current of field winding, respectively. i_{gi}, i_S represent for current of the loop "i" and stator winding "s" in pole "g." M_{fS}, $M_{f\,gi}$, L_f represent for mutual inductance and self-inductance of field windings [5].

107.5 Simulation and Analysis

The inner interturn faults of stators are familiar in normal operation, especially in synchronous generators with large capacity. The windings of synchronous generators with large capacity are in the same channel, which increase the probability of interturn faults.

In can be seen that the analysis of reason and theory of interturn faults has great value, which proves the matter to seek the discipline of the interturn faults. The simulation used PSCAD software to make models of generator and simple grid, simulating the operation of full-load generation. The simulation focused on one-winding and two-winding faults and looked for the discipline of faults through interturn faults. A 10 kW generator model was used in the simulation, which was made of PMVR, field system, etc.

Table 107.1 Parameters of generator

Parameters	Range
Kinetic energy of generator	$E_{WMS} > 0$
Original active power (pu)	$0 \leq P \leq 1.0$
Original reactive power (pu)	$0 \leq Q \leq 1.0$
Active power on a node (pu)	$\Sigma P = 1.0$
Reactive power on a node (pu)	$\Sigma Q = 1.0$
Direct axis transient reactance	$X'd > 0, Xd > X'd > X''d$
Quadrature axis transient reactance	$Xq = X'q > X''q (T'q0 = 0)$
Damping reactor	$XL \leq X''d$

107.5.1 Simulation of Interturn Faults in Stator on Full-Load Generator

The chapter uses the typical parameters of generator in PSCAD. And parameters are shown in Table 107.1.

The model of magnetic field was the IEEE-1968 model, which could represent the typical generator. The full-load operation means generator operates with rated load, which is the most familiar working condition.

One of the features when the full-load generator is on operation is that the current is heavy. The long-time operation of full-load generator has great harm to the isolate winding, which makes it meaningful to study interturn faults under the condition when the generation is full load.

107.5.2 One-Winding Fault of Phase A on Stator

When one-winding short circuit happens on phase A of the generator, the simulation occurs, as shown in Figs. 107.2 and 107.3.

As shown in Figs. 107.2 to 107.3, current of three phases is heavier when the generator is on normal operation.

There would be an obvious change when a fault happens. On the contrary, field current has an obvious change and the amplitude is about 0.1 A, for which the field current is chosen to be the criterion of faults.

107.5.3 Two-Winding Fault of Phase A on Stator

When two-winding short circuit happens on phase A of the generator, the simulation occurs, as shown in Figs. 107.4 and 107.5.

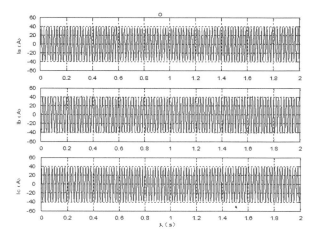

Fig. 107.2 Current wave of three phases when one-winding short circuit happens

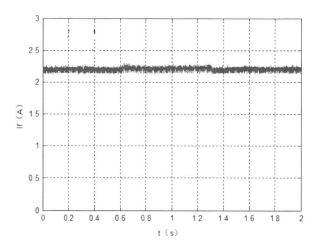

Fig. 107.3 Field current wave when one-winding short circuit happens

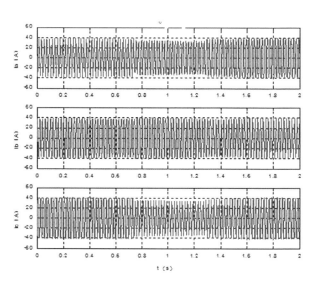

Fig. 107.4 Current wave of three phases when two-winding short circuit happens

Fig. 107.5 Field current wave when two-winding short circuit happens

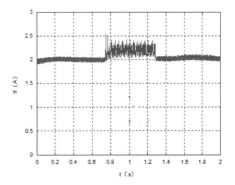

As shown in Figs. 107.4 to 107.5, current of three phases are heavier when the generator is on two-winding short circuit. The change of field current is obvious and the amplitude is about 0.4 A.

107.6 Conclusion

The chapter used PSCAD software to prove multi-loop theory could be used to study inner faults of motors, checking the validity that current variation and voltage variation of stator could be used to analyze the stator faults.

The math models of stator windings are with commonality when used to analyze faults, which are adapted to be used in analysis of non-salient pole machine and salient pole machine. The whole simulation has some meanings to the analyzation of real motor faults.

References

1. Ishida, K., & Dokai, K. (1992). Development of a 500kV transmission lines arrester and its characteristics. *IEEE Transactions on Power Delivery, 7*(3), 1265–1274.
2. Sakshaug, E. C. (1989). Metal oxide arresters on distribution systems fundamental consideration. *IEEE Transmission on Power Delivery, 4*(4), 2076–2089.
3. Flisowski, Z., Mazzetti, C., & Wlodek, R. (2004). New approach to the selection of effective measures for lightning protection of structures containing sensitive equipment. *Journal of Electrostatics, 60*(4), 287–295.
4. Mathier, L., Perreault, L., & Bobe, B. (1992). The use of geometric and gamma-related distributions for frequency analysis of water deficit. *Stochastic Hydrology and Hydraulics, 6*(4), 239–254.
5. Frank, J., & Masse, J. (1951). The Kolmogorov-Smirnov test for goodness of fit. *Journal of the American Statistical Association, 46*(253), 68–78.

Chapter 108
An Improved Edge Flag Algorithm Suitable for Hardware Implementation

Lixiang Wang and Tiejun Xiao

Abstract The traditional edge marking algorithm cannot fill the elongated polygon and a polygon with local points correctly. After doing a lot of research and analysis about polygon fill algorithms, this paper presents a new improved algorithm, which is suitable for hardware implementation, to meet the need for high-quality graphic display in the embedded system. The new algorithm makes full use of the characteristic that the local point or elongated point is accessed repeatedly when it meets local points and elongated points. We can define a measurement variable named FLAG, which is used to mark the boundary point of the polygon. The flag of the present point will add one when it is accessed. This method can conveniently and simply distinguish singular points and elongated points from ordinary points. What's more, the improved algorithm solves the previously mentioned problems effectively. In the new algorithm, we only use the addition operation so it is easy to be implemented by the hardware.

108.1 Introduction

With the wide application of embedded systems, polygon filling has been extensively used in the field of embedded systems gradually. The traditional display of the embedded graphics depended largely on the microprocessor. However, to satisfy the high-quality and high-efficiency requirements, using only software to do the complex process of polygon filling has become a bottleneck in the graphics display, so it is a feasible scheme to meet the high demand with the hardware. The problem of polygon filling [1] is one of the basic issues in computer graphics. Polygon fill algorithm mainly includes edge marking algorithm [2, 3], scan line fill algorithm [4], and seed fill algorithm. For the seed fill algorithm [5], it has to call

---

L. Wang (✉) · T. Xiao
School of Computer Science and Telecommunication Engineering, Jiangsu University, Zhenjiang 212013, China
e-mail: lxwang2012@163.com

stack time and time again, so the efficiency of the filling is reduced greatly. For the scan line fill algorithm, it has to build edge tables and rank them when filling the polygon, so the efficiency is also restricted. When scan line fill algorithm and edge marking algorithm are implemented in software, their efficiency is quite good; however, the edge marking algorithm doesn't need to build edge tables and maintain them, so its efficiency is greater by one to two orders of magnitude than that of the scan line fill algorithm. But the traditional edge marking algorithm will not fill normally when the polygon has elongated points and singular points. So a lot of papers have already been submitted to provide many solutions to improve it, such as Xiaohua Wang who proposed an improved edge marking algorithm in 2004 that successfully resolved the problem of the singular points. Unfortunately the introduction of the edge table, which is used in the scan line fill algorithm, made the data structure complex. Guodong Ye [6] presented a designed edge marking algorithm aiming at elongated polygon, but the efficiency of this algorithm isn't high while using the operation of intersection. This paper presents an improved algorithm that uses a measurement variable as the flag of boundary, whose initial value is 0. In the process of rasterizing the boundary, the flag of the current point will add one when it is accessed. The method distinguishes the normal boundary points from singular points and elongated points is very easy and it not only keeps the advantage of the traditional edge marking algorithm but also solves the anomalies when filling the special point effectively.

108.2 The Traditional Edge Marking Algorithm

The traditional edge marking algorithm is very easy to be understood as follows: first, marking the boundary and, second, filling the polygon, pixel by pixel in accordance with the order from the left point to the right for each scanning line intersecting with the boundary of the polygon. It is judged whether the pixel point is in the interior of the polygon or not. If the pixel point is in the interior of the polygon, fill it with filling color; otherwise, fill it with background color. But an abnormal phenomenon will appear when the polygon has local points and elongated points. In this process, a Boolean variable INSIDER marks whether the point is in the polygon or not. If the initial value of the INSIDER is FALSE, then the value of the INSIDER will be negated when the scanning line encounters the boundary point, and if the value of the INSIDER is TRUE, then the color of the point is set with filling color.

108.2.1 Singular Point

The general method to judge whether a local point is a singular point is as follows: assuming an intersecting point of A, if the two sides of another vertex are located on

Fig. 108.1 The map of the local point and elongated points

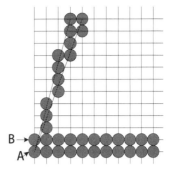

Fig. 108.2 Filling of local point in polygon

the same side of the intersection of A, the intersection is a singular point. Otherwise, it isn't a local point. In Fig. 108.1, point A is a singular point; when the scanning line encounters point A, the value of the INSIDER is TRUE, and then the pixels after point A will be filled, resulting in abnormal filling. Such a polygon (10, 10), (50, 120), (90, 10), (50, 70) which has local points and is filled by traditional edge marking algorithm is shown in Fig. 108.2. And so, we make mistake for it strongly.

108.2.2 Elongated Point

The problem of the elongated polygon is that when the slope of the two edges of the polygon is so close, the polygonal area is too thin and narrow to fit in this region at any point. In Fig. 108.1, point B is an elongated point. Then the pixels will be filled after point B, resulting in abnormal filling. Such an elongated polygon (50, 25), (50, 225), (90, 225), (90, 125), (55, 125) is shown in Fig. 108.3. The points near the highest point of the polygon should not be filled, but they are filled by traditional edge marking algorithm. Obviously, we also make mistake for it.

Fig. 108.3 Filling of elongated polygon

108.3 The Improved Edge Marking Algorithm and Flow Diagram

Based on the above descriptive analysis, the improved edge marking algorithm is explained in this section. The reason of the abnormal filling phenomenon is that the traditional algorithm can't make the distinction between the ordinary points and the local points and elongated points. It presents an easy method to distinguish them. And what's more, in the system we use a memory to store the flags of the boundary points.

As shown in Fig. 108.4, the main processes are as follows: Firstly, we can get the pixels of the polygon. Secondly, when scanning the boundaries of the polygon, we can get the maximum and minimum values of the transverse and longitudinal coordinates of the vertices of the polygon; they are defined as xmin, xmax, ymin, and ymax. Thirdly, in the process of rasterizing, read the flag of the memory, get the flag of the current point, and add one before storing it again in the memory. So when we begin to scan the polygon, we must initialize the scanning line-scan_y with ymin, and scan_y increases by 1 from ymin to ymax. If y is less than y_max, then set the value of the INSIDER to FALSE and then initialize x with x_min. In the next step, if the value of the flag is greater than one, then set the color of the point with filling color and the value of the INSIDER remains unchanged; otherwise, if the value of the flag of the current pixel is 1, then judge the values of the prior and next pixels. If both of them are 1, then the value of the INSIDER remains unchanged; otherwise, invert the value of the INSIDER. If the value of the INSIDER is TRUE, then set the color of the current pixel with filling color; otherwise, set the color of the current pixel with background color. Take point A as an example. When the scanning line encounters A, because the value of flag of A is two, greater than one, we can set the color of A with filling color and the value of the INSIDER remains unchanged; the points at the back of A are not filled.

Fig. 108.4 The flow chart of the improved algorithm

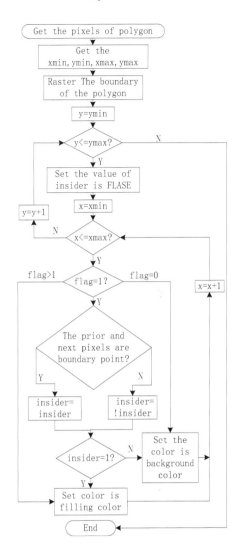

108.4 Result

The improved algorithm has been verified by the software. The software tool is VC++ 6.0. The filling effect of the algorithm, which has been improved, is shown in Figs. 108.5 and 108.6.

From the effect of the polygon filling with improved algorithm, we can see that the improved algorithm is correct. Take point A as an example. When the scanning line meets with A and B, as shown in Fig. 108.1, the value of the INSIDER remains unchanged, so the colors of the point at their back are filled with background color and their colors are filled with filling color because their flags are greater than 1.

Fig. 108.5 Filling of elongated polygon with improved algorithm

Fig. 108.6 Filling of local point in polygon with improved algorithm

108.5 Conclusion

The traditional edge marking algorithm is simple and easy to be implemented by hardware, and it has the advantages of needless to build, keep and order the edge tables. What we can see from the implementation of this new improved algorithm is that not only does it avoid the complex data structure called edge table, which is always used in the scan line fill algorithm, but also it avoids abundant intersection operation. All of them can take up a lot of CPU time, and in this algorithm, we only use the addition operation, so a conclusion can be made that this improved algorithm put forward by this paper is suitable for hardware implementation.

References

1. Wu, Z., et al. (2003). Singular point distinguishing algorithm for area filling. *Journal of Computer-Aided Design & Computer Graphics, 15*(8), 979–983 (In Chinese).
2. Wang, X., & Yan, B. (2004). An improved algorithm of edge marking fill. *Computer Applications, 24*(6), 182–183 (In Chinese).
3. Hao, X. (2006). The common problems and solutions of boundary labeling method in the course of realization. *Journal of Xi'an University of Engineering Science and Technology, 20*(10), 643–645 (In Chinese).
4. Zhang, Z., Liu, X., Zhang, Z., et al. (2009). Improved method for polygon scan conversion. *Computer Engineering and Applications, 45*(4), 193–195 (In Chinese).
5. Zhang, Z., Ma, S., & Li, W. (2009). New regional filling algorithm based on seed. *Computer Engineering and Applications, 45*(6), 201–202 (In Chinese).
6. Ye, G., Lin, G., Zhu, C. (2009). A designed edge marking fill algorithm for elongated polygon. *2009 First International Workshop on Database Technology and Applications* (pp 22–24).

Chapter 109
A Handheld Controller with Embedded Real-Time Video Transmission Based on TCP/IP Protocol

Mingjie Dong, Wusheng Chou, and Yihan Liu

Abstract Cross-platform video transmission is of vital importance in industrial applications. In this paper, we introduce a method for transmitting video from the computer with Windows system to the ARM11 board with embedded Linux system using the Ethernet based on the TCP/IP protocol. The ARM11 board is used as the server to receive video information using its Qt GUI, while the computer on the bank is used as the client that receives video information from the remote-operated underwater vehicle showing with its MFC (Microsoft Foundation Classes) interface and then sends the video information to the handheld controller. The image gained from the computer MFC is JPG format, and after coding, the images are transmitted to the server on the handheld controller continuously. Then the Qt GUI receives the data and decodes the JPG images before displaying them on the screen. The transmission is based on TCP/IP protocol and an image parsing protocol made by us. After testing, the video image can successfully conduct real-time transmission and can meet the industry application.

109.1 Background Information

The project comes from the national 863 project—miniature underwater submarines. Because of the complex underwater environment of the nuclear power plant, the ROV (remote operated vehicle) must be controlled on the bank, and the information of the reactor pool gained by the camera attached to the ROV will be transmitted to the control box made of an IPC (Industrial Personal Computer) with Windows operation system on the bank. Considering the big size of the control box, we need to develop a portable handheld controller which can replace the control box to a certain extent. The video image information will be transmitted to the

---

M. Dong (✉) • W. Chou • Y. Liu
Intelligent Technology and Robotics Research Center, School of Mechanical Engineering and Automation, Beijing University of Aeronautics and Astronautics, Beijing 100191, China
e-mail: buaadmj@gmail.com

system box from the camera on the ROV, and then the control box will transmit the video image information to the handheld controller directly. The latter is just what we will talk on this paper.

The video transmission is based on the TCP/IP protocol [1]. The control box will be the client, while the handheld controller will be the server during the video transmission. The transmission is through Ethernet using network cable, so the transmission is very fast and can be real time after a certain optimization.

109.2 Introduction of the Hardware Platform

The handheld controller is made of an ARM11 development board with an S3C6410 platform. The S3C6410, whose CPU can be up to 667 MHz, is a 32-bit RISC microprocessor which is designed by Samsung to provide a cost-effective, low-power capabilities and high-performance processor solution for mobile devices [2]. During the research, a Linux operation system with kernel number 2.6.38 is transplanted to it. After cross-compiling using arm-Linux-gcc cross-compiler with version number 4.5.1, the program written on the host machine can run successfully on the development board [3]. Given its excellent property, the development board is right for video transmission (Fig. 109.1).

109.3 Software Application Platform Building

The software application platform is composed of two different parts, the software on the Windows system and the software on the embedded Linux system.

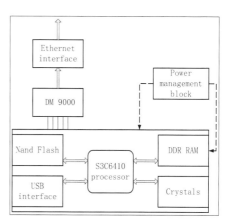

Fig. 109.1 The hardware platform of the handheld controller

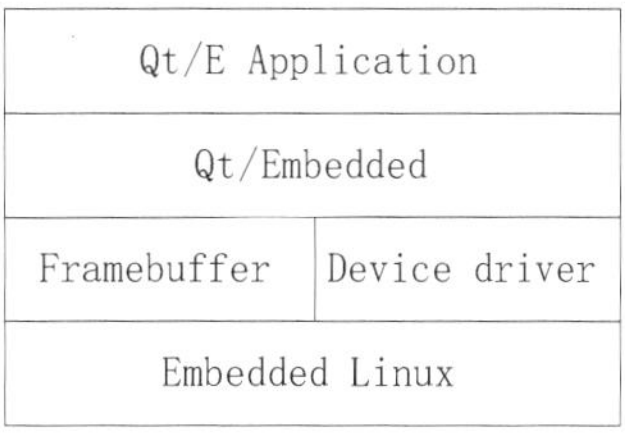

Fig. 109.2 Qt/E mechanism structure diagram

Qt/E Application	
Qt/Embedded	
Framebuffer	Device driver
Embedded Linux	

109.3.1 Software on the Windows System

The software on the Windows system refers to the platform using MFC built on the control box with the Windows system. This part shows the video image transmitted from the camera attached on the ROV. The MFC program uses the SDK functions of the image capture card to display the images. Meanwhile, it is used as the client of the network video transmission through socket based on TCP/IP protocol.

109.3.2 Software on the Embedded Linux System

In the ARM11 development board, a Linux system is transplanted with the kernel version number 2.6.38, and then we successfully setup the cross-compiler arm-linux-gcc with the version number 4.5.1 on it. With all above settled, we set up Qt/E (QT embedded) on the development board in order to use its convenient GUI to decode the JPG image and display the video information on the screen.

Qt/E is an open-source software development kit provided by the software developer Trolltech, with C++ program language as its development tools. Qt/Embedded has a lot of simple class libraries and interface modification tools. Especially, it has unique signals and slots mechanism [4]. It is across platforms and can be easily transplanted. The simple Qt/E mechanism is showed in Fig. 109.2.

109.4 Application of TCP/IP Protocol

Based on the TCP/IP protocol, we use socket as the application programming interface. The socket programming is based on the system call of the socket. At first, we use the function socket () to create a new socket, and then we connect the socket address with the socket we just created using bind (). In order to build the socket connection between the client and the server, we should use connect () and accept () functions. The function connect () is used to build connection, while the function accept () is used by the server, waiting for the connection request from the client. Meanwhile, the server uses the function listen () to monitor whether there is a

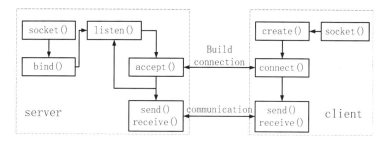

Fig. 109.3 The socket's working procedure

client request. Once a connection between the client side and the sever side is built successfully, we can exchange data from the client and the server [5]. The socket's working procedure is in Fig. 109.3 [6].

109.4.1 Socket on the Server Side

The server is built on the development board with embedded Linux operation system using the Qt/E GUI. We use QTcpServer, a Qt class, to build a sever class. Then, we use the function listen () to monitor the port we set at the beginning [7]. Because of the high encapsulation of Qt, we can save a lot of trouble. Once the server side gets the connection request from the client, the server will accept the request and the Qt function readyread() will be called, and then in our program, the function dataReceived() written by us will run according to our settled signals and slots mechanism. The data from the client will be gained successfully.

109.4.2 Socket on the Client Side

On the client side, as we use the MFC in the visual C++6.0 environment, we should use the Windows socket API to build the socket and build the connection with the server in the Linux operation system. At first, we use socket (AF_INET, SOCK_STREAM, 0) [8] to create a new socket based on the TCP/IP protocol. Then we give the address variables a certain value to make sure the socket address is connected with the socket we created. Lastly, the client will send a connection request to the server using the function connect (). When the server accepts the request, the connection is finished.

109.5 Protocol Development of the Video Transmission

For the data to be transmitted in the form of binary, so after capturing the images information from the camera attached to the ROV, we will change the image format from JPG to binary streams. In order to get the images streams coded in the client side and decoded in the server side securely and to accept the whole image streams in the development board smoothly, we should set up our own video transmission protocol.

109.5.1 Image Acquisition and Coding on the Client Side

Every time we capture an image, the image will be saved in the form of binary with a temporary name given by us. Then we make a protocol for the whole image transmission, using the format "AA:file name*length:EE" in order to identify that we get a whole image all the times. The file name is the temporary name we give after an image is captured, while the length is its length of binary form. Using a while structure loop, the images will be sent to the server continuously.

109.5.2 Image Receiving and Decoding on the Server Side

In the server written by us using Qt GUI, a lot of data will be received. At first, we estimate whether a set of data the server received is a whole image through the image transmission protocol we made. If the data meet the format "AA:file name*length:EE", the video transmission protocol made by us, we can make sure that the data the server received is a whole image transmitted from the client, and then the image will be displayed on the screen. Just as before, using a while structure loop, we decode and display the images continuously. Because of the fast flow of the continuous images using network cable and the excellent property of the development board, we can watch the video information smoothly (Fig. 109.4).

109.6 Experimental Verification

After building the environment on both the Windows system and the embedded Linux system, we programmed in the client side and the server side, respectively, based on the video transmission protocol made by us and the TCP/IP protocol. Then we test the program, and the result is very satisfying. The video information captured from the client side can be transmitted to the server side on the ARM11 development board smoothly.

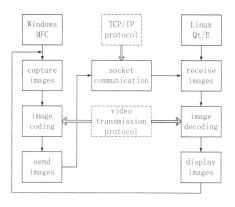

Fig. 109.4 The video transmission procedure

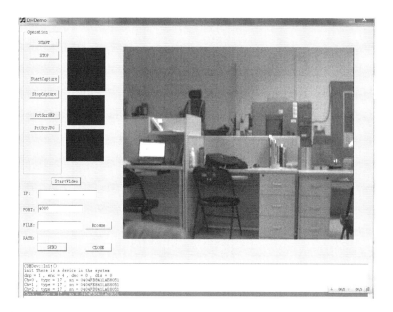

Fig. 109.5 The video information running in the client side on the control box

In order to test whether the video transmission is stable and to gain the transmission speed, we set up the function frame_test() in the Qt GUI. By calling the Qt class libraries with the function, we find that the transmission is very stable and the speed can be up to 24 frames per second. It is real time and can meet the industry application perfectly. The effect picture of the video transmission is shown in Figs. 109.5 and 109.6.

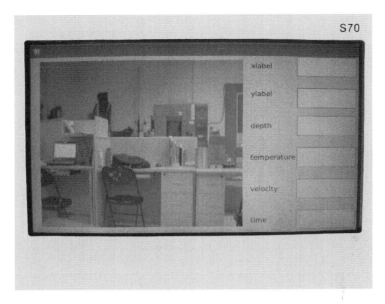

Fig. 109.6 The video information running in the server side on the handheld controller

109.7 Conclusion

Summarily, this paper describes an effective method to transmit video image information across the different operation system based on the TCP/IP protocol. The images transmission protocol we made is simple and useful, while the cost of the whole handheld controller is very low and the video transmission is real time; it can easily meet the needs of the project.

References

1. Wilder, F. (1993). *A guide to the TCP/IP protocol suite* (pp. 3–21). Boston, MA: Artech House.
2. Friendly ARM. (2011). *Tiny6410 hardware manual*. Retrieved August 5, 2011, from http://www.arm9.net
3. Liu, Z. (2012). Application of embedded database in room environment supervisory system. *Modern Electronics Technique, 35*(10), 2–3.
4. Bangwei, Y., & Deng, H. (2011). Control interface of embedded digital monitoring system based on Qt/embedded. *Part and Application, 35*(24), 1–2.
5. Anand Kumar, M., & Karthikeyan, S. (2011). Security model for TCP/IP protocol suite. *Journal of Advance in Information Technology, 2*(2), 1–5.
6. Zhang, Y.-G., Liu, C.-C., Liu, W., & He, F.-Z. (2006). Remote monitoring and control system based on socket and multithread. *Control Engineering of China, 13*(2), 2–3.

7. Blanchette, J., & Summerfield, M. (2008). *C++ GUI programming with Qt4* (2nd ed., pp. 295–303). Beijing, China: Publishing House of Electronics Industry.
8. Tian, L. (2012). Network communication technology based on Winsock. *Hubei: Software Guide, 11*(1), 1–4.

Chapter 110
Evaluating the Energy Consumption of InfiniBand Switch Based on Time Series

Huifeng Wang, Zhanhuai Li, Xiaonan Zhao, Qinlu He, and Jian Sun

Abstract Recently, energy consumption has emerged as a critical factor in designing storage system. In order to test the energy consumption of InfiniBand switch (IB switch), we establish an energy consumption model for IB switch and formulate the test cases. Using the method, you can obtain the energy consumption of the IB switch scientifically and efficiently. Empirical results illustrate the correctness of the energy consumption and reflect the distribution laws of the energy consumption of the IB switch clearly. The scheme can solve the problem of testing and analyzing the energy consumption of the IB switch efficiently. It has positive practice significance to reduce the cost of storage system.

110.1 Introduction

With the development of modern information technology, energy consumption accounts for a significant fraction of cost of data centers. As demonstrated by the successful emergence of the Green500 list [1], energy consumption has become as important as performance. Storage subsystems alone represent roughly 10–25 % of the power consumed by the data center [1]. Energy consumption of storage system can become a greater problem. So it is very meaningful to study on the energy consumption in order to reduce the cost of storage system.

In recent years, high-speed communication protocols and their supporting hardware have emerged, which can provide very low latency and very high bandwidth. As a result, they have become increasingly popular in the areas of high-performance computing (HPC) and enterprise computing where communication is critical to application performance [2]. InfiniBand switch (IB switch) is one of the

H. Wang (✉) · Z. Li · X. Zhao · Q. He · J. Sun
School of Computer Science and Technology, Northwestern Polytechnical University,
Xi'an Shaanxi 710129, China
e-mail: wanghuifeng12@163.com

central devices, which is capable to interconnect a large amount of nodes (large port count). The entire data transfer between servers and storage devices through IB switch. The energy consumption of IB switch is one of the most important sources the storage system consumes. Calabretta et al. [3] study on the relationship between energy consumption and latency. Fu et al. [4] propose a novel router architecture to allow each of its modules to adjust frequency according to traffic loads in order to gain energy efficiency. Vishwanath et al. [4, 5] explore buffer usage of routers and develop a buffer adapting algorithm for letting SRAM and DRAM buffers sleep while not being used.

We can see that the existing studies focus mainly on how to reduce the energy consumption of the storage system. There is little study on how to test the energy consumption of switch. This paper provides a solution to test the energy consumption of the IB switch and make a detailed analysis about the test results. Everyone can use the solution to obtain the distribution of the energy consumption of the IB switch. The test scheme we designed makes the test case ordered. The former test is the condition of the latter test case. The order not only can finish the test the energy consumption of the IB switch but also can make the test data easily and scientifically. The research on energy consumption of InfiniBand switch has positive practice significance on the construction of storage system. In order to describe easily, we abbreviate the InfiniBand switch as IB switch.

The rest of this paper is organized as follows. In Sect. 110.2, we introduce the test content and test methodology. In Sect. 110.3, we describe our test environment and make a detailed analysis on the experiment data of energy consumption of IB switch. We conclude and describe our future work in Sect. 110.4.

110.2 Methodology

In this section, we mainly introduce test contents and present the methodology we use. The paper mainly discusses the energy consumption distribution of the IB switch and investigates whether the workload on servers can make some effect on the energy consumption of IB switch. In order to collect the data correctly, we design a brief solution to test the energy consumption of the IB switch.

110.2.1 Test Topology Graph

We briefly introduce the topology graph about connection between the devices. Figure 110.1 shows that servers connect to the IB switch through IB line and the power the IB switch uses must pass the power analysis tool to provide power supply for IB switch. The power analysis tool has three ports to connect with other devices. The first port of the power analysis tool ties to power supply. The second port is connected to the terminal pc which is used to collect data of the energy consumption. The third port ties to the IB switch.

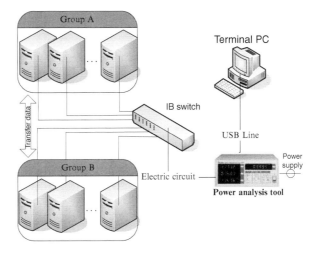

Fig. 110.1 The topology graph of the test system

110.2.2 Energy Consumption Model

In order to make a quantitative analysis, we establish an energy consumption model for IB switch, which relates to the construction and working process of the IB switch. It describes as follows:

$$E_{sw} = E_{base} + E_{workload} + \sum E_{port} \quad (110.1)$$

The energy consumption of IB switch derives from three parts, which are the base energy consumption, the energy consumption of transferring data, and the energy consumption of IB cards. The three parts are described as E_{base}, $E_{workload}$, and ΣE_{base}, respectively.

Although the model roughly describes the distribution of the energy consumption of IB switch, we need to study the proportion of every part to the total energy consumption of IB switch. The proportion is very helpful to evaluate the total energy consumption of IB switch.

110.2.3 Test Case

In order to make the test scientifically and efficiently, we design the test cases as described in Table 110.1. The design not only can save test time but also can make the test data more accurate. The former test case is the precondition and the basis of the latter test case. Using the results of former test case, we can only calculate the subtraction of the consecutive test cases. It need not test the starting value of every test case and also can make the change of two consecutive test cases to not too largely influence the experimental precision.

Table 110.1 Test cases

Test case	Abbreviation	Description
1	Basis test	Test the energy consumption of the idle IB switch
2	Full-switched test	Test the energy consumption of IB switch in full-switched
3	IB cards test	Test the energy consumption of the IB cards
4–8	Workloads test	Test the influence of workloads of servers to the energy consumption of IB switch

We design eight test cases to complete the test task. The test cases are in order. If you conduct the test cases in order, you can easily obtain the energy consumption of IB switch and make a detailed analysis.

Test case 1 is the basis test for the energy consumption of IB switch. The method is to power on the IB switch which is not connecting any IB card. The result of the test case is the energy consumption of the idle IB switch. Other test cases draw their conclusions through comparing with the result of the basis energy consumption.

The mission of test case 2 is to test the energy consumption of IB switch in full-switched. In test case 2, the internal ports transfer data mutually all the time. The result of the full-switched test reflects the energy consumption of switching data.

In test case 3, we connect 20 IB cards to the IB switch and stop the fully switching the data. The purpose of this test case is to collect the energy consumption of the IB cards.

Test cases 4–8 are a series of test cases. Its task is to inspect the influence of workloads of servers to the energy consumption of IB switch. In these five test cases, we connect some servers to the IB switch through IB cards. These servers are divided into groups. At any test case, we make any group switch data between two servers through the IB switch.

Through conducting basis test and full-switched test, we can obtain the energy consumption of full-switched in the IB switch. We can gain the energy consumption of the IB cards by comparing test case 1 and test case 3. Through conducting test cases 4–8, we can obtain the discipline about the energy consumption of workloads to the IB switch.

110.3 Experiment and Analysis

The IB switch we study is used by Inspur Corporation. The switch has 324 ports. The server connected to the IB switch is the model of AS300N. The test equipment of energy is Chroma Digital Power Meter 66202. The tool can easily collect power of the tested object in real time. In order to describe easily, we abbreviate average power as ap.

Fig. 110.2 The result of test case 1

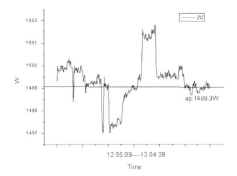

As described in Sect. 110.2.3, the energy consumption of IB switch derives from three parts. We conduct the eight test cases to calculate the E_{base}, $E_{workload}$, and ΣE_{base}, respectively, to obtain the distribution law of the energy consumption of the IB switch.

110.3.1 Test Case 1

We power on the IB switch which is not connecting any IB card to test the E_{base}. Figure 110.2 shows the result of test case 1, which is the energy consumption of the idle IB switch. We get the E_{sw1} which is the average power of test case 1. It is about 1499.3w. The E_{sw1} is equal to E_{base}. It is very simple but indispensable and very important because it is the standard basis of the other test cases. Other test cases draw their conclusions through comparing with the result of the basis energy consumption.

110.3.2 Test Case 2

The difference between test case 1 and test case 2 is that whether the IB switch is full-switched or not. We conduct test case 2 with the similar method of test case 1. The E_{sw2}, which is the ap of test case 2, is about 1757.6w. Based on the energy consumption model, we can get the largest energy consumption of switching data as follows:

$$E_{sw2} = E_{base} + E_{workload\_max} \quad (110.2)$$
$$E_{workload\_max} = E_{sw2} - E_{base} = E_{sw2} - E_{sw1} = 258.3w \quad (110.3)$$

Under normal circumstances the energy consumption is usually smaller than the value.

Fig. 110.3 The result of test cases 4–8

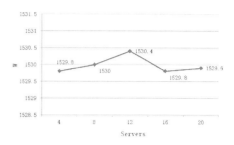

110.3.3 Test Case 3

In test case 3, we connect 20 IB cards to the IB switch. We obtain that the E_{sw3}, which is the ap of test case 3, is about 1528.6w. So we can get the energy consumption of each IB card as follows:

$$E_{sw3} = E_{base} + \sum_{1}^{20} E_{port} = E_{sw1} + \sum_{1}^{20} E_{port} \qquad (110.4)$$

$$E_{port} = \frac{E_{sw3} - E_{sw1}}{20} = 1.465w \qquad (110.5)$$

We can see that the energy consumption of each IB card is a small proportion of the total energy consumption of IB switch. However, when we insert all the IB cards to the IB switch, the energy consumption of IB cards cannot be ignored.

110.3.4 Test Cases 4–8

Figure 110.3 demonstrates the energy consumption of the IB switch when transferring data between the servers connected to the IB switch. In each experiment we add a group to transfer data. Each group has four servers and transfers data through the IB switch. We can find out the ap is a little changeable in the serials of experiments. The ap is about 1530w. It reflects that the energy consumption of transferring data is a small proportion of the total energy consumption of IB switch. In other words, transferring data has little impact on the energy consumption of the IB switch.

According to the empirical results, we can evaluate the energy consumption of IB switch. If n IB cards connect to the IB switch, we can obtain the E_{sw} as follows:

$$E_{sw} = E_{base} + \sum_{1}^{n} E_{port} = 1499.3 + 1.465n \qquad (110.6)$$

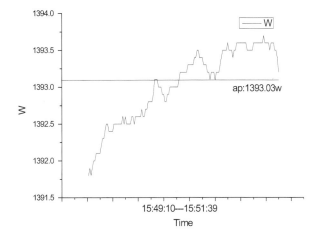

Fig. 110.4 The result of energy consumption IB switch provided by the Chinese Academy of Science

We also can evaluate the largest energy consumption of the IB switch as follows:

$$E_{sw} = E_{sw2} + \sum_{1}^{324} E_{port} = 1757.6 + 1.465 * 324 = 2232.26w \quad (110.7)$$

Using the method, we also test the IB switch provided by the Chinese Academy of Science. The energy consumption of the idle IB switch is about 1196.3w. When inserting six IB cards to the IB switch, the energy consumption grows to 1210w. As described in Fig. 110.4, when conducting full-switched test case, the ap of the IB switch is about 1393.03w. Empirical results show the same distribution law of the energy consumption of the IB switch. So adopting the solution to study on the energy consumption of IB switch has positive practice significance to reduce the cost of storage system.

110.4 Conclusion and Future Work

The series of experiments draw generally two conclusions. Firstly, the energy consumption of IB switch is mainly from three aspects, respectively, the energy consumption of normal run, the energy consumption of IB cards, and the energy consumption of communicating between internal ports. Secondly, the energy consumption of transferring data between servers connected to the IB switch is a small proportion of the total energy consumption of IB switch.

In the future, we will change the place where the IB ports connect to the IB switch to observe the law whether the port the server connect to the switch can have some effect on energy consumption. We plan to collect a long-term data of energy consumption of IB switch to observe the law of the IB switch energy consumption.

Acknowledgements This research program has been supported by the Ministry of Science and Technology of RPC (863 Program:2013AA01A215), the NPU Fundamental Research Foundation under Grant No.JC20120209, the National Key Technology Research and Development Program of the Ministry of Science and Technology of China under Grant No.2011BAH04B05, and the National Natural Science Foundation of China under Grant No.61033007.

References

1. Prada, L., García, J., Calderón, A., García, J. D., & Carretero, J. (2013). A novel black-box simulation model methodology for predicting performance and energy consumption in commodity storage devices. *Simulation Modelling Practice and Theory, 34*, 48–63.
2. Liu, J., Poff, D., & Abali, B. (2009). Evaluating high performance communication: A power perspective[C]. *Proceedings of the 23rd International Conference on Supercomputing* (pp. 326–337). Yorktown Heights, NY: ACM.
3. Calabretta, N., Luo, J., Lucente, S. D., & Dorren, H. (2012). Experimental assessment of low latency and large port count OPS for data center network interconnect[C]. *Transparent Optical Networks (ICTON), 2012 14th International Conference on* (pp. 1–4). Coventry: IEEE.
4. Fu, W., & Song, T. (2012). A frequency adjustment architecture for energy efficient router [C]. In *ACM SIGCOMM Computer Communication Review* (Vol. 42, pp. 107–108). Helsinki, Finland: ACM.
5. Vishwanath, A., Sivaraman, V., Russell, C., Zhao, Z., & Thottan, M. (2011). Adapting router buffers for energy efficiency[C]. *Proceedings of the Seventh Conference on Emerging Networking Experiments and Technologies* (pp. 1–12). New York: ACM.

Chapter 111
Real-Time Filtering Method Based on Neuron Filtering Mechanism and Its Application on Robot Speed Signals

Wa Gao, Fusheng Zha, Baoyu Song, Mantian Li, Pengfei Wang, Zhenyu Jiang, and Wei Guo

Abstract In order to implement the real-time filtering and tracking of robot signals with high efficiency, a novel real-time filtering method based on neuron filtering mechanism is developed in this paper. By considering the ubiquity of resonance in mammal and combining the mechanism of neural information processing, the derived details and the feasible parameter criterion under minimum error variance condition are given. For illustration, the application on quadruped robot is discussed. The quadruped robot feet speed signals are processed by developed real-time filtering method and Kalman filtering algorithm, respectively, and the computation time of both methods is tested. Experiment results show that the performance of developed real-time filtering method is better than that of Kalman filtering algorithm, not only in filtering and tracking performance but also in filtering speed. The novel real-time filtering method based on neuron filtering mechanism can effectively implement the real-time filtering and tracking with regard to robot signals.

111.1 Introduction

The mechanism of biological neuron filtering is being a subject of intense study since several years. Since the work of Hubel et al. [1], the filtering properties of single neurons in auditory cortex have been known and attract the interests of professionals [2, 3]. Recently, it is well known that biological neurons produce temporal filtering property and exhibit responses to vibration stimuli in neuron [4, 5]. It is a real-time

---

W. Gao · F. Zha · M. Li (✉) · P. Wang · Z. Jiang · W. Guo
State Key Laboratory of Robotics and System, Harbin Institute of Technology,
Harbin 150006, China
e-mail: skymoon.hit@gmail.com

B. Song
Department of Mechanical Design, Harbin Institute of Technology, Harbin 150006, China

information transferring process in mammal. Accordingly, it can inspire us to develop a novel filtering approach from the physiological standpoint.

Besides, vibration is ubiquitous in mammal and resonance phenomena can be widely found in vibrations such as the beating of a heart, the conveying of a sound, and the breathing of mammals [6, 7]. Resonance, which refers to the tendency of a system to oscillate with greater amplitude at some frequencies than at others, provides filtering ability by generating vibrations of a specific frequency or picking out specific frequencies from a complex vibration containing many frequencies [8, 9]. Hence, it is available to probe real-time filtering method starting from vibration.

In this paper, we are interested in developing a real-time filtering method by using biological neuron filtering mechanism. The paper is organized as follows. Firstly, the derived details and the feasible parameter criterion under minimum error variance condition are presented. Secondly, experiments of quadruped robot feet signals are given, and comparisons between developed real-time filtering method and Kalman filtering algorithm are discussed. Then, the paper is completed with some concluding remarks.

111.2 Real-Time Filtering Method Based on Neuron Filtering Mechanism

111.2.1 Methods

Generally, the single-degree-of-freedom vibration system can be seen as a typical and basic unit in vibration domain, and its mathematical expression is

$$f'' + \alpha f' + \beta f = F \qquad (111.1)$$

where f is the system displacement output, F is the external excitation, and α, β are the system parameters, respectively. Denote $F = \beta u$, Eq. 111.1 can be rewritten as

$$f'' + \alpha f' + \beta f - \beta u = 0 \qquad (111.2)$$

where u represents the system input. We can solve Eq. 111.2 by the following form:

$$\begin{cases} f' = x \\ x' = \beta u - \beta f - \alpha x \end{cases} \qquad (111.3)$$

where x represents the system velocity. Equation 111.3 represents the state function of single-degree-of-freedom vibration system. Define $n \in N$, and N represents the set of integers. Denote the sampling instant t_n and the sampling interval $\Delta t = t_n - t_{n-1}$. When the sampling interval Δt approaches zero, Eq. 111.3 can be discretized as Eq. 111.4.

$$\begin{cases} f_{n+1} = x_n \\ x_{n+1} = \beta u_n - \beta f_n - \alpha x_n \end{cases} \quad (111.4)$$

Then, we can obtain the numerical solution of Eq. 111.4 as follows:

$$\begin{cases} f_{n+1} = f_n + \Delta f_n \\ x_{n+1} = x_n + \Delta x_n \end{cases} \quad (111.5)$$

where

$$\begin{cases} \Delta f_n = (K_1 + K_2)/2 \\ \Delta x_n = (L_1 + L_2)/2 \end{cases} \text{ and } \begin{cases} K_1 = x_n \\ K_2 = x_n + hL_1 \\ L_1 = \beta u_n - \beta f_n - \alpha x_n \\ L_2 = (1 - \alpha h)L_1 - \beta h K_1 \end{cases} \quad (111.6)$$

where $h > 0$ is the step of numerical solution.

Information in neuron is transferred by synapses. The generations of temporal filters are shown among different synapses, and there exist information exchange processes while neural information transferring [10, 11]. As referred in ref. [12], we assume that Eq. 111.5 is a neural information delivery system, Δf_n and Δx_n are information segments transferred by different synapses. Hence, Δf_n and Δx_n shall exchange to complete the information transferring process. Thus,

$$\begin{cases} f_{n+1} = f_n + \Delta x_n \\ x_{n+1} = x_n + \Delta f_n \end{cases} \quad (111.7)$$

Define

$$\begin{cases} a = -\beta h^2/2 \\ b = h - \alpha h^2/2 \end{cases} \quad (111.8)$$

Substituting Eqs. 111.6 and 111.8 into Eq. 111.7 yields

$$\begin{cases} f_{n+1} = (1 - \beta b)f_n + (a - \alpha b)x_n + \beta b u_n \\ x_{n+1} = a f_n + (1 + b)x_n - a u_n \end{cases} \quad (111.9)$$

Equation 111.9 is the kernel filtering function of developed filtering method. α, β, a, and b are the parameters. Its derivation mainly depends on the principle of neuron filtering mechanism, and it is quite different from common filtering theory.

111.2.2 Preferences Under Minimum Error Variance Condition

Considering the filtering process and the observing process, we have

$$\begin{cases} X_{n+1} = \Phi_n X_n + B_n U_n \\ Z_{n+1} = H_{n+1} X_{n+1} + V_{n+1} \end{cases} \quad (111.10)$$

Denote

$$\Phi = \begin{bmatrix} 1 - \beta b & a - \alpha b \\ a & 1 + b \end{bmatrix}, B = \begin{bmatrix} \beta b \\ -a \end{bmatrix} \quad (111.11)$$

where $X_{n+1} = [f_{n+1}\ x_{n+1}]^T$ is the state sequence, Z_{n+1} is the observation sequence, Φ is the state transfer matrix, $H = [1\ 0]$ is the observation matrix, and V is the white noise caused by observation equipment with the expectation $E[V] = 0$ and the variance value $\sigma[V] = r$. Then, we can obtain the system step prediction equation:

$$\begin{cases} \hat{X}_{n+1,n} = \Phi_{n+1,n}\hat{X}_n + B_{n+1,n}U_n \\ \hat{Z}_{n+1,n} = H_{n+1}\hat{X}_{n+1,n} \end{cases} \quad (111.12)$$

where $\hat{X}_{n+1,n}$ and $\hat{Z}_{n+1,n}$ are the step predicted state value and the step observed value, respectively. Define the observed error as $\tilde{Z}_{n+1,n}$, and $\tilde{Z}_{n+1,n} = Z_{n+1} - \hat{Z}_{n+1,n}$. It can be used to upgrade the step predicted state value $\hat{X}_{n+1,n}$. Thus, we have

$$\hat{X}_{n+1} = \hat{X}_{n+1,n} + \tilde{Z}_{n+1,n} = \hat{X}_{n+1,n} + G_{n+1}\left(Z_{n+1} - H_{n+1}\hat{X}_{n+1,n}\right) \quad (111.13)$$

where G_{n+1} is the undetermined filtering gain matrix.

According to the linear minimum variance estimation method, the system error variance matrix function can be defined as follows:

$$J_{n+1} = E\left[\tilde{X}_{n+1}\tilde{X}^T_{n+1}\right] \quad (111.14)$$

where $\tilde{X}_{n+1}$ is the state error, and $\tilde{X}_{n+1} = X_{n+1} - \hat{X}_{n+1}$. Hence,

$$J_{n+1} = [I - G_{n+1}H_{n+1}]J_{n+1,n}[I - G_{n+1}H_{n+1}]^T + G_{n+1}J_{n+1}G^T_{n+1} \quad (111.15)$$

where $J_{n+1,n} = \Phi_{n+1,n}J_n\Phi^T_{n+1,n}$ is the step error variance matrix.

Minimizing the system error variance matrix J_{n+1}, we can obtain the optimized gain matrix $G = J_{n+1}H^T r^{-1}$. When the filtering process is stable, the error variance is constant value, i.e., $J_{n+1} = J$. Thus,

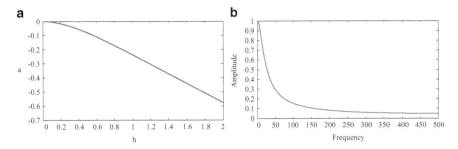

Fig. 111.1 (a) The relationship between parameter *a* and *h*. *h* is the step of numerical method. (b) The amplitude-frequency figure of developed real-time filtering method when the parameter combination is selected as Table 111.1

Table 111.1 The parameter combination

a	b	α	β
−0.002	0.003	10	350

$$J = (I - Jr^{-1})\Phi J \Phi^T \qquad (111.16)$$

Substituting Eqs. 111.8 and 111.11 into 111.16, we can yield the feasible parameter criterion as follows:

$$2a^3h^2 + 4a^2(h+1) + 2ah^2(h+2) + h^4 = 0 \qquad (111.17)$$

where h is the step of numerical solution. The relationship between parameter a and h can be seen clearly in Fig. 111.1a. According to the value of h, the acceptable parameter combinations under minimum error variance condition can be obtained by Eqs. 111.8 and 111.17, and h depends on different scenarios.

The data in Table 111.1 is one of the feasible parameter combinations which meet Eqs. 111.8 and 111.17, and the corresponding amplitude-frequency figure is shown in Fig. 111.1b. The parameter combination is selected by the motion properties of quadruped robot, and the details will be given in later section. Then, the above-proposed real-time filtering method (consists of Eqs. 111.8, 111.9, and 111.17) will be substantially testified.

111.3 Experiments

To validate the developed real-time filtering method, we compare it with the Kalman filtering algorithm by using speed signals of quadruped robot feet which are sampled by accelerometers. And then, we test the computation time of both methods in LPC2148.

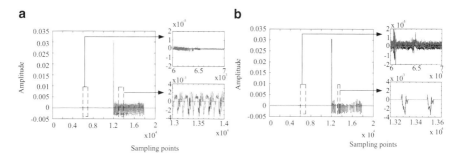

Fig. 111.2 (**a**) The sampled quadruped robot feet speed signals. Different color lines represent different robot feet. Figures on the *right* show local enlarge parts of the *left* figure. (**b**) The comparisons between the developed real-time filtering method and the Kalman filtering algorithm. The *black*, the *blue*, and the *red* represent the measured speed signal, the filtering result of Kalman filtering algorithm, and the filtering result of developed real-time filtering method, respectively. Figures on the *right* show local enlarge parts of the *left* figure

The measured quadruped robot feet speed signals are shown by different color lines in Fig. 111.2a. The sampling points are 20,000, and the sampling frequency is 200 Hz. The interference noises caused by robot natural vibration and signal mensuration process can be seen clearly from local enlarge figures, and the frequencies of interference noises are uncertain. Towards the sampled quadruped robot feet speed signals, the useful signals generated by the motion of quadruped robot feet are quite small, and the frequencies are less than 10 Hz due to the dynamic balance and self-weight of quadruped robot. It is an extreme narrow band-pass width of frequency. Hence, in order to acquire the useful signals and track accurately, h is demanded to select a small value. Here we choose h = 0.1. The corresponding parameter combination obtain by Eqs. 111.8 and 111.17 is listed in Table 111.1, and the amplitude-frequency figure is shown in Fig. 111.1b.

For the motor behaviors of robot feet are similar, we take one foot as an example. The filtering comparisons are shown in Fig. 111.2b. The black, the red, and the blue represent the measured robot foot speed signal, the filtering result of developed real-time filtering method, and the filtering result of Kalman filtering algorithm, respectively. Obviously, the filtering and tracking performance of developed real-time filtering method is better than that of Kalman filtering algorithm, and the wave form of filtered signal is much smoother. It indicates that the developed real-time filtering method implements filtering and tracking functions at the same time, while the Kalman filtering algorithm tends to track.

We also compare the computation time of two filtering methods in LPC2148 whose main frequency is 60 MHz. For 5,000 sampling points, the comparison result is shown in Table 111.2. It indicates that the computation time of developed real-time filtering method is about 3 times than that of Kalman algorithm under the same condition.

Table 111.2 The computation time in LPC2148 (milliseconds)

Developed real-time filtering method	Kalman filtering algorithm
2.75	7.75

From Fig. 111.2 and Table 111.2, the real-time filtering method based on neuron filtering mechanism, which shows better performance than Kalman filtering algorithm towards robot signals, meets higher real-time and better tracking and filtering requirements.

111.4 Conclusion

A real-time filtering method based on neuron filtering mechanism was proposed and applied to quadruped robot speed signals. Filtering function and preferences under minimum error variance condition are given, and the filtering and tracking performance are discussed compared with Kalman filtering algorithm. Besides, the computation time of the two methods are measured in LPC2148, and the results show that the filtering speed of developed real-time filtering method is several times faster than that of Kalman filtering algorithm.

Acknowledgements This work is partially supported by the National Natural Science Foundation of China (Nos. 60901074, 51075092, 61175107, 61005076), the National High Technology Research and Development Program ("863" Program) (No. 2007AA042105) of China, the Natural Science Foundation of Heilongjiang Province in China (No. E200903), and the State Key Laboratory of Robotics and System (HIT) (No. SKLRS200801B).

References

1. Hubel, D. H., Henson, C. O., Rupert, R., & Galambos, R. (1959). Attention units in the auditory cortex. *Science, 129*, 1279–1280.
2. Schnupp, J. (2006). Auditory filters, features, and redundant representations. *Neuron, 51*, 278–280.
3. Romanyshyn, Y., & Pukish, S. (2011). Filtering signals in models of neurons and neurthods in MEMS Design (MEMSTECH), Proal networks. Perspective Technologies and Meceedings of VIIth International Conference on. IEEE, (p.191).
4. Hrocholle-Bossavit, B., & Quenet, B. (2009). Neural model of frog ventilatory rhythmogenesis. *Biosystems., 97*, 35–43.
5. Xie, J. L., Wang, Z. J., & Shi, H. B. (2010). Effect of Synaptic Plasticity on Correlation between Neural Spike Trains. Information Engineering and Computer Science (ICIECS), 2nd International Conference on. IEEE, (pp.1–4).
6. Izhikevich, E. M., Desai, N. S., Walcott, E. C., & Hoppensteadt, F. C. (2003). Bursts as a unit of neural information: selective communication via resonance. *Trends in Neurosciences, 26*, 161–167.
7. Daniel, I. J. (2001). *Engineering vibration* (pp. 28–60). Upper Saddle River, NJ: Prentice Hall.

8. Thompson, W. T. (1996). *Theory of vibrations* (pp. 35–55). Cheltenham, England: Nelson Thornes Ltd.
9. Zhang, J. J., & Jin, Y. F. (2012). Stochastic resonance in FHN neural system driven by non-Gaussian noise. *Acta Physica Sinica, 61*, 13 (In Chinese).
10. Edwards, C. J., Alder, T. B., & Rose, G. J. (2005). Pulse rise time but not duty cycle affects the temporal selectivity of neurons in the anuran midbrain that prefer slow AM rates. *Journal of Neurophysiology, 93*, 1336–1341.
11. Xie, X. P., Song, D., Wang, Z., Marmarelis, V. Z., & Berger, T. W. (2006). Interaction of short-term neuronal plasticity and synaptic plasticity revealed by nonlinear systems analysis in dentate granule cells. Engineering in Medicine and Biology Society, 28th Annual International Conference of the IEEE, (pp.5543–5546).
12. Forture, E. S., & Rose, G. J. (2001). Short-term synaptic plasticity as a temporal filter. *Trends in Neurosciences, 24*, 381–385.

Chapter 112
Multiple-View Spectral Embedded Clustering Using a Co-training Approach

Hong Tao, Chenping Hou, and Dongyun Yi

Abstract It is a challenging task to integrate multi-view representations, each of which is of high dimension to improve the clustering performance. In this paper, we aim to improve the clustering performance of spectral clustering method when the manifold for high-dimensional data is not well defined in the multiple-view setting. We abstract the discriminative information on each view by spectral embedded clustering which performs well on high-dimensional data without a clear low-dimensional manifold structure. We bootstrap the clusterings of different views using discriminative information from one another. We derive a co-training algorithm to obtain a most informative clustering by iteratively modifying the affinity graph used for one view using the discriminative information from the other views. The approach is based on the assumption that the clustering from one view should agree with the clustering from another view. Comprehensive experiments on four real-world multiple-view high-dimensional datasets are presented to demonstrate the effectiveness of the proposed approach.

112.1 Introduction

In many important data mining applications, an instance may have multiple representations (views) from different feature spaces [1]. For example, a document can be translated to multiple languages, an Internet webpage can be represented as page-text as well as the hyperlinks pointing to it [2]. The phenomenal impact of multiple-view data in many applications has raised interest in the so-called multiple-view learning. Although each single view of the data might be sufficient for a given learning task, the complementary information of different views is ignored by learning from each view separately. So the main challenge of multiple-view

H. Tao (✉) • C. Hou • D. Yi
Department of Mathematics and Systems Science, National University of Defense Technology, Changsha 410073, China
e-mail: taohong08@sina.com

learning is to develop algorithms that can integrate complementary information of different views to improve the learning performance [3].

Co-training is the first algorithm to deal with multi-view data. It assumes each view of the data is sufficient for learning and the two views are conditionally independent. In the original co-training algorithm, a separate classifier for each view is learned using any labeled examples. The most confident predictions of each classifier on the unlabeled data are then used to iteratively construct additional labeled training data. This process should slowly drive the two classifiers to agree with each other on labels [1].

In this paper, we focus on multi-view clustering particularly in multi-view spectral clustering (SC). The available literature for this topic is growing with encouraging results [4–7]. However, the success of traditional SC methods is largely dependent on the manifold assumption, and this assumption does not always hold on high-dimensional data. When the data do not exhibit a clear low-dimensional manifold structure (e.g., high-dimensional and sparse data), the clustering performance of SC degrades. The spectral embedded clustering (SEC) [8] solves this problem in the single-view setting, but little work takes this issue into consideration in the multiple-view setting.

We propose a new approach named co-trained multi-view spectral embedded clustering (CoSEC) to solve the problem mentioned above. We assume that the clustering from one view should agree with the clustering from the other view and bootstrap the clusterings of different views using information from one another by co-training. In particular, we use the cluster assignment matrix obtained by the SEC algorithm on one view to modify the affinity graph used for the other view. By iteratively applying this approach, the clusterings of the two views tend to each other. And then we extend the proposed co-training framework for more than two views. The cluster assignment matrix on each view got by SEC algorithm can reflect the local and global structure information of the data. Then the co-training iterations retain the information needed for clustering and throw away the within-cluster details which might be confusing. These two factors result in a better clustering performance.

The rest of this paper is organized as follows. In Sect. 112.2, we will briefly review the SEC approach. Our proposed CoSEC framework is then presented in Sect. 112.3. Experiments on four real-world datasets are displayed in Sect. 112.4, and the conclusion remarks are given in Sect. 112.5.

112.2 SEC Revisited

SEC is a variant of SC methods to enhance the clustering performance on single-view high-dimensional data, motivated by the observation that the true cluster assignment matrix for high-dimensional data can be always embedded in a linear space spanned by the data. We assume $X = [x_1, x_2, \ldots, x_n] \in \mathbb{R}^{d \times n}$ is the high-dimensional data of only one view and Y is the corresponding cluster assignment matrix. Thus, there exist $W \in \mathbb{R}^{d \times c}$ and $b \in \mathbb{R}^{c \times 1}$ such that $Y = X^T W + \mathbf{1}_n b^T$. This equation usually holds for the high-dimensional and small-sample-size problem, which is usually the case in many real-world applications [9].

Suppose A is a symmetric matrix with each entry A_{ij} representing the affinity of a pair of data points; the normalized Laplacian graph L is then defined by $L = I_n - D^{-1/2} A D^{-1/2}$, where D is a diagonal matrix with the diagonal elements as $D_{ii} = \sum_j A_{ij}$, $\forall i$. Denote the relaxed cluster assignment matrix as $U \in \mathbb{R}^{n \times c}$ and the trace operator of a matrix A as $tr(A)$. SEC expects the learned U is close to a linear space spanned by the data X; thus, the optimization problem is [8, 10]

$$\min_{U^T U = I_c, W, b} tr(U^T L U) + \mu \left(\|X^T W + \mathbf{1}_n b^T - U\|^2 + \gamma tr W^T W \right) \quad (112.1)$$

where μ and γ are two trade-off parameters to balance three terms. In Eq. 112.1, the first term reflects the smoothness of data manifold, while the second term characterizes the mismatch between the relaxed cluster assignment matrix U and the low-dimensional representation of the data.

For simplicity, we assume the data is centered, i.e., $X\mathbf{1}_n = \mathbf{0}$. Set the derivatives of the objective function with respect to b and W to zeros; we have

$$b = \frac{1}{n} U^T \mathbf{1}_n \text{ and } W = \left(XX^T + \gamma I_d \right)^{-1} XU \quad (112.2)$$

Substitute W and b in Eq. 112.1 by Eq. 112.2; the optimization problem Eq. 112.1 becomes

$$\min_{U^T U = I_c} tr\left(U^T \left(L + \mu \tilde{L} \right) U \right) \quad (112.3)$$

where $\tilde{L} = H_n - X^T \left(XX^T + \gamma I_d \right)^{-1} X$, and $H_n = I - \frac{1}{n} \mathbf{1}_n \mathbf{1}_n^T$ is the centering matrix. The global optimal solution U^* to Eq. 112.3 can be relaxed as the eigenvector of $L + \mu \tilde{L}$ corresponding to the c smallest eigenvalues. Based on $Y \in \mathbb{B}^{n \times c}$, the discrete-valued cluster assignment matrix Y can be obtained by K-means or spectral rotation [11].

112.3 Co-training for Spectral Embedded Clustering

In this section, we will present our CoSEC algorithm, which aims to improve the clustering performance on the high-dimensional data without a clear manifold structure.

We cannot make use of the semi-supervised co-training directly because there is no labeled data in unsupervised learning problems. However, the motivation still remains: to restrict the classification (in our problem, clustering) in one view to be consistent with those in other views [6].

For a Laplacian matrix with exactly k number of connected components, the first k eigenvectors of it are the cluster assignment vectors. That is to say, these k eigenvectors only contain discriminative information about the different clusters, leaving out the details within the same cluster. In the case that the Laplacian matrix is fully connected, the eigenvectors are no longer the cluster assignment vectors, yet they still contain discriminative information which can be used for clustering [12]. In addition, the eigenvectors obtained by SEC capture local and global discriminative information because SEC balances the smoothness of data manifold (local) and the mismatch between the relaxed clustering assignment matrix U and the low-dimensional representation of the data (global). In multi-view setting, we exploit the eigenvectors obtained by SEC on one view to modify the graph structure in the other view, and vice versa.

How to modify the graph structure in one view using discriminative information from the other view? We consider each column $\mathbf{a}_i$ of the affinity matrix $A \in \mathbb{R}^{n \times n}$ as an n-dimensional vector that indicates the affinities of the ith point with all the points in the graph. The eigenvectors of the Laplacian matrix are vectors in the space spanned by these n affinity vectors. As above analyzed, the first k eigenvectors have the discriminative information for clustering, so projecting the affinity vectors along these directions (eigenvectors) can reserve the information needed for clustering and drop the within-cluster details that might confuse us. Then project them back to the original n-dimensional space, we get the modified graph. The inverse projection is easily finished by multiplying the transpose of the projection matrix thanks to its orthogonality [6].

Let us denote the data matrix, the affinity matrix, and the relaxed clustering assignment matrix on the vth view as $X^{(v)}$, $A^{(v)}$, and $U^{(v)}$ ($v = 1, 2$), respectively. Then the modified affinity matrixes on both views are $S^{(1)} = sym(U^{(2)}U^{(2)T}A^{(1)})$ and $S^{(2)} = sym(U^{(1)}U^{(1)T}A^{(2)})$, where $sym(S)$ is the symmetrization operator on a matrix S. We symmetrize $S^{(v)}$ because the projection of affinity matrix $A^{(v)}$ on the eigenvectors does not yield a symmetric matrix. And we add a rank-1 matrix to $sym(S)$ that has all its entries equal to the absolute value of the minimum negative entry of $sym(S)$. This makes sure that the corresponding Laplacian matrix is positive semi-definite at each iteration. We iteratively repeat this process by using $S^{(1)}$ and $S^{(2)}$ as new affinity matrixes to conduct SEC on each view to get new relaxed cluster assignment matrixes.

For the data have more than two views, we take the affinity matrix $A^{(v)}$ of a view and project it onto the union of subspaces spanned by top k discriminative eigenvectors of the other views, i.e., $S^{(v)} = sym((\sum_{i \neq v} U^{(i)}U^{(i)T})A^{(v)})$ for all the views [6]. Algorithm 1 gives a detailed description of the CoSEC algorithm.

Algorithm 1: CoSEC Algorithm

Input:
Data matrix $X^{(v)}$ and affinity matrix $A^{(v)}$ for each view, $v = 1, 2, \ldots, l$
The number of iteration *iter*, the number of clusters c
Output: Assignment to c clusters
Initialize: Compute the matrixes $L_0^{(v)} + \mu^{(v)}\widetilde{L}_0^{(v)}$, solve Eq. 112.3 by SEC, and obtain the optimal $U_0^{(v)}$, $v = 1, 2, \ldots, l$
for $i = 1$ to *iter* **do**
1: $S^{(v)} = sym((\sum_{j \neq v} U_{i-1}^{(v)} U_{i-1}^{(v)T}) A_{i-1}^{(v)})$, $v = 1, 2, \ldots, l$.
2: Use $S^{(v)}$ as the new affinity matrix, i.e., $A_i^{(v)} = S^{(v)}$, and compute the matrix $L_i^{(v)} + \mu^{(v)}\widetilde{L}_i^{(v)}$ and conduct SEC to obtain $U_i^{(v)}$, $v = 1, 2, \ldots, l$.
end for
3: Row-normalize $U_{iter}^{(v)}$.
4: Let $V = U_{iter}^{(v*)}$, where $v*$ is believed to be the most informative view. If there is no prior knowledge on the view informativeness, matrix V can also be set to column-wise concatenation of all $U_{iter}^{(v)}$s.
5: Based on V, compute the discrete cluster assignment matrix Y by using K-means clustering or spectral clustering.

112.4 Experiments

In this section, several experiments are conducted on four real-world datasets to compare our CoSEC approach with a number of baselines. The baselines are:

- Single view: The view that achieves the best SEC performance using a single view of the data, which is called the most informative view.
- Feature concatenation (FC): Concatenating the features of each view, and then running SEC using the joint view representation of the data.
- CCA-based feature extraction (CCA): Applying CCA for feature fusion from multiple views of the data [13] and then running SEC using these extracted features.
- Multi-view partitioning via tensor methods (TensorSC) [7]: Building up an affinity tensor from multiple affinity matrices and obtaining a joint optimal subspace by tensor decomposition, partitioning the subspace to obtain the cluster labels.
- Co-trained multi-view spectral clustering [6] (CoSC): This is most closely related to our approach. The only difference is that CoSC uses SC to obtain the relaxed cluster assignment matrix.

The four real-world datasets used in our experiments are the 3sources dataset, the UCI Handwritten digits dataset (Digits), the Internet advertisement dataset (AD), and the WebKB dataset. Some datasets are resized, and Table 112.1 summarizes the details of the datasets used in the experiments.

Table 112.1 Dataset description

Dataset	Size	Views	Classes
3sources	169	3	6
Digits	2,000	2	10
AD	3,264	2	2
WebKB	1,051	2	2

We employ the clustering accuracy (Acc) to evaluate the performance for all the clustering algorithm. Acc discovers one-to-one relationship between clusters and the true classes. It measures the extent to which cluster contains examples from the corresponding category.

We adopt the Gaussian similarity with cosine distance to construct the graph for spectral partition. That is to say, the entry of the affinity matrix is defined as

$$A_{ij} = \exp(-(1 - \cos\theta)/(2\delta^2)) \quad (112.4)$$

Usually this local similarity is not sensitive to δ, so we put $\delta = 0.5$.

For fair comparison, when conducting SEC, we set the parameter γ as 1, and the parameter μ as $\{10^{-10}, 10^{-7}, 10^{-4}, 10^{-1}, 10^0, 10, 10^2, 10^5, 10^8, 10^{11}, 10^{14}\}$ in all approaches except TensorSC and CoSC. For CoSEC, we set the parameter μ the same value on each view though it may be more practical to set different values according to the essence of data on different views.

We use the spectral rotation to calculate the discrete assignment matrix Y based on the relaxed continuous solution. To reduce statistical variety, we independently repeat spectral rotation for 50 times with random initialization for all methods, and then we report the mean clustering accuracy and standard deviation corresponding to the best parameters.

'We first study the sensitivity of the clustering performances of CoSEC with respect to parameter μ in Fig. 112.1. CoSEC favors a small value on 3sources data and Handwritten digits data, and its performances on these datasets are relatively stable when a small value is set for μ, while it prefers an intermediate value of μ for WebKB dataset and a larger one for the Internet advertisement data. These indicate that the 3sources data and Handwritten digits data have a clear manifold structure. Setting a large value for μ implies that the linearity regularization is more important for SEC, so the Internet advertisement dataset probably has a strong linearity relationship between the data matrix X and the cluster assignment matrix Y. For WebKB data, the balanced combination of data manifold and linearity regularization helps to bring about a better clustering performance.

For the comprehensive study of performances of various clustering methods, the results for the four datasets are shown in Table 112.2. The numbers in the brackets are the standard deviations of the clustering accuracy obtained with 50 different runs of spectral rotation with random initializations.

As it can be seen, our proposed CoSEC outperforms all the baselines on all datasets. For 3sources data, all methods are run first using any two views and then using all three views, and the best results are reported. On 3sources data, all the

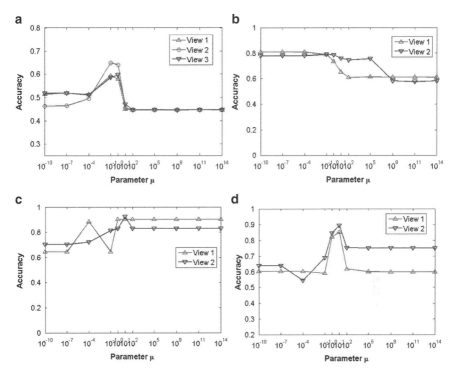

Fig. 112.1 Acc in different views of CoSEC with different μ on the four datasets. (**a**) 3 sources, (**b**) Handwritten digits, (**c**) AD, (**d**) WebKB

Table 112.2 Clustering accuracy (*Acc*) results on four datasets. Numbers in parentheses are the std. deviations

Method	Single view	FC	CCA	TensorSC	CoSC	CoSEC
3sources	0.6523 (0.0058)	0.6256 (0.0046)	0.3275 (0.0129)	0.2522 (0.0139)	0.5339 (0.0189)	0.6749 (0.0131)
Digits	0.7418 (0.0371)	0.6983 (0.0319)	0.7558 (0.0065)	0.7543 (0.0056)	0.7133 (0.0300)	0.8093 (0.0020)
AD	0.9033 (0)	0.8227 (0.0061)	0.8400 (0)	0.8401 (0.0005)	0.6496 (0.0392)	0.9233 (0.0012)
WebKB	0.8173 (0)	0.6100 (0.0043)	0.7175 (0.0053)	0.7187 (0.0059)	0.7093 (0.0005)	0.8943 (0)

baselines except CoSEC perform worse than single-view SEC. On Handwritten digits dataset and WebKB dataset, CoSEC outperforms all the baselines by a significant margin. FC performs worst on both datasets. For the Internet advertisement data, the clustering accuracy of the best single view is slightly lower than that of CoSEC and higher than those of the other baselines. CoSC performs worst on this dataset.

112.5 Conclusion

In this paper, a CoSEC was proposed. It aims to improve the clustering performance of spectral clustering on multi-view high-dimensional datasets when the data manifold is not well defined. Based on the key assumption that the true underlying clustering is same for all views, we apply the idea of co-training to the unsupervised clustering. The central idea of our approach is to use the discriminative information on one view to modify the graph structure used for the other view. The discriminative information obtained by SEC on each view is a balance of the global and the local discriminative information. And the co-training iterations retain the information needed for clustering and throw away the within-cluster details. These two factors result in a better clustering performance. Empirical evaluation on the four different kinds of high-dimensional data shows the effectiveness of our CoSEC approach.

References

1. Blum, A., & Mitchell, T. (1998). Combining labeled data with co-training [C]. *Proceedings of the Workshop on Computational Learning Theory* (pp. 92–100). San Francisco: Morgan Kaufmann.
2. Long, B., Yu, P.S., & Zhang, Z. (2008). A general model for multiple view unsupervised learning [C]. *Proceedings of the SIAM International Conference on Data Mining* (pp. 822–833). Atlanta, GA: SIAM.
3. Tzortzis, G. F., & Likas, C. L. (2010). Multiple view clustering using a weighted combination of exemplar-based mixture models. *IEEE Transaction on Neural Networks, 21*(12), 1925–1938.
4. de Sa, V. R. (2005). Spectral clustering with two views [C]. *Proceedings of the Workshop on Learning with Multiple Views, 22th International Conference on Machine Learning* (pp. 20–27). Bonn, Germany: ACM.
5. Zhou, D., & Burges, C.J.C. (2007). Spectral clustering and transductive learning with multiple views [C]. *Proceedings of the 24th International Conference on Machine Learning* (pp. 1159–1166). Corvallis, OR: ACM.
6. Kumar, A., & Daumé, H. (2011). A co-training approach for multiview spectral clustering [C]. *Proceedings of the 28th International Conference on Machine Learning* (pp. 393–400). Bellevue, WA: Omnipress.
7. Liu, X., Ji, S., Glänzel, W., & De Moor, B. (2013). Multiview partitioning via tensor methods. *IEEE Transactions on Knowledge and Data Engineering, 25*(5), 1056–1069.
8. Nie, F., Xu, D., Tsang, I.W., & Zhang, C. (2009). Spectral embedded clustering [C]. *Proceedings of the International Joint Conference on Artificial Intelligence* (pp. 1181–1186). Westerville, OH: Odyssey Press.
9. Ye, J. (2007). Least squares linear discriminant analysis [C]. *Proceedings of the 24th International Conference on Machine Learning* (pp. 1087–1093). Corvallis, OR: ACM.
10. Shi, J., & Malik, J. (2000). Normalized cuts and image segmentation. *IEEE Transactions on Pattern Analysis and Machine Intelligence, 22*(8), 888–905.
11. Yu, S. X., & Shi, J. (2003). Multiclass spectral clustering [C]. *Proceedings of the 9th IEEE International Conference on Computer Vision* (pp. 313–319). Nice, France: IEEE Computer Society.

12. Luxburg, U. (2007). A tutorial on spectral clustering. *Statistics and Computing, 17*(4), 395–426.
13. Chaudhuri, K., Kakade, S.M., Livescu, K., & Sridharan, K. (2009). Multi-view clustering via canonical correlation analysis [C]. *Proceedings of the 26th International Conference on Machine Learning* (pp. 129–136). Montreal, Canada: ACM.

Chapter 113
Feedback Earliest Deadline First Exploiting Hardware Assisted Voltage Scaling

Chuansheng Wu

Abstract In this paper, we examine the merits of hardware/software co-design of a feedback dynamic voltage scaling algorithm and a new processor are capable of executing instructions in the frequency and voltage conversion. We study several energy-aware feedback schemes based on earliest-deadline-first scheduling, dynamic adjustment of the behavior of the system, for different workload characteristics. An infrastructure for investigating several hard real-time dynamic voltage scaling schemes, including our feedback dynamic voltage scaling algorithm, is implemented on an NEC 530 embedded board. System structure and algorithm overhead is evaluated for different dynamic voltage scaling schemes. Feedback dynamic voltage scaling algorithm saves at least more energy frequently than the previous dynamic voltage scaling algorithm, with an additional 18 % energy reduction peak savings.

113.1 Introduction

In order to reduce the power dissipation of CPU, the dynamic voltage scaling technique has received widespread support in recent years to extend battery life. Dynamic Voltage Scaling dynamically scales the processor core voltage up or down depending on the computational demand of the system. Switch to reduce the speed of the transistor with low power supply voltage also allows a lower clock frequency. Assuming that the voltage and frequency are linearly related, lower voltage and frequency result in reduced cubic power consumption [1].

We developed several energy-aware feedback schemes for dynamic voltage scaling algorithm feedback earliest deadline based on priority scheduling, dynamic adjustment according to different load characteristics of real-time system [2]. Feedback dynamic voltage scaling algorithms have been proposed in our previous work,

---

C. Wu (✉)
Software College, University of Science and Technology Liaoning, Anshan 114051, Liaoning, China
e-mail: gykwcs@163.com

the simulation evaluation. Several of our improved algorithms are developed in this paper feedback scheme considering practical design and implementation issues of the actual embedded system structure. We are interested in studying the performance of the dynamic voltage scaling algorithm in an embedded environment where the overhead and the actual energy consumption can be measured quantitatively. The real-time scheduler itself, when integrated with a dynamic voltage scaling algorithm, may execute at several different CPU frequencies, which also requires accurate modeling of the system overhead.

113.2 Earliest-Deadline-First Scheduling with Dynamic Voltage Scaling Support

In order to evaluate energy-saving performance of dynamic voltage scaling algorithm in the embedded environment, we consider the scheduling problem with the earliest-deadline-first policy in hard real-time system. The scheduling framework is composed of two parts: (1) the earliest-deadline-first and (2) dynamic voltage scaling scheduling. These two components are independent of each other; therefore, earliest-deadline-first scheduling with dynamic voltage scaling algorithm is different. The earliest deadline is particularly attractive for dynamic voltage scaling algorithm because of its dynamic characteristics, which allows the dynamic voltage scaling scheme, using relaxation. Our dynamic voltage scaling scheduling is based on the feedback control incremental adjustments to the behavior of the system to reduce energy consumption.

A cycle uses the framework of fully preemptive and independent task model. Each task T_i is defined by a triple (C_i, P_i, c_i), where C_i is the period of T_i, P_i is the measured worst-case execution time of T_i, and c_i is the actual execution time of T_i. Each task's relative deadline, d_i, is equal to its period, and all tasks are released at time zero. The periodically released instances of a task are called jobs. T_j is used to denote the jth job of task T_i. Its release time is $C_i * (j-1)$ and its deadline is $C_i * j$. P_{ij} is used to represent the actual execution time of job T_j. The hyperperiod H of the task set is defined as the least common multiplier (LCM) among all the tasks' periods. The schedule is repeated in each extended end.

113.3 Feedback Dynamic Voltage Scaling Algorithm

Dynamic voltage scaling algorithm for feedback on each task calls for a real execution time (work) based on feedback from a previous call to execution time. Then, a task execution budget is divided into two parts, as shown in Fig. 113.1. Frequency of the real-time CA is proportional to the minimum of the expected. On the contrary, the maximum frequency of the rest of the execution time is scalable, $C_a + C_B = $ WCET (worst-case execution time).

Fig. 113.1 Task splitting

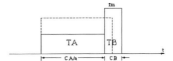

All future tasks are deferred as far as possible with the use of the maximum (worst) plan and schedule to achieve the task K. Therefore, currently available relaxation of SK shows the scaling factor and the corresponding minimum frequency. Through feedback schemes, the algorithm is able to capture changes in actual execution time. The current task preemption is expected in the future through the scheduling time slot allocation. This is the implementation of a backward scanning to fill idle times and finish the task ahead of the scheduled time slot (algorithms in detail see). Due to the even more greedy approach than any of the previous schemes, the algorithm was reported to exhibit additional energy savings in simulation experiments, particularly for medium utilization systems, which are quite common [3]. More substantial savings have been observed for execution time in pulsating PID feedback that provides new opportunity for positive scaling. NEC 530 of the embedded development board is in the process of implementation. We propose a feedback scheme by refining the following two feedback mechanisms.

113.3.1 Simple Feedback

Some periodic real-time workload is in a certain time interval of relatively stable behavior. The actual execution time of their different works remained almost unchanged or changes only in a very small range interval. For such workloads, we use a very simple feedback mechanism by computing the moving average of previous jobs' actual execution times and feed it back to the dynamic voltage scaling scheduler. We try to avoid the overhead of more complex feedback mechanisms, like the PID feedback controller as described in the next section, because a simple feedback usually in these situations provides a performance that is good enough. Quantitative comparison of the overhead of dynamic voltage scaling algorithm for our PID feedback and several other dynamic voltage scaling algorithms also makes us believe that a complex feedback dynamic voltage scaling scheme can reduce the number of expansion of its energy saving potential [4].

The simple feedback mechanism chooses the value of CA as the controlled variable. Each job Tij's actual execution time Pij is chosen as the set point. CA is assigned to be 50 % WCET for the first job of each task, which means half of the job's execution is budgeted at a low frequency and half of it is reserved at the maximum frequency. The maximum frequency portion guarantees the deadline requirements, even if the worst-case execution time is used in full. Each time a job completes, its actual execution time is fed back and aggregated to anticipate the next job's CA. Let CAij denote the CA value for Tij. The (j + 1)th job of the task is assigned a CA value according to:

$$CAi(j+1) = (CAij + ci(j+1) - ci(j - N + 1))/N \quad (113.1)$$

where N is a constant that represents the number of items used in the moving average calculation.

113.3.2 PID Feedback

Multiple input-multiple output function is very difficult not only to accurately control the behavior of the system but also to increase the complexity of the algorithm. Therefore, we improve the original PID feedback dynamic voltage scaling mechanism with the use of the following simplified design.

Instead of using CAi (i = 1...n) as the controlled variable for each task Ti and creating n different feedback controller for n different tasks, we now define a single variable r as the controlled variable for the entire system as:

$$r = \frac{1}{n}\sum_{i=1}^{n} \frac{C_{Aij} - c_{ij}}{c_{ij}} \quad (113.2)$$

$$e(t) = r - o \quad (113.3)$$

where J is the TI's new job index in the sampling point. Our goal is to make R approximately 0 (i.e., the set point). System error is E (t PID) feedback to control the controlled variable R. PID feedback controller is now defined as:

$$\Delta r_j = K_p e_i(t) + \frac{1}{k_i}\sum IW e_i(t) + K_d \frac{e_i(t) - e_i(t - DW)}{DW} \quad (113.4)$$

$$r_{j+1} = r_j + \Delta r_j$$

For each rj, we adjust the CA value for task Ti by CAi(j + 1) = rj cij + cij. The transfer function Gr between r and CA can be derived by taking derivative of both sides of Eq. (113.2):

$$Gr(s) = Mrs \quad (113.5)$$

where $M_r = \frac{1}{n}\sum_{l=1}^{n} \frac{1}{c_i}$. The block diagram of the model is shown in Fig. 113.2. Its transfer function is:

Fig. 113.2 Control loop model

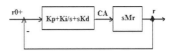

$$\frac{G_p(s)G_r(s)}{1+G_p(s)G_r(s)} = \frac{MK_{ps} + MK_i + MK_d s^2}{1 + MK_{ps} + MK_i + MK_d s^2} \quad (113.6)$$

According to control theory, a system is stable if and only if all the poles of its transfer function are in the negative half-plane of the s-domain. From Eq. (113.6), we can infer the pole of our system as follows:

$$\frac{-MK_p \pm \sqrt{MKP_p^2 - 4MK_d(MK_i + 1)}}{2MK_d} \quad (113.7)$$

Note that $-MK_P + \sqrt{MK_p^2 - 4MK_d(MK_i + 1)}$ is still less than 0 when $MK_p^2 - 4MK_d(MK_i + 1) > 0$. Hence, all the poles in the s domain are in the negative half plane, in order to ensure the stability of the system.

113.4 Experimental Evaluation

Through dynamic voltage scaling algorithm for our feedback in embedded architecture practical evaluation, we evaluate our algorithm's potential for energy savings in a real system and simulation environment.

113.4.1 Platform and Methodology

The embedded platform used in our experiment is a 495LP embedded board running on a diskless MontaVista Embedded Linux variant, which is based on the 2.4. 21 stock kernel but has been patched to support dynamic voltage scaling on the PPC 405LP. This board provides the hardware support required for dynamic voltage scaling and allows software to scale voltage and frequency via user-defined operation points ranging from a high end of 266 MHz at 1.8V to a low end of 33 MHz at 1V. The board has been modified to 50 % reduced capacitor, which allows the dynamic voltage regulating switch to occur more rapidly, i.e., switch is composed of a maximum of 0 μs duration from 1 to 1.8 V.

This set of pairs was constrained by a need to have a common PLL multiplier of 16 relative to the 33MHz base clock and a divider of two or any multiple of 4. To change the multiplier brings the extra spending for switching, which we hope to eliminate in the research [5]. Dynamic power management (DPM) facilities are

Table 113.1 Dynamic voltage scaling switching overhead

Activity	Sync. DVS	Sync. DVS	Signal handler
Overhead (µs)	117–162	8–20	0.07–0.6

used as an enhanced Linux kernel support for the dynamic voltage scaling function. The DPM operating point defines the stable frequency/voltage (and the related system parameters) that we experimentally determined.

113.4.2 Synchronous vs. Asynchronous Switch

We first evaluate the overhead of different dynamic voltage scaling techniques, through the expansion of DPM test plate support and operating system. A unique dynamic voltage scaling function supported by the NEC 530 embedded motherboard is that the frequency switching can be synchronous or asynchronous. The traditional method of processor frequency and voltage conversion is synchronous switching; the application must stop at the transition interval [6]. On the contrary, asynchronous switching of frequency and voltage allows the process to continue. Figure 113.2 describes the core of the voltage and current changes in the asynchronous switch of the PPC 530 processor [7].

Table 113.1 reports the overhead for synchronous and asynchronous switching in a time range bounded by two extremes: (a) exchange switch adjacent frequency/voltage level between and (b) between the lowest and highest frequency/voltage grade. In addition, the overhead of the following signal processing is also measured for a series of minimum and maximum processor frequencies with each asynchronous switch. The results show that the synchronous dynamic voltage regulating switch has about an order of magnitude greater overhead than asynchronous switching. The timer interrupt handler that is triggered only increases the overall cost by a little amount during each asynchronous switch.

113.4.3 Dynamic Voltage Scaling Scheduler Overhead

The overhead of our feedback dynamic voltage scaling algorithm is compared with several other dynamic voltage scaling algorithms. On the embedded development board, we first measured the execution time of the scheduling algorithm of the dynamic voltage scaling in different frequencies, as shown in Table 113.3. The overhead was obtained by measuring the amount of time when a task issues a yield system call till another task was dispatched by the scheduler. The table shows that the static dynamic voltage scaling has the lowest overhead among the four and our PID feedback dynamic voltage scaling has the highest one [8]. This is not surprising since static dynamic voltage scaling uses a very simple strategy to select the

113 Feedback Earliest Deadline First Exploiting Hardware Assisted Voltage Scaling

Table 113.2 Overhead of dynamic voltage scaling-earliest-deadline-first scheduler

CPU freq. (MHz)	Static	DVS	Idling over look-ahead	Egad [spec] CiD-feedback
33	217	487	2,296	3,652
44	170	366	1,714	2,943
66	100	232	1,112	1,728
133	52	120	546	801
266	36	76	229	472

Table 113.3 Task set, times in msec

	Task set 1		Task set 2		Task set 3	
Task	Period (Ci)	WCET (Pi)	Period (Ci)	WCET (Pi)	Period (Ci)	WCET (Pi)
1	2,400	400	600	80	90	12
2	2,400	600	320	120	48	18
3	1,200	200	400	40	60	6

frequency and voltage falling short in finding the best energy saving opportunities. Cycle-conserving dynamic voltage scaling, look-ahead dynamic voltage scaling and our PID feedback dynamic voltage scaling use more sophisticated and aggressive algorithms for lower energy consumption, albeit at higher overheads. The trade-off between cost and performance needs to be carefully checked.

Next, we evaluate if feedback dynamic voltage scaling algorithm, although having the biggest cost among the four, provides the best energy-saving effect in the actual embedded environment. We measured the actual energy consumption of these dynamic voltage scaling algorithms when executing three medium utilization task sets depicted in Table 113.2 using both synchronous and asynchronous dynamic voltage scaling switchings. As a baseline for comparison, we also implemented a naive dynamic voltage scaling scheme, in which the maximum frequency is selected when a task is scheduled and the minimum frequency is selected when the system is idle.

The first task set in Table 113.3 is harmonic, i.e., the time is an integer multiple of the shortest cycle that facilitates scheduling. This will allow the scheduling algorithm to demonstrate the extreme behavior, usually outperforming any other choice of period. The second and third of the task sets are nonharmonic with longer and short periods, respectively. The actual execution time is half that of the WCET for each of the experimental tasks.

The naive dynamic voltage scaling algorithm is used as a base of comparison for each dynamic voltage scaling algorithm subsequently. In the task set, a static dynamic voltage scaling can reduce energy consumption by about 29 % over the naive scheme. Dynamic voltage scaling cycle can save 47 % energy. The look-ahead real-time dynamic voltage scaling can save more than 50 %, and our feedback method saves 54 % energy compared to the naive dynamic voltage scaling. This clearly shows great potential in energy saving for real-time scheduling.

113.5 Conclusion

We evaluated it as well as several other real-time dynamic voltage scaling algorithms on an NEC 530 embedded platform. We compared the energy consumption and scheduling overhead between different dynamic voltage scaling schemes. The experimental results show that the positive feedback dynamic voltage scaling algorithm in our energy consumption reached additional savings 24 % over the look-ahead dynamic voltage scaling algorithm and AGR-2 algorithm and up to 64 % energy savings over the naive dynamic voltage scaling scheme when considering the scheduling overhead.

References

1. Jejurikar, R., & Gupta, R. (2012). Procrastination scheduling in fixed priority real-time systems. *Proceedings of the Language Comcilers and Tools for Embedded Systems, 9*(2), 101–111.
2. Kang, D., Crago, S., & Suh, J. (2011). A fast resource synthesis technique for energy-efficient real-time systems. *IEEE Real-Time Systems Symposium, 35*(5), 204–207.
3. Lu, C., Stankovic, J. A., Abdelzaher, T. F., Tao, G., Son, S. H., & Marley, M. (2009). Performance specifications and metrics for adaptive real-time systems. *Proceedings of the IEEE Real-Time Systems Symposium, 23*(4), 1011–1013
4. Brock, B., & Rajamani, K. (2011). Dynamic power management for embedded systems. In *IEEE international SOC conference* (pp. 23). London: WET publishing.
5. Chandrakasan, A., Sheng, S., & Brodersen, R. W. (2012). Low-power CMOS digital design. *IEEE Journal of Solid-State Pircuits, 27*(3), 473–484
6. Govil, K., Chan, E., & Wasserman, H. (2012). Comparing algorithms for dynamic speed-setting of a low-power CPU. *First International Conference on Mobile Computing and Networking, 63* (4), 483–494
7. Gruian, F. (2012). Hard real-time scheduling for low energy using stochastic data and DVS processors. In *Proceedings of the international, symposium on low-power electronics and design ISLPED'02* (pp. 77–86). New York: Addison-weslet Publishing.
8. Dirk Grunwald, Philip Levis, Charles B. Morrey, Michael Neufeld & Keith I. Farkas (2012). Policies for dynamic clock scheduling. In *Symposium on operating systems design and implementation* (pp. 235–247). Paris: Springer.

Chapter 114
Design and Realization of General Interface Based on Object Linking and Embedding for Process Control

Jiguang Liu, Jianbing Wu, and Zhiguo He

Abstract Based on the analysis of existing problems of interface software development process of industrial control, the importance of building the general interface system based on OPC (Object Linking and Embedding for Process Control) was proposed. The data model was given with database technology. On the basis of the data model, the configurable general interface system based on OPC was implemented. Versatility and configurability is the most important feature of the interface system. By simple modification of configuration information, the interface system will meet the needs of different projects. The application results show that the interface system greatly reduces the development cycle of the related software, improves the reliability and stability of the application system, and reduces costs of system operation and maintenance.

114.1 Introduction

In conventional control system, information interchange between hardware device and software of control system was done through device driver. Different control equipment manufacturers often use different communication protocol. To access production data on field device, users have to write particular communication interfaces to talk with those peripherals. Besides, industrial field devices are abundant in variety, and hardware products are updated constantly. Therefore, it is difficult to develop a suit of industrial control interface software for production information interchange which is universally applicable to all industrial devices. Consequently, "Information islands" are formed in industry field, which constrained further development and application in production fields.

J. Liu (✉) • J. Wu • Z. He
School of Mathematics and Computer Science, Panzhihua University,
Panzhihua 617000, China
e-mail: liujig@gmail.com

To solve the existing problems in conventional control system, the industry launched OPC technology; the technology as the newest, the most widely used soft bus standards in the industry control domain, has been supported by basic automation manufacturers in general and was widely used in industrial field. Currently, although there are many interface applications based on OPC technology, but these interface systems had a common characteristic of high coupling between itself and application, it caused excessive customizability, poor flexibility, extensibility and independence of the interface system. The OPC interface given in [1–4] belonged to such applications. For these interface systems mentioned above, due to the lack of abstraction of the common characteristics of the interface systems, therefore, it was impossible to develop a reusable generic interface system. In these applications, the application developers wrote some important parameters (such as information of basic automation device, data acquisition point information) into interface program in the form of hard coding; the slight changes of external environment parameters will lead to the entire program to recompile and release, resulting in poor interface system adaptability to changes in the external environment. To buy mature industrial control software can eliminate the coupling between the interface system and application system, but industrial control software is generally not low cost, and different process control project must repeat to purchase the same software, resulting in project cost go straight up. In OPC technology maturing today, therefore, building a general interface system based on OPC is of great significance to solve the problem of "Information islands" of automation, improve the level of factory process control, reduce the cost of implementation of the process control system, and have rapid implementation of the production process control system.

114.2 System Design Objective and Essential Function

In order to meet various needs of different external system and automation system, the interface system must be configurable. For various external system and automation system, it can meet different application requirements with a simple modify configuration information. This interface system's most outstanding characteristic is universality and configurability, which is the biggest difference between this interface system and other system based on OPC. Guided by the principle of universality and configurability, a design objective of the interface system was given as follows:

- The system can simultaneously connect to multiple OPC servers. (Generally, hardware from different manufacturers exist simultaneously in industry field.)
- The system is capable of dynamic configuration for data collecting point.
- The system has the general purpose interface to external system, which facilitates data exchange between basic automation system and external system and which also triggers process tracking logic of external system.
- The system must have producibility.

Interface system is a basis for solving "islands of automation" problem, achieving production data sharing, further expanding and applying of production fields. Therefore, interface system must realize the following two basic functions:

- It implements bidirectional process information exchange between any automation system based on OPC technology and external system.
- To establish triggering mechanism for tracking and monitoring production process, provide the conditions for tracking and monitoring production process.

114.3 Data Model Design

Data model is the core part of interface system design and is a basis of implementing universalization and configurability of interface system. Excellent data model can add good extensibility, flexibility, and adaptability to the interface system. On the analysis of domestic and foreign application experience of OPC, a data model as shown in Fig. 114.1 was obtained. Follow-up discussion centers around the data model which is a basis of implementation of the interface system. The following paragraphs introduce the key entities of the model.

114.3.1 OPC_SERVER

Usually, OPC servers from several control equipment manufacturers exist simultaneously in industry field. The total number of the servers may be changed at any time. To accomplish production data acquisition from control equipment in use, the

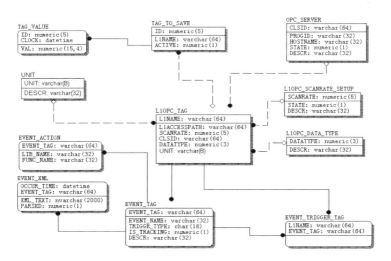

Fig. 114.1 Structure diagram of system data model

interface system must have the capability of dynamic connecting to OPC servers. This entity stores information of all OPC servers in the industry field, and the administrative staff can modify, add, or reduce the data in the table according to actual situations when increasing or decreasing or changing OPC server in the field. As the interface system starts up, it dynamically accesses the information of all OPC servers and connects to each OPC server. Consequently, the system has good extensibility and adaptability.

114.3.2 L1OPC_SCANRATE_SETUP

Every production data has a definite occurrence cycle. User can define scanning period according to characteristic of production data. This entity is used to hold user-defined rate of scanning on field data point. Users can increase or adjust scan rate at any time based on actual conditions. The latest scan rate can be acquired from configuration table when system starts. To facilitate setting up a mechanism of tracking and triggering of the production process, the interface system categorized the production data into four lists according to the characteristics of the production data. The categories are cycle data, event data, trigger data, and download data which are described, respectively, as follows.

114.3.2.1 Cycle Data

Cycle data is a large amount of analog signal acquired successively in the industry field, such as temperature, pressure, and flow. Data points whose scan rate is above 0 in configuration table all belong to cycle data point. Cycle data is characterized by volatility and continuity. Its value typically fluctuates frequently in a certain range. On-spot operators can quickly find whether the device or production process stays normal through monitoring this type of data. The asynchronous data access method was adopted to acquire cycle data. In consideration of the frequent fluctuation of the cycle data, in order to improve performance but also saving memory, the definition of opc-dead band was provided for interface system. OPC server of automation system will send varying data to the interface system only when automation system variation range of cycle data is above definition of the dead band.

114.3.2.2 Event Data

Event data is a signal triggered by change of state, position, behavior, or features of field device or materials. For example, continuous cast slab cutting signal and billet entering the furnace signal all belong to digital signal. In configuration table, it belongs to event data if its SCANRATE is 0. Event data is a kind of important production data which is a key to establish triggering mechanism for tracking and monitoring production process. When an event occurs in the production site, the system will accomplish an invocation of interface of external system according to interface information of external system stored in EVENT_ACTION. Meanwhile, it transmits the production process information to external system. That completes the tracking and monitoring of the production process. Asynchronous mode is adopted to acquire the event data.

114.3.2.3 Trigger Data

Trigger data is a set of production process data closely bound to event data, such as pouring weight of big casting ladle, pouring weight of middle casting ladle, and pouring time. In configuration table, it belongs to trigger data if its SCANRATE is -1. A single event signal can be bound to multiple trigger data. The configuration information of trigger data bound to specific event is stored in EVENT_TRIGGER_TAG. When the event occurs, the system gets the trigger data according to the relevant configuration information in EVENT_TRIGGER_TAG and transmits it with event data to an external system in a form of package. The external system can track and monitor production process after parsing the received data. It should be pointed out that trigger data must bind to event and it may have no practical meaning to get this type of data alone. Therefore, the system reads trigger data synchronously only when an event occurs.

114.3.2.4 Download Data

Download data refers to the data transmitted to automation system by external system through the interface system. This type of data generally is the control information of production. In configuration table, it belongs to download data if its SCANRATE is -2. Interface system do not read but write this type of data into automation system if necessary. To avoid industrial accidents or plant accident, automation system must verify the legitimacy of this type of data when used, and valid data can be used.

114.3.3 L1OPC_TAG

This entity is used to save data acquisition point information of all OPC servers which include access path and data type. Configuration information is obtained from this entity and is added to the OPC Group object. Since the regular change of field data point, management staff can increase or decrease or modify data point in the light of actual conditions without leading to modification of system source code. This makes the system more universal, flexible, and easily extensible.

114.3.4 EVENT_ACTION

This entity is used to save logical program interfaces of external system corresponding to an event. When some event of L1 system happens, interface system will invoke program interface of external system on the basis of configuration of this event to perform business logic and realize the tracking and monitoring of production process.

114.4 Implementation of System

114.4.1 EVENT_ACTION

Data access method of OPC is either synchronous access or asynchronous access [3]. The synchronous access means OPC client is in a wait state after sent a request to an automation OPC server. The invocation of OPC client can be returned, and other process will be executed after a requested data from the OPC server was returned to the client. The asynchronous access means the client will return immediately and process other task after sending a data request to OPC server. OPC server will call the callback function of OPC client immediately and send the requested data to OPC client when a request is received. The implementation of synchronization is simpler than that of asynchronous. Synchronization can be adopted when data quantity are less in interaction between OPC server and client. But a large amount of data exchange will necessarily lead to performance and efficiency of system degradation. It may even influence the proper use of basic automation system. The implementation process of asynchronous mode is more complex. Its advantage is high performance of system so that a block caused by a large volume of data request from multiple clients can be avoided as well as CPU and net resources can be saved at maximum.

There are different formulations about which is better of using synchronization for data-gain or asynchronous for data-gain. The characteristic of production data decides which mode should be used. Asynchronous should be adopted for access of cycle data and event data and synchronization for trigger data.

114.4.2 System Implementation

System implementation is actually converting data model to program. According to data model and knowledge of object-oriented programming, data model is converted to OPC server object management class, OPC server object class, OPC Group object class, OPC Item object class, EventManager object class, and connection point object class. Each of the abovementioned classes is an encapsulation

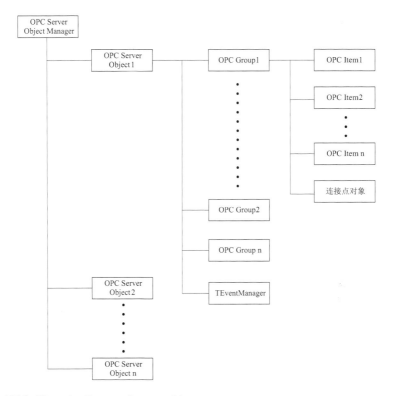

Fig. 114.2 Hierarchy diagram of system object

of OPC interface. To improve the efficiency of program developing and debugging, each class has corresponding management function, for example, connection and disconnection of server and activating and deactivating of OPC Group. Object hierarchy of system is shown as Fig. 114.2. The functions and implementations of each class are described as follows.

114.4.2.1 OPC Server Management Object

This class is a container class of OPC server object which is responsible for creating all OPC server objects. The class of OPC server management object traverses the configuration information of OPC_SERVER data table and creates each connection to OPC server object when the system starts up. If creating OPC server object failed, the log will be recorded by the system so that relevant staff can examine the failure. The operators can conduct the management of a specific OPC server such as start-up and shutdown.

114.4.2.2 OPC Server Object

This is an encapsulation of OPC interface server in the system which is responsible for executing connection to field OPC server. This object is created by OPC server management object and performs the connection to remote OPC server. It will traverse the information of configuration table of L1OPC_SCANRATE_SETUP and create OPC Group object if connecting successfully. OPC server object is a container class of OPC Group object which can conduct management of creating, deleting, activating, deactivating, etc., on OPC Group.

114.4.2.3 OPC Group Object

The data transmission of OPC is conducted with the group as the unit [4]. This is an encapsulation of OPC interface group and is created by OPC server object. This object will get all data acquisition point information belonging to this group from L1OPC_TAG and add it to the group item by item. OPC Group is organized by scan rate in interface system and the threads are arranged by OPC Group in OPC server. Therefore, it is not suitable to set excessive scan rate in L1OPC_SCANRATE_SETUP, otherwise might result in performance decline of OPC server. The dead band parameter setting is also supported by OPC Group object.

114.4.2.4 OPC Item Object

This object is an encapsulation of OPC interface Item which is responsible for management and maintenance information about OPC Item. Each OPC Item object is subordinate to the OPC Group object which has the same scan rate as OPC Item. OPC Item represents data connection with OPC server and general corresponding to registers on device. There is a correspondence between the OPC Item object and the records in L1OPC_TAG. The management staff can increase or decrease or modify L1OPC_TAG information to adapt to data acquisition requirement in industry field and the latest configuration information will be used when system starts up. The management functions such as activating, deactivating, and value writing can be conducted by this object.

114.4.2.5 Event Management Object

This is essentially an encapsulation of OPC Group which belongs to OPC server object. This object is a container of event object and facilitates flexible managing event object for the management. The capacity of retrieving event object through index or event name is provided by the system.

114.4.2.6 Event Object

Event object is also an encapsulation of OPC interface Item. Due to special features of event, an individual encapsulation is needed for event object. Event calls external interface function according to a configuration in EVENT_ACTION and sends process data of automation system to the external system. Thus completing the tracking and monitoring of the production process. It is the important function of event object. Event corresponds to event data point in L1OPC_TAG. Therefore, increase or decrease of event object can be realized by modification of configuration. Through this, the dynamic expansion of system function can be realized.

114.4.2.7 OPC Connection Point Object

There are two different mechanism of OPC asynchronous access: advisory connection mechanism and connection object mechanism [4] which respectively corresponds to OPC DA 1.0 [5] specification and OPC DA 2.0 [6] specification. The interface system uses connection object mechanism to realize asynchronous mechanism. OPC connection object is an encapsulation of OPC connection object mechanism. This object which belongs to each OPC Group object is a key to realize asynchronous communication. OPC Group adopted asynchronous communication mechanism will register OnDataChange function with OPC server of automation system. The OnDataChange function registered by client will be invoked automatically by OPC server of automation system when it detects data change of a certain group. This realized pushing change to client. Asynchronous communication is adopted for cycle data and event data of interface system. To avoid frequently invoked OnDataChange function by server caused by subtle change of cycle data, dead band is defined for each OPC group. Only the range of data change is above the dead band; a notification of data change will be received by client.

114.5 Conclusion

This universal interface system has been applied in many projects of process control in Panzhihua Iron and Steel Co. There are some achievements of this successfully developed system:

- Facilitating the design and integration of industry control software, shortening the product development period.
- Standardize interface specification of automation system and external system, facilitating debugging of process tracing system, and improve the efficiency of software development.
- Configurability of system enhances extensibility and adaptability of interface system. Meanwhile, producibility interface system also improves stability and reliability of application system.

References

1. Zhou, J., & Zhou, Y. (2004). Application of middleware OPC technology in industrial control system. *Computer Engineering, 30*(23), 165–167 (in Chinese).
2. Chen, J., & Yuan, N. (2003). Research and application of OPC data access specifications. *Techniques of Automation and Applications, 22*(8), 61–64 (in Chinese).
3. Su, M., & Wang, Z. (2006). Research and realization of OPC data access sever. *MicroComputer, 22*(3-1), 11–13 (in Chinese).
4. Tan, J., Jiang, S., Wu, Z., & Shao, H. (2011). The design of remote monitoring system of gas drainage based on OPC. *Coal Engineering, 7*, 128–130 (in Chinese).
5. OPC common definitions and interfaces version 1.0. OPC Foundation, 1998.
6. OPC data access custom interface specification 2.05A. OPC Foundation, 2002.

Chapter 115
A Stateful and Stateless IPv4/IPv6 Translator Based on Embedded System

Yanlin Yin and Dalin Jiang

Abstract In order to solve intercommunication problem between IPv4 network and IPv6 network more flexibly, this paper has proposed an improved IPv4-IPv6 translator based on embedded system. By using an optimized address mapping regulation, it can support both stateful and stateless translation method. In addition, a lightweight SIP-ALG and Modbus-ALG have been designed to assist the translator to process the datagram, which may take address and domain information at the seven layers of OSI model. The results show the translator can work well between sensor network and Internet, and the mixed use of stateful and stateless method has much less memory usage than stateful method and nearly the same process delay as stateless method.

115.1 Introduction

Since the last five IPv4 address spaces had been allocated completely in 2011 by ICANN/IANA and a growing number of users have the requirement for Internet, IPv4 addresses will be exhausted very soon [1]. At the same time, with the rise of the Internet of Things (IOT), which is considered as the third wave of development of global information industry that comes after computer and Internet, a great many of sensors and equipments need billions of IP addresses to accomplish communications. IPv6, as the core of Next Generation Internet (NGN), with exhaustless addresses, more secure and flexible, will take over IPv4 step by step eventually and undoubtedly.

However, the transition from IPv4 to IPv6 is considered as a long-term strategy; therefore, the intercommunication between these two networks becomes the main issue. Typical transition technologies include dual stack technology, which is

Y. Yin (✉) • D. Jiang
College of Electronics Information and Control Engineering, Beijing University of Technology, Beijing 100124, China
e-mail: alin4187@163.com

implemented at the early stage and still has to consume IPv4 address when a new node adds in; tunnel technology, which can only support IPv4 to IPv4 communication over IPv6 or IPv6 to IPv6 communication over IPv4; and translation technology, which is considered as the most feasible solution so far and will be discussed throughout this paper.

There are four types of IPv4/IPv6 translation cases which can be divided into eight scenarios depending on whether the IPv6 side or the IPv4 side initiates communication [2]. Translation technology is based on stateful method which refers to the translator that keeps every mapping information in its memory or stateless method which refers to the translator that only uses predetermined rules for address mapping [3, 4]. RFC6145 suggests that stateless translation supports end-to-end address transparency and has better scalability compared with stateful translation. In the early year, Aoun and Davies proposed to move the Network Address Translator-Protocol Translator (NAT-PT), which is a stateful translation mechanism illustrated in RFC2766, to historic status because of its complicated implementation and insecurity [5]. Yet RFC4996 also mentioned that in some circumstances, an IPv6-IPv4 protocol translation solution may be a useful transitional solution. In recent years, some stateless translation methods based on IVI [6], of which IV represents IPv4 and VI represents IPv6, have been proposed one after the other such as IVI/MAP-T/MAP-E [7]. But all these researches and experiments of stateless translation methods were under personal computer environment, and they may not be appropriate for sensor network which considered as the key to the IOT. Considering the scalability and utilization of IPv4 addresses, this paper proposed and designed an improved NAT-PT translator which supported both stateful and stateless translation method based on embedded system in order to fit the particularity of the sensor network.

115.2 Translation Model

Translation model includes two parts, address translation and protocol translation, which sometimes probably go along with the functional application layer gateway (ALG) [8]. In fact, the main difference between stateful and stateless translation method is the address translation. For the stateful method, like original NAT-PT, it keeps every entry of address mapping in the memory. Therefore, it is wasted storage and time when looking up corresponding entry in mapping table. For the stateless method, like original IVI, it uses a number of certain rules for address mapping. However, the multiplexing number of its each IPv4 address is suggested up to 256, and it consumes at least 2^{64} available IPv6 addresses for stateless usage [9]. Hence, the improved NAT-PT translator should balance both pros and cons of these two methods. In this paper, we will only discuss the top half IPv4-IPv6 translation scenarios mentioned in RFC6144 [2] because of the particularity of sensor network, as is shown in Fig. 115.1.

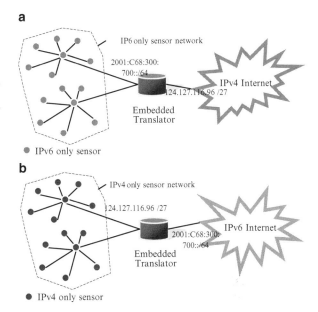

Fig. 115.1 Panel (**a**) shows the interoperation between an IPv6-only sensor network and the IPv4 Internet in bidirection, panel (**b**) shows the interoperation between an IPv4-only sensor network and the IPv6 Internet in bidirection

To support both the stateful and stateless translation is a tough thing, thus we should do some work on the IPv4-IPv6 address mapping and describe how it works, and then illustrate the protocol translation algorithm and application level gateway (ALG) solution.

115.2.1 Net Address Translation

115.2.1.1 Address and Mapping Regulation

The main idea to support both stateful and stateless translation is to utilize IPv6 address mapping regulation to separate them. Figure 115.2 described the address mapping regulation for the improved NAT-PT translator, which is optimized from IVI. Notice that the IPv6 prefix must be smaller than /64 and has to use 1/16 of its address space for stateless implement. Each segment defined in the address format is explained as follows:

- *Prefix*: a 64-bit IPv6 prefix assigned by the Internet service provider (ISP) for global use or defined by the network itself for local use.
- *Flag* (*F*): a 4-bit field to decide which kind of translation method to be used, stateful or stateless. 4-bit all zero stands for stateful and 4-bit all one stands for stateless; other combinations are reserved.
- *Reserve* (*R*): a 4-bit field reserved for future use, default is all zero.

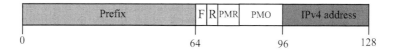

Fig. 115.2 Address mapping regulation

- *Port multiplexing ratio* (*PMR*): an 8-bit field, in the power of two, is to reveal the multiplex number of each IPv4 address. The theoretical maximum multiplex ratio is $2^{16}:1$ (PMR value is $0 \times 0f$), while a sensor can only use one port to communicate, and the minimum is 1:1.
- *Port multiplexing offset* (*PMO*): a 16-bit field to determine the start position of a port range, which is decided by the PMR.
- *IPv4 address*: a 32-bit IPv4 address for sensors in IPv6-only network to map.

In order to illustrate conveniently, we use the combination of colon hexadecimal and dotted decimal to represent an IPv6 address (e.g., use 2001::255.255.255.255 instead of 2001::FFFF:FFFF). For example, 2001:C68: 300:700:F00A:0100: 124.127. 116.125 is an IPv6 address in which 2001:C68:300:700::/64 is assigned by the ISP and the prefix 2001:c68:300:700:F000::/68 means it uses the stateless address translation. The PMR, $0 \times 0A$, reveals that the address multiplex ratio is 1:1m024 while the available port number for a single sensor is 64. The PMO, 0×0210, shows that these 64 port numbers start from 0×0210 to 0×0250 (512–592 in decimal).

1. How It Works

    (a) Interoperation between IPv6 sensor network and IPv4 Internet

    Firstly, we will discuss about the case that interoperation happened between an IPv6-only sensor network and the IPv4 Internet shown in Fig. 115.1a. Indeed, this is the situation at the beginning of IPv4-IPv6 transition when the IPv4 network is still the backbone network. In this case, both stateful method and stateless method can be used. We assume that the translator has been assigned an IPv4 prefix 124.127.116.96/27; one-half is for stateless use and the other half is for stateful use. When a data packet passes through the translator in the direction of IPv6 to IPv4, it will go through the following steps:

    - *Step 1*: Check the 4-bit flag field in the source address of the IP header. If it is set to $0 \times f$, then go to step 2; else if it is set to 0×0, then go to step 5; else go to step 7.
    - *Step 2*: Check the source port field of UDP or TCP header, if it is in the region of [PMO, PMO+$2^{16\text{-PMR}}$], then go to step 3; else go to step 7.
    - *Step 3*: Extract the 96–127 bits from the IPv6 destination address as the new IPv4 destination address; go to step 4.
    - *Step 4*: Extract the 96–127 bits from the IPv6 source address as the new IPv4 source address; go to step 6.

- *Step 5*: Do stateful translation by checking source address in mapping table, if you find the entry, then use the corresponding address and port as the new address and port; else create a new mapping entry, using the mapped IPv4 address and port. Go to step 6.
- *Step 6*: Do protocol translation with ALGs and send the translated packet.
- *Step 7*: Drop the packet.

When a data packet passes through the translator in the direction of IPv4 to IPv6, it will go through the following steps:

- *Step 1*: Read the translator's configuration and check the IPv4 destination address; if it is for stateless use, then go to step 2; else if it is for stateful use, go to step 5; else go to step 7.
- *Step 2*: Get the PMR from translator's configuration, and get the source UDP or TCP port as P from the IPv4 packet, calculate PMO, PMO = INT(P/PMR)* PMR, and go to step 3.
- *Step 3*: Fill the new IPv6 source address with prefix, F, R, PMR, PMO, and IPv4 source address extracted from the old IPv4 packet; go to step 4.
- *Step 4*: Fill the new IPv6 destination address with prefix, F, R, PMR, PMO, and IPv4 destination address extracted from the old IPv4 packet; go to step 6.
- *Step 5*: Do stateful translation by checking destination address in mapping table; if you find the entry, then use the corresponding address and port as the new address and port; else create a new mapping entry, using the mapped IPv6 address and port. Go step 6.
- *Step 6*: Do protocol translation with ALGs and send the translated packet.
- *Step 7*: Drop the packet.

(b) Interoperation between IPv4 sensor network and IPv6 Internet

Next, we will talk about the case that interoperation happened between an IPv4-only sensor network and the IPv6 Internet shown in Fig. 115.1b. This is the situation that happens in the late stage of IPv4-IPv6 transition when legacy sensors or equipments want to access the IPv6 Internet. Unfortunately, stateless transition cannot be used due to the philosophical logic: IPv4 network needs to communicate with all of the IPv6 Internet, not just a small subset, and stateless can only support a subset of the IPv6 addresses. But we can use stateful method to accomplish address translation; the process is similar as previous illustration.

115.2.2 Protocol Translation

Protocol translation between IPv4 and IPv6 is a critical part; this is because besides the wide difference in address space and protocol header, the architecture of these two protocols has an extreme variation. In this paper, the protocol translation algorithm is based on Stateless IP/ICMP Translation Algorithm (SIIT) [10], as is shown in Fig. 115.3.

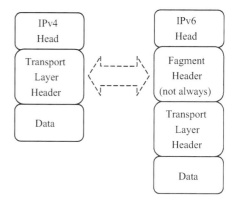

Fig. 115.3 Protocol translation diagram

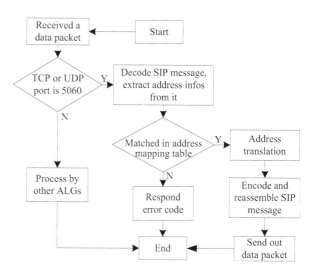

Fig. 115.4 Lightweight SIP-ALG workflow

115.2.3 Application Level Gateway

It is inevitable that sensors or equipments are running Layer7 network applications that take address information, leading to communication problem. Thus, ALG should be set to solve the problem. Typical ALG is DNS-ALG and FTP-ALG which have much discussed. Considering that a number of sensors, maybe advanced and intelligent, are using Modbus protocol over TCP/IP stack to signaling [11], or utilizing Session Initiation Protocol (SIP) to control voice and video transportation, this paper has proposed a lightweight SIP-ALG and a Modbus-ALG solution for the translator. Figure 115.4 shows the workflow of lightweight SIP-ALG in stateful method, and Modbus-ALG is similar.

115.3 Architecture and Design

The translator is based on the embedded system, which uses ARM Cortex-A8 development board (1 GHz CPU, 256M RAM) as the hardware, uses Linux as the operating system, and takes advantage of netfilter framework for developing. Figure 115.5 shows the system architecture. The Linux device drivers provide some methods to access raw data packet from the hardware devices. The netfilter framework gets the packet from the device drivers and deals with it. Firstly, the NAT module generates new IP source and destination addresses by using stateful or stateless method. Then, the PT module generates a new IP or ICMP header. At last, ALG modules do some further processing when the packet takes address information at the application layer. All the OS kernel-level developments, including device driver development and translation modules development, are using C language. However, the user space development mostly is based on shell or PHP script.

115.4 Testing

115.4.1 Testing of Connectivity

We wrote a shell script (using ping command) running on Linux-based sensors and hosts to test the connectivity between nodes in different networks, including eight cases, and the statistics is shown in Table 115.1. Notice that the success rates of cases 1, 2, 4, 5, 6 are under 100 %, probably because the compatibility of IPv6 stack in some sensors is not very good. Generally, the translator works well.

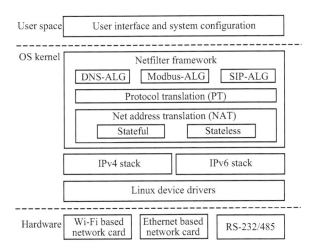

Fig. 115.5 Architecture of the translator

Table 115.1 Results of connectivity testing

Method	Case	Total	Success	Rate (%)
Stateful	v6 sensor Pings v4 host	10,000	9,989	99.89
	v4 host Pings v6 sensor	10,000	9,973	99.73
	v4 sensor Pings v6 host	10,000	10,000	100
	v6 host Pings v4 sensor	10,000	9,999	99.99
Stateless	v6 sensor Pings v4 host	10,000	9,980	99.80
	v4 host Pings v6 sensor	10,000	9,987	99.87
	v4 sensor Pings v6 host	10,000	10,000	100
	v6 host Pings v4 sensor	10,000	10,000	100

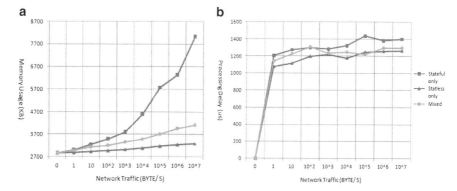

Fig. 115.6 Panels (**a**) and (**b**) show the relationship between network traffic and memory usage or processing delay, respectively, in three modes (stateful, stateless, and mixed)

115.4.2 Testing of Pressure

In this test, we wrote a shell script running on the translator to monitor memory usage, processing delay while network traffic is changing under different translation methods (MTU is set to 1,460). The statistical results are shown in Fig. 115.6, ignoring the flow direction. We can infer that the mixed use of stateful and stateless method has a lot less memory usage than stateful method and nearly the same process delay as stateless method.

115.5 Conclusion

This paper has proposed an improved IPv4-IPv6 translator that can support both stateful and stateless translation for sensor network accessing Internet. Then we implemented it on the ARM Cortex-A8 based on embedded system and did some system testings. Although stateful method is resource wasting and time-consuming, it is still the only solution under some certain circumstances. Surely, this translator is useful but maybe not perfect in architecture, translation algorithms, connectivity of software modules, table search algorithm, C code efficiency, etc. And these will be optimized gradually in our future researches.

References

1. IPv4 address report (online). (2013). Retrieved from http://www.potaroo.net/tools/ipv4/.
2. Baker, F., Li, X., Bao, C., & Yin, K. (2011). Framework for IPv4/IPv6 translation, RFC 6144.
3. Bagnulo, M., Matthews, P., & van Beijnum, I. (2011). Stateful NAT64: Network address and protocol translation from IPv6 clients to IPv4 servers, RFC6146.
4. Li, X., Bao, C., & Baker, F. (2011). IP/ICMP translation algorithm, RFC6145.
5. Aoun, C., & Davies, E. (2007). Reasons to move the network address translator—protocol translator (NAT-PT) to historic status, RFC4966.
6. Li, X., Bao, C., Chen, M., Zhang, H., & Wu, J. (2011). The China Education and Research Network (CERNET) IVI translation design and deployment for the IPv4/IPv6 coexistence and transition, RFC6219.
7. Li, X., & Bao, C. (2013). IVI/MAP-T/MAP-E: Unified IPv4/IPv6 stateless translation and encapsulation technologies (online). Retrieved from http://www.cnki.net/kcms/detail/34.1228.TN.20130228.1703.003.html.
8. Holdrege, M., & Srisuresh, P. (2001). Protocol complications with the IP network address translator, RFC3027.
9. Zhu, Y. C. (2008). Stateless mapping and multiplexing of IPv4 addresses in migration to IPv6 Internet (pp. 2248–2252). In: *2008 I.E. global telecommunications conference*. New York: Institute of Electrical and Electronics Engineers Inc.
10. Nordmark, E. (2000). Stateless IP/ICMP translation algorithm (SIIT), RFC2765.
11. Modbus messaging on TCP/IP implementation guide V1.0b (online). (2006). Retrieved from http://www.modbus.org/specs.php.

Chapter 116
A Novel Collaborative Filtering Approach by Using Tags and Field Authorities

Zhi Xue, Yaoxue Zhang, Yuezhi Zhou, and Wei Hu

Abstract Traditional collaborative filtering is widely used in social media and e-business, but data sparsity and noise problems have not been solved effectively yet. In this chapter, we propose a novel approach of collaborative filtering based on field authorities, which achieves genre tendency of items by mapping tags to genres and simulates a fine-grained word-of-mouth recommendation mode. We select the nearest neighbors from sets of experienced users as field authorities in different genres and assign weights to genres according to genre tendency. Our method can solve sparsity and noise problems efficiently and has much higher prediction accuracy. Experiments on MovieLens datasets show that the accuracy of our approach is significantly higher than traditional user-based kNN CF approach in both MAE and precision tests.

116.1 Introduction

Collaborative filtering (CF) is the most popular technology in current recommender systems. The basic idea is to recommend items to active users based on the opinions of other users who have similar tastes. CF approach achieves a great success in research and practice, such as Google News, Netflix, and Amazon [1].

A typical collaborative filtering approach, k-Nearest Neighbor (kNN) [2, 3], for example, is a way to find the "nearest neighbors" of the active user. The items will be recommended to the active user only if they are most liked by the neighbors. However, because of the limitation of data, this approach has data sparsity and noise problems, which cause failures in neighbor searching. In order to solve these problems, a CF approach based on expert opinions has been proposed [4] (ECF). Rather than applying a nearest neighbor algorithm to the user-rating data, predictions

Z. Xue (✉) • Y. Zhang • Y. Zhou • W. Hu
National Laboratory of Information Science and Technology, Department of Computer Science and Technology, Tsinghua University, Beijing 100084, China
e-mail: raphaelxue@gmail.com

are computed by using a set of expert neighbors from an independent dataset. Compared to traditional kNN method, this approach can effectively solve the problems of data sparsity and noise but leads to several other problems, such as prediction accuracy declination, and more than that, it is hard to find external experts normally.

In this chapter, we present a novel collaborative filtering algorithm based on field authorities. We define field authorities as experienced user sets in different areas. The genre information of items is from an experience-decision dataset, and genre weights are calculated by user-generated tags. We use MovieLens dataset to experiment and find the following results: (1) the prediction accuracy of our approach is significantly higher than traditional user-based CF and ECF; (2) there is no need to use external data, so our approach has good expansibility.

116.2 Genre Resolution Model Based on Tags

In order to describe the reality more fine-grained, we model the genres of items by using tags marked on the content. Using tags to model genres can solve the credibility problem of items for us, because the nominal and the real are always not consistent and undoubtedly the most authoritative judgments are from the choice of users.

The performance of a single tag can be very unstable, but if we observe the statistical significance from the macrostructure, we can find obvious corresponding relations between tags and genres. We use the associated characteristics between tags and genres and can judge the tendency of each item on different genres.

As shown in Fig. 116.1, we use this relationship to associate the tag and the genres, by mapping the number of times to the genres. It is reasonable because we can treat the tagging behavior as a vote to the item. From statistical sense, a few individual, local views will be erased by the majority of user mainstream view, and individual differences in statistical sense are irrelevant. By focusing on the vote results, we can achieve maps from tags to genres and get the frequency of it, which is shown in Fig. 116.2.

For tag_i, the number of labeled times to $genre_j$ is F_{ij}, and the number of labeled times to $item_m$ is f_{im}:

$$F_{ij} = \sum_{genre_j \in item_m} f_{im}$$

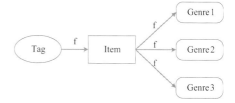

Fig. 116.1 The relationship of tag, item, and genre

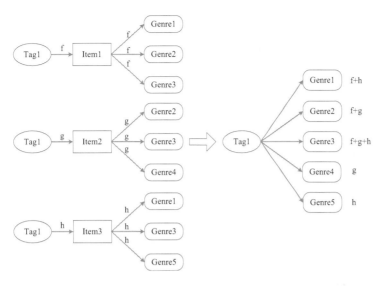

Fig. 116.2 The vote from the tag to genres

Fig. 116.3 Calculate item tendency via tag tendency

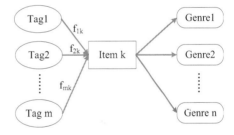

Then we can calculate the ratio of tag_i on $genre_j$:

$$ratio_{ij} = \frac{F_{ij}}{\sum_j F_{ij}}$$

Now we have the tendency of each tag on different genres. We calculate the tendency of each item on different genres by using the tag tendency as a bridge, shown in Fig. 116.3.

We calculate the frequency of $item_k$ on different genres using the matrix shown in Fig. 116.4.

For $genre_j$, the equivalent frequency FG_{jk} on $item_k$ is:

$$FG_{jk} = \sum_{i=1}^{m} f_{ik} \cdot ratio_{ij}$$

Fig. 116.4 The matrix for calculating frequency

	$genre_1$	$\cdots$	$genre_j$	$\cdots$	$genre_n$
tag_1	$f_{1k} \cdot ratio_{11}$	$\cdots$	$f_{1k} \cdot ratio_{1j}$	$\cdots$	$f_{1k} \cdot ratio_{1n}$
$\vdots$	$\vdots$		$\vdots$		$\vdots$
tag_i	$f_{ik} \cdot ratio_{i1}$	$\cdots$	$f_{ik} \cdot ratio_{ij}$	$\cdots$	$f_{ik} \cdot ratio_{in}$
$\vdots$	$\vdots$		$\vdots$		$\vdots$
tag_m	$f_{mk} \cdot ratio_{m1}$	$\cdots$	$f_{mk} \cdot ratio_{mj}$	$\cdots$	$f_{mk} \cdot ratio_{m1}$

Finally we get Tendency$_{jk}$ between genre$_j$ and item$_k$:

$$Tendency_{jk} = \frac{FG_{jk}}{\sum_j FG_{jk}}$$

In order to facilitate the comparison between different items, we have to normalize the tendency of different genres on the same item.

$$Tendency_{jk}' = \frac{Tendency_{jk}}{\max_j Tendency_{jk}}$$

116.3 Field Authority Nearest Neighbors

On the basis of genre resolution model, we design our collaborative filtering approach based on field authority (FACF), and our core tasks are to predict ratings for blank "user-item" pairs. Figure 116.5 shows the algorithm process. Traditional CF usually predicts rating using user-based [5–7] or item-based [1, 8] mode. Since our algorithm is based on field authorities, we use the user-based mode in the calculating process.

For user u and item i, we predict r_{ui} step by step as follows. First we have got the genre tendency of item i. In each genre involved, we use FACF to calculate a prediction rating and use genre tendency as weight to sum the ratings of different genres to get the final result r_{ui}.

The most important aspect of FACF is to find field authority nearest neighbors. Here we require that field authorities are in the top 100 sorted by rating numbers and at least rate 10 movies in a certain genre. For user u and Fa$_k$, which stands for field authorities on genre k, we calculate the similarities between u and member v of Fa$_k$, and there are K field authorities who are the most similar with user u as his nearest neighbors. We use the cosine similarity here:

$$w_{ui} = \frac{\sum_{i \in I}(r_{ui} - \bar{r}_u) \cdot (r_{vi} - \bar{r}_v)}{\sqrt{\sum_{i \in I}(r_{ui} - \bar{r}_u)^2 \cdot \sum_{i \in I}(r_{vi} - \bar{r}_v)^2}}$$

Fig. 116.5 FACF algorithm process

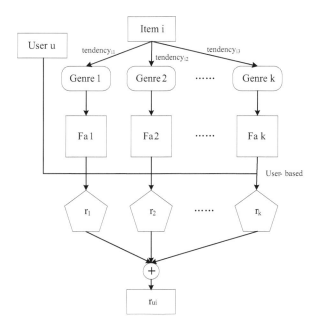

where I stands for the set that items are co-rated by u and v; $\bar{r}_u$ and $\bar{r}_v$ are the average ratings in current genre for u and v.

We can predict rating by using the similarity as a weight. This is done by means of a similarity-weighted average of the ratings input from each field authority [9]:

$$\hat{r}_{ui} = \bar{r}_u + \frac{\sum_{v \in S(u,K) \cap N(i)} w_{uv} \cdot (r_{vi} - \bar{r}_v)}{\sum_{v \in S(u,K) \cap N(i)} |w_{uv}|}$$

where $\hat{r}_{ui}$ is the prediction rating for genre k. $\bar{r}_v$ is the average rating for Fa_k member v on item i. $\bar{r}_u$ is the average rating of user u. After we get the prediction rating of each genre, we use weighted summation of predicted ratings to get the final prediction result:

$$r_{ui} = \sum_{e=1}^{k} r_e \cdot tendency_{ie}$$

In summary, the core idea of FACF is to predict ratings for user-item pairs from different genre perspective and then use user's preferences of each genre as weights to get the final result. The results are able to take the preferences between the user and field authorities in common into account and reflect the item tendency on different genres quite suitable.

116.4 Experiments

We test our algorithm on MovieLens dataset. We can verify FACF performance in prediction accuracy obviously by contrasting with kNN CF and localized ECF. The MovieLens dataset is a subset of MovieLens community, which contains 544 thousand ratings on 4,988 movies rated by 1,000 users. Each user rates 20 ratings and each movie has 1 rating at least.

In order to evaluate the accuracy of predictions, we divide our data set into 80 % training – 20 % testing sets and calculate the average results of a 5-fold cross-validation.

116.4.1 Mean Absolute Error

For each user-item pair in the test set, r_{ui} is its actual rating, and $\hat{r}_{ui}$ is its prediction rating. MAE is defined as follow:

$$MAE = \frac{\sum_{u,i \in T} |r_{ui} - \hat{r}_{ui}|}{|T|}$$

To test the performance of our algorithm under different conditions, we take the number of nearest neighbors K as a variable. We also run the same experiment on user-based kNN CF (U-kNN) and ECF as a comparison.

Figure 116.6 shows the MAE results of three approaches. FACF has the best performance on MAE test. We can see details in Table 116.1. FACF has a reduction of MAE from 3.6 to 10.8 % compared to U-kNN and from 2.4 to 6.0 % compared to ECF. It is strong evidence that FACF is significantly more accurate than U-kNN and ECF.

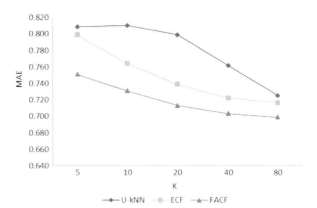

Fig. 116.6 MAE of U-kNN, ECF, and FACF

Table 116.1 MAE and reductions of U-kNN, ECF, and FACF

	U-kNN	ECF	FACF	Reduction (1-FACF/U-kNN)	Reduction (1-FACF/ECF)
K=5	0.809	0.799	0.751	7.1 %	6.0 %
K=10	0.811	0.764	0.731	9.8 %	4.3 %
K=20	0.799	0.739	0.713	10.8 %	3.5 %
K=40	0.762	0.723	0.703	7.7 %	2.6 %
K=80	0.726	0.717	0.699	3.6 %	2.4 %

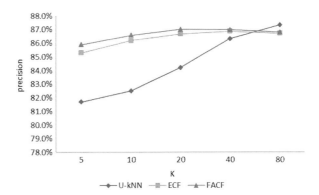

Fig. 116.7 Precision when threshold = 3

116.4.2 Precision of Recommendation Lists

Even though MAE is a good indicator, users do not consider it the way as MAE does in real life. We are more concerned with whether users feel good to what we recommend, like a top-N recommendation list [10]. Here we observe the prediction accuracy by setting a threshold to distinguish a movie between "recommended" and "not recommended" [4].

We set a threshold τ as a criterion. When our predicted rating is higher than τ, we think this movie is recommendable, otherwise is not recommendable. So if the actual rating on this movie is also higher than τ, we consider it as a successful recommendation. By calculating the success rate, we can compare the performance of user-based kNN CF, ECF, and FACF.

Figures 116.7 and 116.8 show the precision under different threshold conditions. When threshold $\tau=3$, the precision of FACF has been around 87 %, which is much higher than that of U-kNN and a little higher than that of ECF. When threshold $\tau=4$, the precision of FACF is 3 % higher than both U-kNN and ECF. We can see that when we need a high standard recommendation by using a tough condition, FACF can give us a much better result.

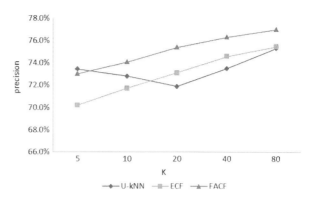

Fig. 116.8 Precision when threshold = 4

116.5 Conclusion

We propose a novel CF approach by using tags and field authorities. By mapping tags to genres of each item and introducing field authorities based on genre information, we simulate the word-of-mouth recommendation mode properly. We perform experiments on MovieLens dataset. Compared to kNN CF and localized ECF, (1) we have a much higher accuracy than user-based kNN CF and ECF; (2) there is no need to use external data, so our algorithm has a larger scope than ECF. Besides, because of using the concept of field authorities, the inherent characteristics, such as high degree of participation and consistent tendency, can ameliorate the data sparsity and noise problems existing in traditional collaborative filtering naturally.

References

1. Linden, G., Smith, B., & York, J. (2003). Amazon.com recommendations: Item-to-item collaborative filtering. *IEEE Internet Computing, 7*(1), 76–80.
2. Beyer, K., Goldstein, J., Ramakrishnan, R., & Shaft, U. (1998). When is "Nearest Neighbor" meaningful? In *ICDT'99* (pp. 217–235). LNCS 1540. Berlin: Springer.
3. Hall, P., Park, B., & Samworth, R. (2008). Choice of neighbor order in nearest-neighbor classification. *Annals of Statistics, 36*(5), 2135–2152.
4. Amatriain, X., Lathia, N., Pujol, J., Kwak, H., & Oliver, N. (2009). The wisdom of the few: A collaborative filtering approach based on expert opinions from the web. In *Proceedings of the 32nd international ACM SIGIR conference on research and development in information retrieval* (pp. 532–539). New York, NY: ACM.
5. Schafer, B., Frankowski, D., Herlocker, J., & Sen, S. (2007). Collaborative filtering recommender systems. In *The adaptive web* (pp. 291–324). LNCS 4321. Berlin: Springer.
6. Koren, Y. (2008). Tutorial on recent progress in collaborative filtering. In *Proceedings of the 2008 ACM conference on recommender systems* (pp. 333–334). New York, NY: ACM
7. Su, X., & Khoshgoftaar, T. (2009). A survey of collaborative filtering techniques. *Advances in Artificial Intelligence, 2009*, 1–19.

8. Sarwar, B., Karypis, G., Konstan, J., & Riedl, J. (2001). Item-based collaborative filtering recommendation algorithms. In *Proceedings of the 10th international conference on world wide web* (pp. 285–295). New York, NY: ACM.
9. Resnick, P., Iacovou, N., Suchak, M., Bergstrom, P., & Riedl, J. (1994). GroupLens: An open architecture for collaborative filtering of netnews. In *Proceedings of the 1994 ACM conference on computer supported cooperative work* (pp. 175–186). New York, NY: ACM.
10. Deshpande, M., & Karypis, G. (2004). Item-based top-N recommendation algorithms. *ACM Transactions on Information Systems, 22*(1), 143–177.

Chapter 117
Characteristics of Impedance for Plasma Antenna

Bo Yin and Feng Yang

Abstract Impedance analysis is very important for antenna design. In this chapter, the internal impedance of the plasma antenna is analyzed by building the model of high-frequency electromagnetic waves acting with plasma. At the same time, a model of surface current for plasma antenna is developed in accordance with the eigenvalue equation of guided mode, and the radiation resistance of plasma antenna is analyzed according to the method of Poynting vector. From the results, we find that the internal impedance and the radiation resistance of the plasma antenna are affected distinctly by the plasma density and electron-neutral collision frequency. The internal resistance could be reduced, and the radiation resistance would be added efficiently by increasing the plasma density and decreasing the collision frequency.

117.1 Introduction

More and more people are interested in plasma antenna which is based on plasma elements instead of metal conductors in recent years. Plasma antenna's behavior is determined by a circular plasma column in which a surface wave is propagating along it. A plasma antenna may work immediately once it is energized through using of an RF source, and will stop instantly with the source removed. When de-energized, the plasma antenna becomes a dielectric tube filled with inert gas, and

B. Yin (✉)
School of Electronic Engineering, University of Electronic Science and Technology of China, Chengdu 611731, China

College of Electronic Engineering, Chongqing University of Posts and Telecommunications, Chongqing 400065, China
e-mail: byin0520@163.com

F. Yang
School of Electronic Engineering, University of Electronic Science and Technology of China, Chengdu 611731, China

it reflects little return wave signals to radar [1]. Reconfiguration is another advantage of plasma antenna in which its radiation characters can be changed conveniently by electrical rather than mechanical control [2].

In the past years, many scientific or technical documents concerning plasma antenna have been reported. Kumar [3] investigated the radiation properties of a plasma column as a reconfigurable plasma antenna by controlling the operating parameters, such as drive frequency, input power, and argon gas. Zhu [4] presented the characteristics of the AC-biased plasma antenna by experimental observations. Wu [5] analyzed radiation pattern of plasma antenna through a model for a plasma antenna of beam-forming, and results indicated a good performance in the aspects of beam-forming, beam-scanning, and radiation efficiency. However, from these studies it is very difficult to infer the information of impedance of plasma antenna. As we all know, the impedance of a plasma antenna is important for impedance matching and efficiency analysis in the antenna design. In this chapter, the internal impedance and radiation resistance of plasma antenna are analyzed through the theory of impedance and the method of Poynting vector, respectively. Since physics property of plasma varies as the signal frequency due to a disperse material, the plasma density and the collision frequency have obviously influence on the impedance of plasma antenna, and results show that the internal resistance could be reduced and the radiation resistance be increased efficiently by adjusting the plasma density and the collision frequency properly.

117.2 The Internal Impedance of a Plasma Antenna

Plasma is a collection of free charged particles moving in random directions which consists of electrons, ions, and neutrons; in general, it is neutral in the steady state. When the density of ionized gas is very high enough, the electromagnetic wave cannot go deep into the plasma; the skin depth is quite small, and then plasma exhibits properties of a conductor [6]. The electric conductivity determines the ohmic power dissipation, which is an important mechanism for electron heating in discharges.

A plasma column antenna is modeled in Fig. 117.1, assume that its density is homogeneous in axial and radial directions, and the length and radius of the plasma column are 1.2 and 0.0125 m, respectively. Then the wave vector propagating in the plasma column is a complex quantity, and it can be expressed as

$$k = \omega\sqrt{\mu\varepsilon_p} = k_0\sqrt{\varepsilon_r} = \beta - j\alpha \qquad (117.1)$$

where $\varepsilon_r = 1 - \omega_{pe}^2/\omega(\omega - jv)$ is the plasma relative dielectric constant [7], ω is the frequency of propagation signal, and v_m is the electron-neutral collision frequency. The quantity $\omega_{pe}^2 = n_e e^2/m_e \varepsilon_0$ is the electron plasma frequency, where n_e and m_e are the plasma density and quality, respectively [7]. As the relative

117 Characteristics of Impedance for Plasma Antenna

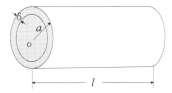

Fig. 117.1 Structure of plasma column antenna

permittivity of plasma is complex, so it can be expressed as $\varepsilon_r = \varepsilon_1 - j\varepsilon_2$. Then the attenuation constant and phase constant are

$$\beta = \omega \sqrt{\frac{\mu_0 \varepsilon_0}{2}} \sqrt{\varepsilon_1 + \sqrt{\varepsilon_1^2 + \varepsilon_2^2}} \qquad (117.2)$$

$$\alpha = \omega \sqrt{\frac{\mu_0 \varepsilon_0}{2}} \sqrt{-\varepsilon_1 + \sqrt{\varepsilon_1^2 + \varepsilon_2^2}} \qquad (117.3)$$

So we can obtain skin depth of the plasma column $\delta = 1/\alpha$.

For a plasma column with certain density, it shows conductive properties like a conducting column. So the internal impedance of plasma antenna can be counted as the following:

When $\delta < a$,

$$R_1 = R_s = \frac{l}{\sigma S} = \frac{l}{\sigma S} = \frac{l}{2\pi a \delta \sigma} \qquad (117.4)$$

When $\delta > a$,

$$R_1 = R_s = \frac{l}{\sigma S} = \frac{l}{\sigma S} = \frac{l}{\pi a^2 \sigma} \qquad (117.5)$$

The internal impedance of the plasma antenna is plotted in Fig. 117.2a, b at different plasma densities, when the electron-neutral collision frequency is 5×10^8 Hz. From Fig. 117.2a, the internal resistance of the plasma antenna becomes bigger slowly as the signal frequency rises, and increases rapidly in the low-frequency band. And if we enhance the plasma density, the internal resistance can be decreased observably, which is the results of the plasma conductivity increasing with the increase of the plasma density. In Fig. 117.2b, the internal reactance of plasma antenna increases linearly with the increase of signal frequency. And for a higher plasma density, the internal reactance becomes smaller, and the slope is smaller too.

It is the collision frequency that is relevant to plasma's conductive and dielectric properties, which impact antennas' characteristics with using plasma as a conducting column. The internal impedance of the plasma antenna is plotted in Fig. 117.3a, b at electron-neutral collision frequencies, when the plasma density is $n_e = 1 \times 10^{18}$ m^{-3}. As shown in Fig. 117.3a, the internal resistance of plasma

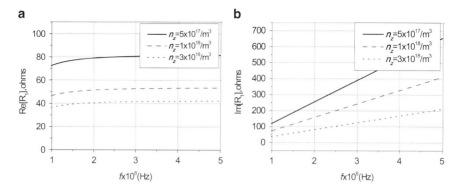

Fig. 117.2 (**a**) Internal resistance and (**b**) internal reactance of the plasma antenna at different plasma densities

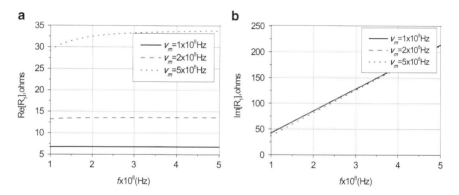

Fig. 117.3 (**a**) Internal resistance and (**b**) internal reactance of the plasma antenna at different collision frequencies

antenna becomes bigger as the signal frequency increases, and arises rapidly in the low-frequency band. At same time, if we enhance the collision frequency of plasma, the internal resistance can be increased evidently. This is easily understood on the basis of the collision model for plasma. Electromagnetic (EM) waves propagating in plasma will cause oscillation of electrons. If no collision, the plasma will show characteristics of a lossless medium. However, electrons will collide with each other and with neutral particles and their kinetic energy is generally exchanged between particles. And the energy transfer is critically dependent on the electron's oscillation cycle when the collision occurs [8]. So the higher collision frequency is, the greater plasma resistance will be. In Fig. 117.3b, the results indicate that the internal reactance of the plasma antenna increases linearly with the increase of signal frequency, and the internal reactance is almost not affected by collision frequency of plasma.

117 Characteristics of Impedance for Plasma Antenna

117.3 The Radiation Resistance of a Plasma Antenna

When frequencies of incident EM waves are far lower than the plasma frequency, the surface wave will appear on the interface between the plasma and the surrounding dielectric just like traveling on the surface of a metal column. However, the wave vector of surface wave should vary with the plasma density. For a uniform density, we may obtain the plasma surface wave dispersion relation by Helmholtz's wave equation and the boundary condition [9, 10], that is,

$$\varepsilon_r T_0 I_1(T_p a) K_0(T_0 a) + T_p K_1(T_0 a) I_0(T_p a) = 0 \quad (117.6)$$

where $T_p^2 = k^2 - \varepsilon_r k_0^2$ and $T_0^2 = k^2 - k_0^2$, a is the radius of the plasma column, $k_0 = \omega/c$ is the wave number in free space, $I_i(\cdot)$ and $K_i(\cdot)$ denote modified Bessel function of the first and second kind, respectively, and ε_r is the plasma relative dielectric constant.

It is through the mechanism of the radiation resistance that power is transferred from the guided wave of the plasma antenna to the free-space wave. The greater the radiation resistance is, the higher will be the power radiated for a given electric current. In order to find the radiation resistance of plasma antenna, the Poynting vector is formed in terms of the electric field and magnetic field radiated by the antenna. By integrating the Poynting vector over a closed surface (usually a sphere of very large radius), the total power radiated by the source is found. As approximation, the current distribution of monopole for plasma column is

$$I(z) = I_0 \left(e^{jk_p z} - e^{jk_p(2l-z)} \right) \quad (117.7)$$

and the current in the fed end of plasma column is

$$I_1 = I_0 \left(1 - e^{j2k_p l} \right) \quad (117.8)$$

where k_p is the propagation constant of surface wave and can be got from Eq. 117.6. The far-zone field of electric current element is given by [11]

$$dE_\theta = j \frac{60\pi I dz}{\lambda R} \sin\theta e^{-jk_0 R} \quad (117.9)$$

where λ and k_0 are the wave length and the wave number in free space, respectively, and R is the distance between field point and source point.

So the far-zone electric field of the plasma antenna with a length l is

$$E_\theta = \int_0^l j \frac{60\pi I_z}{\lambda r_1} \sin\theta e^{-jk_0 r_1} dz \quad (117.10)$$

Fig. 117.4 Plasma antenna placed along the z axis

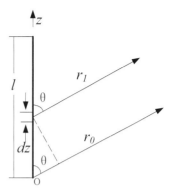

As shown in Fig. 117.4, a plasma column antenna, $l = 1.2$ m and $a = 0.0125$ m, is placed along the z axis. r_0 is the distance between the origin to the field point, and r_1 is the distance between the electric current element (Idz) to the field point in far region. Consider $l \ll r_0$, so that we take $1/r_0 \approx 1/r_1$. Due to $r_0 \| r_1$ in the far field, as the first approximation, we can take $r_1 = r_0 \cos\theta$, then

$$E_\theta = j\frac{60\pi}{\lambda r_0} e^{-jk_0 r_0} \sin\theta \int_0^l I_z e^{jk_0 z \cos\theta} dz \qquad (117.11)$$

The power radiated by the plasma antenna is

$$P_r = \oiint_\Omega |\vec{S}| dA \qquad (117.12)$$

where Ω is a sphere surface of its radius in the far-field region. Poynting vector $\vec{S} = \frac{1}{2}\mathrm{Re}\left[\vec{E} \times \vec{H}^*\right]$.

Finally, the radiation resistance can be obtained by the principle of the equivalent circuit.

$$R_r = P_r/I_1^2 \qquad (117.13)$$

When the electron-neutral collision frequency is 5×10^8 Hz, the radiation resistance of the plasma antenna is plotted in Fig. 117.5a at different plasma densities. In the figure, we get that the plasma antenna displays resonance characteristic similar to that of the metal one, the curve is moved to the right, and its peak values become bigger with the increase of concentration for plasma. For a plasma column, the increasing plasma density will cause a good conductivity, and this leads to a stronger ability to radiate EM wave. On the other hand, the peak value of radiation resistance is approximately periodic, which will help us select a frequency band to realize a high-efficiency plasma antenna. Figure 117.5b shows that the radiation resistance of the plasma antenna varies with different electron-neutral

117 Characteristics of Impedance for Plasma Antenna

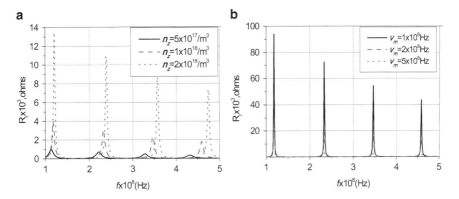

Fig. 117.5 Radiation resistance for the plasma antenna (**a**) at different plasma densities and (**b**) at different collision frequencies

collision frequencies, when $n_e = 1 \times 10^{18}$ m^{-3}. The results tell us that the plasma antenna displays resonance characteristics too and the peak values of the curve with the increase of collision frequencies of plasma. As mentioned previously, the power is consumed in the plasma column, which reduces the EM power radiation, so the radiation resistance becomes smaller.

117.4 Conclusion

Since high-frequency electromagnetic waves act with plasma and charged particles collide with each other, a plasma column not only shows the ability of EM power radiation but also demonstrates ohmic loss characteristics. In this chapter, the internal impedance and radiation resistance of plasma antenna are analyzed through the theory of impedance and the method of Poynting vector respectively. The results indicate that the internal resistance could be reduced and the radiation resistance be increased efficiently by adjusting the plasma density and the collision frequency properly; accordingly, the efficiency of the antenna is improved too. It is very useful for the analysis and design of plasma antenna.

References

1. Cerri, G., De Leo, R., Mariani Primiani, V., & Russo, P. (2008). Measurement of the properties of a plasma column used as a radiated element. *IEEE Transactions on Instrumentation and Measurement, 57*(2), 242–247.
2. Alexe, I., Anderson, T., Farshi, E., Karnam, N., & Pulasani, N. R. (2008). Recent results for plasma antennas. *Physics of Plasmas, 15*(5), 057104.

3. Kumar, R., & Bora, D. (2010). A reconfigurable plasma antenna. *Journal of Applied Physics, 107*(5), 053303.
4. Zhu, A., Chen, Z., Lv, J., & Liu, J. (2012). Characteristics of AC-biased plasma antenna and plasma antenna excited by surface wave. *Journal of Electromagnetic Analysis and Applications, 4*(7), 279–284.
5. Wu, X. P., Shi, J. M., Chen, Z. S., & Xu, B. (2012). A new plasma of beam-forming. *Progress in Electromagnetics Research, 126*, 539–553.
6. Yin, B., Yang, F., Wang, B., & Hao, H. G. (2011). Mutual impedance of plasma antennas, 2011. In *7th International Conference on Wireless Communications, Networking and Mobile Computing* (pp. 1–4). Wuhan, China: IEEE Press.
7. Lieberman, M. A., & Lichtenberg, A. J. (2005). *Principles of plasma discharges and materials processing* (pp. 46–88). New York, NY: Wiley.
8. Yuan, C. X., Zhou, Z. X., & Sun, H. G. (2010). Reflection properties of electromagnetic wave in a bounded plasma slab. *IEEE Transaction on Plasma Science, 38*(12), 3348–3355.
9. Rayner, J. P., Whichello, A. P., & Cheetham, A. D. (2004). Physical characteristics of plasma antennas. *IEEE Transactions on Plasma Science, 32*(1), 269–281.
10. Rayner, J. P., & Cheetham, A. D. (2010). Travelling modes in wave-heated plasma sources. *IEEE Transactions on Plasma Science, 38*(2), 62–72.
11. Balanis, C. (2005). *Antenna theory-analysis and design* (3rd ed., pp. 31–69). New York, NY: Wiley.

Chapter 118
A Low-Voltage 5.8-GHz Complementary Metal Oxide Semiconductor Transceiver Front-End Chip Design for Dedicated Short-Range Communication Application

Jhin-Fang Huang, Jiun-Yu Wen, and Yong-Jhen Jiangn

Abstract A 5.8-GHz transceiver front-end applied in dedicated short-range communication (DSRC) systems which is developed in public traffic transportation to improve the safety is fabricated on a chip using TSMC 0.18-μm CMOS process. The proposed prototype includes an asymmetric T/R switch, a current-reused LNA, and a class A power amplifier (PA) on the low-voltage operation in order to minimize the power consumption. Measured results achieve the power gain of 11 dB, the NF of 4.9 dB, the third-order intercept point (IIP3) of −5.4 dBm, and the power consumption of 3.9 mW in the receiving (Rx) mode. On the other hand, the power gain of 12.4 dB, the output 1 dB compression point (OP_{-1dB}) of 11.4 dBm, the PAE of 14.7 % at P_{-1dB}, the IMD3 of −15.8 dBc at 1 dB compression level, the output power of 2.6 dBm with a 50 Ω load, and power consumption of 116.3 mW are obtained in the transmitting (Tx) mode. The overall chip area is 1.5 (1.32 × 1.14) mm^2. This RF CMOS transceiver front-end includes all matching circuits and biasing circuits, and no external components are required.

118.1 Introduction

For mobile wireless communication, handheld sets of small size and light weight are more attractive, and the battery endurance plays an important role in the market. In order to minimize the required power consumption, operating the circuit at a reduced supply voltage is apparently an effective approach. ITS communication system can effectively improve the mobile safety and traffic efficiency in vehicle transportation. The DSRC protocol is defined in the physical layer of ITS and

---

J.-F. Huang (✉) • Y.-J. Jiangn
Department of Electronic Engineering, National Taiwan University of Science and Technology, Taipei 106, Taiwan
e-mail: jfhuang@mail.ntust.edu.tw

J.-Y. Wen
National Communications Commission, Taipei 106, Taiwan

supports both public safety and private operations in vehicle to roadside communications and provides a high-speed radio link between the roadside unit (RSU) and onboard unit (OBU). Meanwhile, the DSRC protocol has been developed worldwide and practically applied for electronic toll collection (ETC) system. Furthermore, the 5.8-GHz band is located in the unlicensed industrial, scientific, and medical (ISM) bands and is widely applied for medium distant communication applications.

A fully integrated 0.25-μm SiGe-BiCMOS transceiver for DSRC applications is presented, but this chip is more expensive than CMOS one [1]. In addition, larger power consumption is unpractical for OBU's application under the supply voltage of 3.3 V. A series-shunt T/R switch is integrated with transceiver front-end amplifier, but its power-handling capability is too low for Tx path [2]. A fully integrated transceiver front-end is proposed with the T/R switch, LNA, and PA devices, yet this chip is unpractical due to larger power consumption [3]. A fully integrated high-efficiency linear CMOS class E PA for 5.8-GHz ETC applications is presented [4]. Class E PA has high efficiency, but it is only suitable for nonlinear modulation communication systems. For DSRC systems, it is not suitable. Hence, with those considerations, a low-voltage 5.8-GHz CMOS transceiver front-end chip design for DSRC applications is presented in this paper.

118.2 Transceiver Front-End Circuit Design

Figure 118.1 shows the structure of the proposed prototype which includes a T/R switch, an LNA, and a PA. The T/R switch is a key block for a time division duplex (TDD)-based radio system, and it connects radio transmitter and receiver alternatively to a shared antenna. It receives the radio signal from antenna to the LNA or transmits the radio signal from the PA to antenna. In Rx mode, the LNA appropriately amplifies the weak radio signal from the antenna through the T/R switch but not adding too much noise to it. In Tx mode, the PA amplifies the radio signal from the up-converter to appropriate signal power level through the T/R switch to the antenna.

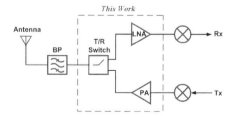

Fig. 118.1 The structure of the proposed transceiver front-end

Fig. 118.2 The proposed asymmetric T/R switch circuit

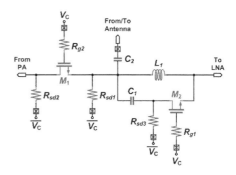

118.2.1 An Asymmetrical T/R Switch Circuit

The proposed T/R switch is an asymmetrical architecture shown in Fig. 118.2. In the Tx path, a series switch, M_1, is used, whereas a shunt switch, M_2, is employed in the Rx path. A digital control signal, V_C, is applied to the gates of the switches to select Tx or Rx mode operation. Series gate bias resistors, R_{g1} and R_{g2}, are used, so the gate potential is bootstrapped to the source and drain. The size of M_1 is determined based on the trade-off of the on-resistance, R_{on} which affects insertion loss, and C_{sd} in cutoff region, which affects isolation. The on-resistance can be evaluated as

$$R_{on} = \frac{1}{\mu_n C_{ox} W (V_{gs} - V_{TH})/L}. \quad (118.1)$$

It is desired to keep R_{on} small to reduce the insertion loss. This can be achieved by choosing large mobility, μ; increasing transistor aspect ratio, W/L; and keeping $V_{gs} - V_{TH}$ large where V_{TH} is the threshold voltage.

One criterion uses NMOS transistors rather than PMOS transistors in the design. The other rule designs transistors with minimum allowable channel length, L. Because the minimum value of L is limited by the process, low R_{on} eventually requires large W. However, broadening a transistor definitely increases its junction and parasitic capacitances proportionally. The source and drain voltages ($V_{S/D}$) of M_1 and M_2 are biased by the inverted V_C through large bias resistors ($R_{sd1} - R_{sd3}$). In Tx mode, V_C is set at 1.8 V, so V_{SD} of M_1 and M_2 is at 0 V. The Rx path presents a parallel resonant tank with L_1 and C_1 shortened through M_2. The large bias resistor R_{sd3} causes the source terminal of M_2 to be floating. This keeps the impedance between M_2's source and drain small in spite of large signal swings at the antenna node. As a result, the quality factor (Q) of the L_1C_1 tank remains sufficiently high to effectively block out leakage power from the Tx branch.

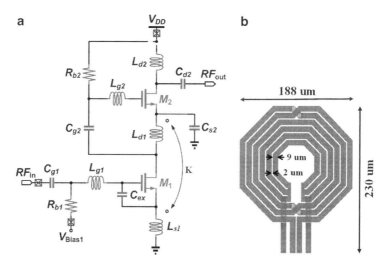

Fig. 118.3 The proposed current-reused LNA: (**a**) schematic and (**b**) transformer layout

118.2.2 A Current-Reused Low-Noise Amplifier

To reduce power consumption, a two-stage cascade amplifier is folded into a single-stage cascode amplifier. Applying this idea obtains the proposed current-reused LNA circuit which comprises an input matching network matched to 50, and an on-chip transformer-degenerated cascode amplifier shown in Fig. 118.3a. Figure 118.3b illustrates this transformer layout. The two inductors L_{d1} and L_{s1} forming this transformer are connected at the source terminals of M_1 and M_2 for reducing chip area. The transformer has turn ratio of 1:3, 9-μm metal width, 2-μm space, 30-μm inner length, 188-μm outer length, and a chip area of 188 × 230 μm². This LNA provides higher gain than the common cascode amplifier since when operating in a 5.8-GHz frequency band, C_{g2} is shortened and the LNA acts as a cascade amplifier. On the contrary, when operating in lower frequency band, C_{g2} is open and the LNA acts as a cascode amplifier. Obviously, the power consumption of this circuit will keep the same as a cascode amplifier. The input series gate matching inductor actually comes from the L_1 used in Fig. 118.2. The chip area can then be reduced. The capacitive coupling C_{g2} is needed to achieve DC isolation between the active devices of both transistors M_1 and M_2. L_{g2} is added to optimize interstage matching with C_{g2}, the effective parasitic capacitance of the active device M_2 and the on-chip transformer.

Neglecting the effect of the bias resistor, R_{b1}, the LNA input impedance Z_{in} is solved by writing Kirchhoff's voltage law in its phasor form across its input loop:

$$Z_{in} = \frac{g_{m1}L_{s1}}{C_{ex}} + j\left[\omega(L_{s1} + L_{g1}) + \left(-\frac{1}{\omega C_{g1}} - \frac{1}{\omega C_{ex}}\right)\right], \quad (118.2)$$

Fig. 118.4 Schematic of the proposed power amplifier circuit

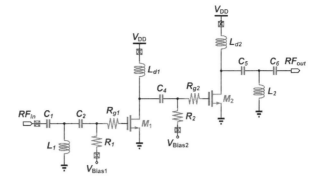

where g_{m1} is the transconductance of the transistor M_1 and the added C_{ex} contains C_{gs1}, the parasitic capacitance between the gate and source nodes.

For input matching, $Z_{in} = R_s = 50\,\Omega$, from Eq. 118.2, the input matching occurs at the frequency ω_c as

$$\omega_c = \sqrt{\frac{1}{L_{s1}+L_{g1}} \cdot \frac{C_{g1}+C_{ex}}{C_{g1}+C_{ex}}}. \tag{118.3}$$

At this frequency, the input impedance becomes

$$R_s = \mathrm{Re}(Z_{in}) = \frac{g_{m1}L_{s1}}{C_{ex}}, \tag{118.4}$$

where $\mathrm{Re}(Z_{in})$ means the real part of Z_{in}.

118.2.3 Class A Power Amplifier

The proposed PA circuit is shown in Fig. 118.4. It is a two-stage cascade common source amplifier. Both the drive stage and the power stage are biased on class A operation to achieve high linearity. To evaluate the PA function, power-added efficiency, *PAE* includes information on the driving power for the PA and is more commonly used than power conversion efficiency. *PAE* is defined as

$$PAE = \frac{P_{out} - P_{in}}{P_{DC}}, \tag{118.5}$$

where P_{in} and P_{out} are input power and output power, respectively, at the frequency of interest. P_{DC} is the DC supply power.

118.3 Measured Results

The proposed transceiver front-end is fabricated in TSMC 0.18-μm CMOS process. The die photograph is shown in Fig. 118.5, and the chip area including pads is 1.5 (1.32 × 1.14) mm². Measurements have been performed with a GSG probe bench with an HP 8510C network analyzer, an Agilent 8975A NF analyzer, and an Agilent E4407B spectrum analyzer. The power consumption is 3.9 mW with 1−V supply voltage in the Rx mode and 116.3 mW with 1.8−V supply voltage in the Tx mode.

Figure 118.6a, b shows the measured power gain and noise figure, respectively, in the Rx mode. The measured power gain is 11 dB somewhat below the simulated value about 6 dB at 5.8 GHz, but their data curves are pretty matched. The measured NF is 4.9 dB at 5.8 GHz. Figure 118.7a, b shows the measured P_{-1dB} in the Rx mode. The measured P_{-1dB} is −15 dBm. Figure 118.8 shows the measured IIP3 in the Rx mode. The measured IIP3 is −5.4 dBm. The P_{-1dB} of

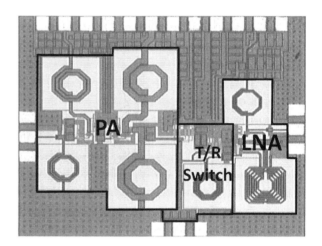

Fig. 118.5 Die photomicrograph of the transceiver front-end with a chip area of 1.32 × 1.14 (1.50) mm²

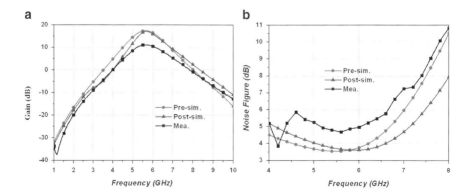

Fig. 118.6 Measured power gain and noise figure vs. frequency in the Rx mode

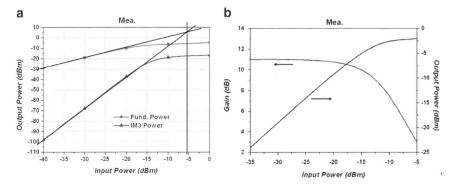

Fig. 118.7 (a) Measured power gain and output power vs. input power in the Rx mode with $P_{-1dB} = -15$ dBm and (b) measured IIP3 in the Rx mode with an IIP3 value of -5.4 dBm

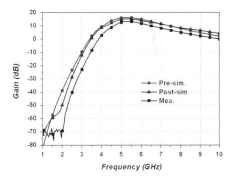

Fig. 118.8 Measured power gain vs. frequency in the Tx mode with a power gain of 12.4 dB at 5.8 GHz

output power at the linear region of operation can also be verified by $P_{-1dB} = \text{IIP3} - 9$ (dBm).

Figure 118.8 shows the measured power gain in the Tx mode with a value of 12.4 dB at 5.8 GHz. Figure 118.9a shows that the measured result exhibits an OP_{-1dB} of 11.4 dBm with a PAE of 14.7 % in the Tx mode. The maximum PAE of 21 % is achieved at an output power of 13.7 dBm, where the power gain is greater than 12 dB.

Figure 118.9b depicts the measured IMD3 and output power as a function of input power in the Tx mode, and the measured IMD3 is -15.8 dBc at $P_{-1dB} = -15$ dBm. The measured output power at single port with no buffer circuits is 2.6 dBm at the frequency f_o of 5.8 GHz with input power of -10 dBm. The measured performance of the proposed transceiver front-end is summarized and compared to recently other reported works in Table 118.1. The proposed transceiver front-end achieves the smallest chip area, the lowest NF, and the lowest power consumption while attaining very good performances compared to other features.

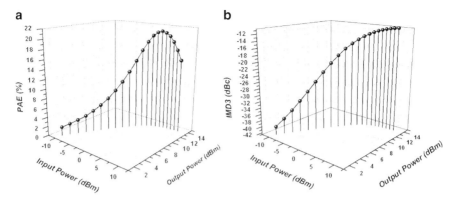

Fig. 118.9 (a) Measured PAE and output power vs. input power in the Tx mode with a measured PAE of 14.7 % at P_{-1dB} of 11.4 dBm and (b) measured IMD3 and output power vs. input power in the Tx mode with a measured IMD3 of -15.8 dBc at $P_{-1dB} = -15$ dBm

Table 118.1 Comparison of some transceiver front-ends in recent publications

	[5] 2005	[6] 2010	[7] 2006	This work
Process (μm)	BiCMOS 0.35	CMOS 0.13	CMOS 0.18	CMOS 0.18
f_{RF} (GHz)	5.2	5.8	5.5	5.8
Chip area (mm^2)	2.86	6.24	NA	1.5
Rx mode				
V_{DD} (V)	N/A	1.2	3.0	1
DC current (mA)	9.8	52[a]	40[b]	3.9
Conver. gain (dB)	21.9	48	NA	11
NF (dB)	N/A	5	NA	4.9
IIP3 (dBm)	N/A	1.1	N/A	−5.4
Tx mode				
V_{DD} (V)	N/A	3.3	3.0	1.8
Current (mA)	59.1	150	40[b]	64.6
Power gain (dB)	12.1	NA	NA	12.4
OP_{-1dB} (dBm)	N/A	−41	NA	11.4
PAE @ P_{-1dB} (%)	N/A	9.2	N/A	14.7

[a]52 mA includes Rx RF front-end, Rx baseband analog circuits, and Rx LO buffers. 150 mA includes Tx RF front-end, Tx baseband analog circuits, and Tx LO buffers
[b]Both Tx/Rx modes

118.4 Conclusion

A 0.18-μm 5.8-GHz CMOS fully integrated transceiver front-end for DSRC applications was presented. A T/R switch of asymmetric topology is advantageous for handling high power since it takes into account the asymmetrical power level in the Tx and Rx branches of a typical transceiver. The T/R switch in Rx side is merged with the LNA architecture consuming very small chip area and is almost with no power consumption. On-chip T/R switch is built in the transceiver front-end chip, allowing for reduced signal loss in the whole architecture. This integrated CMOS transceiver front-end includes all matching circuits and biasing circuits, and no external components are required. Our work can provide a compact, low-power, and low-cost solution to DSRC payloads.

Acknowledgements The authors would like to acknowledge the fabrication support and chip fabrication provided by the National Chip Implementation Center (CIC). Thanks are also given to Dr. Ron-Yi Liu for his layout guidance and Taiwan Mobile-Phone Inc. for the financial support.

References

1. Sasho, N., Minami, K., Fujita, H., Takahashi, T., Iimura, K., Abe, M., et al. (2008). Single chip 5.8GHz DSRC transceiver with dual-mode of ASK and Pi/4-QPSK[C]. *Proceedings of IEEE Radio and Wireless Symposium* (pp. 799–802). Orlando, FL.
2. Yamamoto, K., Heima, T., Furukawa, A., Ono, M., Hashizume, Y., Komurasaki, H., et al. (2001). A 2.4-GHz-band 1.8-V operation single-chip Si-CMOS T/R-MMIC front-end with a low insertion loss switch. *IEEE Journal of Solid-State Circuits, 36*(8), 1186–1197.
3. Jou, C.-F., Huang, P.-R., & Cheng, K.-H. (2003). Design of a 0.25-μm transceiver front-end[C]. *Proceedings of IEEE International Symposium on Electronics, Circuits, and Systems* (Vol. 3, pp. 1090–1093). Sharjah, United Arab Emirates.
4. Suh, Y., Sun, J., Horie, K., Itoh, N., & Yoshimasu, T. (2009). Fully-integrated novel high efficiency linear CMOS power amplifier for 5.8 GHz ETC applications[C]. *Proceedings of Asia Pacific Microwave Conference* (pp. 365–368). Singapore.
5. Kanaya, H., Koga, F., Seki, K., & Yoshida, K. (2005). Impedance matching circuit for wireless transceiver amplifier based on transmission line theory[C]. *Proceedings of Asia Pacific Microwave Conference* (pp. 19–22). Suzhou, China.
6. Kwon, K., Choi, J., Choi, J., Hwang, Y., Lee, K., & Ko, J. (2010). A 5.8 GHz integrated CMOS dedicated short range communication transceiver for the Korea/Japan electronic toll collection system. *IEEE Transactions on Microwave Theory and Techniques, 58*(11), 2751–2763.
7. Nagata, M., Masuoka, H., Fukase, S.-I., Kikuta, M., Morita, M., & Itoh, N. (2006). 5.8 GHz RF transceiver LSI including on-chip matching circuits[C]. *Proceedings of IEEE Bipolar Circuits and Technology Meeting* (pp. 263–266). Maastricht Dutch.

Chapter 119
A 5.8-GHz Frequency Synthesizer with Dynamic Current-Matching Charge Pump Linearization Technique and an Average Varactor Circuit

Jhin-Fang Huang, Jia-Lun Yang, and Kuo-Lung Chen

Abstract A 5.8-GHz frequency synthesizer is implemented in TSMC 0.18-μm CMOS process. This paper proposes a dynamic current-matching charge pump linearization technique and uses a current-switching differential Colpitts VCO to lower the phase noise and an averaged varactor circuit to increase the linearity of the VCO tuning range. At the supply voltage of 1.8 V, measured results achieve the locked tuning frequency from 5.55 to 5.94 GHz, corresponding to 6.8 % and the phase noise of −105.83 dBc/Hz at 1 MHz offset frequency from 5.8 GHz. The overall power consumption is 21.6 mW. Including pads, the chip area is 0.729 (0.961 × 0.761) mm^2.

119.1 Introduction

Frequency synthesizer is an important component used in wireless transceiver front-end to perform signal up- and down-conversion. The integer-N frequency synthesizer is considered to be well understood and less complicated to design. Several multiband frequency synthesizers have been published [1–3]. A 5-GHz frequency synthesizer utilizes injection-locked frequency divider (ILFD) in the first divider stage to save power found, but it may cause the frequency synthesizer becoming unlocked due to the narrow locked range of ILFD and consuming more chip area due to the inductor of ILFD [1]. A 1-V frequency synthesizer for

---

J.-F. Huang (✉) • J.-L. Yang
Department of Electronic Engineering, National Taiwan University of Science and Technology, Taipei 106, Taiwan, China
e-mail: jfhuang@mail.ntust.edu.tw

K.-L. Chen
National Communications Commission, Taipei 106, Taiwan, China

low-voltage applications is presented, but it consumes large chip area of 0.988 mm² and much power of 27.5 mW [2]. A locking detector to detect the locking situation of the PLL is adopted, but it is more complex as illustrated in [3]. In the phase/frequency detector (PFD) and charge pump (CP) circuits, the nonlinearity is mainly attributed to the up/down current mismatch and the gain (slope) variation around the region of phase error; while in the dividers, the circuit timing jitters modulate the zero-crossing points of a signal and cause the system to exhibit nonlinear behavior. Hence, considering those factors of power consumption, phase noise, tuning range and chip area, a low-phase noise, wide tuning range, and small chip area, frequency synthesizer is proposed and fabricated in TSMC 0.18-μm CMOS process.

119.2 Architecture of Frequency Synthesizer

The proposed frequency synthesizer consists of a PFD, a CP, an off-chip 3rd-order passive loop filter and a VCO in the feed-forward path and a programmable frequency divider in the feedback path as shown in Fig. 119.1. An accurate VCO with low noise is essential in designing a quality frequency synthesizer. The PFD detects the phase error between the reference signal F_{REF} and the feedback signal F_{DIV}. The digital output signals of PFD control the VCO through the CP and filter circuits. The locked frequency synthesizer output frequency f_{VCO} is expressed as follows:

$$f_{VCO} = 4 \times N_{MMFD} \times F_{REF}, \qquad (119.1)$$

where N_{MMFD} is the programmable divider ratio and F_{REF} is the reference frequency.

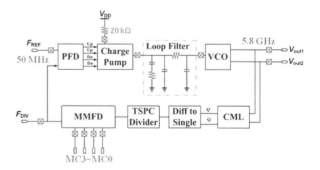

Fig. 119.1 Architecture of the proposed frequency synthesizer

119.3 Frequency Synthesizer Functions

119.3.1 G_m-Boosting Voltage-Controlled Oscillator

The cross-coupled VCO in CMOS has attracted considerable interest due to its easy start-up and good phase noise characteristics. Colpitts VCO features superior phase noise because noise current from active devices is injected into the tank during minima of the tank voltage when the impulse sensitivity is low. Unfortunately, the conventional Colpitts VCOs suffer from poor start-up characteristics; i.e., higher power consumption is needed to ensure reliable start-up.

In order to resolve the poor start-up characteristic, and improve the phase noise, the current-switching differential Colpitts VCO shown in Fig. 119.2 where the differential oscillator is built with symmetry around the resonator tank is modified from [4]. The phases of gate voltages of M_1 and M_3 and M_2 and M_4 are the same as the phases of drain voltages of M_2 and M_1, respectively. Connecting them together will have the effect of G_m-boosting scheme. The negative resistance $-2/g_m$ where g_m denotes the transconductance of each transistor generated by the cross-coupled pMOS transistors is to compensate for the loss with the LC-tank. The differential outputs of the VCO connect to the common source amplifiers which function as analog buffers. This proposed balanced VCO consists of two single-ended LC-tanks and two pairs of pMOSFETs so that their gate-source voltages become small. To enhance the start-up oscillation condition of the balanced Colpitts VCO, the pMOSFET core is chosen to reuse the dc current. Therefore, the proposed balanced VCO uses four MOSFETs, and the consumed current still remains very small; therefore, low power dissipation can be achieved.

The capacitance $C_{var1, 2}$, realized from a high-Q MOSFET capacitor, in parallel with the inductors forms the LC-tank resonator which determines the oscillating frequency. The oscillating frequency is given by

$$f_o = \frac{1}{2\pi\sqrt{L_1 C_{vnet}}}. \tag{119.2}$$

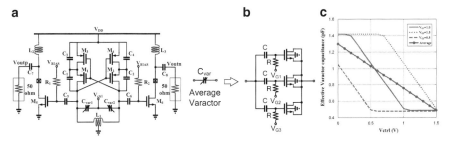

Fig. 119.2 The gain-boosting Colpitts VCO: (**a**) the overall circuit; (**b**) an averaged varactor circuit by using distributed bias voltages of 1.0, 1.5, and 0.5 V; and (**c**) linear property of the averaged capacitance

where C_{vnet} is the effective capacitance of the VCO. Three MOS varactors are connected in parallel with dc bias voltages V_{G1}, V_{G2}, and V_{G3}. Rather than using a fixed dc bias voltage (i.e., $V_{G1} = V_{G2} = V_{G3}$) in the traditional varactor design, the averaged varactor circuit uses distributed voltage values for V_{G1}, V_{G2}, and V_{G3} which are 1.0, 1.5, and 0.5 V, respectively, as shown in Fig. 119.2b [5]. The tuning capacitance of the combination of the three varactors is approaching linear shown in Fig. 119.2c and is therefore insensitive to the varactor errors. The resonant frequency of the LC-tank will become more stable and the nonlinearities of the varactors are averaged, but at the sacrifice of the chip area of 0.15 mm^2. The averaged varactor increases the linearity of VCO tuning range. Therefore, the phase noise is lowered and this feature is applied to these VCO varactors.

119.3.2 Phase Frequency Detector

Important techniques to design PFD operating at high frequency with minimum dead zone are adopted to reach the minimum phase offset and to reduce the dead zone in the circuit. With these considerations, a domino-logic PFD shown in Fig. 119.3 is presented [6]. On the reset path, the reset delay cell is inserted to reduce the minimum UP and DN pulse widths and then to improve the dead zone in the PFD. When the input and output frequencies are sufficiently close, the PFD operates as a phase detector, performing phase lock. The loop locks when the phase difference drops to zero and the charge remains relatively idle.

119.3.3 Dynamic Current-Matching Charge Pump

A charge pump consists of two switched current sources that pump charge into or out of the loop filter according to logic inputs. The main objective is to design a reasonably sized CP circuit to achieve the performance of a large-sized one for area efficiency and minimizing the unwanted transient corruptions. Figure 119.4 illustrates the proposed CP PLL driven by a PFD, such an implementation senses the transitions at the input and output, detects phase or frequency differences, and activates the charge pump accordingly [7].

This CP is based on a switches-in-source architecture. The feature is that two extra feedback transistors, M_{fbN} and M_{fbP}, are added to compensate for the channel-length modulation effect of the up/down current mirrors via negative feedback. The amount of compensation is dynamically adjusted according to the CP output voltage (V_{CP}). The technique reduces the current error between charge up and charge down due to the mismatch of charge-sharing effects when the switches are turned ON and hence improves the in-band phase noise.

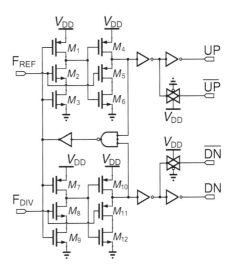

Fig. 119.3 The proposed PFD circuit with minimum dead zone

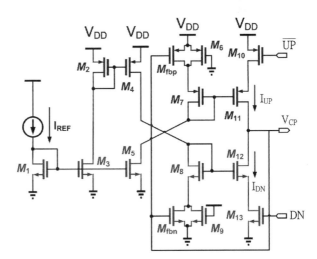

Fig. 119.4 Dynamic current-matching charge pump circuit

119.3.4 Low-Pass Filter

The low-pass filter eliminates noise from the tuning voltage. The VCO output signal is altered by the control data inputted into the filter which follows some parameter specs in the PLL, including reference frequency (F_{REF}), divide ratio (N), loop bandwidth (K), charge pump current (I_{CP}), slope of VCO (K_{VCO}), and phase margin (ϕ_P). The filter suppresses spurs introduced by the reference frequency.

Fig. 119.5 The third-order low-pass loop filter circuit

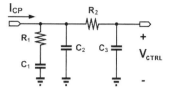

Figure 119.5 shows the schematic of the third-order low-pass loop filter which provides more attenuation of spurs by placing a series resistor R_2 and a shunt capacitor C_3. $V_{CTRL}(s)/I_{cp}(s)$ will become

$$Z(s) = \frac{sR_1C_1 + 1}{s^3R_1R_3C_1C_2C_3 + s^2(R_1C_1C_2 + R_1C_1C_3 + R_3C_2C_3 + R_3C_1C_3) + s(C_1 + C_2 + C_3)}. \tag{119.3}$$

Resistor R_1 and capacitor C_1 generate a pole at the origin and a zero at $1/(R_1C_1)$. C_2 and the combination of R_2 and C_3 generate extra poles at frequencies higher than frequency synthesizer bandwidth of interest to reduce the feedthrough at reference frequency and decrease spurious harmonics of the reference frequency. The component parameters used in this 3rd low-pass loop filter with consideration of CP currents, VCO tuning range gain, frequency divider ratio, loop filter bandwidth, phase noise, etc., are listed as follows: $F_{REF} = 50$ MHz, $K = 350$ kHz, $I_{CP} = 100$ μA, $K_{VCO} = 216$ MHz/V, phase margin $= 62°$, $C_1 = 271$ pF, $C_2 = 18$ pF, $C_3 = 1$ pF, $R_1 = 6.8$ kΩ, and $R_2 = 13.6$ kΩ.

119.3.5 Multi-Modulus Frequency Divider

Figure 119.6a shows the programmable MMFD circuit which contains 2/3 true-single-phase-clock (TSPC) divider and traditional logic gate circuits, and Fig. 119.6b shows the 2/3 TSPC divider architecture which contains two traditional DFFs [8]. The MMFD has to treat frequency division over large continuous range and can be programmable.

In this design, the 16-modulus divider is chosen to deal with all of integer divide ratio values from 16 to 31. The divide ratio of MMFD is defined in Eq. 119.4. When the frequency synthesizer oscillates at 5.8 GHz, the divide ratio must be 29, and then the control code, MC0-MC3, is set to be (1, 1, 0, 1):

$$N_{Divide\_Ratio} = 16 + (2^3 \times MC3) + (2^2 \times MC2) + (2^1 \times MC1) \\ + (2^0 \times MC0). \tag{119.4}$$

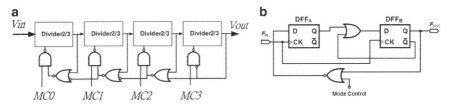

Fig. 119.6 (**a**) The MMFD circuit containing 2/3 dividers and logic gates, and (**b**) the schematic of the proposed 2/3 divider circuit

119.4 Measured Results

The proposed frequency synthesizer was implemented in TSMC 0.18-mm CMOS process. Figure 119.7 shows the die microphotograph including the wire-bound pads. The overall chip area is 0.729 (0.961 × 0.761) mm^2 including measured pads. Under the supply voltage of 1.8 V, the power consumption is 21.6 mW.

Measurements have been performed with an Agilent E4446A spectrum analyzer and an HP 8110A 150-MHz function generator which provides 50 MHz for reference frequency used to perform the measurement. The plot of the tuning characteristic of the locked frequency synthesizer versus the controlled voltage varying from 0 to 1.8 V is shown in Fig. 119.8a. From it we can find the much linearity of using average varactor circuit. The VCO output frequency is tunable from 5.5 to 5.94 GHz by varying the controlled voltage V_{ctrl}. Figure 119.8b shows the measured output power of −13.33 dBm of the prototype after locking at 5.805 GHz. The reference spur is about 50 dBc and appears exactly at the reference frequency of 50 MHz.

Figure 119.9 shows both the measured phase noises of free-running VCO and the phase-locked VCO. According to the measured results, the phase noise of the locked VCO is −105.83 dBc/Hz at 1 MHz offset frequency from 5.805 GHz. The measured performances of the proposed frequency synthesizer are summarized in Table 119.1 in comparison with other recently published papers. The proposed prototype achieves the highest output frequency, the widest tuning range, the least chip area, and lower phase noise, comparing to the other three references. The phase noise is −111 dBc/Hz @ 1 MHz, but with a power consumption of 27.5 mW and a larger chip area of 0.988 mm^2 shown in [2]. Our prototype only needs power of 21.6 mW and chip area of 0.729 mm^2.

119.5 Conclusion

In this paper, a 5.8-GHz frequency synthesizer was fabricated in TSMC 0.18-mm CMOS process. To improve phase noise, a cross-coupled Colpitts VCO with an average varactor circuit was adopted. The CP employed dynamic current-matching

Fig. 119.7 Die micrograph of the proposed frequency synthesizer with a chip area of 0.761 × 0.961 mm^2

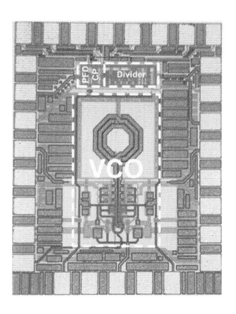

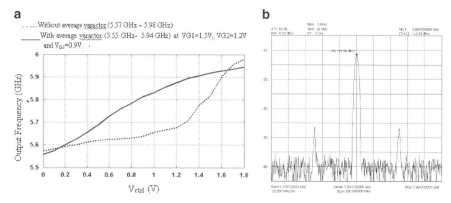

Fig. 119.8 (**a**) Measured tuning range versus varying the control voltage, with average varactor circuit; the tuning frequency range is much linear, and (**b**) output spectrum of the locked frequency synthesizer with $V_{DD} = 1.8$ V and $V_{ctrl} = 0.9$ V

Fig. 119.9 Measured phase noises of VCO and locked PLL

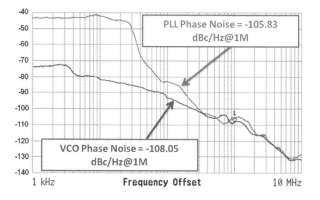

Table 119.1 Performance comparison of proposed PLL with previously published papers

	[1] (2009)	[2] (2004)	[3] (2009)	This work
Technology (μm)	0.18	0.18	0.18	0.18
Supply voltage (V)	1.8	1	1.8	1.8
Tuning range (GHz) (%)	5.15~5.35 (3.8)	5.45~5.65 (3.6)	5.1~5.35 (3.8)	5.55~5.94 (6.8)
Phase noise @ 1 MHz offset (dBc/Hz)	−104	−111	−104	−105.83
KVCO (MHz/V)	200	75	111	216
Loop bandwidth (kHz)	200	N.A.	200	350
Power consump (mW)	18	27.5	18	21.6
Chip area (mm^2)	1.05	0.988	1.045	0.729

circuit to compensate for the channel-length modulation effect. The measured phase noise and power consumption are −105.83 dBc/Hz and 21.6 mW, respectively, and the locked tuning frequency is from 5.55 to 5.94 GHz and the chip area is only 0.729 mm^2.

Acknowledgements The authors would like to thank Prof. Ron-Yi Liu for his layout guidance and the National Chip Implementation Center (CIC) for the chip fabrication and technical supports. We also thank the Taiwan Mobile Phone Company for the financial support.

References

1. Deng, P.-Y., & Kiang, J.-F. (2009). A 5-GHz CMOS frequency synthesizer with an injection-locked frequency divider and differential switch capacitors. *IEEE Transactions on Circuits and Systems, 56*(2), 320–326.
2. Leung, G.-C., & Luong, H.-C. (2004). A 1-V 5.2-GHz CMOS synthesizer for WLAN applications. *IEEE Journal of Solid-State Circuits, 36*(11), 1873–1882.
3. Chiu, W.-H., Huang, Y.-H., & Lin, T.-H. (2009). A 5GHz phase-locked loop using dynamic phase-error compensation technique for fast settling in 0.18-μm CMOS. In *IEEE Symposium on VLSI Circuits* (pp. 128–129).
4. Li, X., Shekhar, S., & Allstot, D.-J. (2005). Gm-boosted common-gate LNA and differential Colpitts VCO/QVCO in 0.18-μm CMOS. *IEEE Journal of Solid-State Circuits, 40*(12), 2609–2619.
5. Wu, T., Hanumolu, P.-K., Mayaram, K., & Moon, U.-K. (2009). Method for a constant loop bandwidth in LC-VCO PLL frequency synthesizers. *IEEE Journal of Solid-State Circuits, 44*(2), 427–435.
6. Huang, J.-F., Mao, C.-C., & Liu, R.-Y. (2011). The 10 GHz wide tuning and low phase-noise PLL chip design. In *IEEE International Security and Identification Conference*, Xiamen, China (pp. 1–4).
7. Lin, T.-H., Ti, C.-L., & Liu, Y.-H. (2009). Dynamic current-matching charge pump and gated-offset linearization technique for delta-sigma fractional-N PLLs. *IEEE Transactions on Circuits and Systems I, 56*(6), 877–885.
8. Huang, J.-F., Shih, C.-W., & Liu, R.-Y. (2011). A 5.8-GHz frequency synthesizer chip design for worldwide interoperability for microwave access application. *Microwave and Optical Technology Letters, 53*(12), 2931–2935.

Chapter 120
Full-Wave Design of Wireless Charging System for Electronic Vehicle

Yongxiang Liu, Yi Ren, and Yi Wang

Abstract This chapter studies magnetic resonance based on wireless power transmission (WPT) system for electronic vehicle (EV). In this system, the two resonant coils mounted on the bottom of the vehicle and on the ground were simultaneously analyzed by the method of moments (MoM), an accurate and efficient full-wave electromagnetic analysis method. Then, compared with traditional WPT in ideal circumstance, the different performance of WPT in wireless charging system of EV is studied. Finally, a new design of the WPT integrated with circumstance is proposed, which achieves 90 % energy transmission efficiency at the resonant frequency of 13.56 MHz with the distance between two resonant coils varying within 15–25 cm.

120.1 Introduction

Higher power delivery through electromagnetic wave transmission is a fantastic technology after Tesla's hypothesis, which has been a topic of continued interests since the last several decades. This technology is not achieved until the electromagnetic inductance-based wireless power transmission (WPT) occurred [1]. The inductance-based WPT can only work in short range as several centimeters, which is far apart from industry's requirements. Meanwhile, another WPT was proposed as the radio-frequency WPT, which can transmit higher power energy by microwave in long distance. However, the ultra-high-gain antenna array is required which is usually not easy to achieve and very expensive. Besides that, many other

---

Y. Liu
Electric Power Research Institute, Chongqing 404100, China
e-mail: l_yx123@qq.com

Y. Ren (✉) • Y. Wang
Chongqing University of Posts and Telecommunications, Chongqing 404100, China
e-mail: renyi@cqupt.edu.cn; wangyi@cqupt.edu.cn

shortages hampered this technology to apply and develop [2]. Recently, André Kurs proposed a new WPT based on nonradiative near-field magnetic resonant which is called magnetic resonant WPT [3]. Compared to the former two WPT systems, this new one worked in middle range, typically 2 m with energy transmit efficiency as high as 40 %. The high performance generates great interests in industry and scientific research, which is thought to have broad application prospects.

Nowadays, many potential applications of magnetic resonant WPT are proposed, typically in EV, mobile communication system, implant medical devices, etc. Specially, considering the shortage of fossil energy in the future and the convenience of WPT in EV, many research institute and company pay much more attention on its application research in EV [4]. To author's best knowledge, most of those researches relied on experimental tests and equivalent circuit theory analyses, in which the electromagnetic circumstance is usually simplified. Typically, most of the WPT design in EV system is not considering the affluence of the metallic car chassis and the ground. However, according to the knowledge of electromagnetic theory, the circumstance will strongly affect the performance of WPT. As a result, the design considering electromagnetic environmental integration is required for stability and higher-efficiency performance of magnetic resonant WPT in EV application.

In this work, the method of moments (MoM) [5, 6], an accurate and efficient full-wave electromagnetic analysis method, is applied to simulate the performance of WPT in our design. Furthermore, a new structure of WPT design is proposed, in which the chassis and the surface of the ground is set as perfect electronic conductor (PEC) and the circuits of WPT is mounted in the cavity on PEC ground. Our simulation results show that the power delivery efficiency is higher than 90 % when the distance of transmit and receive ports varied by 15–25 cm, which is the distance between the ground and the majority of car's chassis.

120.2 Wireless Power Transmission System

120.2.1 Power Transmission with Nonradiative Near-Field System

A typical nonradiative near-field WPT system consists of 2 or 4 circles, and the 4 circles system is considered much more efficient than the 2 circuits system. Therefore, in this paper, the 4 circles WPT system is analyzed as shown in Fig. 120.1.

Usually, this structure is analyzed with the equivalent circuit theory, and the details can be found in reference [3]. This method can only work in ideal circumstance where the WPT is applied without considering the influence of electromagnetic circumstances. Namely, this model cannot be applied in the EV circumstances analysis, where the chassis and ground will affect WPT strongly. Therefore, the best

120 Full-Wave Design of Wireless Charging System for Electronic Vehicle

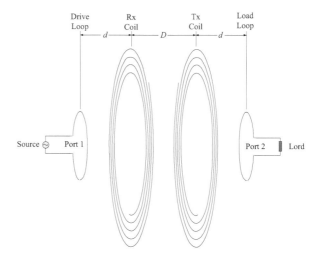

Fig. 120.1 Circles WPT system

way will be the full-wave electromagnetic simulation method. Finally, in this work, MoM is introduced to simulate the WPT in EV application.

120.2.2 Method of Moments

MoM is deduced from the time-harmonic Maxwell's equations and the boundary conditions, which is thought as the most accurate and efficient method in electromagnetic simulation method. Usually, MoM can be classified as several kinds of forms for the different applied circumstance and integral function core. Given the WPT is usually constructed as PEC, the electric field surface integral equation (EFIE) is applied in this simulation. According to the boundary integral equations, the EFIE is expressed as

$$(j\omega A + \nabla \Psi)_{\tan} = E_{\tan}^{\text{sou}}, \quad r \in S \tag{120.1}$$

where A is the magnetic vector potential, Ψ is the electronic scalar potential, ω is the angular frequency, and $E_{\tan}^{\text{sou}}$ is the imposed voltage source on port 1. A and Ψ can be expressed as

$$A(r) = \mu_0 \int_S J(r') \frac{e^{-jk|r-r'|}}{4\pi|r-r'|} ds' \tag{120.2}$$

$$\Psi(r) = \frac{1}{\varepsilon_0} \int_S \sigma(r') \frac{e^{-jk|r-r'|}}{4\pi|r-r'|} ds' \tag{120.3}$$

where $J(r')$ and $\sigma(r')$ are the induced current and charge, respectively. Usually, $J(r')$ and $\sigma(r')$ can be contacted with the charge consistent law as $J + j\omega\sigma = 0$. Finally, $J(r')$ is the wanted solution which can be solved by the matrix method. The readers can find more details about MoM in reference [5]. Furthermore, the application property of MOM in the designing of WPT is proved in reference [7].

120.3 The Designed New WPT System

The purpose of this work is introducing MoM to design a new WPT system in EV application, and the key point is that the design should be integrated with circumstance in which the influence of the background should be considered. As this purpose, a typical model of the PWT is shown in Fig. 120.2. Here, in order to express clearly, only the transmit port on the ground is shown in Fig. 120.2, where the ground is modeled as the PEC surface will be easily achieved in the engineering. It is noted that the receive port on car's chassis is symmetrical with the transmit port in our design. In our design, the size of rectangular PEC ground is set to 4.6 × 1.8 m, which is the typical size of a small car's chassis. We proposed that the WPT should be mounted in the backed cavity which is in the middle of the PEC plate. In order to display more clearly, the WPT with backed cavity is enlarged in Fig. 120.3.

Fig. 120.2 The structure of the WPT in EV application (transmit port)

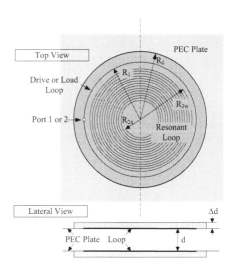

Fig. 120.3 The detailed structure of WPT with backed cavity

Fig. 120.4 The energy transmission efficiency versus frequency and d. (**a**) $d = 15$ cm; (**b**) $d = 20$ cm; (**c**) $d = 25$ cm

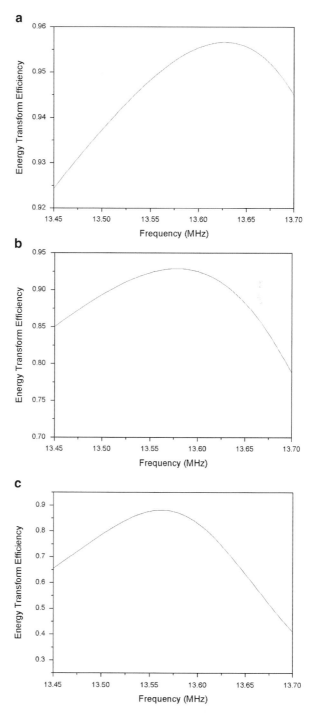

This WPT is constructed as drive loop, load loop, and resonant loop. All of the loops are mounted in the backed cavity. The drive loop and resonant loop are mounted in the same plane and the same thing is the load loop and resonant loop in Rx. Given the versatility, the distance d between Tx and Rx should vary by 15–25 cm, which is the distance of most small car's chassis to the ground. Therefore, the design should make sure that the resonant frequency is fixed at 13.56 MHz and the efficiency is higher than 85 % in the variable range. Therefore, the appropriate parameter should be set to satisfy this target. In this work, the genetic algorithm (GA) [8] combined with MoM is applied in parameter optimization. Given the limited space of this literature, GA will not be detailed here. Finally, the optimized parameter is $R_1 = 16.2$ cm, $d_1 = 1$ mm, $R_{2n} = 5$ cm, $R_{2w} = 15$ cm, $d_2 = 3$ mm, $R_d = 20$ cm, $\Delta d = 1.5$ cm, and $n = 14.515$ (n is the number of turns of the resonant loop). Finally, with the optimized parameters, the simulated efficiency with different d is shown in Fig. 120.4. It is obvious that the resonant frequency is fixed at 13.56 MHz and the efficiency is higher than 85 %, which achieves the designing requirements.

120.4 Conclusion

In this work, the MoM was introduced to design the WPT system in EV application. This new structure consisted of four loops which are mounted in the backed cavity on a large PEC plate, and the PEC plate was used to offset the effect of the ground and chassis. Furthermore, the structure was improved by setting the drive loop around the resonant loop to decrease the sensitivity of the resonant frequency within the distance. Finally, the MoM and GA were combined to optimize the related parameters. The optimized results showed that the designed structure can fix the resonant frequency at 13.56 MHz and the energy transmission efficiency can be kept higher than 85 % when the distance varied within 15–25 cm.

References

1. PowerMat Inc. (2009). Retrieved from http://www.powermat.com.
2. McSpadden, J., & Mankins, J. (2002). Space solar power programs and microwave wireless power transmission technology. *IEEE Microwave Magazine, 3*(4), 46–57.
3. Kurs, A., Karalis, A., Moffatt, R., Joannopoulos, J. D., Fisher, P., & Soljacic, M. (2007). Wireless power transfer via strongly coupled magnetic resonances. *Science, 317*(18), 83–91.
4. Tan, L.-L., Huang, X.-L., Huang, H., Zou, Y.-W., & Li, H. (2011). Transfer efficiency optimal control of magnetic resonance coupled system of wireless power transfer based on frequency control. *SCIENCE CHINA, Technological Sciences, 54*(6), 1428–1434.
5. Harrington, R. F. (1968). *Field computation by moment methods* (pp. 76–124). Malabar, FL: R. E. Krieger.
6. Rao, S. M., Wilton, D. R., & Glisson, A. W. (1982). Electromagnetic scattering by surfaces of arbitrary shape. *IEEE Transactions on Antennas Propagation, 30*(2), 409–418.

7. Moshfegh, J., Shahabadi, M., & Rashed-Mohassel, J. (2011). Conditions of maximum efficiency for wireless power transfer between two helical wires. *IET Microwaves, Antennas & Propagation, 5*(5), 545–550.
8. Dong, Y.-F., Gu, J.-H., Li, N.-N., Hou, X. D., & Yan, W. L. (2007, August 19–22). Combination of genetic algorithm and ant colony algorithm for distribution network planning. *2007 International Conference on Machine Learning and Cybernetics* (Vol. 2, pp. 999–1002), Hong Kong.

Chapter 121
A Hierarchical Local-Interconnection Structure for Reconfigurable Processing Unit

Yujia Zou, Leibo Liu, Shouyi Yin, Min Zhu, and Shaojun Wei

Abstract Reconfigurable computing is being widely used in Computation-intensive applications. With the rapid development of applications, we have higher requirements for the computational efficiency of reconfigurable computing. In order to improve the computational efficiency, the array size gradually increased for applications that are more complex. With the upgrade of the array size, the hardware overhead of traditional interconnection structure used for reconfigurable processing unit (RPU) increases significantly. This paper proposed a new interconnection structure called hierarchical local interconnection for RPU. Comparing to traditional full-mesh structure used in MorphoSys, the hierarchical local interconnection greatly enhanced the area efficiency while retaining the flexibility of interconnection. When the array scale is 8×8, hardware overhead of new structure is 28.6 % of the traditional structure.

121.1 Introduction

In the past 2 decades, the reconfigurable computing technology developed rapidly. Now it is more and more popular in computation-intensive applications, such as media processing [1] and communication [2]. In order to handle the more complex applications, the structure and size of reconfigurable processor is becoming more and more complex. With the improvement of the array size, the proportion of interconnection structure is gradually increasing and the interconnection structure is also changing gradually.

Y. Zou · L. Liu (✉) · S. Yin · M. Zhu · S. Wei
Research Centre for Mobile Computing, Tsinghua University, Beijing 100084, China

Institute of Microelectronics, Tsinghua University, Beijing 100084, China

Tsinghua National Laboratory for Information Science and Technology, Beijing 100084, China
e-mail: liulb@tsinghua.edu.cn

This paper proposes a method with which we can maintain the area efficiency of the interconnection, increasing the array scale and computing capability. When the array scale is 8 × 8, the ratio of processing unit hardware overhead and interconnection hardware overhead in our structure is 3.22 compared to 0.536 in MorphoSys. When the array scale increases to 128 × 128, the ratio in our structure can maintain 0.345 compared to 0.002 in MorphoSys.

121.2 Related Work

The RaPiD [3] of Washington University and the Garp [4] of Berkeley used a one-dimensional interconnection. This structure is simple for wiring and widely used for early reconfigurable processor. The MorphoSys [5] of University of California Irvine and the ADRES [6] of IMEC used a traditional full-mesh structure. The full-mesh structure is more flexible than the one-dimensional interconnection. Good flexibility is conducive for mapping algorithm on the array. However, with the expansion of the array scale, traditional full-mesh structure will bring more hardware overhead. With the rapid development of applications, larger array is required to enhance the computing capacity. If the traditional full mesh structure is used in a large reconfigurable architecture, the interconnection hardware overhead will be significantly enhanced. The area efficiency of the array will drop sharply. Therefore, we must develop a new interconnection structure for large-scale reconfigurable processor, which is flexible enough for algorithmic mapping and with less hardware overhead.

REMUS [7] is a reconfigurable processor mainly used for multimedia system, which can be dynamically configured as the H.264, AVS, or MPEG2 video decoder, as well as other multimedia applications.

The architecture of REMUS processor is shown in Fig. 121.1a, which consists of an ARM, two reconfigurable processing units (RPUs), an entropy decoder (EnD), and some assistant modules. The ARM is a RISC CPU with two tightly coupled memories to accelerate the specific loaded codes and is used as the host processor mainly to handle control application and generate the configuration information for RPU.

RPU is a coarse-grained dynamic reconfigurable computing module and is the core element for computation-intensive task in the system (Fig. 121.1b). Each RPU consists of four units of reconfigurable computing array (RCA). Each RCA consists of an 8 × 8 PE array (Fig. 121.2), an internal memory for the temporarily storage, and some control modules. In RCA, reading external data may require multiple cycles, or when the RCA is processing, the input and output data need cache. We design input FIFO and output FIFO for data cache. The depth of FIFO is designed by simulation results of the algorithm mapping.

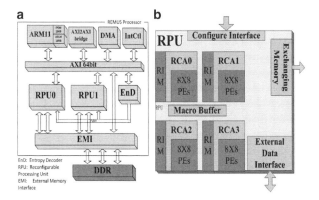

Fig. 121.1 (a) REMUS architecture, (b) architecture of RPU

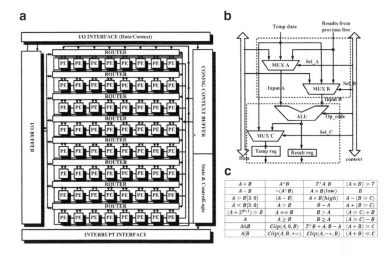

Fig. 121.2 (a) PE 8 × 8 arrays, (b) architecture of PE, and (c) PE functions

121.3 Analysis of Interconnect Structure

121.3.1 Traditional Full-Mesh Structure

The traditional full-mesh structure has a great deal of flexibility, which is adopted in MorphoSys [5]. The traditional full-mesh structure contains three parts: (1) the interconnection between input FIFO to processing unit, (2) the interconnection between processing units, and (3) the interconnection between processing unit to output FIFO. The basic traditional full-mesh structure is shown in Fig. 121.3.

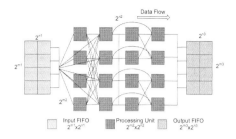

Fig. 121.3 Traditional full-mesh structure

We assume that the width of the input FIFO is 2^{m1} data, the depth of data required for a single operation is 2^{n1} rows. The data required by a task may be larger than the width of input FIFO. So we suppose that the scale of the arrays is $2^{m2} \times 2^{n2}$, and the direction of data flow is shown as Fig. 121.2. The scale of output FIFO is $2^{m3} \times 2^{n3}$. The processing unit may obtain the data from any layer of the input FIFO. The output data from the processing unit may be transferred to any layer of the output FIFO.

In order to illustrate the interconnection of the hardware scale more clearly, we give a simple example: there are two rows of data interconnected by the full-mesh interconnection structure, the first row is the source with 2^m data and the second row is the target with 2^n data. Firstly, the MUX overhead of transfer one from 2^m data to target address is $2^{m-1} + 2^{m-2} + \ldots + 2^0 = 2^m - 1$. Then repeat the operation $(2^n - 1)$ times, making 2^n data transfer from the source to the target. By multiplication principle, we can derive that all the MUX overheard is $(2^m - 1)*2^n$.

Then we calculate the hardware scale of traditional full-mesh structure in three parts:

1. The interconnection between input FIFO and processing unit:

$$(2^{m1} \times 2^{n1} - 1)(2^{m2} \times 2^{n2}) \qquad (121.1)$$

2. The interconnection between processing units:

$$(2^{m2} - 1)2^{m2}(2^{n2} - 1 + 1)/2 \times (2^{n2} - 1)$$
$$= (2^{m2} - 1)(2^{n2} - 1) \times 2^{m2} \times 2^{n2-1} \qquad (121.2)$$

3. The interconnection between processing unit and output FIFO:

$$(2^{m2} \times 2^{n2} - 1)(2^{m3} \times 2^{n3}) \qquad (121.3)$$

We assume that m and n increase in equal proportion, so the hardware scale is in direct proportion to $2^{m*}2^n$.

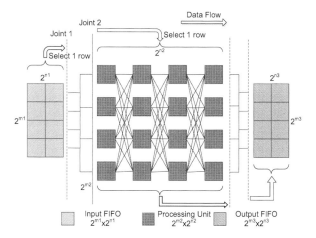

Fig. 121.4 Hierarchical local-interconnection structure

121.3.2 New Local-Connection Structure

Although the traditional full-mesh structure has such a great deal of flexibility, the hardware overhead and configuration information of the traditional full-mesh structure is too large for reconfigurable array. The huge hardware overhead of traditional full-mesh structure leads to the number of RPUs decrease in the same area of a chip. It may not meet the scale of processing units required by the application.

Under the premise of ensuring the flexibility of reconfigurable array as much as possible, we develop a hierarchical local-interconnection structure for reconfigurable array. Generally, the adjacent data in DFG are stored in adjacent memory. The input data position of processing unit array can adjust by passing through the idle processing unit in array. By using the localization of data in the algorithm, hierarchical local interconnection meets the requirement of the algorithm mapping.

The structure of our hierarchical local interconnection is shown in Fig. 121.4. In the hierarchical local interconnection, one row is selected from 2^{n1} rows of the input FIFO connected to Joint 1, the structure between Joint 1 and Joint 2 is full mesh, and one row is selected from 2^{n2} rows of processing unit connected to Joint 2, thus completing the interconnection between input FIFO and processing unit array. In the same manner, we complete the interconnection between the processing unit array and output FIFO. In processing unit array, only the two adjacent rows are full-mesh structure.

The calculation of hardware overhead of the hierarchical local-interconnection structure is a little different from 3-D traditional full-mesh structure. Firstly, selecting one row with 2^{m1} data from 2^{n1} rows needs $(2^{n1} - 1)*2^{m1}$ MUX. Secondly, achieving full mesh between 2^{m1} and 2^{m2} needs $(2^{m1} - 1)*2^{m2}$ MUX; finally, selecting one row with 2^{m2} from 2^{n1} rows needs $(2^{n1} - 1)*2^{m2}$ MUX. So the hardware overhead of hierarchical local-interconnection structure is in three parts:

1. The interconnection between input FIFO and processing unit:

$$(2^{n1} - 1)2^{m1} + (2^{m1} - 1)2^{m2} + (2^{n2} - 1)2^{m2} \qquad (121.4)$$

2. The interconnection between processing units:

$$(2^{m2} - 1)2^{m2}(2^{n2} - 1) \qquad (121.5)$$

3. The interconnection between processing unit and output FIFO:

$$(2^{n2} - 1)2^{m2} + (2^{m2} - 1)2^{m3} + (2^{n3} - 1)2^{m3} \qquad (121.6)$$

Then, we will compare the traditional full-mesh structure to hierarchical local-interconnection structure.

121.3.3 Area Optimization

In the section of area optimization, we compare the three parts of interconnection between two structures. Firstly, we compare the interconnection between input FIFO and processing unit in two structures. In order to compare the area overhead of two structures, we use Eq. 121.1/Eq. 121.4 $\cong 2^{m1+n1+m2+n2}/(2^{m1+n1} + 2^{m1+m2} + 2^{n2+m2})$; under normal circumstances, n1 is smaller (because of the width of the input FIFO is larger, the number of rows in FIFO is less in one circulation); we derived the Eq. 121.1/Eq. 121.4 $\cong 2^{n1} \times 2^{m1+n2}/(2^{m1} + 2^{n2})$. Secondly, we compare the interconnection between processing units in two structures. We derived Eq. 121.2/Eq. 121.5 = 2^{n2-1}. In the same way, we could derive Eq. 121.3/Eq. 121.6 = $2^{n3} \frac{2^{m3+n2}}{2^{m3}+2^{n2}}$. When the scale of array is 16 × 16, that is, n2 = 4, the hardware overhead of hierarchical local interconnection is 1/8 of the traditional full-mesh structure. The hierarchical local-interconnection structure greatly reduce the hardware overhead of array.

121.3.4 Configuration Optimization

Since each MUX needs to be configured, the configuration of two interconnection structures is different. The configuration of input terminal, output terminal, and processing units is, respectively, (m1 + n1)*2^{m2}*2^{n2}, (m3 + n3)*2^{m2}*2^{n2}, and (m2 + n2)*2^{m2}*2^{n2} in traditional full-mesh structure. The configuration of input terminal, output terminal, and processing units is, respectively, n1*2^{n2} + m1*2^{m2}, n3*2^{n2} + m3*2^{m2} and m2*2^{m2}*2^{n2} in hierarchical local-interconnection structure.

Comparing the configuration of two structures, we could derive the equation as follows:

1. The input terminal:

$$\frac{(m1+n1)2^{m2+n2}}{n1 \times 2^{n2} + m1 \times 2^{m2}} \approx 2^{n2} \qquad (121.7)$$

2. The output terminal:

$$\frac{(m3+n3)2^{m2+n2}}{n3 \times 2^{n2} + m3 \times 2^{m2}} \approx 2^{n2} \qquad (121.8)$$

3. The array:

$$\frac{(m2+n2)2^{m2+n2}}{m2 \times 2^{m2+n2}} = \frac{m2+n2}{m2} \qquad (121.9)$$

When the array size is 16×16, m2 = n2 = 4, the configuration of input terminal and output terminal in hierarchical local interconnection is 1/16 of full-mesh structure, and the configuration of processing units reduces to 50 %.

121.4 Experimental Results and Analysis

The relationship between the scale of array and area in traditional full-mesh structure and hierarchical local-interconnection structure is shown, respectively, in Tables 121.1 and 121.2. Based on the architecture of MorphoSys and our structure, we extended the array scale. Then, we synthesize the circuit by Synopsys Design Compiler under TSMC65nmLP processing technique and get the data in Tables 121.1 and 121.2.

Comparing the data in Tables 121.1 and 121.2, the hardware overhead of hierarchical local interconnection is 28.6 % of the traditional structure when the array scale is 8×8. The hardware overhead of interconnection in traditional full-mesh structure is double of the processing unit. When the array scale expanded to 32×32, the hardware overhead of interconnection in traditional full-mesh structure is about 32 times of the processing unit. That area efficiency is too low to be used. On the other hand, even if the array scale expanded to 128×128, the hardware overhead of hierarchical local interconnection is only triple of the processing unit. The area efficiency is still reasonable.

Figures 121.5 and 121.6 show the relationship between area percentage and array size. With the growth of the array size, the area percentage of interconnection in MorphoSys increases rapidly to more than 80 %, which makes the area efficiency

Table 121.1 Full-mesh area percentage vs. array size

Array scale	Storage	Interconnection	Processing unit
8 × 8	249974 μm^2	540160 μm^2	289456 μm^2
	23.20 %	50.00 %	26.80 %
16 × 16	9.30 %	80.00 %	10.70 %
32 × 32	2.70 %	94.10 %	3.20 %
64 × 64	0.70 %	98.50 %	0.80 %
128 × 128	0.20 %	99.60 %	0.20 %

Table 121.2 Hierarchical local-interconnection area percentage vs. array size

Array Scale	Storage	Interconnection	Processing unit
8 × 8	249974 μm^2	90000 μm^2	289456 μm^2
	39.70 %	14.30 %	46.00 %
16 × 16	36.10 %	22.10 %	41.80 %
32 × 32	31.10 %	32.90 %	36.00 %
64 × 64	24.80 %	46.40 %	28.80 %
128 × 128	18.10 %	60.90 %	21.00 %

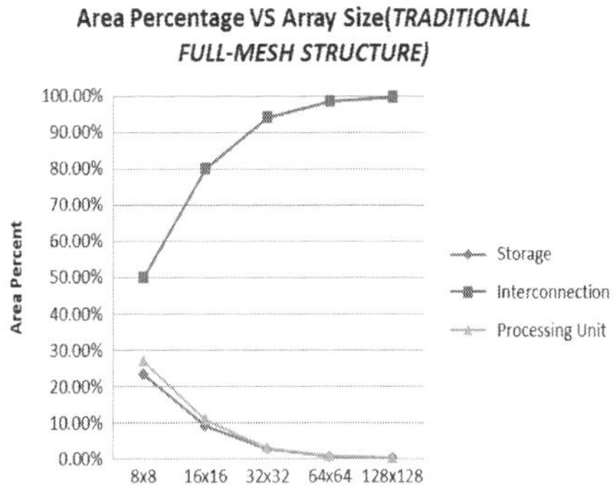

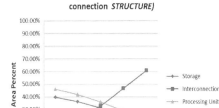

Fig. 121.5 Area percentage vs. array size (traditional full-mesh structure)

Fig. 121.6 Area percentage vs. array size (hierarchical local-interconnection structure)

of array reduce drastically. On the contrary, the area percentage of interconnection increases linearly with the array size; even if the array size expands to 128 × 128, the area percentage of our interconnection structure is 60 %.

121.5 Conclusion

In this paper, we proposed a novel interconnection structure called hierarchical local interconnection for RPU. Comparing to traditional full-mesh structure, which is adopted in MorphoSys, the hierarchical local interconnection greatly enhanced the area efficiency while retaining the flexibility of interconnection. When the array scale is 8 × 8, hardware overhead of new structure is 28.6 % of the traditional structure. Even if the array size expands to 128 × 128, hierarchical local-interconnection structure can still make the array maintain reasonable area efficiency.

Acknowledgements This work is supported in part by the China National High Technologies Research Program (No. 2012AA012701), the Tsinghua Information S&T National Lab Creative Team Project, the International S&T Cooperation Project of China grant (No. 2012DFA11170), the Tsinghua Indigenous Research Project (No. 20111080997), the Special Scientific Research Funds for Commonweal Section (No. 200903010), the Science and Technology Project of Jiangxi Province (No. 20112BBF60050) and the NNSF of China grant (No. 61274131).

References

1. Veredas, F. J., Scheppler, M., Moffat, W., & Mei, B. (2005). Custom implementation of the coarse-grained reconfigurable ADRES architecture for multimedia purposes. In *International Conference on Field Programmable Logic and Applications, 2005, IEEE* (pp. 106–111).
2. Ebeling, C., Fisher, C., Xing, G., Shen, M., & Liu, H. (2004). Implementing an OFDM receiver on the RaPiD reconfigurable architecture. *IEEE Transactions on Computers, 53*(11), 1436–1448.
3. Ebeling, C. Cronquist, D. and Franklin P. (1997), Configurable Computing: The Catalyst for High-Performance Architectures," Proc. IEEE Int'l Conf. Application-Specific Systems, Architectures, and Processors, pp. 364–372
4. Hauser, J. R., & Wawrzynek, J. (1997). Garp: A MIPS processor with a reconfigurable coprocessor. In *Proceedings of the 5th Annual IEEE Symposium on FPGAs for Custom Computing Machines, IEEE* (pp. 12–21).
5. Singh, H., Lee, M. H., Lu, G., Kurdahi, F. J., Bagherzadeh, N., & Filho, E. M. C. (2000). MorphoSys: An integrated reconfigurable system for data-parallel and computation-intensive applications. *IEEE Transactions on Computers, 49*(5), 465–481.
6. Novo, D., Moffat, W., Derudder, V., & Bougard, B. (2005). Mapping a multiple antenna SDM-OFDM receiver on the ADRES coarse-grained reconfigurable processor. In *IEEE Workshop on Signal Processing Systems Design and Implementation, 2005, IEEE* (pp. 473–478).
7. Zhu, M., Liu, L., Yin, S., & Wang, Y. (2010). A reconfigurable multi-processor SoC for media applications. In *Proceedings of 2010 IEEE International Symposium on Circuits and Systems (ISCAS), IEEE* (pp. 2011–2014).

Chapter 122
High Impedance Fault Location in Distribution System Based on Nonlinear Frequency Analysis

Jinqian Zhai, Di Su, Wenjian Li, Feng Li, and Guohong Zhang

Abstract A methodology is presented to detect and locate high impedance faults (HIFs) in radial distribution system by means of nonlinear frequency analysis. The proposed technique is based on the analysis of the feeder responses to power line carrier signals, which are periodically injected at the outlet of transformer. The effectiveness of the method has been verified through simulation studies. The results demonstrated that the proposed method has the potential to be applied in practice to resolve HIF real-time monitoring problem.

122.1 Introduction

As to power line fault location, most of the research aims at finding the positions of transmission line faults; the locations of faults on distribution systems have started receiving much attention [1]. The detection and location algorithms in power transmission systems are not useful in power distribution systems. Fault location in distribution system is much more difficult than in a transmission network. Due to the complex topology of downstream network, the calculated value is far from being as accurate as needed for a fast and reliable service restoration; so any method that helps to locate faults, as soon as possible, is welcomed. For fault such as short circuit with low impedance fault, several techniques have been proposed by many authors [2–4]. But there is no appropriate solution to high impedance fault (HIF) in distribution system [5–7].

A great number of HIFs can be detected today, but cannot be localized due to lack of communication among distribution feeder sections. HIF are inherently nonlinear and always result in distorted currents. According to nonlinear characteristics of HIF, nonlinear frequency analysis method [8] is proposed to detect the

---

J. Zhai (✉) · D. Su · W. Li · F. Li · G. Zhang
ZhengZhou Power Supply Company, ZhengZhou 450000, China
e-mail: jinqianzhai@163.com

position of HIF in distribution system without turning off the electric power. In outlet side of substation, high-frequency signal is injected onto power line through coupling equipment; this high-frequency signal flows through power line, which was received by receiving device installed in tower. By wireless network, the device sends the received signal to control center. In monitoring center, by analyzing all receiving signals, using nonlinear frequency analysis method, the position of HIF in distribution system is found. The purpose of this work is to evaluate the applicability of nonlinear frequency analysis to HIF location in distribution system.

122.2 Modeling of HIF

The electric arc has a voltage/current nonlinear relation and might show an asymmetric behavior of the positive half cycle with respect to the negative one. The nonlinear HIF model is shown in Fig. 122.1 [9].

The model includes two dc sources: V_p and V_n which present the arcing voltage between the trees and line [10]. During the positive half cycle, the current flows through V_p and during the negative, through V_n. When the phase voltage is greater than the positive DC voltage V_p, the fault current flows toward the ground. The fault current reverses when the line voltage is less than the negative DC voltage V_n. The values of the phase voltage between V_n and V_p have no fault current flows. Two resistances—R_p and R_n—present the resistance of trees and/or earth resistance.

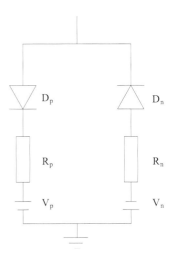

Fig. 122.1 Simplified two-diode fault model of HIFs

122.3 Nonlinear Frequency Analysis Approach for Locating HIF

122.3.1 Derivation of the Approach

For the power transmission system with HIF, the equivalent circuit of power transmission line is shown in Fig. 122.2 [11].

For system without HIF, from the current and voltage laws, the following equations are obtained:

$$\frac{u_{n-1}(s) - u_n(s)}{L_n s + R_n} = \frac{u_n(s)}{L_{n+1} s + R_{n+1} + Z_{load}} + (C_n s + G_n) u_n(s) \tag{122.1}$$

The expansion equation is shown below:

$$C_n L_n L_{n+1} s^3 u_n(s) + [L_n C_n R_{n+1} + R_n C_n L_{n+1} + G_n L_n L_{n+1} + L_n C_n Z_{load}] s^2 u_n(s)$$
$$+ [R_n R_{n+1} C_n + L_n G_n R_{n+1} + R_n G_n L_{n+1} + L_{n+1} + L_n + (R_n C_n + L_n G_n) Z_{load}] s u_n(s)$$
$$+ [R_n + (1 + G_n R_n) R_{n+1} + (1 + G_n R_n) Z_{load}] u_n(s)$$
$$- L_{n+1} s u_{n-1}(s) - (R_{n+1} + Z_{load}) u_{n-1}(s) = 0$$

$$\tag{122.2}$$

So, the mathematical model of transmission line system can be written in time domain as

$$A\dddot{U}(t) + B\ddot{U}(t) + C\dot{U}(t) + DU(t) = U_s(t) \tag{122.3}$$

where A, B, C, and D are the system parameter matrices, respectively. $U = (u_1, \cdots, u_n)'$ is the voltage vector, and $U_s = (L_2 \dot{u}_s + R_2 u_s, 0, \cdots, 0)'$ is the external input vector acting on the system. Obviously, this system is a linear system.

Assume that HIF is located at J-th section with $J \in \{2, \cdots, n\}$. There must be a change to the circuit parameter R at the J-th section of practical transmission line, that is,

$$\Delta R_J = R_{J-fault} - R_{J-normal}, \quad \Delta u = U(\Delta R_J), \quad \text{so } NU = \left(\overbrace{0, \cdots, 0}^{J-2}, -\Delta u, \Delta u, \overbrace{0, \cdots, 0}^{n-J} \right)'$$

$$\tag{122.4}$$

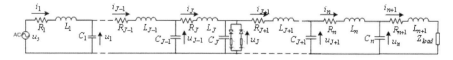

Fig. 122.2 The equivalent circuit of power line system with one high impedance fault

In this case, the power line system can be described as

$$A\dddot{U}(t) + B\ddot{U}(t) + C\dot{U}(t) + DU(t) = NU + U_s(t) \quad (122.5)$$

The system described by eq. (122.5) is a typical locally nonlinear multi-degree of freedom (MDOF) system [12]. By applying the nonlinearity detection approach in [12] to model (122.5), a nonlinear frequency analysis technique for locating HIF of transmission line is proposed. The proposed technique can be described as follows [12]:

1. Excite power line system separately using two different input voltages $u_s^{(q)}(t)$, $q = 1, 2$, and measure the corresponding voltage response at each receiving site of power line to obtain $u_i^{(q)}(t)$, $q = 1, 2$, $i = 1, \cdots, n$.
2. Calculate the FFT spectrum of $u_s^{(q)}(t)$, $q = 1, 2$ and $u_i^{(q)}(t)$, $q = 1, 2$, $i = 1, \cdots, n$ to produce $U_s^{(q)}(j\omega)$, $q = 1, 2$ and $U_i^{(q)}(j\omega)$, $q = 1, 2$, $i = 1, \cdots, n$.
3. Evaluate the functions of $E^{i,i+1}(j\omega)$, $i = 1, \cdots, n-1$ from the results obtained in Step (2) as follows:

$$E^{i,i+1}(j\omega) = [1\ 0]\begin{bmatrix} U_s^{(1)}(j\omega), U_{i+1}^{(1)}(j\omega) \\ U_s^{(2)}(j\omega), U_{i+1}^{(2)}(j\omega) \end{bmatrix}^{-1}\begin{bmatrix} U_i^{(1)}(j\omega) \\ U_i^{(2)}(j\omega) \end{bmatrix} \quad (122.6)$$

4. Evaluate $\overline{E}^{i,i+1}$ for $i = 1, \cdots, n-1$ as

$$\overline{E}^{i,i+1} = \frac{\int_{\omega_1}^{\omega_2} |E^{i,i+1}(j\omega)|d\omega}{\max_{i \in (1,\cdots,n-1)}\left[\int_{\omega_1}^{\omega_2} |E^{i,i+1}(j\omega)|d\omega\right]} \quad (122.7)$$

where $[\omega_1, \omega_2]$ is a frequency band within the frequency range of the input spectrum $U_s^{(q)}(j\omega)$, q = 1,2.
5. Examine $\overline{E}^{i,i+1}$ for $i = 1, \cdots, n-1$. If $\overline{E}^{i,i+1}$ is found to change sharply in the J-th section, it can be concluded that HIF is located at the J-th section of power line system.

122.3.2 Implementation of the Approach in Distribution System Using Power Line Carrier Method

In order to test the feasibility of the proposed technique in distribution system, a relatively simple distribution system is selected; it is shown in Fig. 122.3. The proposed technique is based on the analysis of the feeder responses to high-frequency signals, which are periodically injected at the feeder inlet. The detection

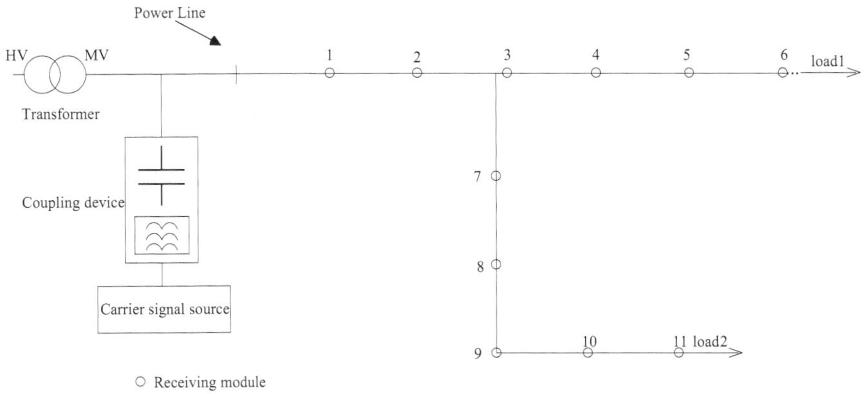

Fig. 122.3 The schematic of distribution line monitoring module

procedure requires exciting the distribution line networks twice using two sinusoidal high-frequency input signals. Several receiving devices are installed along with the feeder, which are used to receive the high-frequency signals and filter out the power frequency signals. The responses to two separate exciting high-frequency signals are used as the information collected in Step (1) of the technique. The procedure in distribution system is as follows:

1. Suppose the receiving module 1–6 is in the main feeder, the receiving module 7–11 is in the branch.
2. Judge the index value of $\overline{E}^{i,i+1}$ and $\overline{F}^{i,i+1}$.

If $\overline{E}^{i,i+1}$ is found to change sharply in the J-th section, and $\overline{F}^{i,i+1}$ is found to change randomly, then there is HIF in the main feeder.

If $\overline{E}^{i,i+1}$ is found to change sharply in the J-th section, and $\overline{F}^{i,i+1}$ is found to change sharply in the K-th section, then that illustrates the branch is in the J-th section, and there is HIF in the K-th section of branch.

If $\overline{E}^{i,i+1}$ and $\overline{F}^{i,i+1}$ are both found to change randomly, there is no HIF in distribution network.

122.4 Simulation Studies

In order to verify the effectiveness of the proposed approach, simulation studies using the Matlab/Simulink facilities for distribution networks were conducted. 35 kV power line level was used for the simulation studies. Six inspecting devices were installed in the main feeder of the distribution network, e.g., receiving module 1–6; five inspecting devices were installed in the branch feeder, e.g., receiving module 7–11. To simplify analysis, only one HIF is supposed to exist in the distribution networks. For power line system parameters, each section of line length

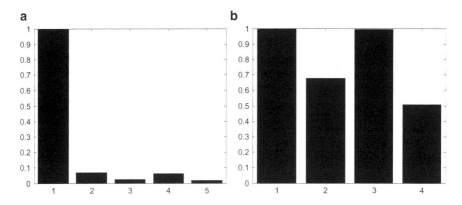

Fig. 122.4 Illustration of $\overline{E}^{i,i+1}$ and $\overline{F}^{i,i+1}$ in Table 122.1(**a**) is illustration of $\overline{E}^{i,i+1}$, (**b**) is that of $\overline{F}^{i,i+1}$

is 1 km, lien type is ACSR LGJ-70. Simulation study is used in the configuration system shown in Fig. 122.4. The used HIF model parameters are as follows: $R_p = 1250\Omega$, $R_n = 500\Omega$, $V_p = 5000V$, $V_n = 7000V$. The load parameter is 15,000 Ω.

122.4.1 Case Study 1

When fault is in the main feeder between 1 and 2, from the five-step procedure, the proposed technique was implemented in the main feeder as follows:

1. Two sinusoidal voltage inputs,
   $u_s^{(q)}(t) = \alpha_q \sin(\omega_u t)$, $q = 1, 2$,
   where $\omega_u = 2\pi \times 50000$, $\alpha_1 = 12$, $\alpha_2 = 20$, were applied to excite the system and to generate two sets of output responses on the distribution system.
2. Calculate the FFT spectrum of $u_s^{(q)}(t)$, $q = 1, 2$ and $u_i^{(q)}(t)$, $q = 1, 2$, $i = 1, \cdots, n$ to produce $U_s^{(q)}(j\omega)$, $q = 1, 2$ and $U_i^{(q)}(j\omega)$, $q = 1, 2$, $i = 1, \cdots, n$.
3. Evaluate function of $E^{i,i+1}(j\omega)$, $i = 1, \cdots, n-1$ from the results obtained in Step (2).
4. Evaluate $\overline{E}^{i,i+1}$ for $i = 1, \cdots, n-1$; the results obtained are given in Table 122.1 and illustrated in Fig. 122.4a
5. From Table 122.1, it is shown that the index value of $\overline{E}^{i,i+1}$ changes sharply from $\overline{E}^{1,2}$ to $\overline{E}^{2,3}$ from large to small. The index value of $\overline{E}^{i,i+1}$ is very small after $\overline{E}^{2,3}$. So the HIF happened in the second section in the main feeder.

Again the above five-step procedure is used in the branch line. Here, the input voltage has changed, because the branch line is in the third section in the main

Table 122.1 $\overline{E}^{i,i+1}$ and $\overline{F}^{i,i+1}$ evaluated for a case where HIF is located at the Sect. 122.2 of main feeder

	$\overline{E}^{1,2}$	$\overline{E}^{2,3}$	$\overline{E}^{3,4}$	$\overline{E}^{4,5}$	$\overline{E}^{5,6}$
	1.0000	0.0686	0.0235	0.0623	0.0201
	$\overline{F}^{1,2}$	$\overline{F}^{2,3}$	$\overline{F}^{3,4}$	$\overline{F}^{4,5}$	
	1.0000	0.6758	0.9930	0.5061	

Table 122.2 $\overline{E}^{i,i+1}$ and $\overline{F}^{i,i+1}$ evaluated for a case where HIF is located at the Sect. 122.2 of branch line

	$\overline{E}^{1,2}$	$\overline{E}^{2,3}$	$\overline{E}^{3,4}$	$\overline{E}^{4,5}$	$\overline{E}^{5,6}$
	0.0411	1.0000	0.0332	0.0332	0.0091
	$\overline{F}^{1,2}$	$\overline{F}^{2,3}$	$\overline{F}^{3,4}$	$\overline{F}^{4,5}$	
	1.0000	0.1031	0.2080	0.0304	

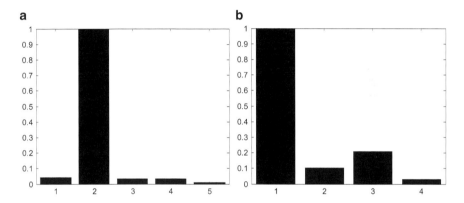

Fig. 122.5 Illustration of $\overline{E}^{i,i+1}$ and $\overline{F}^{i,i+1}$ in Table 122.2, (**a**) is illustration of $\overline{E}^{i,i+1}$ and (**b**) is that of $\overline{F}^{i,i+1}$

feeder, so $u_3(t)$ is source signal of the branch line. The value of index $\overline{F}^{i,i+1}$ is shown in Table 122.1 and illustrated in Fig. 122.4b; it is shown that the index value of $\overline{F}^{i,i+1}$ changes randomly. According to the evaluation standard of Sect. 122.3.2, the HIF is found in the second section of main feeder, which is obviously correct.

122.4.2 Case Study 2

In this case, supposed HIF happened in the branch line. First, fault is in the branch line between 7 and 8. The five-step procedure is implemented in the main feeder and branch line, respectively. As the same case study 1, due to the branch line is in the Sect. 122.3 of the main feeder, $u_3(t)$ is used to the signal source of branch line. The results obtained are given in Table 122.2 and illustrated in Fig. 122.5.

From Table 122.2 or Fig. 122.5a, it is shown that the index value of $\overline{E}^{i,i+1}$ changes sharply from $\overline{E}^{2,3}$ to $\overline{E}^{3,4}$ from large to small; the index value is very small after $\overline{E}^{3,4}$, not associated with the index value before $\overline{E}^{2,3}$. From Table 122.2 or Fig. 122.5b, it is shown that the index value of $\overline{F}^{i,i+1}$ change sharply from $\overline{F}^{1,2}$ to $\overline{F}^{2,3}$, the index value is very small after $\overline{F}^{2,3}$. According to the topology of distribution line and the evaluation standard of Sect. 122.3.2, the conclusion is that the branch line is in the third section of the main feeder, and the HIF is located in the second section of the branch line, which is correct.

122.5 Conclusion

In this paper, the application of a nonlinear frequency analysis-based approach is proposed to detect and locate the HIF in distribution system. The power carrier signal technology has been suggested for the practical implementation of the proposed technique. Numerical simulation studies have been conducted. The results verified the effectiveness of the new technique and demonstrated the potential to apply the technique in practice to resolve the important HIF location problem. Further research will be focused on laboratory tests to make necessary preparations for future experimental studies on real distribution systems.

References

1. Saha, M.M., Das, R., Verho, P., & Novosel, D. (2002). Review of fault location techniques for distribution systems[C]. *Power systems and communications infrastructures for the future* (pp 1–6), Beijing.
2. Choowong-Wattanasakpubal, & Teratum-Bunyagul. (2010). Algorithm for detecting, indentifying, locating and experience to develop the automate faults location in radial distribution system[J]. *JEET, 5*(1), 36–44.
3. Campoccia, A. Silvestre, M.L.D., Incontrera, I., Sanseverino, E.R., & Spoto, G. (2010). An efficient diagnostic technique for distribution systems based on under fault voltages and currents[J]. *Electric Power Systems Research, 80*(10), 1205–1214.
4. Seung-Jae Lee, Myeon-Song Choi, Sang-Hee Kang, Bo-Gun Jin, Duck-Su Lee, Bok-Shin Ahn, et al. (2004). An intelligent and efficient fault location and diagnosis scheme for radial distribution systems[J]. *IEEE Transactions on Power Delivery, 19*(2), 524–532.
5. Flauzino, R. A., Ziolkowski, V., Silva, I.N., de Souza, & D.M.B.S. (2009). Hybrid intelligent architecture for fault identification in power distribution systems[C]. *Power & Energy Society General Meeting. PES'09* (pp. 1–6). Calgary, AB.
6. Elkalashy, N.I., Lehtonen, M., Darwish, H.A., Taalab, A.M.I., Izzularab, & M.A. (2007). DWT-based extraction of residual currents throughout unearthed MV networks for detecting high-impedance faults due to leaning trees[J]. *ETEP., 17*(6), 597–614.
7. Borghetti, A., Corsi, S., Nucci, C.A., Paolone, M., Peretto, L., & Tinarelli, R. (2006). On the use of continuous-wavelet transform for fault location in distribution power systems[J]. *Electrical Power and Energy Systems., 28*(9), 608–617.

8. Lang, Z. Q., & Billings, S. A. (1996). Output frequency characteristics of non-linear system [J]. *International Journal of Control, 64*(16), 1049–1067.
9. Aboul-Zahab, E.M., Eldin, E.-S.T., Ibrahim, D.K., & Saleh, S.M. (2008). High impedance fault detection in mutually coupled double-ended transmission lines using high frequency disturbances[C]. *12th Middle-East Power System* (pp. 412–419). Aswan.
10. Ibrahim, D.K., El Sayed, T.E., El-Zahab, E.E.-D.A. & Saleh S.M. (2010). Unsynchronized fault-location scheme for nonlinear HIF in transmission lines[J]. *IEEE Transactions on Power Delivery, 25*(2), 631–637.
11. Lonngren, K. E., & Bai, E. W. (1996). Simulink simulation of transmission lines[J]. *IEEE Transactions on Circuits and Device Magazine., 12*(3), 10–16.
12. Lang, Z.Q. & Peng, Z.K. (2008). A novel approach for nonlinearity detection in vibrating systems[J]. *Journal of Sound and Vibration, 314*(64), 603–615.

Chapter 123
Early Fault Detection of Distribution Network Based on High-Frequency Component of Residual Current

Jinqian Zhai, Di Su, Wenjian Li, Feng Li, and Guohong Zhang

Abstract A methodology is presented to detect incipient faults in distribution networks by means of DWT and energy detection algorithm. The proposed technique is to extract the characteristic of incipient fault by DWT method, that is, to extract the d5 coefficient of wavelet decomposition of residual current and residual voltage. Compare energy value with normal situation using an energy detection algorithm; incipient faults are detected. The proposed technique has been investigated by ATP/EMTP. Simulation results show that this technique is effective and robust, and the proposed method has the potential to be applied in practice to resolve incipient fault real-time monitoring problem.

123.1 Introduction

Incipient faults in power lines are normally characterized as the faulty phenomena with the relatively low fault currents, such as high impedance faults, insulator leakage current faults, and intermittent/transient faults [1]. Unlike low impedance short circuits, which involve relatively large fault currents and are readily detectable by conventional overcurrent protection, these faults represent little threat of damage to power system equipment [2]. But with time, they may lead to a catastrophic failure (i.e., a permanent damage beyond repair). So, early detection of power line faults would undoubtedly be a great benefit to the utilities, enabling them to avoid catastrophic failures, unscheduled outages, and thus loss of revenues.

Various methods have been proposed by researchers and protection engineers. Among them, harmonic analysis [3], randomness detection [4], artificial neural networks [5], Hilbert-Transform based [6], and wavelet transform [7–11] are used to extract the feature of incipient fault signals in distribution line. But, due to

J. Zhai (✉) • D. Su • W. Li • F. Li • G. Zhang
Zhengzhou Power Supply Company, Zhengzhou 450000, China
e-mail: jinqianzhai@163.com

high-time resolution and low-frequency resolution for high frequencies and high-frequency resolution and low-time resolution for low frequencies, wavelet transform can achieve a better solution. Recently, a power line condition monitoring system [12] is proposed; according to the project's goal, it is directed to develop a new power line sensornet, along with sensors to be scattered along the line, for prediction of incipient faults and momentary line contact (such as incipient failure of insulators and tree limb contact). However, due to the complexity of fault mechanism and environments, a proven approach for incipient failure detection in distribution system is not yet available and needs high maintenance cost.

The focus of the work reported in this paper is to develop an efficient online system that uses measured voltage and current values over a period to diagnose power line incipient faults. Because it is difficult to collect a large amount of incipient fault data, we propose a deterministic method, rather than a training-based intelligent system method. The proposed algorithm is as follows: The incipient fault feature is extracted by DWT. The energy value and average energy value is computed by using the detail d5 coefficients of the residual current signal. Comparing the energy value with the set value, incipient fault in distribution network can be detected.

123.2 Incipient Faults of Distribution Line

The term "incipient faults" refers to certain pre-fault "symptoms" or electrical activities taking place prior to a power system failure or blackout. Power line incipient faults are the primary causes of catastrophic failures in distribution network. For underground cable, these faults develop in the extruded cables from gradual deterioration of the solid insulation due to the persisting stress factors. For overhead lines, incipient faults are associated with degraded equipment (insulators, arresters, transformer insulation and bushings, etc.) and the gradual intrusion of tree limbs as they grow into the overhead power line.

Unlike short circuit faults, incipient faults in distribution networks do not draw sufficient currents from the line to trigger the protective devices. Incipient faults may present intermittent, asymmetric, and sporadic spikes, which are random in magnitude and could involve sporadic bursts as well, and exhibit complex, nonlinear, and dynamic characteristics [13].

123.3 Proposed Technique Principles

The detail coefficients in the high frequency are used to analyze incipient fault signal; the energy value above set value is used to characterize incipient fault signal in distribution network. The flow chart of incipient fault detection in distribution networks is shown in Fig. 123.1.

123 Early Fault Detection of Distribution Network Based on High-Frequency... 1085

Fig. 123.1 Flow chart of incipient fault based on DWT

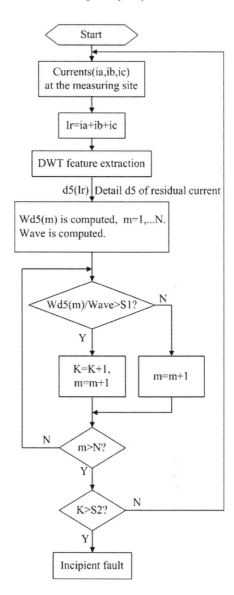

At the measuring site, three phase currents are measured. The corresponding residual currents are computed and they are extracted using DWT. The sampling frequency of the raw signals is 100 kHz. In this algorithm, only the wavelet transform coefficients on scales 3 to scales 5 will be considered, because the incipient fault which have been detected has a frequency range from 2 to 10 kHz [14], which is verified by a large number of field experiments, and scales 3–5 correspond to a frequency range of 3.125–12.5 kHz, d3, d4, and d5 including the frequency bands 25-12.5, 12.5–6.25, and 6.25-3.125 kHz are investigated, where the sampling frequency is 100 kHz. It is obvious that detail d5 is included in the frequency band of incipient fault. So the detail d5 is selected as the research frequency band. The residual current detail d5 coefficient over one measuring period is divided into N section for the fault detection purpose. The sum over absolute value for every section is computed, which is described as $W_{d5}(m)$, $m = 1, \cdots, N$ as

$$W_{d5}(m) = \sum_{n=(m-1)*p+1}^{m*p} |d5\_Ir(n)| \qquad (123.1)$$

where $W_{d5}(m)$ means the detector in the discrete samples according to $d5\_Ir$, which is the detail level d5 of the residual current with incipient fault, n is used for carry out a sliding window covering 5 ms, P is sample number in a window and N is a number of window samples.

The average value of the absolute sum over one measuring period is computed, which is shown as follows:

$$W_{ave} = \frac{1}{N*P} \sum_{i=1}^{N*P} |d5\_Ir(n)| \qquad (123.2)$$

The ratio value for $W_{d5}(m)$ to W_{ave} is obtained as

$$J = \frac{W_{d5}(m)}{W_{ave}} \qquad (123.3)$$

When the ratio J is above the setting value S1, the counter K is triggered once. During one measuring period, when the counter K is above the setting value S2, the incipient fault of distribution system is detected.

123.4 Simulated System

123.4.1 Test Power System

The 10-kV distribution system is supplied with power by a 110-kV grid via 40-MVA transformer as shown in Fig. 123.2; the resonant-earthed 10-kV system consists of several overhead lines and cables as radial feeders, 4-feeder distribution network simulated using ATP/EMTP, in which the processing is created by ATPDraw. L1 and L4 are overhead lines, 15 and 10 km, respectively. L2 and L3 are cables, 5 and 7 km, respectively. The feeder overhead line and cable are represented using the frequency-dependent JMarti model.

The polluted insulator fault is simulated according to the model of reference [15]; transient faults are generated by a fast electronic switch to simulate Peterson arc mechanism [6]. They are supposed at 5 km far away the measuring site.

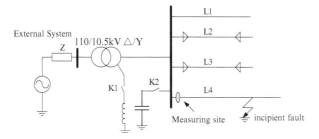

Fig. 123.2 Configuration of simulated system

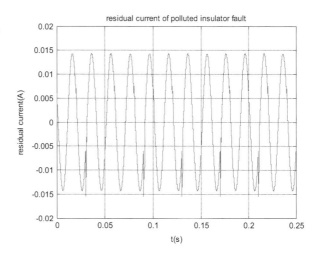

Fig. 123.3 Residual current of polluted insulator fault

123.4.2 Simulation Results

Here, simulated system is an isolated neutral system; polluted insulator fault signals were simulated with a sampling rate of 100 kHz for a duration of 0.25 s; instantaneous arc fault signals were simulated with a sampling rate of 100 kHz for a duration of 0.5 s. Herein, the setting values S1 and S2 are set to 100 and 15, respectively. Each value is set according to a large amount of simulation experiments; S1 < 100 and S2 < 15 account for little incipient fault, and we do not care for them.

The residual current for the fault case, which occurs on polluted insulator far away 5 km from the measuring site, is depicted in Fig. 123.3. After the discrete wavelet analysis, the details d5 are investigated. According to the proposed algorithm, the simulation time is divided into 50 time segments. The value of $W_{d5}(m)$ and W_{ave} are calculated, the value of $W_{d5}(m)$ and W_{ave} are calculated by Eq. 123.1

Fig. 123.4 Detector $W_{d5}(m)$ of residual current details shown in Fig. 123.3

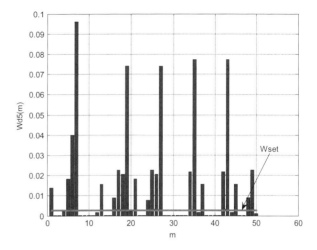

Fig. 123.5 Residual current of instantaneous arc fault

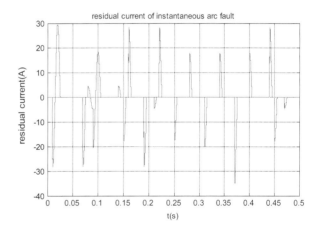

and Eq. 123.2, respectively, W_{set} is obtained from W_{ave} multiply by the setting value S1, which is shown in Fig. 123.4. In Fig. 123.4, it is shown that K = 22, it accounts for among 50 time segments, 22 time segment W_{set} is above the setting value S2. So this method can effectively detect the occurrence of polluted insulator in distribution system.

Similar to the above fault case, Fig. 123.5 depicts the residual current of instantaneous arc fault in distribution network. From Fig. 123.5, although the residual current of instantaneous arc fault is very large, it is intermittent signal; intermittence appears as a series of transients. The value K, which is the number of the ratio above the setting value S1, are obtained from Fig. 123.6. It is 43, which are above the setting value S2. Perhaps, instantaneous arc fault will access to the brink of collapse. These results clearly again demonstrate the effectiveness of the proposed technique.

Fig. 123.6 Detector $W_{d5}(m)$ of residual current details shown in Fig. 123.5

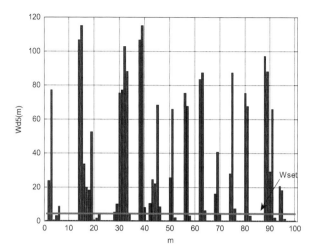

123.4.3 Discussion

From the aforementioned results, the proposed technique has a good performance for incipient fault signals in distribution network, especially for incipient fault signals such as instantaneous arc fault. The proposed technique has a distinctive indicator of incipient fault. But, for polluted insulator fault signal, due to very large resistance, the amplitude of fault feature is small, although it is implemented by this approach, it accounts for the approach is relevant with the fault resistance. Still, the proposed technique is a good approach for implementing these kinds of faults.

123.5 Conclusion

Successful detection of abnormalities would be a great benefit to the utilities, enabling them to detect severe faults at an early stage of their development, and consequently preventing unscheduled outages due to failures in distribution line. An ideal incipient fault detection should capture the degrading path of equipment and detect the root causes. The ultimate goal is to improve the overall distribution network reliability and reduce the operational costs strategically.

For incipient fault detection, the key issue is difficult to extract the characteristic of fault information. In this manuscript, a dwt-based method is proposed to extract the characteristic of incipient fault. The basic idea is extract the d5 coefficient of wavelet decomposition of residual current and residual voltage. Together with an energy detection algorithm, a scheme for incipient fault detection has been proposed. Simulation results verify the effectiveness of the proposed approach.

References

1. Sidhu, T. S., & Zhihan, X. (2010). Detection of incipient faults in distribution underground cables. *IEEE Transactions on Power Delivery, 25*(3), 1363–1371.
2. Wester, C. G. (1998). High impedance fault detection on distribution systems. In *42nd Annual Conference on Rural Electric Power Conference*, St. Louis, MO (pp. c5-1–5).
3. Kim, C. J., Shin, J. H., Yoo, M.-H., & Lee, G. W. (1999). A study on the characterization of the incipient failure behavior of insulators in power distribution line. *IEEE Transactions on Power Delivery, 14*(2), 519–524.
4. Benner, C. L., & Russell, B. D. (1997). Practical high-impedance fault detection on distribution feeders. *IEEE Transactions on Industry Applications, 33*(3), 635–640.
5. Al-Dabbagh, M., & A1-Dabbagh, L. (1999). Neural networks based algorithm for detecting high impedance faults on power distribution lines. In *Proceedings of International Joint Conference on Neural Networks*, Washington, DC (Vol. 5, pp. 3386–3390).
6. Cui, T., Dong, X., Bo, Z., & Juszczyk, A. (2011). Hilbert-transform-based transient/intermittent earth fault detection in noneffectively grounded distribution systems. *IEEE Transactions on Power Delivery, 26*(1), 143–151.
7. Lovisolo, L., Moor Neto, J. A., Figueiredo, K., de Menezes Laporte, L., & dos Santos Rocha, J. C. (2012). Location of faults generating short-duration voltage variations in distribution systems regions from records captured at one point and decomposed into damped sinusoids. *IET Generation, Transmission and Distribution, 6*(12), 1225–1234.
8. Butler, K. L. (1999). An expert system based framework for an incipient failure detection and predictive maintenance system. In *Intelligent System Application to Power Systems Conference*, Orlando, FL (pp. 321–326).
9. Miri, S. M., & Privette, A. (1994). A survey of incipient fault detection and location techniques for extruded shielded power cables. In *The 26th Annual Southeastern Symposium on System Theory*, Athens, OH (pp. 402–405).
10. Kim, C. J., Lee, S.-J., & Kang, S.-H. (2004). Evaluation of feeder monitoring parameters for incipient fault detection using Laplace trend statistic. *IEEE Transactions on Industry Applications, 40*(6), 1718–1724.
11. Apostolos, N. M., Andreou, G. T., & Labridis, D. P. (2012). Enhanced protection scheme for smart grids using power line communications techniques-part II: Location of high impedance fault position. *IEEE Transactions on Power Delivery, 3*(4), 1631–1640.
12. Yang, Y., Divan, D., Harley, R. G., & Habetler, T. G. (2006). Power line sensornet—A new concept for power grid monitoring. In *Power Engineering Society General Meeting*, IEEE (pp. 1–8).
13. Mousavi, J., & Rasoul, M. (2005). *Underground distribution cable incipient fault diagnosis system*. Ph.D. dissertation, TEXAS Digital Library, Texas A&M University, College Station, TX.
14. Ebron, S., Lubkeman, D. L., & White, M. (1990). A neural network approach to the detection of incipient faults on power distribution feeders. *IEEE Transactions on Power Delivery, 5*(2), 905–914.
15. Tsarabaris, P. T., Karagiannopoulos, C. G., & Theodorou, N. J. (2005). A model for high voltage polluted insulators suffering arcs and partial discharges. *Simulation Modelling Practice and Theory, 13*(2), 157–167.

Chapter 124
A Complementary Metal Oxide Semiconductor D/A Converter with R-2R Ladder Based on T-Type Weighted Current Network

Junshen Jiao

Abstract The mathematical expression and physical implementation are analyzed for a D/A converter and illuminated the T-type network framework of a binary digital-to-analog transform by dividing current means in this paper. Based on it, slice of half-dividing current is suggested by way of the symmetry of the drain and the source terminals in CMOS transistor. The paper puts forward a novel CMOS D/A converter based on T-type weighted current network with R-2R ladder. It has the merits of low power consumption and easy making of integration. Simulation result reveals a monotonic characteristic of the D/A converter.

124.1 Introduction

With the digital technology, especially the rapid development of computer technology, modern control, communication, and testing, the signal processing is widely adopted in digital computer technology in order to improve the performance of the system. Digital-to-analog (D/A) converter provides the interface between the analog world and digital signal processing systems. D/A converter is widely used in computer, automatic control, measurement, and many other areas. It is an indispensable device in modern communication.

D/A converter of traditional MOSFET architecture is a binary-weighted current source, which is composed of identical complementary metal oxide semiconductor (CMOS), and it is current steering. It has been used in a wide range of applications for conversion [1–3]. The architecture designing allows high-speed data converter, but a serious drawback is that it has consumed current source due to the high number of units, wherein D/A converter doubles the number of bits and a large

---

J. Jiao (✉)
Department of Electronic Engineering Technology, Tongling University,
Anhui 244061, China
e-mail: jiaotlu@sina.com

silicon area. In addition, consumption of large areas of the current source array is difficult to match all the MOS transistors [4–6].

This study evaluates the R-2R ladder D/A converter to present that the current-or inverse-mode framework is best suited for low-power operation. Then, methods to characterize the current-mode converter are put forward. The capability of an 8-bit D/A converter fabricated using CMOS process is finally brought forward to display the characterization techniques.

124.2 Mathematical Expression and Physical Implementation of Binary D/A

A n-bit binary digital quantity $D(b_{n-1}, b_{n-2} \cdots, b_1, b_0)$ can be given as follows:

$$A = b_{n-1} \times 2^{n-1} + b_{n-1} \times 2^{n-1} + \cdots + b_1 \times 2^1 + b_0 \quad (124.1)$$

where $b_i \in \{0,1\}$; thus, there are 2^n possible values for A. As long as the number n is large enough, the obtained A is probably regarded as an analog quantity. If signals are represented by voltage, the converted analog voltage signal can be expressed as

$$V_A = K \times V_{REF} \times A = K \times V_{REF} \times \left(d_{n-1} \times 2^{n-1} + \cdots + d_1 \times 2 + d_0 \times 2^0\right)$$
$$= 2^{n-1} \times K \times V_{REF} \times \left(d_{n-1} + d_{n-2} \times \frac{1}{2} + \cdots + d_1 \times \frac{1}{2^{n-2}} + d_0 \times \frac{1}{2^{n-1}}\right) \quad (124.2)$$

where K is a constant and V_{REF} is the normal reference voltage. It is known that it is difficult to gain voltage signals, but the current signal facility is to be added by tying wires. Thus, it is guessed that each item in summing is represented by current. These currents are found weighted and switched by $d_{n-1}d_{n-2} \cdots d_1d_0$, respectively. D/A converter with weighted current signal can be given as

$$I_{n-1} : I_{n-2} : \cdots : I_1 : I_0 = 1 : \frac{1}{2} : \cdots : \frac{1}{2^{n-2}} : \frac{1}{2^{n-1}} \quad (124.3)$$

It can be replaced as T-type resistor network with R-2R, as shown in Fig. 124.1.

In Fig. 124.1, the current passing through the resistor R is never changed whether $S_{n-1}S_{n-2}\ldots S_1S_0$ is connected to ground ($d_i = 0$) or virtual ground ($d_i = 1$) because of summing amplifier input $V-$ approach 0. Each branch current is always the same in T-type resistor network with R-2R. It should be noted that $V-$ is a virtual ground. It can get net equivalent resistance R from A–A, B–B, C–C, D–D, and E–E port. So,

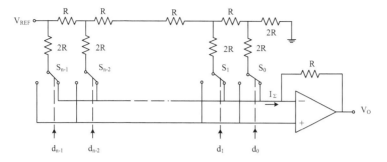

Fig. 124.1 Binary D/A converter T-type resistor network with R-2R

I = V_{REF}/R, each branch current is I/2, I/4, I/8, I/16...I/2n−1 and I/2n, respectively.

A main error source of the R-2R ladder D/A converter is the mismatch in switch on resistance, but the advances in CMOS fine-line technology have greatly improved the matching accuracy of switches [7, 8]. Based on these technological aspects, the R-2R ladder is revisited.

124.3 Current Division Principle of MOS Transistor

In order to testify MOS transistor division of current, it makes use of N-MOS transistor, which also uses P-MOS. N channel MOS transistor cross-sectional diagram can be given in Fig. 124.2.

If the conductive channel voltage is $V(x)$, electron diffusion and drift give birth to arbitrary position current. Assume $I(x)$ is the current of inversion layer [9, 10]. In this way,

$$I(x) = I_{drift}(x) + I_{diff}(x) \tag{124.4}$$

It is in proportion with $I_{diff}(x)$, channel charge density (Q_C), electron mobility (μ), channel field strength (dv_c/d_x), and the channel width (W). $I_{drift(x)}$ can be written as

$$I_{drift}(x) = -W\mu Q_C \frac{dV_C}{dx} \tag{124.5}$$

The diffusion current is in proportion with μ, thermal voltage (KT/Q), and derivative of Q_C. $I_{diff}(x)$ can be given as

Fig. 124.2 N channel MOS transistor section

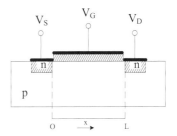

$$I_{diff}(x) = W\mu \frac{KT}{q}\frac{dQ_C}{dx} \qquad (124.6)$$

Thus,

$$I(x) = W\mu\left(-Q_C\frac{dV_C}{dx} + \frac{KT}{q}\frac{dQ_C}{dx}\right) \qquad (124.7)$$

The current along the channel is constant because the channel length is L. Consequently,

$$I \times L = W\int_0^L -\mu\left(Q_C\frac{dV_C}{dx} - \frac{KT}{q}\frac{dQ_C}{dx}\right)dx \qquad (124.8)$$

It is known that $I_D = -I$, so

$$I_D = \frac{W}{L}\int_0^L \mu\left(Q_C\frac{dV_C}{dx} - \frac{KT}{q}\frac{dQ_C}{dx}\right)dx \qquad (124.9)$$

Equation 124.9 clearly reveals the MOS transistor drain and gate symmetry characteristics. I_D is in proportion with W/L in the same substrate.

Consequently, the principles of half-dividing current can be used in MOS transistor replacement in Fig. 124.1. In spite of V-I identities of MOS transistor having nonlinear connection, symmetric outcomes can be shown in Eq. 124.10.

$$\frac{I_{D1}}{I_{D2}} = \frac{W_1/L_1}{W_2/L_2} \qquad (124.10)$$

Equation 124.10 shows that the ratio I_{D1}/I_{D2} is constant and independent of the input current and the terminal voltages V_G, V_1, and V_2. Also, for an equal division of the input current, the transistors T_1 and T_2 should have the same size. The resistors in the R-2R ladder can be replaced by CMOS transistors and still preserve the current division principle, despite the nonlinear current–voltage connection in CMOS transistors. Principle of half-dividing current for MOS is shown in Fig. 124.3.

Fig. 124.3 Principle of half-dividing current for MOS

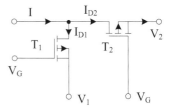

124.4 Half-Dividing Current Component of MOS with R-2R Ladder

MOS D/A conversion current component is shown in Fig. 124.4. It is composed of four patches PMOS transistors. T_2 assumes the R branch function. 2R branch is connected in series by T_1 and T_3 (or T_4). The input current is divided into equal parts. T_3 or T_4 not only takes in hand switching function but also occupies the matching role. Therefore, the entire unit layout is compact and has good match.

Each unit consists of T_3 and T_4 tube of two complementary as the switch transistor. If $d_i = 1$, weighted current can be exported by I_{OUT}, T_4 transistor is close. If $d_i = 0$, weighted current connects virtual ground terminal of the amplifier. Output current accords with the Eq. 124.3.

124.5 8-Bit D/A Converter of MOS with R-2R Ladder

An 8-bit D/A converter fabricated using CMOS process is shown in Fig. 124.5 with R-2R ladder. The resistor R in Fig. 124.5 is replaced by the slice PMOS transistor operating in the linear region and the four slice transistors form the slice cell for 1-bit conversion. Those unit transistors are driven by the digital input d_i and also operate as switches.

124.6 Simulation Result

Based on the 0.35-μm CMOS slice and power voltage of 3 V, it makes use of HSPICE simulation in order to test the proposed architecture linearity. T_1 and T_2 parameters are W/L = 2 μm/20 μm and T_3 and T_4 parameters are W/L = 1 μm/ 0.7 μm.

Simulation results are shown in Figs. 124.6 and 124.7. DNL and INL have skip courses because of the output of the amplifier limiting the cause in the initial moment. Despite the initial skips, DNL has no excess of the range of ±1 LSB, which reveals a monotonic characteristic of the D/A converter. The INL stays within the limit of ±0.9 LSB.

Fig. 124.4 Half-dividing current component of MOS with equivalent R-2R network

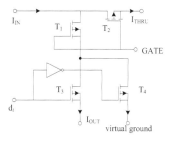

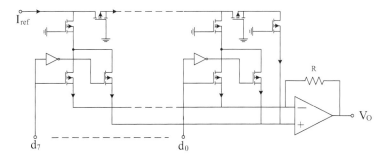

Fig. 124.5 8-bit D/A converter of MOS with R-2R ladder

Fig. 124.6 DNL simulation result

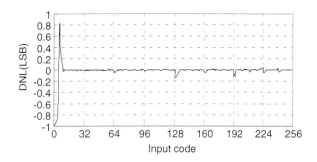

Fig. 124.7 INL simulation result

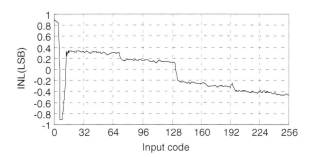

124.7 Conclusion

By analyzing the mathematical expressions of D/A converter, a T-type resistance network is introduced by using the dividing current. Based on this principle, CMOS transistor as a resistor ladder is proposed in this paper. Compared with the traditional resistance network, the architecture has good linearity. In this network, the CMOS parameter transistors can be precisely controlled in size. It can maintain the ratio between the MOS transistor. So, it has important theoretical and application value in the field of electronic technology.

Acknowledgements The work was supported by Educational Commission of Anhui Province of China (No. 2007JYYM443).

References

1. Yin, G. M., Eynde, F. O., & Sansen, W. (1992). A high-speed CMOS comparator with 8-bit resolution. *IEEE Journal of Solid-State Circuits, 27*(2), 208–211.
2. Lee, S. C., & Cho, M. H. (2002). 10-Bit 200 ms/s CMOS D/A converter employing high-speed limiter. *IET Journal of Electronics Letters., 38*(23), 1407–1408.
3. Borremans, M. A. F., & Steyaert, M. S. J. (2001). A 10-bit 1-Gsample/s Nyquist current-steering CMOS D/A converter. *IEEE Journal of Solid-State Circuits, 36*(1), 315–324.
4. Zhou, Y. J., & Yuan, J. (2003). An 8-bit 100-MHz CMOS linear interpolation DAC. *IEEE Journal of Solid-State Circuits, 38*(10), 1758–1761.
5. Ripley, D., Balteanu, F., & Gheorghe, I. (2004). Quad-band GSM/GPRS/EDGE polar loop transmitter. *IEEE Journal of Solid-State Circuits, 39*(12), 2179–2188.
6. Vleugels, K. (2001). A 2.5-V sigma-delta modulator for wideband communication applications. *IEEE Journal of Solid-State Circuits, 36*(12), 1887–1898.
7. Tseng, W. H., Wu, J. T., & Chu, Y. C. (2011). A CMOS 8-bit 1.6-gs/s DAC with digital random return-to-zero. *IEEE Transactions on Circuits and Systems II: Express Briefs., 58*(1), 1–5.
8. Deveugele, J., & Steyaert, M. S. J. (2006). A 10-bit 250-MS/s binary-weighted current-steering DAC. *IEEE Journal of Solid-State Circuits, 41*(2), 320–329.
9. Woo, J. K., & Shin, D. Y. (2009). High-speed 10-bit LCD column driver with a split DAC and a class-AB output buffer. *IEEE Transactions on Consumer Electronics, 55*(3), 1431–1438.
10. Marche, D., Savaria, Y., & Gagnon, Y. (2008). Laser fine-tuneable deep submicron CMOS 14 bit DAC. *IEEE Transactions on Circuits System I, 55*(8), 2157–2165.

Chapter 125
Detecting Repackaged Android Applications

Zhongyuan Qin, Zhongyun Yang, Yuxing Di, Qunfang Zhang, Xinshuai Zhang, and Zhiwel Zhang

Abstract The rapid development of the smartphone brings immense convenience to people. Recently more and more developers publish their own applications (or apps) on the android markets to make profits. The so-called repackaged apps emerge by embedding malicious codes or injecting ads into the existing apps and then republishing them. In this paper, focusing on the shortcomings of existing detection system, we propose an efficient repackaged apps detection scheme based on context-triggered piecewise hash (CTPH). We also optimize the similarity calculation method (edit distance) and filter unnecessary matching process to make the matching more efficient. Experimental results show that there are about 5 % repackaged apps in pre-collected data. The proposed scheme improves the detection accuracy of the repackaged apps and has positive significance to the ecosystem of android markets.

125.1 Introduction

In the past few years, android has developed strikingly which occupies a dominant position in the smartphone markets since its market share exceeds Apple in 2010. The latest data from the research company Strategy Analytics show that android's market share has risen to 70 % to the end of 2012 [1].

Z. Qin (✉) • Z. Yang • X. Zhang • Z. Zhang
School of Information Science and Engineering, Southeast University, Nanjing 210096, China

Information Security Research Center, Southeast University, Nanjing 210096, China
e-mail: zyqin@seu.edu.cn

Y. Di
Communication Department, Nanjing Institute of Artillery Corps, Nanjing 210000, China

Q. Zhang
Computer Department, Nanjing Institute of Artillery Corps, Nanjing 210000, China

Because android is free and open, app developers can publish their own apps in the [1] android markets. Repackaged app developers first download original apps, disassemble, modify configuration file, inject malicious code, insert ads, re-sign with a private key, and then release to android markets again. Further analysis indicates that these repackaged apps are typically used to steal ad revenues and obtain user location, phone number, and other personal information, even to control user's phone remotely.

Some schemes have been proposed to detect the repackaged apps. DEXCD [2], developed by Ian Davis, extracts the opcodes from Java class in Dex file and tries to find a steam match of opcodes to detect the cloned apps. Clint et al. propose DNADroid to detect apps copying [3], which utilizes WALA to construct program dependency graphs (PDGs) [4], applies lossless and lossy filters to discard the method pairs, and computes the similarity by subgraph isomorphism. However, the robust techniques in DNADroid are largely expensive. DroidMOSS [5], presented by Wu et al., computes the fuzzy hashes of apps and compares similarity of all fingerprints; the similarity scoring algorithm is memory-consuming, thus slowing down the process.

In this paper, we proposed an improved repackaged android apps detection system based on CTPH (i.e., fuzzy hash) [6], which uses two small primes to do twice CTPH (T-CTPH) process and generates two fingerprints for each app. The main contributions of this paper are the following: (1) We propose an improved fingerprint-generating algorithm, two small primes are selected as the trigger values for T-CTPH to increase the randomness against possible attacks and improve the accuracy, which can be further used to filter unnecessary matching processes. (2) An efficient edit distance calculation method is proposed to speed the calculation of the similarity between fingerprints, which greatly reduce the memory usage. (3) We have realized our system to detect repackaged apps and found that about 5 % repackaged apps in pre-collected 6,438 samples of four app types.

The remainder of this paper is organized as follows: Sect. 125.2 introduces our approach, including feature extraction, fingerprint generation, and similarity matching. Section 125.3 illustrates the evaluation results based on 6,438 real applications from several android markets. Finally, we make our conclusion in Sect. 125.4.

125.2 System Design

Usually the repackaged apps have the following features: (1) They always have the same app type with the official ones. (2) The size will not have a big difference from the original one. Furthermore, we assume that the signing keys are not leaked. Therefore, the matched pairs (with high similarity) with the same author information are ignored because they are often the different versions of the same application.

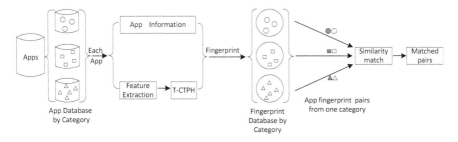

Fig. 125.1 System architecture

125.2.1 System Overview

To detect the repackaged apps, we have designed the following system shown in Fig. 125.1, which contains three parts: (1) sample collection and classification, (2) feature extraction and fingerprint generation, and (3) similarity matching. We first download apps from android markets by app category; for each app we extract the signature information and instruction sequence for fingerprint generation. Then we store fingerprints in their own databases and select any two fingerprints from each database for similarity calculation. Given a comparison threshold, suspicious repackaged app pairs will be fixed.

125.2.2 App Feature Extraction

To extract the app feature, we first uncompress it and use keytool to extract certificate information in *META-INF* directory [7]. As shown in Fig. 125.2, we get the MD5 as the unique information of one app, for it will be different between the original and repackaged apps. Then we leverage existing Dalvik disassembler *baksmali* to disassemble *classes.dex* file [8] and extract instructions by the following rules: (1) depth traversal with the alphabetical order of generated smali files and folders; (2) before releasing, some names of class are modified, we ignore the confused names of classes to reduce the error of instruction extraction; and (3) extracting methods of different classes.

125.2.3 Fingerprint Generation

For an actual APK file, the extracted instruction sequences may be very long. To generate the fingerprints, one common way presented in DroidMOSS is to use fuzzy hash directly, which does not consider the nature that the size of repackaged app will not have a big difference from the original one. Figure 125.3 shows the once

```
administrator@ubuntu:~$ keytool -printcert -v -file \
> /renren/META-INF/CERT.RSA
Owner: CN=renren, OU=renren.com, O=opi, L=beijing, ST=beijing, C=CN
Issuer: CN=renren, OU=renren.com, O=opi, L=beijing, ST=beijing, C=CN
Serial number: 4b85da7d
Valid from: Thu Feb 25 10:03:41 CST 2010 until: Fri Nov 28 10:03:41 CST 2064
Certificate fingerprints:
         MD5:  FB:5C:BF:1E:21:6D:40:74:54:5C:72:17:84:DB:18:48
         SHA1: CE:4A:B1:BB:60:4F:74:67:3A:1B:8B:3B:C7:F7:D4:10:71:F6:E7:93
         Signature algorithm name: MD5withRSA
         Version: 1
```

Fig. 125.2 Certificate information of one application

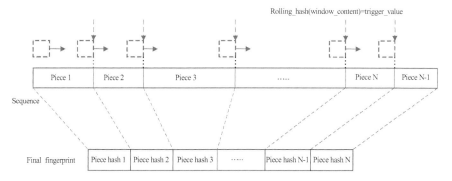

Fig. 125.3 Context-triggered piecewise hash

CTPH process, the sequence is the input and a trigger value is for dividing pieces, then all the piece hashes are calculated and concatenated directly as the final fingerprint.

In this paper, we present an improved approach. Specifically, we remove the compression mapping process in Spamsum [9]; and for sequences of different lengths, we use different primes to calculate the fingerprints; for an instruction sequence, we use two small primes to do twice CTPH processes, respectively. As visually shown in Fig. 125.4, the *original sequence* is the instructions extracted by the rules in Sect. 125.2.2. Suppose the length of the original sequence is N and $S1$ pieces generate after the first CTPH process (e.g., the left in solid line in Fig. 125.4), then the average size of the trigger value tv is about $\lfloor N/S1 \rfloor$ ($\lfloor \ \rfloor$ means round downwards). We represent each 32-bit binary piece hash as an 8-digit hexadecimal number in this paper, so the length of the sequence generated by the first CTPH process is about $8*S1$. If the piece number is S after the second CTPH process, then the trigger tv' of the second CTPH is about $\lfloor 8*S1/S \rfloor$. So we have $tv \approx \frac{N}{S1}$ and $tv' \approx \frac{8*S1}{S}$. Further, $tv * tv' \approx \frac{8*N}{S}$. In order to make the trigger values in similar size, we have $tv = tv'$. That is, $tv = tv' \approx \sqrt{\frac{8*N}{S}}$. And we use primes to trigger pieces to increase the randomness against possible attacks. Given tv (or tv'), we have two adjacent primes $r1$ and $r2$, that is, $r1 <= tv = tv' <= r2$; thus we take $r1$ and $r2$ as the trigger values to generate two fingerprints. In order to balance efficiency and

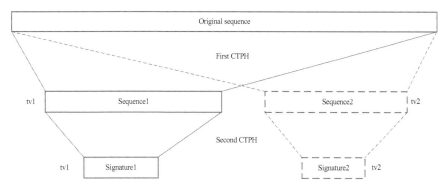

Fig. 125.4 T-CTPH for fingerprint generation

similarity between fingerprints, S is taken as 128, that is, *signature*1 (or *signature*2) is made up about 1,024 characters in formula Eq. 125.1. Here, $r1$ is $tv1$ and $r2$ is $tv2$ in Fig. 125.4. Therefore, the final fingerprint of an app is

$$signature = (tv1, signature1) || (tv2, signature2) \qquad (125.1)$$

Next the fingerprint of each app is calculated and stored in its corresponding type database. If the lengths of two sequences are far away, it is little possible they are repackaged pairs. For this nature, not all the fingerprints are compared, and we just concern the fingerprints triggered by the same prime. For two different apps, the fingerprints are x and y: $x = (x1, xsign1)||(x2, xsign2)$ and $y = (y1, ysign1)||(y2, ysign2)$. If $x1 = y1$, it will definitely have $x2 = y2$; we compare $(xsign1, ysign1)$ and $(xsign2, ysign2)$ with the method in Sect. 125.2.4, respectively, and put the larger similarity score as the final result; if $x1 = y2$ or $x2 = y1$, we compare $(xsign1, ysign2)$ or $(xsign2, ysign1)$. In other cases, we do not treat them as repackaged apps. With such filtering method, it will be more efficient to detect the repackaged apps in the android market.

125.2.4 Similarity Matching

Our above steps are applied for each app generating the fingerprint. Then we compare the similarity between the fingerprints by adopting an optimized edit distance method. The edit distance is the minimum edit operations to turn one fingerprint into another, including insertion, deletion, and substitution. The conventional way is to use a two-dimensional array to calculate the edit distance of two strings (with lengths of *len*1 and *len*2, respectively) by filling the array circularly. So the final result is *array*[*len*1-1,*len*2-1].

However, if two strings are very long, *len*1**len*2 size of memory will be needed for calculating, slowing the matching speed. In order to speed up the calculation for long strings, we have optimized the method by using three one-dimensional arrays, *array*1, *array*2, and *array*3 (with sizes of *len*1, *len*2, *len*2, respectively), to calculate the edit distance. Here, *array*1 denotes the first column of the conventional two-dimensional array; *array*2 and *array*3 denote two adjacent rows of that two-dimensional array. We fill *array*2 and *array*3 circularly with an iterative method and exchange *array*2 and *array*3 continuously to denote the two adjacent rows. At last, if *len*1 is odd, the edit distance is *array*2[*len*2-1]. Otherwise, the edit distance is *array*3[*len*2-1]. This process is described in Algorithm 1, and it just needs about *len*1 + *len*2*2 memory space.

Algorithm 1 Calculate the edit distance between two apps

Input: Two fingerprints fp1 and fp2
Output: Edit distance between fp1 and fp2
1: *len*1e*strlen*(*fp*1) , *len*2 ←*strlen*(*fp*2)
2: array1[0]=(*fp*1[0]==*fp*2[0] ? 0:1) //initialize array1
3: **for** *i* = 1→=*len*1-1 **do**
4:    *cost*=(*fp*1[*i*]==*fp*2[0] ? 0:1)
5:    *array*1[*i*]=*min*(*array*1[*i*-1]+1,*i*+*cost*)
6: **end for**
7:    array2[0]= array1[0] //initialize array2
8: **for** *i* = 1→=*len*2-1 **do**
9:    *cost*=(*fp*2[*i*]==*fp*1[0] ? 0:1)
10: *array*2[*i*]=*min*(*array*2[*i*-1]+1,*i*+*cost*)
11: **end for**
13: **for** *i* = 1i *len*1-1 **do**
14:    (*i* mod 2==0) ? *array*2[0]=*array*1[*i*]:*array*3[0]=*array*1[*i*]
15: **for** *j* = 1ay*len*2-1 **do**
16:    *cost*=(*fp*1[*i*]==*fp*2[*j*] ? 0:1)
17:    **if** (*i* mod 2==0) **then**
18:       *array*2[*j*]=*min*(*array*2[*j*-1]+1,*array*3[*j*]+1,*array*3[*j*-1]+*cost*)
19:    **else**
20:       *array*3[*j*]=*min*(*array*3[*j*-1]+1,*array*2[*j*]+1,*array*2[*j*-1]+*cost*)
21:    **end if**
22: **end for**
23: **end for**
24: edit_dist=((len1mod 2=0) ? array3[len2-1]:array2[len2-1])
25: **return** edit_dist

After the edit distance between two fingerprints is calculated, we use the following formula to measure the similarity between the two fingerprints [5]:

$$Sim\_Score = \left[1 - \frac{edit\_dist}{\max(len1, len2)}\right] * 100 \quad (125.2)$$

If two apps are signed with different keys and the similarity score exceeds a certain threshold, we treat them as repackaged matching pairs. Note that the choice of threshold largely affects the false-positive and false-negative rate, thus influencing the accuracy of our test results. In our experiments, we choose 70 as the threshold, and it shows a good balance between the false-positive and false-negative rate.

125.3 Evaluation

In our experiment, we collect 6,438 apps from several android markets and store them with different categories. There are four types: social networking, game, system tool, and shopping, which are shown in column 2 of Table 125.1.

From Table 125.1, we find that our scheme (T-CTPH) improves the accuracy of detecting repackaged apps, and the results are closer to the results of manual analysis. In addition, the apps about social networking have the highest rate. This is because the many repackaged apps are used for stealing user's Internet traffic and phone bill by injecting malicious code to control user's phone remotely, which will inevitably require Internet service. Also, games related are rather high. It is the reason that developers can reroute or steal ad revenues by replacing or embedding ads to the games.

In order to detect so many applications, the complexity of the algorithm is very important. In our scheme, the time and space complexity of fingerprint generation are both $O(n)$, which has little influence on the overall efficiency. For the edit distance of two fingerprints, the time and space complexity are $O(n^2)$ and $O(n)$, respectively. Note that before optimizing, they are both $O(n^2)$. Here we reduce the space complexity from $O(n^2)$ to $O(n)$, which largely improve the efficiency. Then we test the optimized similarity algorithm on a computer with a Linux system (Ubuntu 10. 04). The CPU is Intel (R) Pentium 4 running at 2.93 GHz, 512 MB RAM. Table 125.2 shows the consumed time of sequences in different length ranges. We find that the optimized algorithm largely speeds up the calculation and is more significant to reduce the time when the sequences are longer.

To perform a concrete study of the repackaged apps and reveal how one app is repackaged, we show the analysis of repackaged apps detected by our scheme. Usually, advertising SDK needs to add a publisher identifier to AndroidManifest.xml and modify the layout description and the program bytecodes to show ads. As shown in Fig. 125.5, it is a detected example that repackage a normal app (*com. racingstudio. racingmoto*) by including AdMob SDK [10]. We find that the signed keys are different, and they are similarity matching pairs. Further, with manual analysis, the repackaged app (right) always pops up ads in the bottom when running the two games and functions like *setVisibility*, *findViewById*, and *loadAd* are inserted into *onCreate* to display ads in the disassembled files, by which developers can steal the ads revenues.

Table 125.1 The results of different detecting repackaged apps methods and manual analysis

Category	Number	DroidMOSS	T-CTPH	Manual analysis	Percentage
Social	2,557	157	156	155	6.1
Game	2,396	140	140	138	5.8
Tool	838	39	38	38	4.5
Shopping	647	34	34	33	5.1

Table 125.2 Consumed time of before and after optimization

Lengths	0.5k	2k	4k	6k	8k	10k
Before (ms)	10.27	127.41	468.56	1,028.37	1,818.52	2,992.54
After (ms)	4.04	68.24	272.91	604.49	1,032.50	1,443.57

Fig. 125.5 Screenshots of repackaging

125.4 Conclusion

In this paper, we propose an efficient method to detect repackaged apps based on CTPH. We remove the compression mapping step, do twice CTPH process with two small primes to improve the fingerprint accuracy, optimize the similarity algorithm, and filter unnecessary matching processes to make the matching more efficient. Our experimental results show there are about 5 % repackaged apps in pre-collected samples, and it has a positive and practical significance for the ecological system of the android markets.

Acknowledgements This paper is funded by the Information Security Special Projects of National Development and Reform Commission. The authors would like to thank the anonymous reviewers for their insightful comments that helped improve the presentation of this paper.

References

1. Bicheno, S. (2013). *Global smartphone OS market share by region: Q4 2012*. https://www.strategyanalytics.com/default.aspx?mod=reportabstractviewer&a0=8222
2. Davis, I. (2012). *Dex clone detector*. http://www.swag.uwaterloo.ca/dexcd/index.html
3. Crussell, J., Gibler, C., & Chen, H. (2012). Attack of the clones: Detecting cloned applications on Android markets. In *Computer Security–ESORICS 2012* (pp. 37–54). Heidelberg: Springer.
4. IBM T.J. Watson Research Center. (2012). *Watson libraries for analysis (WALA)*. http://wala.sourceforge.net/wiki/index.php/Main_Page

5. Zhou, W., Zhou, Y., Jiang, X., & Ning, P. (2012). Detecting repackaged smartphone applications in third-party android marketplaces. *Proceedings of the Second ACM Conference on Data and Application Security and Privacy* (pp. 317–326). New York, NY: ACM.
6. Kornblum, J. (2006). Identifying almost identical files using context triggered piecewise hashing. *Digital Investigation, 3S*, S91–S97.
7. *Android development guide*: *Signing your applications*. http://developer.android.com/tools/publishing/app-signing.html
8. *Smali-An assembler/disassembler for Android's dex format*. https://code.google.com/p/smali/
9. Andrew, T. *Spamsum README*. http://www.samba.org/ftp/unpacked/junkcode/spamsum/
10. Google Inc. *Admob for android developers*. http://support.google.com/admob/topic/1307236?hl=zh-Hans&ref_topic=1307209

Chapter 126
Design of Wireless Local Area Network Security Program Based on Near Field Communication Technology

Pengfei Hu and Leizhen Wang

Abstract In order to solve wireless local area network (WLAN) security problem due to the open-wide nature of wireless radio and the improvement of computing power, a design of WLAN security program based on near field communication (NFC) is presented in this paper. In this paper, the importance of having access to handshake for WPA2 brute force is explained. The proposed design protects the four-way handshake by taking advantage of NFC short-range character to eliminate the risk of intercept. For implementation, Android system is selected as a mobile device development platform. The design is compatible with the IEEE 802.11i which ensures the massive expansion in the future. Furthermore, the design simplifies operations to improve users' experience without much extra hardware cost and offers an option to the owner of WLAN to control the access physically, which benefits commercialization of NFC. From one perspective, this design can solve the wireless network security problem effectively.

126.1 Introduction

With the popularization of mobile terminals and Internet of Things, wireless communication technology has gained great progress. The near field communication (NFC) has got considerable concern due to its security, low power consumption characteristics. Rapidly growing NFC is expected to be one of the most important trends and continues to gain popularity in the business and IT industry.

At the same time, wireless networking has been experiencing an explosive growth and offers attractive mobility and flexibility to both network users and operators [1] in the age of mobile Internet. The wireless local area network (WLAN) systems like IEEE 802.11 networks become common access networks in

---

P. Hu (✉) · L. Wang
Northeastern University at Qinhuangdao, Qinhuangdao 066004, China
e-mail: hupengfei1993@gmail.com

public and private environments. Due to the wide-open nature of wireless radio, security over a wireless environment is more complicated than in a wired environment. Many attacks make the wireless network insecure [2]. The security problem has already become the Achilles' heel of the further development of the wireless network. To overcome the security challenges, IEEE 802.11i (also called WPA2) has been developed to enhance the security. However, current wireless technologies in use allow hackers to monitor and even change the integrity of transmitted data [3]. The improvement of computing power makes it possible for the effective WPA2 brute-force cracking.

To solve these security challenges, we propose a program to set up a WLAN that combine Wi-Fi with NFC. The proposed program provides a simple but safe mean to set up a WLAN among mobile devices. We realize the importance of capturing handshake packet to hack WPA/WPA2-psk, so we introduce the NFC by taking advantage of NFC short-range character to protect the 4-way handshake to guarantee WLAN security.

The rest of this chapter is organized as follows. In Sect. 2 we introduce some related works about NFC and IEEE 802.11i security threats. After that, our design of WLAN security program based on NFC technology is explained in Sect. 3. In Sect. 4, we give some discussion. This chapter is concluded with a summary and future works in Sect. 5.

126.2 Related Work

126.2.1 Introduction of NFC

NFC is a short-range wireless connectivity technology which is heavily based on *radio frequency identification* (RFID). Extending the ability of the contactless card technology, NFC also enables devices to share information over a distance of a few centimeters with a maximum communication speed of 424 kbps operating in the 13.56 MHz frequency band. NFC has three operating modes: read/write mode, card emulation mode, and peer-to-peer mode. The peer-to-peer mode is an operating mode specific to NFC and allows two NFC devices to communicate directly with each other [4] with the lightweight, binary message data format called *NFC Data Exchange Format* (NDEF). NFC bidirectional communication is an ideal ability for establishing connections with other technologies. This technology can be integrated into an existing system to simplify and speed up the process of monitoring and control of the system [5]. To support the peer-to-peer communication, the *Logical Link Control Protocol* (LLCP) was designed to make peer-to-peer transactions smoother as it enables NFC devices to be equal in communication [6].

Because the communication distance is short, it is almost impossible to intercept. NFC and WLAN do not share the same frequency band [7]. The difference of frequency obstructs the radio interception among the transmission of NFC data.

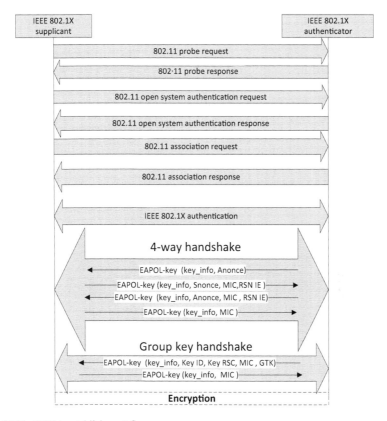

Fig. 126.1 RSNA establishment flow

126.2.2 IEEE 802.11i Security Threats

The IEEE 802.11 committee established the Task Group i (TGi) and ratified IEEE 802.11i on July 2004 [8]. 802.11i incorporates IEEE802.11X as its authentication enhancement and adopts a concept called *robust security network associations* (RSNAs) which is established as shown in Fig. 126.1. The success of authentication is witnessed by the fact that the access point (AP) and the client station (STA) own the same *Pairwise Master Key* (PMK) at the end. The four-way handshake provides a new temporary key called the *Pairwise Transient Key* (PTK) for confidentiality, data authentication, and anti-replay attack [8]. All the messages during the four-way authentication handshake are sent as EAPOL-Key frames.

Considering the fact that WPA2 has its own advanced system, there is no specific effective attack method till now except for brute-force attack or dictionary attacks. But the security risk cannot be ignored with computing power booming. From theory to practice, the high-speed brute force for WPA/WPA2-psk based on distributed multi-core CPU and GPU get a performance optimization in recent years. The development of cloud computing provides the optimal balance between cost and benefit as well. Both of them let WPA/WPA2-psk cracking available.

To make the matter worse, people will be prone to choose easy key for convenience in home or cafe. In such environment, privacy information and sensitive data are in transmission every day without enough protection. The high-speed brute force for WPA/WPA2-psk based on distributed multi-core CPU and GPU can crack 8 bits digital key in tens of seconds [9]. As for the required data to generate and verify, the key is broadcast with normal traffic and is really obtainable for the attacker. And the weakness is that the PMK was derived from the concatenation of the passphrase, SSID, length of the SSID, and nonces [3]. Once the four-way authentication handshake has been captured, the attacker has enough information required to do a dictionary attack to subject the passphrase. While PTK is a keyed-HMAC function based on the PMK, the challenges still exist in WPA2.

Dictionary and brute-force attacks are typically done automatically with tools. We must find a way to jump out the infinite loop that computing power's gain means our wireless network security loss. We have realized that capturing the four-way authentication handshake is the basement of the cracking. So what we need to do is to find a way of avoiding the four-way handshake being captured.

126.3 Design of the Program

Through the explanations of the weakness of WPA/WPA2-psk, we conclude that protecting the four-way handshake will be a good way to make sure our WLAN is secured. Meanwhile WLAN must transmit its authentication information in air to meet the flexibility requirement. So our program must be able to prevent the attacker from obtaining the four-way handshake and meet the flexibility requirement. NFC as a short-range wireless connectivity technology exactly fixes these requirements. Consequently, we hold the belief that the security of NFC should be introduced to the handshaking to connect the WLANs.

Considering that most mobile NFC devices available in the market are for Android system, the most popular open-source operating system, we select Android system as the mobile device development platform. The function of NFC was introduced by Google into Android2.3 device. The Android 4.0 system provides API support for NFC tag read/write mode and NFC P2P mode (Android Beam).

The Android SDK provides an NFC API to develop NFC applications which conduct peer-to-peer data exchange [10]. As the Wi-Fi functions are hiding behind the SDK interface, we cannot call the bottom relevant class. So our design pays more attention on the NFC part. The *.nfc* android package provides access to NFC function. On *android.nfc*, several classes could be used to running NFC function [11]:

- *Android.app.Activity*
- *android.content.Intent*
- *android.nfc.NdefMessage*
- *android.nfc.Record*
- *android.nfc.NfcAdapter*
- *android.nfc.NfcEvent*
- *android.nfc.NfcAdapter.CreatNdefMessageCallback*

We can use *NdefMessage* (byte[ ]) to construct an NDEF message from binary data or *NdefMessage* (*NdefRord* [ ]) to construct from one or more *NdefRecord*s and then transform the information from EAPOL-Key frames into NDEF frames. So we construct *NdefRecord* (short, byte [ ], byte [ ], byte [ ]) according to the EAPOL-Key which contains the *key_info*, *Annoce*, *MIC*, and *RSN IE*. The program is designed as follows [12] (shown in the Fig. 126.2):

1. *Initialization and Anti-collision*

    The mobile devices should initialize first. The initiator periodically probes the presence of a target by scanning the surrounding Wi-Fi radio signals and then verifies the signature. If the signature does not match, continue scanning.

2. *Activation and Parameters Selection*

    Once a target has been detected by two devices in range to communicate, the initiator gets into activation state to let the IEEE 802.1X authentication process as normal. Once the authentication phase is completed, the initiator creates an *NdefMessage* that contains the *NdefRecords* which are constructed according to the EAPOL-Key and a set of parameters like *key_info*, *Annoce*, *MIC,* and *RSN IE* are notified or negotiated. LLCP services will be selected after those abovementioned are finished.

3. *Data Exchange*

    As the old EAPOL-Key frames have already exchanged into NDEF, the initiator calls the *setNdefPushMessage()* with the *NdefMessage* created before. Data exchange over the LLCP is in good protection. Once the message is received by the initiator, the initiator checks the replay counter to avoid replay attack. The initiator will load the PTK if the MIC is valid.

4. *Deactivation*

    The initiator can release the NFC session after the success of authentication, via Release-Request/Response messages.

126.4 Discussion

The factor that our supplicant/initiator and authenticator are both on the Android system makes our program similar with the Android Beam or Wi-Fi Direct. The reason/cause is that there is no NFC-enable wireless router availed. In order to implement this program, we have to take the NEC-enable Ultrabook or tabletPC on

Fig. 126.2 Approach schema

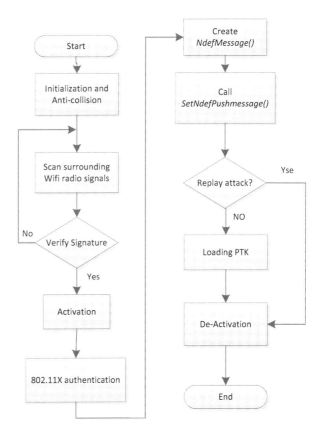

the Android platform as access point to be the authenticator. In our ideal condition (shown in Fig. 126.3), there is a machine like point-of-sale (POS) or wireless router ingrate NFC, so we can finish our payment conveniently and join the WLANs in safety by simple touch.

Base on the description of the program, we can get several points as what are written below:

1. We found that the importance of hacking WPA/WPA2-psk is capturing handshake via the wide-open nature of wireless radio. So we take advantage of the short-range character from NFC to protect the four-way handshake and use NFC to share WPA/WPA2-psk link setup parameters to eliminate the risk of intercept.
2. Due to the compatibility with IEEE 802.11i without changing its framework and protocol, our program could be massively expanded easily if the NFC-enable router appears. Along with the Internet of Things walking into our life, wireless telecommunication equipment that supports NFC function will be mass produced.

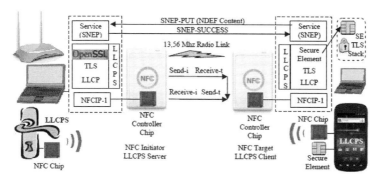

Fig. 126.3 Program overview

3. Considering the convenience and economy, people usually choose WPA/WPA2-psk (key pre-shared) model in their office and home. This key management cannot provide high-level security. In our program, the balance point between cost and security has been found. Our program provides the management of dynamic key to guarantee the WLAN security. People could say no to the trouble of remembering passwords without much extra hardware cost.
4. As the joining WPA2 via application calling is not supported by the Android system, Wi-Fi connection is sort of complicated. In the era of mobile Internet, these obstacles will disappear along with mobile device system upgrade [13]. In our program, the user can join the WLAN just by simple touch. The users' experience gain is improved, and making the power consumption becomes less.
5. Because exchanging the pairing data is allowed, the user's device does not need to store the password before. And the unsafety of temporary authority is disappeared. The owner could control the access to the WLAN physically on condition that the access methods except NFC are restricted. This point will benefit the commercialization of NFC. Businesses could use this way to let their own customers join the WLAN in an open field while closing the door to someone who is not their customer. This option could decrease network cost, improve quality of service, and especially offer a safe network environment to customers.

126.5 Conclusion

This paper introduced NFC, a promising technology, to the wireless network access and tried to solve the problem of wireless network security completely from one perspective. This program made full use of the short-range characteristics of NFC to eliminate the possibility of intercept of handshaking package. The main benefit of our program is compatibility with IEEE 802.11i without changing the network framework. So we can make full use of existing resource to reduce upgrade cost. To compare with the similar technology such as Android Beam, our program used IEEE 802.11i for mutual authentication, key generation, key exchange, encryption, and integrity. Since NFC can establish a connection faster, the cost of time decreases. The user can join the WLAN just by a simple touch to improve user experience. As for future research, we will study the implementation of our program in other systems and further optimize the program.

Acknowledgement This work was supported by the National Natural Science Foundation of China (Grant No.61273203).

References

1. Samiah, A., Aziz, A., & Ikram, N. (2007). An efficient software implementation of AES-CCM for IEEE 802.11i Wireless St. In *31st Annual International Computer Software and Applications Conference (COMPSAC 2007)* (pp. 689–694). IEEE.
2. Chen, J. C., Jiang, M. C., & Liu, Y. W. (2005). Wireless LAN security and IEEE 802.11i. *Wireless Communications, 12*(1), 27–36.
3. Lashkari, A. H., Danesh, M. M. S., & Samadi, B. (2009). A survey on wireless security protocols (WEP, WPA and WPA2/802.11i). In *2nd IEEE International Conference on Computer Science and Information Technology* (pp. 48–52). IEEE.
4. Monteiro, D. M., Rodrigues, J. J., & Lloret, J. (2012). A secure NFC application for credit transfer among mobile phones. In *International Conference on Computer Information and Telecommunication Systems* (pp. 1–5). IEEE.
5. Opperman, C. A., & Hancke, G. P. (2011). A generic NFC-enabled measurement system for remote monitoring and control of client-side equipment. In *3rd International Workshop on Near Field Communication (NFC)* (pp. 44–49). IEEE.
6. Lotito, A., Mazzocchi, D. (2012). OPEN-NPP: an open source library to enable P2P over NFC. In *4th International Workshop on Near Field Communication (NFC)* (pp. 57–62). IEEE.
7. Jie, MA, & Jin-long, E. (2013) Program of establishing connection of WiFi transmission rapidly based on NFC technology. *Computer Engineering, 39*(6), 1–6 (In Chinese).
8. Hori, Y., & Sakurai, K. (2006) Security analysis of MIS protocol on wireless LAN comparison with IEEE802. 11i. In *3rd International Conference on Mobile Technology, Applications & Systems* (p. 11). ACM.
9. Liu, Y. L., Jin, Z. G., Chen, Z., & Liu, J. W. (2007). Design and implement of high-speed brute forcer for wpa/wpa2-psk. *Computer Engineering., 37*(10), 125–127 (In Chinese).
10. Serfass, D., & Yoshigoe, K. (2012). Wireless sensor networks using android virtual devices and near field communication peer-to-peer emulation. *Proceedings of IEEE Southeastcon* (pp. 1–6). IEEE.
11. Android SDK Developer Guide. Available: http://developer.android.com/
12. Urien, P. (2013). LLCPS: A new security framework based on TLS for NFC P2P applications in the Internet of Things. In *Consumer Communications and Networking Conference* (pp. 845–846). IEEE.
13. Arakawa, Y., Sonoda, Y., Tagashira, S., & Fukuda, A. (2012). WiFiTag: Direct link from the real world to online digital contents. In *Seventh International Conference on P2P, Parallel, Grid, Cloud and Internet Computing* (pp. 339–344). IEEE.

Chapter 127
A Mechanism of Transforming Architecture Analysis and Design Language into Modelica

Shuguang Feng and Lichen Zhang

Abstract One of the fundamental challenges in research related to cyber-physical system is accurate modeling and representation of these systems. The main difficulty lies in developing an integrated model that represents both cyber and physical aspects with high fidelity. Among existing techniques, an approach to integrate Modelica with AADL is a suitable choice, as it can encapsulate diverse attributes of cyber-physical systems. AADL modeling language provides a comprehensive set of diagrams and constructs for modeling many common aspects of systems engineering problems, such as system requirements, architectures, components, and behaviors. Complementing these AADL constructs, the Modelica language has emerged as a standard for modeling the continuous dynamics of cyber-physical systems in terms of hybrid discrete event and differential algebraic equation systems. Integrating the descriptive power of AADL models with the analytic and computational power of Modelica models provides a capability that is significantly greater than provided by AADL or Modelica individually. A transformation of AADL into Modelica is developed that will support implementations to transfer efficiently the modeling information between AADL and Modelica models without ambiguity. This chapter proposes an approach to transform the models of AADL into the models of Modelica, to clarify the transformation principles, and to illustrate the important synergies resulting from the integration between these two languages.

127.1 Introduction

Cyber-physical systems (CPS) are becoming an integral part of modern societies [1]. As an application domain, CPS is not new. For example, early automotive-embedded systems in the 1970s already combined closed-loop control of the brake

S. Feng • L. Zhang (✉)
Shanghai Key Laboratory of Trustworthy Computing, East China Normal University, Shanghai 200062, China
e-mail: zhanglichen1962@163.com

and engine subsystems (physical parts) with the embedded computer systems (cyber parts) [2]. Since then, new requirements, functionalities, and networking have dramatically increased the scope, capabilities, and complexities of CPS. This has created needs to bridge the gaps between the separate CPS subdisciplines (computer science, automatic control, mechanical engineering, etc.) and to establish CPS as an intellectual discipline in its own right [3]. The development of a CPS involves many stakeholders who are interested in different aspects of the system [4]. CPS require more advanced modeling techniques to capture physicality including time and space, reliability in terms of probabilistic models, and connectivity in terms of communication links, adaptivity, context awareness, interoperability, and autonomy. This requires a comprehensive integrated modeling framework for specification, modeling of architecture, and tracing their relationships [5].

In order to meet the challenge of cyber-physical system design, we need to realign abstraction layers in design flows and develop semantic foundations for composing heterogeneous models and modeling languages describing different physics and logics. We need to develop new understanding of compositionality in heterogeneous systems that allows us to take into account both physical and computational properties. One of the fundamental challenges in research related to cyber-physical system is accurate modeling and representation of these systems. The main difficulty lies in developing an integrated model that represents both cyber and physical aspects with high fidelity. Among existing techniques, an approach to integrate Modelica [6] with AADL [7] is a suitable choice, as it can encapsulate diverse attributes of CPS. AADL is designed for modeling system architecture. The Society of Automotive Engineers (SAE) released the AADL in November 2004 [8]. AADL can design system architecture and analyzes the time property, reliability, efficiency, and some other properties. According to the principle of MDA (model-driven architecture), a complete model can reduce the risk of consistency and security of a system. Modelica [9] is a multi-domain modeling language. This language is put forward in 1997 with a group of international efforts. Modelica is object-oriented, a causal-modeling, and equation-based language. With an object-oriented property, model reusability can be improved. Modelica describes the physical world in a direct way. In the period of model checking, Modelica can analyze the dynamic change of a variable in continuous time. AADL cannot model the continuous time properties for a system, while Modelica can do this. AADL focuses on design system architecture. In a system architecture, there are many components. The consistency among components can be checked. But in a component, changes with continuous time cannot be visualized. For this reason, we can use Modelica to model these components.

Integrating the descriptive power of AADL models with the analytic and computational power of Modelica models provides a capability that is significantly greater than that provided by AADL or Modelica individually. A transformation of AADL into Modelica is developed that will support implementations to transfer efficiently the modeling information between AADL and Modelica models without ambiguity. AADL and Modelica are two complementary languages supported by two active communities. By integrating AADL and Modelica, we combine the very expressive, formal language for differential algebraic equations and discrete events of Modelica with the very expressive AADL constructs for requirements, structural decomposition, logical behavior, and corresponding crosscutting constructs. In addition, the two communities are expected to benefit from the exchange of multi-domain model libraries and the potential for improved and expanded commercial and open-source tool support.

In this chapter we propose an approach to transform the models of AADL into the models of Modelica, to clarify the transformation principles, and to illustrate the important synergies resulting from the integration between these two languages. Based on this transformation mechanism, the properties of AADL components can be checked with the Modelica tool, which validates the security and consistency of architecture.

127.2 Models of AADL and Modelica

Many CPS applications are systems-of-systems, integrating various mechanical, electronic, and information technology systems. The design of these systems depends more and more on effective solutions that can address heterogeneity and interplay of physical and software elements. In particular, design languages used for specifying CPS should incorporate, in a consistent manner, essential concepts from multiple disciplines, such as mechanical, electronic, and software engineering. Model-driven engineering (MDE) approaches to system development have been adopted in diverse domains, in particular, CPS. This is because the use of models has shown to be promising in addressing the above issues, as well as in handling the increasing complexity of CPSs, reducing their cost of construction, and supporting efficient maintenance and evolution. AADL and Modelica are two complementary languages supported by two active communities. By integrating AADL and Modelica, we combine the very expressive, formal language for differential algebraic equations and discrete events of Modelica with the very expressive AADL constructs for requirements, structural decomposition, logical behavior, and corresponding crosscutting constructs. Although both Modelica and AADL have extension mechanisms, in unifying the two languages, we can better use AADL to model the cyber part of the cyber-physical system. This is because its dynamic analysis mechanism comes from Modelica, which is much more widespread and better supported by tools and which is also more powerful compared to that of AADL (annexes).

127.2.1 AADL Introduction

AADL is a design for architecture. The fundamental element of AADL is component. AADL provides standard components for modeling system architecture.

Table 127.1 AADL components and layers

AADL component layer	Label	Component	Description
Application software	1	Thread	An active component, can be initialized in process
	2	Thread group	An abstraction for thread, data and thread group
	3	Process	Represents the protected address
	4	Data	Represents statics data and data types within a system
	5	Subprogram	Represents executable source text-a callable component with or without parameters
Execution platform	6	Processor	Responsible for scheduling and executing threads
	7	Memory	Storage components for data and executable code
	8	Device	Entities that interface with the external environment of an application system
	9	Bus	Represents hardware and associated communication protocols that enable interactions among other execution platform components
Composite	10	System	Represents a composite of software, execution platform, or system components

These components are divided into three layers: application software, execution platform, and composite. Each layer has its focus. The application software layer is used to build the software architecture. The execution platform is used to build the hardware architecture. The composite is used to integrate the software and hardware architecture. Each layer has its components.

The detailed statements about layers and components are in Table 127.1. AADL components are divided into three categories: application software, execution platform, and composite. Each component has its description for different utilizations.

127.2.2 Modelica Introduction

Modelica is a multi-domain modeling language for the physical world. Modelica has three features: object-oriented modeling, a casual modeling, and equation-based modeling [3]. With these features, Modelica can reduce the complexities of modeling. Models can be built for inheriting from its father. The behaviors in a model can be described with equations and physical principles.

Modelica model is formed with classes. Modelica classes contains: class, model, record, block, function, connector, type, and package. The class "class" is a non-special class while other classes are special classes.

127.3 Transforming Mechanism

The mechanism is based on projecting AADL components into Modelica classes. Not all classes in Modelica can be used. Firstly, analyze the features of each AADL component. Then find a proper class in Modelica that can represent this AADL component. Check the relationship of this component. If the type of component is Type, it can be directly transformed into Modelica class. If the type of component is Implementation, transform this component into Modelica class and replace the relation between Type component and Implementation component with extents in Modelica.

The project of AADL package can be transformed into Modelica project package for the hierarchy similarity of the two kinds of packages.

127.3.1 Project Structure Transformation

The AADL project structure is in a package. The components of a system are contained in a package. Packages are contained in a project.

The Modelica project structure is in a package. In the Modelica project, there is a package. But a package can use classes from other packages with importing these packages.

The project projection between AADL and Modelica can be carried out in a package level, which means transforming an AADL package into a Modelica package. In this level, the private components in AADL package may be public classes in Modelica.

127.3.2 Inheritance Transformation

Modelica is object oriented, so the class has inheritance. In AADL, there is some sense of inheritance between components. The transformation of inheritance relationship is listed in Table 127.2. In AADL, the same kind of components can have an inheritance relationship. In Modelica, the same kind of classes can have an inheritance relationship.

127.3.3 AADL Components Projected into Modelica Classes

AADL components can be projected into Modelica classes. After the analysis of AADL components and Modelica classes, a transformation table is listed in Table 127.3. While mapping components of AADL into Modelica classes, there may be some missed information. According to Table 127.3, AADL components can be transformed into Modelica classes.

Table 127.2 The mapping between AADL and Modelica

Label	Modelica class relationship	AADL component relationship	Description
1	Class	Type ← Implement	In AADL, the *Type* component define features and implement complement implements *Type* component
2	ClassA ← ClassB	TypeA → TypeB, ImplementA ← , ImplementB ←	The relationship of component extension can be represented with Modelica class extension
3	ClassA:a	ImplementA:a	AADL component instantiation can be represented by Modelica class instantiation

Table 127.3 The projection from AADL components into Modelica classes

Label	AADL components	Modelica classes	Projection pair
1	Thread	Model	(1, 5)
2	Thread group	Class	(2, 2)
3	Process	Model	(3, 1)
4	Data	Record and type	(4, [3, 7])
5	Sub program	Model	(5, 1)
6	Processor	Model	(6, 1)
7	Memory	Model	(7, 1)
8	Device	Model	(8, 1)
9	Bus	Connector	(9, 1)
10	System	Model	(10, 1)

Table 127.4 Keywords transformation from AADL into Modelica

Label	AADL keywords	Modelica keywords
1	Extends	Extends
2	Features	Parameter
3	Flows	Connect
4	Properties	Equation
5	Packages	Packages
6	Implementation	Extends
7	Port	Connector
8	Connections	Connect
9	Subcomponents	Declared as variable
10	Modes	Transformed as comments in Modelica

127.3.4 Keywords Transformation

The keywords in AADL are different from those in Modelica. So, we need to transform the keywords in AADL into Modelica. The transformation table is in Table 127.4.

The port in AADL is transformed as connector class in Modelica.

127.4 Conclusion

By integrating AADL and Modelica, we combine the very expressive, formal language for differential algebraic equations and discrete events of Modelica with the very expressive AADL constructs for requirements, structural decomposition, logical behavior, and corresponding crosscutting constructs. In this chapter we propose an approach to transform the models of AADL into the models of Modelica, to clarify the transformation principles, and to illustrate the important synergies resulting from the integration between these two languages. Based on this transformation mechanism, the properties of AADL components can be checked with a Modelica tool, which validates the security and consistency of architecture.

In the future work, we will work on implementing the transformation tool from AADL to Modelica, which makes the transformation from AADL models to Modelica automatic.

Acknowledgments This work is supported by the Shanghai Knowledge Service Platform Project (No. ZF1213). This work is supported by the National High Technology Research and Development Program of China (No.2011AA010101); National Basic Research Program of China (No.2011CB302904); the National Science Foundation of China under grant No.61173046, No.61021004, and No.61061130541; Doctoral Program Foundation of Institutions of Higher Education of China (No. 200802690018); and National Science Foundation of Guangdong Province under grant No. S2011010004905.

References

1. Broy, M. (2012). Cyber physical systems (Part 1). *it—Information Technology, 54*(6), 255–256.
2. Broy, M. (2013). Cyber physical systems (Part 2). *it—Information Technology, 55*(1), 3–4.
3. Lee, E. A. (2008). Cyber physical systems: Design challenges. In *Proceedings of the 11th IEEE Symposium on Object/Component/Service-Oriented Real-Time Distributed Computing (ISORC '08)* (Vol. 100, Part 1, pp. 363–369).
4. Broman, D., Lee, E. A., Tripakis, S., & Törngren, M. (2012). Viewpoints, formalisms, languages, and tools for cyber-physical systems. *Proceedings of the 6th international workshop on multi-paradigm modeling (MPM'12), ACM SIG* (pp. 56–63).
5. Eidson, J., Lee, E. A., Matic, S., Seshia, S. A., & Zou, J. (2012). Distributed real-time software for cyber-physical systems. *Proceedings of the IEEE (Special Issue on CPS), 100*(1), 45–59.
6. Junjie, T., et al. (2012). Cyber-physical systems modeling method based on Modelica. In *2012 I. E. sixth international conference on software security and reliability companion (SERE-C)* (pp. 188–191).
7. Feiler, P. H., Gluch, D. P., & Hudak, J. H. (2006). *The architecture analysis & design language (AADL): An introduction*. Pittsburgh, PA: Software Engineering Institute, Carnegie-Mellon University.
8. Feiler, P. H., Lewis, B. A., & Vestal, S. (2006). The SAE Architecture Analysis & Design Language (AADL) a standard for engineering performance critical systems. In *Computer Aided Control System Design, 2006 I.E. International Conference on Control Applications, 2006 I.E. International Symposium on Intelligent Control* (pp. 1206–1211).
9. Fritzson P, & Modelica, E. V. (1998). A unified object-oriented language for system modeling and simulation. In *ECOOP'98-Object-Oriented Programming* (pp. 67–90). Heidelberg: Springer.

Chapter 128
Aspect-Oriented QoS Modeling of Cyber-Physical Systems by the Extension of Architecture Analysis and Design Language

Lichen Zhang and Shuguang Feng

Abstract Cyber-physical systems have varying quality-of-service (QoS) requirements driven by the dynamics of the physical environment in which they operate. Developing cyber-physical systems is hard because of their end-to-end QoS requirements. Aspect-oriented development method can decrease the complexity of models by separating their different concerns. We can model QoS as a crosscutting concern of cyber-physical systems to reduce the complexity of cyber-physical system development. In this paper, we propose an aspect-oriented QoS modeling method based on AADL. We present our current effort to extend AADL to include new features for separation of concerns, and we make an AADL extension for QoS by aspect-oriented method. Finally, we illustrate QoS aspect-oriented modeling via an example of transportation cyber-physical system.

128.1 Introduction

The very recent development of cyber-physical systems (CPS) provides a smart infrastructure for connecting abstract computational artifacts with the physical world. As new CPS applications start to interact with the physical world using sensors and actuators, there is a great need for ensuring that the actions initiated by the CPS are timely. This will require new quality-of-service (QoS) functionality and mechanisms for CPS [1]. Cyber-physical systems are characterized by their stringent requirements for QoS, such as predictable end-to-end latencies, timeliness, and scalability. Delivering the QoS needs of cyber-physical systems entails the need to specify and analyze QoS requirements correctly.

Cyber-physical systems share characteristics giving rise to tangled concerns in their development and maintenance lifecycle [2]. The characteristics must

L. Zhang (✉) • S. Feng
Shanghai Key Laboratory of Trustworthy Computing, East China Normal University, Shanghai 200062, China
e-mail: zhanglichen1962@163.com

simultaneously support. Distributed, Real-Time, and Embedded "software controllers are increasingly replacing mechanical and human control of critical systems. These controllers must simultaneously support many challenging QoS constraints, including (1) real-time requirements, such as low latency and bounded jitter, (2) availability requirements, such as fault propagation/recovery across boundaries, (3) security requirements, such as appropriate authentication and authorization, and (4) physical requirements, such as limited weight, power consumption, and memory footprint. For example, a distributed patient monitoring system requires predictable, reliable, and secure monitoring of patient health data that can be distributed in a timely manner to healthcare providers." [2]

Fundamental limitations for CPS include [3]:

- Lack of good formal representations and tools capable of expressing and integrating multiple viewpoints and multiple aspects
- Lack of strategies to cleanly separate safety-critical and non-safety-critical functionality, as well as for safe composition of their functionality during human in-the-loop operation
- Ability to reason about, and trade off between, physical constraints and QoS of the CPS [4]

Aspect-oriented programming (AOP) [5] is a new software development technique, which is based on the separation of concerns. Systems could be separated into different crosscutting concerns and designed independently by using AOP techniques. Every concern is called an "aspect."

As the QoS concern needs to be considered as the most parts of the system, it is a crosscutting concern. Crosscutting concerns [6] are concerns that span multiple objects or components. Crosscutting concerns need to be separated and modularized to enable the components to work in different configurations without having to rewrite the code. By using AOP, concerns can be modularized in an aspect and later weaved into the code. The QoS of CPS [7] is very complex; currently QoS research still does not have a completely technical system, and there is no solution meeting all the QoS requirements.

This paper proposes an aspect-oriented QoS modeling method based on Architecture Analysis and Design Language (AADL) [8]. In this paper, we present our current effort to extend AADL to include new features for separation of concerns. We make an in-depth study of AADL extension for QoS. Finally, we illustrate QoS aspect-oriented modeling via an example of transportation cyber-physical system.

128.2 Related Works

Developing cyber-physical systems is hard since it requires a coordinated, physics-aware allocation of CPU and network resources to satisfy their end-to-end QoS requirements. Jaiganesh Balasubramanian et al. make two contributions to address these challenges.

The development of Distributed, Real-Time, and Embedded (DRE) systems is often a challenging task due to conflicting QoS constraints that must be explored as trade-offs among a series of alternative design decisions. Jeff Gray et al. present a model-driven approach for generating QoS adaptation in DRE systems.

Carsten Köllman, Lea Kutvonen, Peter Linington, and Arnor Solberg present an approach for managing several dependability dimensions [9]. They use aspect-oriented and model-driven development techniques to separate and construct QoS independent models and graph-based transformation techniques to derive the corresponding QoS specific models.

QoS-UniFrame classifies quantifiable QoS requirements into static and dynamic. *Static QoS* is design-related, whereas *dynamic QoS* is substantially influenced by the deployment environment.

Bikash Sabata et al. specify QoS as a combination of metrics and policies. QoS metrics are used to specify performance parameters, security requirements, and the relative importance of the work in the system. They define three types of QoS performance parameters: timeliness, precision, and accuracy.

Mohammad Mousavi et al. present an extension to the GAMMA formalism, which they name AspectGAMMA [10], and they show how non-computational aspects can be expressed separately from the computation in this framework.

Dionisio de Niz and Peter H. Feiler discuss their effort to extend the AADL to include new features for separation of concerns [11]. These features include not only constructs to describe design choices but also routines to verify the proper combination of constructs from different concerns.

Lydia Michotte, Thomas Vergnaud, Peter H. Feiler, and Robert B. France extended the AOM approach to support the separation of crosscutting concerns in component architectures using AADL components [12].

AO4AADL is an aspect-oriented extension for AADL [13]. This language considers aspects as an extension concept of AADL components called aspect annex. Instead of defining a new aspect-oriented ADL, they extend AADL, a well-known ADL, with an aspect annex. Therefore, they consider, in their work, that aspects can be specified in a language other than AADL and then integrated in AADL models as annexes.

Ana-Elena Rugina, Karama Kanoun, and Mohamed Kaâniche proposed a four-step modeling dependability method based on AADL [14].

128.3 The Extension of AADL by Aspect-Oriented Method

In its conformity to the ADL definition, AADL provides support for various kinds of nonfunctional analyses along with conventional modeling:

Flow latency analysis: Understands the amount of time consumed for information flows within a system, particularly the end-to-end time consumed from a starting point to a destination.

Fig. 128.1 Property sets of AADL

```
property set Clemson is
MbitPerSec : type units (MPS, GPS => MPS*1000);
Band_width: type aadlinteger units Clemson::MbitPerSec;
Radio_band_width: Clemson::Band_width applies to (all);
Band_width_802_11g: constant Clemson::Band_width => 54 MPS;
Band_width_802_11n: constant Clemson::Band_width => 300 MPS;
Band_width_fast_ethernet: constant Clemson::Band_width => 100 MPS;
end Clemson;
```

Resource consumption analysis: Allows system architects to perform resource allocation for processors, memory, and network bandwidth and analyze the requirements against the available resources.

Real-time schedulability analysis: AADL models bind software elements such as threads to hardware elements like processors. Schedulability analysis helps in examining such bindings and scheduling policies.

Safety analysis: Checks the safety criticality level of system components and highlights potential safety hazards that may occur because of communication among components with different safety levels.

Security analysis: Like safety levels, AADL components can be assigned various security levels.

AADL defines two main extension mechanisms: property sets, as shown in Fig. 128.1, and sublanguages (known as annexes). Annexes and properties allow the addition of complex annotations to AADL models that accommodate the needs of multiple concerns.

In this paper, we extend AADL by aspect-oriented method in the following aspect:

Physical world aspect: Cyber-physical systems are often complex and span multiple physical domains, whereas mostly these systems are computer-controlled.

Dynamic continuous aspect: Cyber-physical systems are mixtures of continuous dynamic and discrete events. These continuous and discrete dynamics not only coexist but also interact, and changes occur both in response to discrete, instantaneous events and in response to dynamics as described differential or difference equations in time.

Formal specification aspect of data: A formal specification aspect of data captures the static relation between the object and data. Formal data aspect emphasizes the static structure of the system using objects, attributes, operations, and relationships based on formal techniques.

Formal specification aspect of information flow and control flow: Formal specification aspect of information flow and control flow aims at facilitating the description and evaluation of various flow properties measures.

Spatial aspect: The analysis and understanding of railway cyber-physical systems' spatial behavior—such as guiding, approaching, departing, or coordinating movements—is very important.

128.4 Case Study: Aspect-Oriented Specification of QoS of Lunar Rover

A lunar rover or moon rover is a space exploration vehicle designed to move across the surface of the moon, as shown in Fig. 128.2.

We use the AADL to model the software part of the lunar rover. The whole system is split into measurement system, control system, and perform system. The data stream of the control system of the lunar rover is shown in Fig. 128.3.

The specification of the delays of control system stream is

```
f1:flow source AttitudeData{Latency => 20 Ms;};
f2:flow source HeadingData{Latency => 20 Ms;};
f3:flow source LocationData{Latency => 20 Ms;};
f1:flow sink SurveyData{Latency => 20 Ms;};
f2:flow sink SurveyData{Latency => 20 Ms;};
f3:flow sink SurveyData{Latency => 20 Ms;};
```

Fig. 128.2 Lunar rover

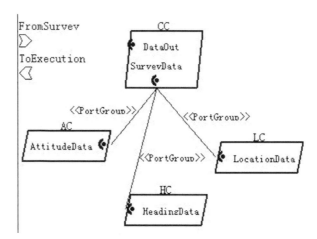

Fig. 128.3 Control system stream

```
Infos (6 items)
i End-to-end flow AC2CC_E2E calculated latency (ASynchronous)40.0 ms is less than expected latency 100.0 ms
i End-to-end flow AC2CC_E2E calculated latency (Synchronous)40.0 ms is less than expected latency 100.0 ms
i End-to-end flow HC2CC_E2E calculated latency (ASynchronous)40.0 ms is less than expected latency 100.0 ms
i End-to-end flow HC2CC_E2E calculated latency (Synchronous)40.0 ms is less than expected latency 100.0 ms
i End-to-end flow LC2CC_E2E calculated latency (ASynchronous)40.0 ms is less than expected latency 100.0 ms
i End-to-end flow LC2CC_E2E calculated latency (Synchronous)40.0 ms is less than expected latency 100.0 ms
```

Fig. 128.4 Control system stream analysis

In the flow implementation, we define the end-to-end delays as follows:

```
AC2CC_E2E:end to end flow AC.f1->AC2CC->CC.f1{
 Latency => 100 Ms;
 };
HC2CC_E2E:end to end flow HC.f2->HC2CC->CC.f2{
 Latency => 100 Ms;
 };
LC2CC_E2E:end to end flow LC.f3->LC2CC->CC.f3{
 Latency => 100 Ms;
 };
```

We use the OSATE tool of AADL to make flow analysis; we obtain the analysis results, as shown in Fig. 128.4.

128.5 Conclusion

This paper proposed an aspect-oriented QoS modeling method based on AADL. Aspect-oriented development method can decrease the complexity of models by separating their different concerns. In model-based development of cyber-physical systems, this separation of concerns is more important given the QoS concerns addressed by cyber-physical systems. These concerns can include timeliness, fault tolerance, and security. Architecture Analysis and Design Language (AADL) is a standard architecture description language used to design and evaluate software architectures for embedded systems already in use by a number of organizations around the world. In this paper, we presented our current effort to extend AADL to include new features for separation of concerns and made an in-depth study of AADL extension for QoS. Finally, we illustrated QoS aspect-oriented modeling via an example of the specification of VANET based on AADL.

The future work focuses on the integration of AADL and formal techniques to specify and verification of QoS of cyber-physical systems.

Acknowledgements This work is supported by Shanghai Knowledge Service Platform Project (No. ZF1213), national high-technology research and development program of China (No. 2011AA010101), national basic research program of China (No. 2011CB302904), the national science foundation of China under grant (Nos. 61173046, 61021004, 61061130541, 91118008), doctoral program foundation of institutions of higher education of China (No. 20120076130003), and national science foundation of Guangdong province under grant (No. S2011010004905).

References

1. Dillon, T., Potdar, V., Singh, J., & Talevski, A. (2011). Cyber-physical systems -providing quality of service (QoS) in a heterogeneous systems-of-systems environment. In *IEEE international conference on digital ecosystems and technologies* (pp. 330–335), Daejeon.
2. Gokhale, A., & Gray, J. (2005). (2005). An integrated aspect-oriented model-driven development toolsuite for distributed real-time and embedded systems. In *Workshop on aspect-oriented modeling workshop, held at AOSD 2005* (pp. 20–26). Chicago, IL: IEEE Computer Society.
3. Wolf, W. (2009). Cyber-physical systems. *Computer., 42*(3), 88–89.
4. Lee, E. A. (2008). Cyber physical systems design challenges. In *11th IEEE international symposium on object oriented real-time distributed computing (ISORC), 2008* (pp. 363–369), Orlando, FL.
5. Kiczales, G., et al. (1997). Aspect-oriented programming. In *Proceedings of ECOOP, LNCS* (Vol. 1241, pp. 220–242). Heidelberg: Springer.
6. Wehrmeister, M. A., Freitas, E. P., Pereira, C. E., et al. (2007) An aspect-oriented approach for dealing with non-functional requirements in a model-driven development of distributed embedded real-time systems. In *Tenth IEEE international symposium on object and component-oriented real-time distributed computing* (pp. 428–432). Santorini Island, Greece: IEEE Computer Society.
7. Frolund, S., & Koistinen, J. (1998). Quality of service specification in distributed object systems. *IEE/BCS Distributed Systems Engineering Journal., 5*(4), 179–202.
8. AE Aerospace. (2009). SAE AS5506A: Architecture analysis and design language V2.0.
9. Köllman, C., Kutvonen, L., Linington, P., & Solberg, A. (2007). An aspect-oriented approach to manage QoS dependability dimensions in model driven development. In *International workshop on model-driven enterprise information systems* (pp. 85–94), Vienna.
10. Mousavi, M. R., Russello, G., Chaudron, M., Reniers, M., Basten, T., Corsaro, A., et al. (2002). Using aspect-GAMMA in the design of embedded systems. In *Proceedings of the seventh IEEE international workshop on high level design, verification and test (HLDVT'02)* (pp. 69–75), Cannes, France.
11. de Niz, D., & Feiler, P. H. (2007). Aspects in the industry standard AADL. In *AOM '07 proceedings of the tenth international workshop on Aspect-oriented modeling* (pp. 15–20), Nashville.
12. Michotte, L., Vergnaud, T., Feiler, P., & France, R. (2008). Aspect oriented modeling of component architectures using AADL. In *Proceedings of the second international conference on new technologies, mobility and security* (pp. 5–7), Auckland.
13. Loukil, S., Kallel, S., Zalila, B., & Jmaiel. M. (2010). Toward an aspect oriented ADL for embedded systems. In *The fourth European conference on software architecture (ECSA 2010), LNCS 6285* (pp. 489–492). Copenhagen: Springer.
14. Rugina, A.-E., Kanoun, K., & Kaaniche, M. (2007). An architecture-based dependability modeling framework using AADL. In *Proceedings of tenth IASTED international conference on software engineering and applications* (pp. 222–227), Dallas, USA.

Chapter 129
Using RC4-BHF to Construct One-way Hash Chains

Qian Yu and Chang N. Zhang

Abstract Cryptographic hash functions play a fundamental role in today's security applications. In general terms, the principal applications of a cryptographic hash function are to verify the integrity of the data, which refers to data authentication or data integrity. The one-way hash chain is an important topic in key management and is also an important cryptographic primitive in many security applications. As one-way chains are very efficient to verify, they are also the primitives to design security protocols for ultra-low-power devices. In this chapter, an RC4-based hash function RC4-BHF is introduced and how to use RC4-BHF to construct efficient one-way hash chains is proposed. The proposed construction for one-way hash chains is efficient and is designed for ultra-low-power devices.

129.1 Introduction

Cryptographic hash functions play a fundamental role in today's security applications. Hash functions take a variable-sized message as input and produce a small fixed-sized string as output [1]. Hash functions are indispensable for a variety of security applications, and the principal application is to verify the integrity of a message, which refers to data authentication or data integrity. In addition, cryptographic hash functions can be used to one-way hash chain, password file generation, intrusion and virus detection, pseudorandom function, and pseudorandom number generator.

Examples of well-known cryptographic hash functions are SHA-family [2, 3], MD4 [4], MD5 [5], and RIPEMD [6]. Analysis has demonstrated that weaknesses are found in some well-known hash functions [2]. Reported a collision in SHA-0 and [7, 8] reported collisions in MD4, MD5, SHA-1, HAVAL-128, and RIPEMD.

Q. Yu (✉) • C.N. Zhang
Department of Computer Science, University of Regina, 3737 Wascana Parkway, Regina, SK, Canada S4S 0A2
e-mail: yu209@cs.uregina.ca

The weaknesses may compromise the security of the applications in which those functions are being used.

Therefore, it is really needed to design some hash functions with totally different internal structures from the broken classes. Furthermore, the emerging ultra-low-power technology sets new challenges for cryptographic algorithms and applications because their resources are limited. The traditional cryptographic hash functions are not well suited to the resource-limited environment, and therefore a hash function designed for ultra-low-power devices is also really needed. RC4-BHF [9] is such a new hash function, which is light-weight, structurally different from the broken hash function classes, and is able to reuse existing RC4 hardware to implement. RC4-BHF is very simple and efficient, designed to run on an eight-bit processor, and rules out most major generic attacks of hash functions, which enables it to run on ultra-low-power devices.

The one-way chain is an important cryptographic primitive in many security applications. In key management, a hash chain is a method to produce many one-time keys or session keys from a single key. For non-repudiation, a hash function can be applied successively to additional pieces of data in order to record the chronology of data's existence [10]. Furthermore, as one-way chains are very efficient to verify, they became the primitive to design security protocols for ultra-low-power devices, such as sensor nodes, as their low-powered processors can compute a one-way function within milliseconds, but would require tens of seconds or up to minutes to generate or verify a traditional digital signature [11], or even impossible to verify the traditional digital signature. Based on RC4-BHF, the proposed construction for one-way hash chains is very efficient and is designed for ultra-low-power devices.

The following is the outline of the rest of the chapter. Sections 129.2 and 129.3 cover the basic knowledge of cryptographic hash function and RC4 stream cipher, respectively. Section 129.4 introduces RC4-BHF and Sect. 129.5 proposes the construction for one-way hash chains. Section 129.6 provides the analysis and Sect. 129.7 provides the conclusion of this chapter.

129.2 Hash Function

A cryptographic hash function H is a transformation that accepts a variable-length block of data M as input and produces a fixed-size hash value $h = H(M)$. The following characteristics are required in a cryptographic hash function:

- The input of H can be of any length.
- The output of H has a fixed length.
- $H(x)$ is relatively easy to compute for any given input, making both hardware and software implementations practical.
- $H(x)$ is a one-way mapping, which means that for any given value h, it is computationally infeasible to find x such that $H(x) = h$.

- $H(x)$ is weak collision resistance, which means that for any given block x, it is computationally infeasible to find $y \neq x$ with $H(y) = H(x)$.
- $H(x)$ is strong collision resistance, which means it is computationally infeasible to find any pair (x,y) such that $H(x) = H(y)$.

Cryptographic hash function is used in a wide variety of applications. In general terms, the principal application of a cryptographic hash function is to verify the integrity of a message, which refers to data authentication or data integrity. Message authentication protects from unauthorized data alteration to assure that the data received are exactly as sent. Two common applications of the cryptographic hash function are message authentication code and digital signature. Following are the security attacks for hash function [12]. We say the hash function is resistant to this attack if it is hard to find one of the attacks, or we say the hash function is resistant to these attacks if it is hard to find all of the attacks [13].

- Collision Attack: can find $M_1 \neq M_2$, but $H(M_1) = H(M_2)$
- Preimage Attack: Given a random y, can find M that $H(M) = y$
- Second Preimage Attack: Given M_1, can find M_2 that $H(M_1) = H(M_2)$

129.3 RC4 Stream Cipher

The stream cipher is an important class of cryptographic algorithms and they encrypt each bit or byte of plaintext one at a time, using a simple time-dependent encryption transformation. Stream ciphers are almost always faster and use far less code than block ciphers, and RC4 is the most widely used stream cipher nowadays because of its high efficiency and simplicity [14, 15]. RC4 has been selected as the encryption algorithm in some security protocols [16, 17] for ultra-low-power devices.

RC4 is a variable key-size stream cipher. It is based on a 256-byte internal state S and two 1-byte indexes i and j. RC4 consists of two algorithms that are key-scheduling algorithm (KSA) and pseudo-random generation algorithm (PRGA) which are described in Algorithm 1 and Algorithm 2, respectively. For a given RC4 base key, KSA generates an initial 256 bytes permutation state. This permutation state is the input to PRGA. PRGA is a repeated loop procedure and each loop generates a 1-byte output as the stream key which to XOR with 1-byte of the plaintext, in the meantime a new 256-byte permutation state S and two 1-byte indexes i and j are getting updated. We call (S, i, j) is an RC4 state.

```
for i from 0 to 255                    i := 0
    S[i] := i                          j := 0
    T[i] := K[i mod keylength]         while GeneratingStreamOutput:
endfor                                     i := (i + 1) mod 256
j := 0                                     j := (j + S[i]) mod 256
for i from 0 to 255                        swap values of S[i] and S[j]
    j := (j + S[i] + T[i]) mod 256         Output k := S[(S[i] + S[j]) mod 256]
    swap values of S[i] and S[j]       endwhile
endfor
                                                    Algorithm 2: PRGA
            Algorithm 1: KSA
```

There are many papers to analyze the security strength of RC4, but none is practical against RC4 with a reasonable key length, such as 128 bytes [15]. So far, the practical attacks (e.g., [17–20]) against RC4 applications remain with WEP attacks, which aim to a key derivation problem in WEP standard [20]. Essentially, this attack is not in RC4 itself, it is in the scheme how to generate the secure keys. This particular problem does not appear to be applicable to other applications using RC4 and can be avoided in WEP by changing the scheme on how to generate the secure keys.

129.4 RC4-BHF

In this section we introduce RC4-BHF, which is an RC4-based hash function. The proposed construction for one-way hash chains is based on it. RC4-BHF includes three steps: padding and dividing step, compression step, and truncation step. Two algorithms KSA^* and $PRGA^*$ are given below in Algorithm 3 and Algorithm 4. KSA and PRGA are already introduced in Algorithm 1 and Algorithm 2. The input of RC4-BHF can be of any length, but equal or less than 2^{64} bits long, and the output is 128-bit or 256-bit long. The internal state of RC4-BHF is 256-byte long.

```
Input: M_k and STATE_k                 Input: len and STATE_Mk
Output: STATE_Mk                       Output: STATE_k
for i from 0 to 255                    for i from 0 to len
    j := (j + S[i] + M_k[i mod 64]) mod 256    i := (i + 1) mod 256
    swap values of S[i] and S[j]           j := (j + S[i]) mod 256
endfor                                     swap values of S[i] and S[j]
                                       endfor
            Algorithm 3: KSA*
                                                   Algorithm 4: PRGA*
```

Step 1: Padding and Dividing Step
The input of the padding and dividing process is the plain message, and the outputs are 512-bit data blocks. The input message is padded as: *Padded Message* := M

129 Using RC4-BHF to Construct One-way Hash Chains

Fig. 129.1 The padding process

|| 1 || 00 ⋯ 0 || L where M is the input message of N bits long, || is the concatenation symbol, the number of zero v is the least non-negative integer in which $N + 65 + v \equiv 0 \mod 512$, and L is a 64-bit data segment indicating the length of message M. The padding process makes the padded message a multiple of 512 bits in length. Figure 129.1 illustrates the padding process. For dividing process, the padded message is divided into 512-bit data blocks, notated by $m_1, m_2, \cdots, m_n$.

Step 2: Compression Step

The first 512-bit data block m_1 and offset integer are input to initialize the internal state S: $State_{m1} = PRGA^*(offset, KSA(m_1))$, then the function $PRGA^*$ modifies the internal state depending on the length len_1 where

$$len_1 = \begin{cases} m_1 \mod 2^5 & \text{if } (m_1 \mod 2^5) \neq 0 \\ \text{offset} & \text{if } (m_1 \mod 2^5) = 0 \end{cases}$$

$State_1 = PRGA^*(len_1, State_{m1})$.

For k (k = 2, 3 ... n), S are updated as below:

$State_{mk} = KSA^*(m_k)$, $State_k = PRGA^*(len_k, State_{mk})$

where $len_k = \begin{cases} m_k \mod 2^5 & \text{if } (m_k \mod 2^5) \neq 0 \\ \text{offset} & \text{if } (m_k \mod 2^5) = 0 \end{cases}$.

The compression process is illustrated in Fig. 129.2. The number of rounds of $PRGA^*$ is controlled by len_k.

Step 3: Truncation Step

The input of the truncation process is 256-byte $State_n$, which is the output of the compression process, and output of the truncation process, which is the final hash value of the hash function, is 128 or 256 bits long.

The detail of the truncation process is that it applies PRGA to generate 512 bytes output in total, and discards the first 256 bytes and only uses the last 256 bytes in calculation. $State_n$ XOR the last 256 bytes of the PRGA output to get a 256 bytes data H. RC4-BHF does not define how to input to applying PRGA so for different applications it can define different input, such as a new random state, or a defined internal state in the compression process.

In order to reduce the overhead, many schemes can be used to reduce the size of the final hash value. Two options have been adopted in RC4-BHF in the truncation process. The first option is to select the least significant bit of each byte of H, and the final hash value is 256-bit value long. The second option is to select the least significant bit of each even or odd number byte of H, and the 128-bit value is the final hash value.

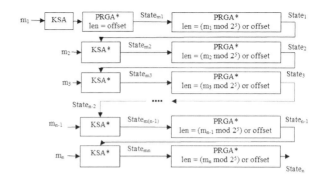

Fig. 129.2 The compression process of RC4-BHF

129.5 Proposed Construction for One-way Hash Chains

The one-way chain is an important cryptographic primitive in many security applications. In key management, some applications require key change frequently. A hash chain is a method to produce many one-time keys or session keys from a single key. For non-repudiation, a hash function can be applied successively to additional pieces of data in order to record the chronology of data's existence [10]. Furthermore, as one-way chains are very efficient to verify, they became the primitive to design security protocols for ultra-low-power devices, such as sensor nodes, as their low-powered processors can compute a one-way function within milliseconds, but would require tens of seconds or up to minutes to generate or verify a traditional digital signature [11], or even impossible to verify due to the ability of the device and the resource required by the security algorithms.

A one-way function $F : K_j = F(K_{j+1})$ can be used to generate a one-way key chain $(K_n, K_{n-1}, \cdots, K_1, K_0)$. Through this function, anybody can compute forward (e.g., computing $K_0, \cdots, K_j$ from a given $K_{j+1} K_0, \cdots, K_j K_{j+1}$ ".) but nobody can compute backward (e.g., computing K_{j+1} for only given $K_0, \cdots, K_j$) in the key chain. RC4-BHF is such a function and can be used to generate the key chain.

The detailed processes of the key chain generation and distribution, as well as of the key self-authentication, are illustrated in the following. Each key generated here is 256-byte long and the key chain is generated by a sender or trusted third party, and a new base key picking up in sequence from the one-way key chain is distributed to receiver(s) when a new key is needed. The sender or trusted third party first generate a random 256-byte permutation state as the last key K_n of the one-way key chain, and generates the rest of the keys by successively applying RC4-BHF $F : K_j = F(K_{j+1})$. Please note that the truncation process is not needed here. By this approach, the key chain is generated in the order of $K_n \rightarrow K_{n-1} \rightarrow \cdots \rightarrow K_1 \rightarrow K_0$. Here we assume n is large enough that the key chain is sufficiently long in relation to the duration of the data transmission. The distribution order is from $K_0, K_1, K_2, \cdots$ to K_n.

In the one-way key chain generated as above, the keys are self-authenticated. The receiver can easily and efficiently authenticate subsequent keys of the one-way

ന key chain using an authenticated key. For example, as soon as a receiver receives a new key K_i, the receiver can authenticate the new received key by its current key K_{i-1}. The new received key is verified once the two keys matches by applying the function $F : K_{i-1} = F(K_i)$. That is, a receiver can use it current key or any previous keys which are from the key chain to verify the new disclosed key.

In many security applications, the integer number is also needed and will need to be update when requested. The sizes of the key and integer number are different, so we cannot use RC4-BHF to generate the integer number chain directly. In the following, we illustrate the detailed processes of the integer number chain generation and distribution, as well as of the integer number self-authentication.

The integer number chain is generated by a sender or trusted third party, and a new integer number picking up in sequence from the one-way integer number chain is distributed to receiver(s) when a new integer number is needed. We assume the biggest of the integer number is $(256^4 - 1)$. The sender or trusted third party first generates two random integer numbers O_n and O_{n-1} as the last two integer numbers of the one-way integer number chain, and generates the rest of the integer numbers by successively applying RC4-BHF $F : O_j = F[(O_{j+1} + O_{j+2}) \bmod 256^4 \times [(O_{j+1} - O_{j+2}) \bmod 256^4 \times O_{i+1} \times O_{i+2}$. Please note that the truncation process will only select the least significant bit of every 8 bytes of H as the output, which is 32-bit long.

By this approach, the integer number chain is generated in the order of $O_n \rightarrow O_{n-1} \rightarrow \cdots \rightarrow O_1 \rightarrow O_0$. Here we assume n is large enough that the integer number chain is sufficiently long in relation to the duration of the data transmission. The distribution order is from $O_0, O_1, O_2, \cdots$ to O_n.

In the integer number chain generated as above, the integer numbers are self-authenticated. The receiver can easily and efficiently authenticate subsequent integer numbers of the one-way integer number chain using two recent integer numbers. For example, as soon as the receiver receives a new integer number O_i, the receiver can authenticate the new received integer number by its current integer number O_{i-1} and most recent integer number O_{i-2}. The new received integer number is verified once the three numbers match by applying RC4-BHF F : $O_{i-2} = F[(O_{i-1} + O_i) \bmod 256^4 \times [(O_{i-1} - O_i) \bmod 256^4 \times O_{i-1} \times O_i]$, that is, the receiver can use any two continuous integer numbers no matter they are current or previous from the integer number chain to verify the new disclosed integer number.

129.6 Analysis

RC4-BHF is based on RC4. The security analysis can be made in the view of the security analysis of RC4 which is well studied as well as the resistance to the major hash function attacks: preimage attack, second preimage attack, and collision attack.

The simulation to analyze the randomness of the RC4 state generation on KSA* and PRGA* shows that the generation of a new RC4 state by KSA* and PRGA* maintains the same randomness of KSA and PRGA. In addition, the truncation process which to generate the final hash value does not reduce the randomness level. Therefore, the generated hash value of RC4-BHF is close to uniform.

The maximum input length of RC4-BHF is 2^{64} bits and the output is fixed 128 or 256 bits. RC4-BHF is relatively easy to compute for any given message. Since the generated hash value of RC4-BHF is close to uniform, it is impossible to find the input through output and it is also computationally infeasible to find any two messages x and y such that $H(x) = H(y)$. Therefore, RC4-BHF is the one-way mapping and strongly collision-free. RC4-BHF satisfies the requirements of a hash function which listed in Sect. 129.2. Since the generated hash value of RC4-BHF is close to uniform, it is hard to find M_1 and M_2 that $M_1 \neq M_2$, but $H(M_1) = H(M_2)$ and to find M_2 when given a message M_1 to make $H(M_1) = H(M_2)$. In addition, RC4-BHF is not reversible, so for a random y we cannot find an M that $H(M) = y$. In conclusion from the above, RC4-BHF is collision-resistant, preimage-resistant, and second preimage attack-resistant. RC4-BHF rules out the major security attacks of hash function. The compression process of RC4-BHF has output size of about 2,048 bits which is much larger than two times of the size of the hash output. Thus, generic attacks such as Kelsey-Schneier second-preimage attack does not work.

For the performance, RC4-BHF is based on the RC4 structure which is very efficient. We conducted a benchmark simulation to compare the relative speeds among RC4-BHF, MD4, MD5, SHA-1, RIPEMD-128, and RIPEMD-160 on a 32-bit processor. Assuming the required time for one time of memory access is two times of the required time for a logic operation or a simple arithmetic operation. The comparison result shows that RC4-BHF is much faster than the other algorithms.

Because the security of RC4-BHF is sound, the proposed construction for one-way hash chains is secure as it is based on RC4-BHF. We have implemented the one-way key chain and one-way integer number chain and both of them work as expected.

129.7 Conclusion

This chapter introduced RC4-BHF, which is an RC4-based hash function, and proposed how to use RC4-BHF to construct efficient one-way hash chains, such as one-way key chain and one-way integer number chain. RC4-BHF is an attempt to use RC4 algorithm to design a cryptographic hash function. The design structure of RC4-BHF is totally different from the broken hash function classes. Moreover, RC4-BHF can be used to ultra-low-power devices, to which most other hash functions cannot be applied. RC4-BHF is very efficient compared to other hash functions, is collision-resistant, preimage-resistant, and second preimage-resistant, and rules out the important generic attacks for hash functions. Based on RC4-BHF, an efficient construction for one-way hash chains is proposed. One-way hash chains are important cryptographic primitives in many security applications and it is an important topic in key management. As one-way chains are very efficient to verify, they are also the primitives to design security protocols for ultra-low-power devices. We believe that RC4-BHF and the proposed construction for one-way hash chains could be applied to many applications, especially to the ultra-low-power devices which have limited resource and capability.

References

1. Stallings, W. (2011). *Cryptography and network security, principles and practice* (5th ed.). Upper Saddle River, NJ: Prentice Hall.
2. SHA-0: A federal standard by NIST. NIST. (1993).
3. FIPS 180–1: Secure hash standard. US Department of Commerce, Washington, DC, Springer. (1996).
4. Ronald, L. (1991). Rivest. The MD4 message-digest algorithm. In *Proceedings of the Crypto'1990, LNCS 537* (pp. 303–311), Springer.
5. Ronald, L. (1992). Rivest. The MD5 message-digest algorithm. RFC 1320, Internet Activities Board, Internet Privacy Task Force.
6. RIPE, Integrity Primitives for secure Information systems, Final report of RACE Integrity Primitive Evaluation. LNCS 1040, Springer. (1995).
7. Wang, X., Yu, H., & Yin, Y. L. (2005). Efficient collision search attacks on SHA-0. In *Proceedings of the Crypto'2005, LNCS 3621* (pp. 1–16), Springer.
8. Wang, X., Yin, Y. L., & Yu, H. (2005). Finding collisions in the full SHA-1. In *Proceedings of the Crypto'2005, LNCS 3621* (pp. 17–36), Springer.
9. Yu, Q., Zhang, C. N., Orumiehchiha, M. A., & Li, H. (2012). RC4-BHF: An improved RC4-based hash function. In *Proceedings of the CIT 2012* (pp. 322–326), IEEE CS Press.
10. Hash chain. http://en.wikipedia.org/wiki/Hash_chain.
11. Hu, Y. C., Jakobsson, M., & Perrig, A. (2005). Efficient constructions for one-way hash chains. In *Proceedings of the ACNS 2005, LNCS 3531* (pp. 423–441), Springer.
12. Stinson, D. R. (2002). *Cryptography, theory and practice* (2nd ed.). Boca Raton: CRC Press.
13. Chang, D., Gupta, K. C., & Nandi, M. (2006). RC4-Hash: A new hash function based on RC4. In *Proceedings of the INDOCRYPT, LNCS 4329* (pp. 80–94), Springer.
14. Karlof, C., Sastry, N., & Wagner, D. (2004). TinySec: A link layer security architecture for wireless sensor networks. In *Proceedings of the 2nd international conference on Embedded networked sensor systems* (pp. 162–175), ACM.
15. Mantin, I. (2001). *Analysis of the stream cipher RC4*. Master's thesis, The Weizmann Institute of Science, Israel.
16. Yu, Q., & Zhang, C. N. (2010). *A lightweight secure data transmission protocol for resource constrained devices. Security and communication networks*. Wiley Press, *3*(5), 362–370.
17. Mitchell, S., & Srinivasan, K. (2004). State based key hop protocol: A lightweight security protocol for wireless networks. In *Proceedings of the MSWiM'04* (pp. 112–118), ACM.
18. Stubblefield, A., Loannidis, J., & Rubin, A. D. (2001). Using the Fluhrer, Mantin, and Shamir attack to break WEP. *AT&T Labs Technical Report*.
19. Fluhrer, S., Mantin, I., & Shamir, A. (2001). Weakness in the key scheduling algorithm of RC4. In *Proceedings of the Workshop in Selected Areas of Cryptography*.
20. IEEE 802.11-1999: Wireless LAN Medium Access Control (MAC) and Physical Layer (PHY) Specifications. (1999).

Chapter 130
Leakage Power Reduction of Instruction Cache Based on Tag Prediction and Drowsy Cache

Wei Li and Jianqing Xiao

Abstract Tag prediction is proposed to reduce the leakage power consumption of instruction cache and the power consumption of branch prediction that represent a sizeable fraction of the total power consumption of embedded processors in this chapter. By extending the architectural control mechanism of the drowsy cache to predict the cache line read in the next access, the tag prediction wakes up the necessary cache line in advance, while the rest of cache line is in the drowsy mode. Empirical results show that the tag prediction reduces the 77 % power consumption compared to the policy adopting branch prediction, and the accuracy of tag prediction is roughly same with the accuracy of BTB prediction. By removing the BTB and adopting the technique of drowsy cache, the tag prediction effectively reduces the power consumption without significant impact on performance of processors.

130.1 Introduction

In modem embedded processor, cache occupies a large portion of the power consumption of the processor chip. On one hand, performance improvements need to use the larger and faster cache that uses the faster transistors consuming the more leakage energy when they are turned off. On the other hand, new applications favor low power and high performance. Figure 130.1 [1] shows the trend of static and dynamic power consumption. As the figure shows, the static power takes up more and more portion of the total power consumption. Because of the popularity of portable device, the power problem is being more serious.

Static power is problem for all the transistors. But it is more important for the cache, because that always need using, especially the instruction cache, cannot be simply turned off. Based on this situation, the drowsy cache [2] is an effective

---

W. Li (✉) • J. Xiao
Xi'an Microelectronic Technology Institute, Xi'an 710054, Shanxi, China
e-mail: lw@stu.xjtu.edu.cn

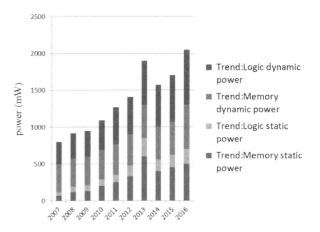

Fig. 130.1 SOC consumer portable power consumption trends

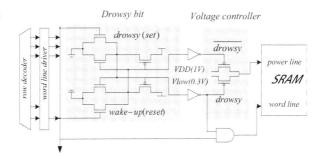

Fig. 130.2 Implementation of the drowsy cache line

method to reduce the leakage power consumption of instruction cache. As shown in Fig. 130.2, the main idea of the drowsy cache is that cache line can be in one of the two modes: in a low-leakage drowsy mode using the lower supply voltage to preserve the data and in a normal mode using the normal supply can be a normal access. To reduce the leakage energy consumption, an algorithm is used to decide which line will be turned on or off in the near future. Then the rest of the cache lines can reduce the power consumption. Therefore, the quality of algorithms decides the power consumption and performance of processors.

There are several wake-up polices including wake-up next set of cache [3], wake-up next block of cache [4, 5], and Just-in-Time Activation [6]. The main idea of these methods that predict which block or line of cache will be read in the next access are based on the principle of spatial locality of programs. The predictor of these methods uses the BTB (branch target buffer) and BHT (branch history table) that can give the target address and the direction of the branch. By using this information, the cache can ahead make the drowsy line or block of cache active. These methods focus on reducing the power consumption of cache and ignore the power consumption of BTB and BHT.

In addition, there is another opposite strategy of no-access [7] based on the principle of cache decay [8] that is used to decide which block or line of cache to switch to drowsy mode when the block or line of cache is not accessed after a period of time called decay time. LRU (Least Recently Used)-assist strategy [9] is also a statistics strategy based on cache decay. This method also uses temporal locality of programs. In executing the program, the cache line accessed is woken up and holds the active mode. When the counter of no-access is overflow, the cache line with no access is switched to drowsy mode. So there are many cache lines in normal mode to consume power. And the method of no-access is more effective for data cache rather than for instruction cache.

In the techniques mentioned previously, some only focus on reducing the power consumption of instruction cache and ignore the extra power consumption of BHT and BTB. Other methods are more appropriate for the data cache rather than for the instruction cache. There are no methods to reduce the power consumption of branch prediction and the leakage power consumption of instruction cache at the same time. In order to reduce the leakage power consumption of instruction cache and reduce the power consumption of branch prediction, this study proposes a method of tag prediction. As Sect. 130.2 presents, the tag prediction extends the structure of drowsy cache to predict the branch target, while the branch prediction is removed. By this way, the extra power dissipation consumed by BTB and BHT can be reduced. As Sect. 130.3 presents, the results show that the tag prediction has almost no loss of performance compared to branch prediction.

130.2 Tag Prediction

130.2.1 Tag Prediction Technique

In order to use the tag prediction to replace the branch prediction, an NCL SRAM (next cache line SRAM) is added to record a number of cache lines. As shown in Fig. 130.3, each cache line needs more than one bit of B/J (Branch or Jump) in the tag word. This bit is used to indicate whether there is a branch or jump instruction in the current cache line. If there is a branch or jump instruction, the target number of next cache line that will be accessed is stored in the NCL SRAM.

The traditional target address buffer modified also can predict the next cache line that will be accessed in the near future. But the BTB uses the full-associated structure or set-associated structure. That means it will consume more dynamic

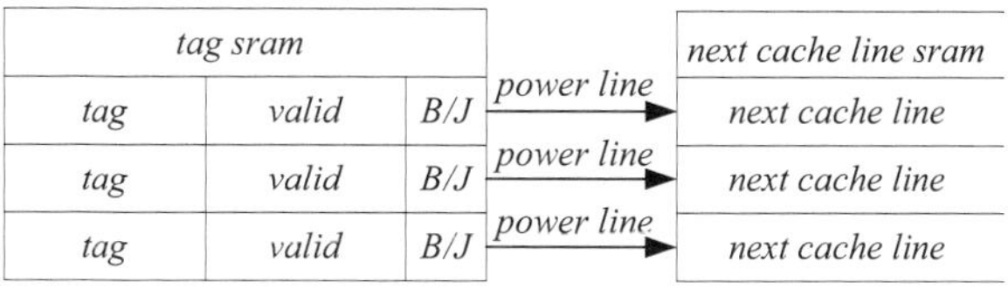

Fig. 130.3 The relationship diagram between tag SRAM and NCL SRAM

tag sram				next cache line sram
tag	valid	B/J	power line →	next cache line
tag	valid	B/J	power line →	next cache line
tag	valid	B/J	power line →	next cache line

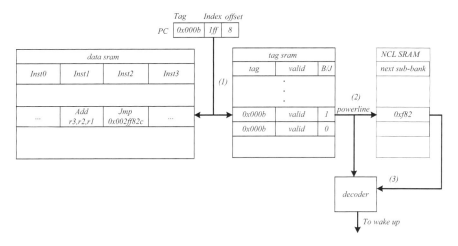

Fig. 130.4 The working scheme of tag prediction in cache line

power. The tag prediction technique can reduce the leakage power and dynamic power consumed by using the BTB technique. The disadvantage is that the information of prediction would be lost when the current line would be replaced.

Figure 130.4 shows how the tag predictor works. Whenever the processor accesses the cache line, the instruction cache checks whether the valid bit and B/J bit are valid. If these bits are valid, the instruction cache sends the address of the next cache line accessed by the processor to the wake-up logic that can early make the corresponding cache line active.

130.2.2 Tag Predictor Design

Figure 130.5 shows the implementation of tag predictor; there are two more bits including L1 (latest access bit 1) and L0 (latest access bit 0) that are used to predict the direction of branch instruction by 2-bit saturating method. B/J is still used to indicate whether there is a branch instruction or a jump instruction in the current cache line. When there is a jump instruction, tag predictor will wake up the NCL SRAM and send the address of cache line to wake-up logic that will make the corresponding cache line active. When there is a branch instruction and the L1 bit is valid, it will make the corresponding cache line also active. The difference between jump instruction and branch instruction is that the direction of branch instruction is decided by 2-bit saturating counter. The state chart of saturating counter is shown in Fig. 130.6. When the direction of prediction is correct, called taken, the state is

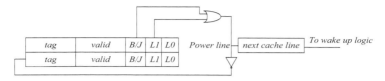

Fig. 130.5 The implementation of tag prediction used to wake up the next cache or the predicted cache line

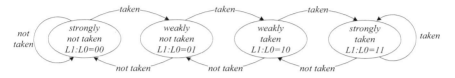

Fig. 130.6 2-Bit saturating counter used to predict the direction of branch

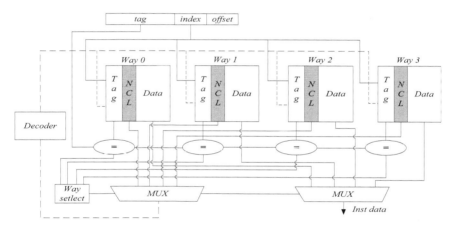

Fig. 130.7 Implementation of 4-way set associative cache

switched to the corresponding taken state. When the direction is not correct, called no-taken, the state is switched to the corresponding no-taken state.

Figure 130.7 shows the implementation of 4-way set associative cache. Each way contains tag SRAM, NCL SRAM, and data SRAM. When the cache line matches, tag prediction checks B/J, L1, and L0 bits to decide which cache line will be accessed by the processor as described above. The solid line indicates the instruction data flow as the common cache dose. The dotted line indicates the wake-up signal flow that is used to wake up predicted cache line.

130.3 Experiments and Analysis

130.3.1 Experimental Setup

In the study, hot-leakage toolsets [10] are used to model an out-of-order speculative processor with a two-level cache hierarchy that is used to evaluate the power consumption and performance. The simulation parameters that roughly correspond to Alpha 21264 is listed in Table 130.1.

There are three different BTB entries: 32-entry, 64-entry, and 128-entry. Instruction cache size is 64 KB and various degrees of associativity are as follows: direct-map, 2-way, and 4-way.

In this study, benchmarks are from SPEC2000 suites that were run on a modified hot-leakage simulator.

Table 130.1 Parameters of out-of-order speculative processor with BTB prediction or tag prediction

Parameter	BTB prediction	Tag prediction
Fetch/issue/decode/commit	4 Instructions	Same with left
Fetch queen/speed	4 Instructions/1x	Same with left
Branch prediction	Bimodal, 2k	No branch prediction
BTB	32-/64-/128-entry	No BTB
RAS	8-entry	Same with left
RUU size	64-entry	Same with left
LSQ size	32-entry	Same with left
Integer ALUs/multi-divs	4/1	Same with left
Floating point ALUs/multi-divs	1/1	Same with left
Memory bus width/latency	8 Bytes/80 and 8 cycles for the first and inter chunks	Same with left
Inst./data TLBs	16-entry/32-entry in each way, 4 KB page size, 4-way, LRU, 30-cycle latency	Same with left
L1 cache	64 KB, 4-way, 32-byte blocks, LRU, 1-cycle latency, write-back	Parameter is the same with left, add the NCL SRAM
L2 unified cache	256 KB, 4-way, 64-byte blocks, LRU, 8-cycle latency	Same with left

Fig. 130.8 Accuracy comparison between BTB prediction and tag prediction

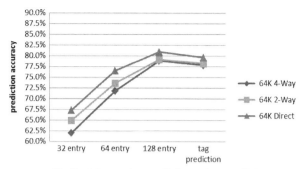

32-,64-,128-entry BTB prediction and tag prediction

130.3.2 Results and Analysis

For more effective evaluation of the quality of the two different wake-up policies, the accuracy of drowsy cache policy is defined as

$$\text{accuracy} = \frac{\text{total\_correct\_predictions}}{\text{total\_wakeups}} \quad (130.1)$$

In the tag prediction, the prediction information is stored in NCL SRAM. In the BTB prediction, the BTB structure is modified to send two addresses. One is the target address of branch. The other is the predicted address used to wake up the cache line in the next access as the most of wake-up methods of BTB prediction dose. The result is shown in Fig. 130.8. The prediction accuracy of tag prediction is quite near the accuracy of BTB using 128-entry. By analyzing the results, the accuracy of tag prediction is little less than the accuracy of BTB prediction; the reasons are that there may be more than one branch instruction in the same cache line and the information of tag prediction may be lost when the replacement of instruction cache line occurs. Nevertheless dynamic and static power is reduced without using the BTB and PHT.

In the following figures, the policy of no-drowsy is the basic reference. The scales of rest policies are given in these figures. Figures 130.9 and 130.10 show the power comparison among different policies considering the power consumption of BTB and PHT and run time of different policies separately. The policies include no-drowsy without using the drowsy technique, no-predict with waking up the drowsy cache on demand, and 32-, 64-, and 128-entry BTB prediction and tag prediction used to wake up drowsy cache ahead. As shown in Fig. 130.9, the leakage power consumption of instruction cache using tag prediction is significantly less than the leakage power consumption of instruction plus the dynamic power of BTB and PHT. The run time using different policies is quite different. As shown in Fig. 130.10, the run time using tag prediction is basically the same with the run time using the 128-entry BTB. And the run time using the rest policies significantly increases because of the low accuracy of prediction.

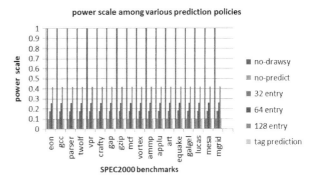

Fig. 130.9 Power comparison among various policies

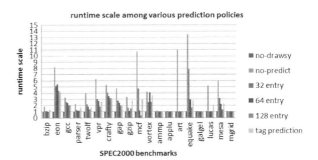

Fig. 130.10 Run-time comparison among various policies

The experimental results show that the tag prediction has about 78 % prediction accuracy just like the branch prediction does. So there is no loss of performance compared to branch prediction. While removing the branch prediction, the tag prediction effectively reduces the power consumption of processors. While the tag prediction reduces the negative performance impact by 76 % compared to the no-prediction policy, the tag prediction reduces 90 % leakage power consumption compared to the no-drowsy policy.

130.4 Conclusion

By using the drowsy policy of tag prediction, the cache line accessed by processor was active and the rest of cache lines were in the drowsy mode. The extra dynamic power consumption of BTB and BHT was eliminated. Compared to the 32-, 64-, and 128-entry BTB prediction, 46, 63, and 77 % power consumption was reduced. The accuracy of tag prediction was near the accuracy of 128-entry BTB prediction at the same time. Though the area of SRAM increased, the area occupied by BTB and BHT can be removed. In the design of embedded processors, the cache is generally small, so increased area used for NCL SRAM is quite small. So this technique is very appropriate for embedded processors using the drowsy technique. During the investigation of drowsy instruction caches in the study, tag prediction of instruction can significantly reduce the dynamic and static power consumption of processors.

References

1. ITRS Organization. (2008). International technology roadmap for semiconductors 2008 updates. Retrieved from http://public.itrs.net/.
2. Flaunter, K., Kim, N. S., et al. (2002). Drowsy caches: simple techniques for reducing leakage power. In *SIGARCH, proceedings of the 29th annual international symposium on computer architecture* (pp. 148–157). Washington: IEEE Computer Society.
3. Zhang, C., Zhou, H. W., et al. (2006). Architectural leakage power reduction method for instruction cache in ultra deep submicron microprocessors. In *The 11th Asia-Pacific computer systems architecture conference* (pp. 588–594). Berlin: Springer.
4. Hu, J., et al. (2003). Exploiting program hotspots and code sequentiality for instruction cache leakage management. In *International symposium on low power electronics and design (ISLPED'03)* (pp. 25–27). Berlin: Springer.
5. Chung, S. W., & Skadron, K. (2006). Using branch prediction information for near-optimal I-Cache leakage. In *The 11th Asia-Pacific computer systems architecture conference* (pp. 24–37). Berlin: Springer.
6. Kim, N. S., Flautner, K., et al. (2004). Single-VDD and single-VT super-drowsy techniques for low-leakage high performance instruction caches. In *International symposium on low power electronics and design (ISLPED'04)* (pp. 54–57). Berlin: Springer.
7. Kim, N. S., Flautner, K., et al. (2004). Circuit and microarchitectural techniques for reducing cache leakage power. *IEEE Transaction on VLSI Systems., 12*(2), 167–184.
8. Kaxiras, S., Hu, Z., & Martonosi, M. (2001). Cache decay: Exploiting generation behavior to reduce cache leakage power. In *ISCA 2001* (pp. 240–251). Goteborg, Sweden: IEEE Computer Society.
9. Zhang, C. Y., Zhang, M. X., et al. (2006). LRU-assist: An efficient algorithm for cache leakage power controlling. *Acta Electronica Sinica., 34*(9), 1626–1630 (In Chinese).
10. Zhang, Y., Parikh, D., et al. (2003). *Hotleakage: An architectural, temperature-aware model of subthreshold and gate leakage*. Virginia, USA: Department of computer sciences, University of Virginia.

Part VI
Network Optimization

Chapter 131
The Human Role Model of Cyber Counterwork

Fang Zhou

Abstract The essence of cyber counterwork mainly reflects as the counterwork process between people, in order to solve effectively the cyber counterwork test problem about the cognitive level and decision level. In this paper, firstly, the dynamic adaptive cyber attack and defense "observation, orient, decision, act" (OODA) loop process models are established, which are based on the traditional military command and control operational process model. Secondly, establishing the cyber attacker and defender role models during the cyber counterwork process, which are mainly from role's identities, role's function, role's capability, and role's relationship utilizing the multi-attribute group description method. At last, establishing capability evaluation index system for each role and evaluating the capability through Delphi method. The human role model can provide theoretical basis for configuring attacker and defender role during the cyber cognitive and decision level test process.

131.1 Introduction

With the development of information technology, cyber counterwork will be the primary information warfare form and become an important part of the concept of full spectrum operation [1–3]. Cyberspace operation is closely related to the land, sea, air, and aerospace operation domain and provides guarantee for military operations within above domains. Presently, the warfare form is translated from physical and transport layer to decision and cognitive layer. And the human function is becoming more and more important, expressed as the man-centered and man–machine combination.

F. Zhou (✉)
The Key Laboratory of Information System Engineering, The 28th Research Institute of China Electronics Technology Group Corporation, Nanjing 210007, China
e-mail: 326zhoufang@163.com

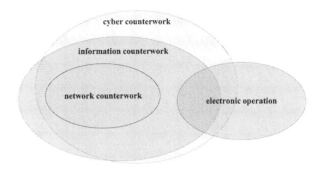

Fig. 131.1 The relation between cyber, network, electronic, and information counterwork

In this paper, we firstly research the concept and characteristic of cyber counterwork. Then the cyber attack and cyber defense process models of dynamic adaptive are established, and several human role models are presented. Those models can provide theoretical basis and guidance for cyber counterwork test.

131.2 The Concept of Cyber Counterwork

Cyber counterwork is defined as the operation action that exploits cyberspace, adopts the reconnaissance, attack, defense and control confrontation, and applies to the information domain; it will also have impact on the physical domain and cognitive domain [4, 5]. The relation between the cyber and network counterwork is inclusive; the relation between electronic and cyber counterwork is intersectant but not inclusive; the relation between cyber and information counterwork is intersectant and convergent. Figure 131.1 shows the above relations.

Similar to the traditional counterwork forms, cyber counterwork mainly includes reconnaissance, attack, protection, and control forms and involves physical domain, information domain, cognitive domain, and social domain [6].

131.3 The Human Role of Cyber Attacking

131.3.1 The OODA Loop Pocess Model of Attacking

Considering the attacking side of cyber counterwork, we build the OODA loop process model with dynamic adaptive OODA loop theory, as shown in Fig. 131.2.

During the observation stage, the main task is to gather objective system information with electronic reconnaissance or network detection tools, such as network topology structure, open ports, OS type and version, system service, and other information.

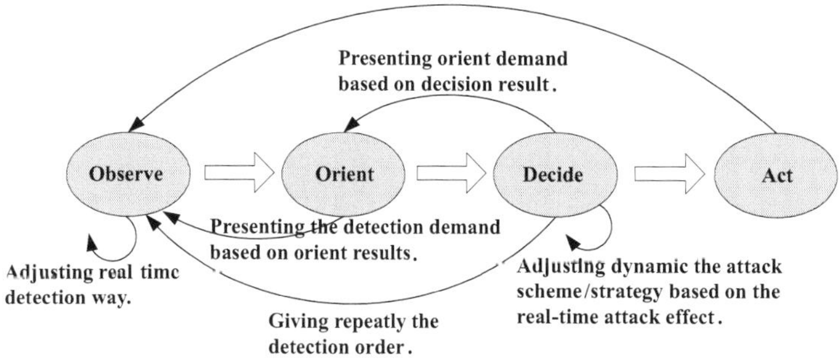

Fig. 131.2 The OODA loop process model of attacking

During the orient stage, the main task is to analyze objective system type, mine system vulnerability, analyze the captured packets, and restructure network or system.

During the decision-making stage, the main task is to make the attack plan that includes determining the attack targets, setting attack way, and choosing an attack weapon.

During the action stage, the main task is to implement the attack and provide real-time feedback information about attack result.

Because of the complexity and uncertainty of cyber counterwork, there is one or more OODA embedded loops, which will be generated during the attack process and which are manifested in three aspects: (1) There are also the observation, orient, decision-making, and action stages; (2) during the orient stage, the detection demand will be repeatedly presented based on orient results; and (3) during the decision-making stage, the detection and orient demand will be presented once the detected information or orient results cannot effectively support the decision making, and the attack plan shall be adjusted based on the attack effect. So the OODA loop of attacking is the iterative, embedded, and repeatedly rising process.

131.3.2 The Human Role Model of Attacking

1. *The Definition and Function of Human Role*
   Based on the analysis about attacking process [7, 8], we propose four roles: detecting controller, attack situation analyzer, attack decision maker, and attack performer. These human roles are defined as follows with multi-property group description method.

$$\text{Role} = \{\text{name, obligation, relation, capability}\}$$

The term "name" is the identifier of human role, such as detecting controller and the attack situation analyzer. The term "obligation" denotes the role's function,

Table 131.1 The human role's function of attacking

Detecting controller	✔ Detecting objective system and gathering network topology, the port, protocol, service, OS type and version, vulnerabilities, and defense measures information using reconnaissance and detection tool
	✔ Adjusting detection way based on information gathered
Attack situation analyzer	✔ Analyzing and evaluating the vulnerabilities of objective system based on information gathered
	✔ Mining the key node of objective system and business relations among nodes
Attack decision maker	✔ Making the attack plan or strategy, which includes determining attack targets, setting attack way, and choosing attack weapon
	✔ Adjusting dynamically the attack plan or strategy based on attack effect
Attack performer	✔ Manipulating attack tool to implement attack activities and providing feedback information about attack effect

Table 131.2 The capability evaluation index sytem of detecting controller

The capability of detecting controller (C_{obs})	The capability of sensitive information gathering (C_{11})	The quantity of sensitive information gathering (C_{111})
		The efficiency of sensitive information gathering (C_{112})
	The capability of vulnerability scanning (C_{12})	The quantity of vulnerability scanning (C_{121})
		The efficiency of vulnerability scanning (C_{122})

describing the assignment undertaken by each role. The term "capability" denotes the skill needed in accomplishing the task, such as scanning system capability, breaking code capability, exploiting vulnerabilities capability, raising privilege capability, and avoiding detection capability. The term "relation" denotes the associated relationship among roles. According to the above definition, we define the role's function, as shown in Table 131.1.

2. *The Capability Level of Human Role*

According to the role's function category, firstly, we propose the measure factor set of each kind of role and establish the evaluation index system and measure criterion of role's ability. Secondly, the Delphi evaluation method is used to determine the weight of each evaluation index in this paper. The basic idea of this method is making full use of expert experience and knowledge to review the evaluation index, and index weight is determined by using the method of geometric mean.

The capability will be evaluated by the capability of sensitive information gathering and the capability of vulnerability scanning. The evaluation indexes of these two capabilities are shown in Table 131.2.

The capability of attack situation analyzer can be evaluated by the time and the success rate of mining the key node of objective system and business relations among nodes. Table 131.3 shows the evaluation index system.

131 The Human Role Model of Cyber Counterwork

Table 131.3 The human role's function of defense

Role's type	Role's function
Defense supervisor	✔ Utilizing system security tools to monitor system running state and collect system or system security equipment log file information
Defense situation analyzer	✔ Generating system security situation, conducting threat estimation and prediction, mining the potential attack or abnormal behavior
Defense decision maker	✔ Making defense plan according to the analysis results ✔ Adjusting dynamically security defense strategy or plan
Defense performer	✔ Preventing illegal operation and intrusion behavior, scanning and remedying system vulnerability, data backup, and fault recovery

Table 131.4 The capability evaluation index system of defense performer

The capability of system recovering (C_{Da1})	The time of system recovering (T_{Ds})
	The degree of system recovering (P_{Dd})
The capability of attack blocking (C_{Da2})	The time of attack blocking (T_{Da})
	The number of attack threat (N_{Dn})
The capability of honeypot technology (C_{Da3})	The time of responsing attack (T_{Dr})
	The time of tracking attack threat (T_{Dt})
	The number of decoying attack threat source (N_{Dd})

Table 131.5 The capability evaluation index system of attack performer

The capability of password decoding (C_{a1})	The time of decoding (T_{ad})
The capability of system penetrating (C_{a2})	The access privilege obtained (P_a)
	The time of system penetrating (T_{ap})
	The concealment of system penetrating (P_{ac})
The capability of system palsying (C_{a3})	The duration time of palsying (T_{as})
	The difficultiy of system recoving (D_a)

The capability evaluation index system of attack decision maker can be evaluated by the time and the effectiveness of making attack plan. Table 131.4 shows the detail index system.

The capability of attack performer is evaluated by three indexes: the access privilege obtained, the duration time of attack, and the concealment of attack. Table 131.5 shows the detail index system.

3. *The Evaluation Model of Attacker's Capability*

   In this paper, AHP (analytical hierarchy process) and the gray evaluation method are used to build the above human roles' capability evaluation model. The basic idea is that the weights of each level index are obtained using the AHP method, to improve the effectiveness, reliability, and viability of index computed. Then the weights of underlying indexes are quantified using the gray evaluation method. Under this basis, the gray weight vector above each index and the gray class are determined. The specific implementation process of the evaluation method is shown in Fig. 131.3.

Fig. 131.3 The flow chart of ability evaluation

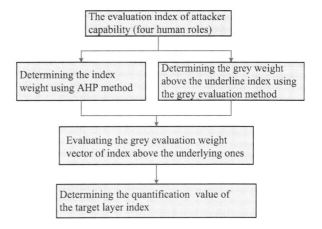

Firstly, denote the first level index as $C_{1i}(i = 1,2)$, and the second level index as C_{1ij}. The computing process with AHP method includes constructing the judgment matrix, calculating the weight vector, and testing the consistency.

The weight vector of the first level index is denoted as $W = (w1, w2)$, and the weight vector of underline level index is denoted as $W1 = (w11, w12)$, $W2 = (w21, w22)$.

Then the underlying index weights are determined with gray evaluation method. The essence of this method is through partial information known to generate and extract valuable information.

The central triangle whitening weight function is superior to the traditional whitenization weight function on the crossover phenomenon, clustering coefficient, and endpoint selection aspects. Therefore, the gray evaluation method of central triangle whitening weight function is used to solve the problem of quantization based on the attacker role of factors. The specific steps are as follows:

(a). Determining the gray class

The value of each index is divided into 5 gray classes, such as "very low," "low," "medium," "high," and "very high." Then selecting the closest center point as K1, K2, K3, K4, and K5, these point values are set to 2, 4, 6, 8, and 10. And the index values are divided in the 4 gray intervals [K1, K2], [κ2, K3], [K3, K4], and [K4, K5].

(b). Determining the index value

Assuming that N experts are invited to rate index, the expert number is denoted as n (n = 1,2,...N). Through establishing the evaluation standards and rules above each index, each evaluation expert independently scores on the underline index according to the grading standards, obtaining the evaluation sample matrix of the role of ability.

(c). Building the center whitenization weight function
The central point albinism function $f_{jm}(x)$ of observation value(x) of index j that belongs to the gray m (m = 1,... 5) can be described as follows:

$$f_{jm}(x) = \begin{cases} \dfrac{x - k_{i-1}}{k_i - k_{i-1}}, & x \in (k_{i-1}, k_i) \\ \dfrac{k_{i+1} - x}{k_{i+1} - k_i}, & x \in (k_i, k_{i+1}) \end{cases}$$

(d). Calculating gray evaluation indexes
The gray evaluation coefficient l_{1jm}, l_{2jm} of the second level indexes C_{11j}, C_{12j} that belongs to gray m is computed as follows:

$$l_{1jm} = \sum_{n=1}^{N} f_{1jm}(d_{1jmn}), \quad l_{2jm} = \sum_{n=1}^{N} f_{2jm}(d_{2jmn})$$

Where $d_{1jm}n$ denotes the score of the n expert about C_{11j}, C_{12j} indexes and at the same time the sums of l_{1jm} and l_{2jm}, the gray evaluation weight coefficient of index of evaluation C_{1ij} is shown as follows:

$$l_{1j} = \sum_{m=1}^{5} l_{1jm}, \quad l_{2j} = \sum_{m=1}^{5} l_{2jm}, \quad p_{1jm} = {l_{1jm}}/{l_{1j}}, \quad p_{2jm} = {l_{2jm}}/{l_{2j}}$$

Therefore, the gray evaluation weight vector of C_{1ij} belonging to five gray classes is shown as follows:

$$p_{1j} = [p_{1j1}, p_{1j2} \cdots p_{1j5}]^T, \quad p_{2j} = [p_{2j1}, p_{2j2} \cdots p_{2j5}]^T$$

Based on the above weight vector, the gray evaluation weight vector can be obtained through the weight matrix W1,W2 and the evaluation weight vectors P_{1j}, P_{2j} are multiplied; the result will be denoted as V = [v1,v2]:

$$V = W \times P^T = (v_1, \ldots v_5), P = [p_{1j}, p_{2j}]$$

The evaluation of gray level according to the "gray level" is assigned; namely, each evaluation gray grade value is translated into vector S = [2, 4, 6, 8, 10]. The ability quantification value of detecting controller role can be expressed as follows:

$$C_{obs} = V \times S^T$$

The capability of attack situation analyzer, attack decision maker, and attack performer can be computed similarly based on the above method.

131.4 The Human Role of Cyber Defense

131.4.1 The Definition and Function of Human Role

After the analysis about the OODA loop model of defending, we propose four roles: defense supervisor, defense situation analyzer, defense decision maker, and defense performer. The function of each human role is defined in Table 131.3.

131.4.2 The Capability Evaluation Index System of Human Role

According to the function of the above human roles, we propose the tactical and technical performance, represented by a set of capacity values.

The capability of defense supervisor is evaluated by the sensitive information count collected within a certain time (N_{Dinfo}) and the detection efficiency (V_{detec}).

The capability of defense situation analyzer is evaluated by the time (S_{Di}), the success ratio, and the efficiency of identifying attack threat behavior or event (Viden).

The capability of defense decision maker is evaluated by the time (T_{Dd}) and the effectiveness of making defense plan (P_{De}).

The capability of defense performer is evaluated by seven indexes; the detail index is shown in Table 131.4.

131.5 Conclusion

Cyber counterwork research is a new area and there are less research results about this area. Based on the analysis of cyber counterwork characteristic, we respectively establish eight human roles for attack side and defense side during the counterwork process. Then the index system of role's ability is established, evaluating the role's ability through Delphi evaluation method. Role model established in this paper can provide theoretical support for the future development of cyber counterwork experiment, such as providing the theory basis for the attacker and defender capacity requirements.

Acknowledgements This work is supported by advanced researched plan of the general armament department No.9140A04040113DZ38054.

References

1. The President of the Unite States. (2003). *The national strategy to secure cyberspace[R]*. Washington, DC: The whitehouse.
2. United States Army Training and Doctrine Command. (2010). *The United States Army's Cyberspace operations concept capablility plan 2016-2028[R]*. Washington, DC: USDOD.
3. Le May Centerfor Doctrine Development and Education. (2010). *Cyberspace operations[R]*. Washington, DC: The United States AirForce.
4. Guang-xia, Z., & Xin, S. (2012). Study on cyberspace operations[J]. *Command Information System and Technology, 3*(2), 6–10 (In Chinese).
5. Chairman of the Joint Chiefs of Staff. (2006). *National military strategy for cyberspace operations[R]*. Washington, DC: USDOD.
6. Kuhl, M. E., & Sudit, M. (2007). Cyber attack modeling and simulation for network security analysis [C]. *Proceedings of the 2007 Winter Simulation Conference, Springer Berlin Heidelberg, Australia. 2119*(1), 1180–1189.
7. Rowe, N. C. (2004). A model of deception during cyber-attacks on information systems [J]. *IEEE First Symposium on Computing & Processing, 1*(2), 21–30.
8. Kotenko, I. (2007). Multi-agent modelling and simulation of cyber-attacks and cyber-defense for home- land security[J]. *IEEE International Workshop on Intelligent Data Acquisition and Advanced Computing Systems: Technology and Applications., 4*(1), 614–620.

Chapter 132
A Service Channel Assignment Scheme for IEEE 802.11p Vehicular Ad Hoc Network

Yao Zhang, Licai Yang, Haiqing Liu, and Lei Wu

Abstract IEEE 802.11p vehicular ad hoc network (VANET) applies multiple channels, including one control channel (CCH) and six service channels (SCHs); the enhanced distributor channel access (EDCA) mechanism is used to support wireless channel assignment and QoS requirements. But the method of SCH assignment is not proposed in IEEE 802.11p standard. We present a scheme to perform SCH assigning, previous transmission indicators of service channels are detected dynamically by service channel assignment controller set in medium access control (MAC) layer, service packets would be delivered into suitable SCH and EDCA access category (AC) queue according to SCH reservation probability and estimated transmission delay. Saturated throughput of our scheme in SCH is analyzed by theoretical model in different conditions; the results show that it can ensure higher SCH utilization and is an efficient way to improve performance of intelligent transportation system.

132.1 Introduction

Vehicular ad hoc network (VANET) has been considered an essential technology for future intelligent transportation system (ITS); the purpose of VANET is to provide vehicle to roadside unit (RSU) as well as vehicle to vehicle wireless communications. IEEE 802.11p is designed for wireless access in vehicular environments (WAVE) to support VANET, allocated 75 MHz bandwidth of licensed

---

Y. Zhang (✉)
School of Control Science and Engineering, Shandong University, Jinan 250061, China

School of Mechanical, Electrical and Information Engineering, Shandong University (Weihai), Weihai 264209, China
e-mail: zhangyao@sdu.edu.cn

L. Yang • H. Liu • L. Wu
School of Control Science and Engineering, Shandong University, Jinan 250061, China

spectrum at 5.9 GHz is divided into one control channel (CCH) and six service channels (SCHs), CCH is dedicated for broadcast of traffic safety messages, and SCHs are dedicated for transmission of various application messages [1–3]. In MAC layer, enhanced distributor channel access (EDCA) mechanism is used to perform wireless channel competition, EDCA classifies data traffic into different priorities to support varying QoS requirements [4]. To coordinate channel access on CCH and multiple SCHs effectively, a synchronized channel coordination scheme is proposed in IEEE 802.11p protocol; the channel access time is divided into synchronization intervals with a fixed length of 100 ms, consisting of 50 ms CCH interval and 50 ms SCH interval. During CCH interval, all OBUs must monitor the CCH for messages of traffic safety or SCH reservation. During SCH interval, OBUs can optionally switch to preconcerted SCH to perform service messages transmitting. Although this scheme can reduce the transmission delay of safety messages in CCH, SCH utilization cannot be ensured [5].

The rest of this paper is organized as follows: In Sect. 132.2, we present a dynamic SCH assignment scheme. In Sect. 132.3, saturated throughput of our scheme is analyzed in different conditions by theoretical model. Finally, the main conclusions and future research work are summed up in Sect. 132.4.

132.2 Service Channel Assignment Scheme

In our scheme, CCH is classified into four EDCA ACs: AC[3] is for safety-related urgent messages, AC[2] is for vehicles to advertise road traffic messages, AC[1] is for nonurgent traffic messages, and AC[0] is for SCH reservation. SCH is classified into three ACs: AC[2], AC[1], and AC[0] are for high-priority UDP packets, and AC[1] and AC[0] are for low-priority TCP packets. Setting service channel assignment controller in MAC layer, it is intended for detecting previous access delay of each AC queue in SCH interval, estimating transmission delay and packet error rate of SCHs, and then guiding AC queue assignment and service channel reservation.

The algorithm of service channel assignment scheme is as follows:

// n_j: the number of reserving SCH-j (j=1-6)
// $adelay_{ji}$: AC[i]'s access delay in SCH-j (i=0,1,2)
// time( ): function of system time
// $lenAC_{ji}$: AC[i]'s queue length in SCH-j
// $delay_{ji}$: AC[i]'s transmission delay in SCH-j
// pe_j: packet error rate in SCH-j
// X_{ji}: the number of acknowledged (ACK) packets through AC[i] in SCH-j
/Y_{ji}: the total number of packets through AC[i] in SCH-j
// pr_j: SCH-j reservation probability

(continued)

(continued)

//*packet_type*: denotes packet priority, the value of UDP packet is 1, TCP packet is 0
initialize: *(i=0,1,2; j=1-6)*
$X_{ji}=0; Y_{ji}=0; pe_j=0; adelay_{ji}=0; delay_{ji}=0;$
//AC queue assignment during SCH-j interval
$lenAC_{ji}=0;$
when ACK from destination node is received by AC[i] in source nodes
$X_{ji}=X_{ji}+1; lenAC_{ji}=lenAC_{ji}-1;$

$$adelay_{jinew} = Time(receive\ ACK)\text{-}Time(frame\ begins\ to\ compete\ channel); \qquad (132.1)$$

if $adelay_{ji}==0$

$$adelay_{ji} = adelayji_{new}; \qquad (132.2)$$

else

$$adelay_{ji} = (1\text{-}\alpha) * adelay_{ji} + \alpha * adelay_{jinew}; \qquad (132.3)$$

when a packet in AC[i] begins to compete channel
$Y_{ji}=Y_{ji}+1;$
when a packet arrives to SCH assignment controller of SCH-j:

$$delay_{ji} = lenAC_{ji} * adelay_{ji}; \qquad (132.4)$$

switch (*packet_type*)
{case 1 *k=2*;
if $delay_{j1}<delay_{j2}$& $delay_{j1}<delay_{j0}$ *k=1*;
if $delay_{j0}<delay_{j2}$ & $delay_{j0}<delay_{j1}$ *k=0*;
case 0 *k=1*;
if $delay_{j0}<delay_{j1}$ *k=0*;}
data packet⟶AC[k] queue ; $lenAC_{jk}=lenAC_{jk}+1;$}
//SCH reservation through AC[0] during CCH interval
$n_j=0;$
when SCH reservation request is received by SCH assignment controller

(continued)

(continued)

$$pe_j = \left(\sum_{k=0}^{2} X_{jk}\right) / \left(\sum_{k=0}^{2} Y_{jk}\right), \quad pr_j$$

$$= \left[1 - \beta \frac{pe_j}{\sum_{k=1}^{6} pe_k} - (1-\beta)\frac{n_j}{\sum_{k=1}^{6} n_k}\right] / 5 \qquad (132.5)$$

RN=a random number from uniform distribution between [0,1];
low=0; high=pr_1;
for k=1:6
{if $low<RN<high$ { $j=k$; break}
$low=low+pr_k$; $high=high+pr_k$;}
sending SCH-j reservation request through AC[0]
when receiving ACK of SCH-j reservation request
$n_j=n_j+1$;

Our algorithm has two steps:

Step 1: AC queue assignment and access delay detecting during SCH interval
AC[i]'s access delay includes time to compete service channels, packet sending delay, propagation delay, and time to send back ACK; it is detected according to formula (132.1), (132.2), and (132.3). In formula (132.3), α is a predefined factor between 0 and 1. If α is bigger, the new sample of access delay makes more influence on average access delay. When a packet arrives to SCH controller, transmission delay of AC[2], AC[1], and AC[0] is estimated by formula (132.4), and then UDP packet would be mapped into the AC queue that estimated transmission delay is smallest among AC[2], AC[1], and AC[0]. TCP packet would be mapped into the AC queue that estimated transmission delay is the smallest among AC[1] and AC[0].

Step 2: SCH reserving during CCH interval
SCH reservation messages are transmitted through AC[0] during CCH interval; SCH-j reserving probability is calculated by formula (132.5). With the increase of packet error rate or the number of nodes that has reserved SCH-j, reserving probability decreases. In formula (132.5), β is a factor between 0 and 1. If the value of β is bigger, packet error rate has more influence on SCH reserving probability than the number of nodes. For example, let β be 0.75; supposing packet error rate of SCH-1 to SCH-6 is 1.25×10^{-4}, 1.25×10^{-4}, 1.25×10^{-5}, 1.25×10^{-5}, 1.25×10^{-6}, and 1.25×10^{-6}, we calculate SCH reservation probability by formula (132.5), as shown in Table 132.1. Obviously, for service

Table 132.1 SCH reservation probability (n1 = 10, n2 = 30, n3 = 15, n4 = 50, n5 = 25, n6 = 35)

SCH-1	SCH-2	SCH-3	SCH-4	SCH-5	SCH-6
0.1289	0.1233	0.1886	0.1781	0.1917	0.1887

channels of high packet error rate or the large number of users, their reservation probabilities are smaller. Our service channel reservation scheme accomplishes payload balance and adaptive SCH selection.

132.3 Performance Analysis of SCH Assignment Scheme

N is defined as the number of nodes in SCH-j, τ_i is the packet transmitting probability of AC[i] in a time slot, τ is the packet transmission probability, p_{ic} is the packet transmission probability of AC[i], p_{si} is the successful probability of transmitting AC[i] packet in a time slot, p_{tr} is the channel utilization probability, CW_{min} is the minimum back-off windows of EDCA, and m_i is the maximum back-off stage of EDCA. We have [6, 7]

$$\tau_i = \frac{2(1 - 2p_{ic})}{(1 - 2p_{ic})(CW[i]_{min} + 1) + p_{ic}CW[i]_{min}(1 - (2p_{ic})^{m_i})}$$

$$\tau = 1 - \prod_{i=0}^{2}(1 - \tau_i), \quad p_{si} = \frac{N\tau_i \prod_{j \neq i}(1 - \tau_j)(1 - \tau)^{N-1}}{1 - (1 - \tau)^{N-1}}$$

$$p_{ic} = 1 - (1 - \tau)^{N-1}\prod_{j \neq i}(1 - \tau_j), \quad p_{tr} = 1 - (1 - \tau)^N$$

Furthermore, if T_{si} is defined as the average time of transmitting AC[i] packets, T_{ci} is the average collision time, δ is the propagation delay, σ is the time slot length, and $E(L_f)$ denotes the average length of service packets, we can get saturated throughput of AC[i] [8, 9]:

$$S_i = \frac{p_{si}p_{tr}E(L_f)}{(1 - p_{tr})\sigma + \sum_{k=0}^{2}(p_{sk}p_{tr}T_{sk}) + p_{tr}\sum_{l=0}^{2}\left[(1 - \sum_{k=0}^{2}p_{sk})\right]T_{cl}} \quad (132.6)$$

In formula (132.6), supposing RTS/CTS mechanism is used in EDCA, we have

$$T_{si} = RTS + 3SIFS[i] + 4\delta + CTS + T_{frame-head} + T_{frame} + ACK + AIFS[i]$$
$$T_{ci} = RTS + AIFS[i] + \delta + CTS$$

Let packet retry limit of AC[i] be RT_i, AC[i] packet arrival rate λ_i, UDP packet arrival rate λ_H, TCP packet arrival rate λ_L, $p(X = k)$ be the probability distribution of channel competing times, and p_i be the probability of AC[i] transmitting packet successfully. We have

$$p_i = \sum_{k=1}^{RT_i} p_{tr}(p_{si}/N)\left[(1 - p_{tr}(p_{si}/N)\right]^{k-1} \quad (i = 0, 1, 2)$$

$$p(X = k) = \frac{(1 - p_{tr}p_{si})^{k-1} p_{tr}p_{si}}{\sum_{j=1}^{RT_i}(1 - p_{tr}p_{si})^{j-1} p_{tr}p_{si}}$$

$$adelay_i = \frac{1}{\mu_i} = \left\{\left[\sum_{k=1}^{RT_i} kp(X=k) - 1\right]T_{ci} + T_{si}\right\}/(1 - pe_j)$$

$$\lambda_2 = p_{H2}\lambda_H + (1-p_2)\mu_2, \quad \lambda_1 = p_{H1}\lambda_H + p_{L1}\lambda_L + (1-p_1)\mu_1$$
$$\lambda_0 = p_{H0}\lambda_H + p_{L0}\lambda_L + (1-p_0)\mu_0, \quad d_i = lenAC[i] \times \{adelay_i\}$$
$$p_{Hi} = probability\left\{\min\left(d_2, d_1, {}^1d_0\right) = d_i\right\} \quad (i = 0, 1, 2)$$
$$p_{Li} = probability\left\{\min\left(d_1, {}^1d_0\right) = d_i\right\} \quad (i = 0, 1)$$

Hence, saturated throughput of UDP packets and TCP packets during SCH interval are

$$S_h = \sum_{j=1}^{6}\left[(\sum_{i=0}^{2} p_{Hi}S_i) \times N \times T_{SCH}\right], \quad S_l = \sum_{j=1}^{6}\left[(\sum_{i=0}^{1} p_{Li}S_i) \times N \times T_{SCH}\right] \quad (132.7)$$

Total saturated throughput of SCH is

$$S_{SCH} = S_h + S_l \quad (132.8)$$

The values of IEEE 802.11 EDCA parameters are shown in Table 132.2; we analyze SCH saturated throughput by formulas (132.7) and (132.8), and the results are shown in Fig. 132.1. It is clear that:

Table 132.2 IEEE 802.11 EDCA parameters

Parameter	Value	Parameter	Value	Parameter	Value
PHY header	192 μs	ACK length	112 bit	AC[0]-AIFS	7
MAC header	272 bit	RTS length	160 bit	AC[1]-AIFS	3
SIFS	10 μs	CTS length	112 bit	AC[2]-AIFS	2
Slot time	20 μs	Propagation delay	1 μs	m2	1
Retry limit	7	Transmission rate	15 M bit/s	AC[2]-CWmin	7
AC[0]-CWmin	15	m1	3	Maximum queue length	300
m0	5	AC[1]-CWmin	15	pe	1.25e-6

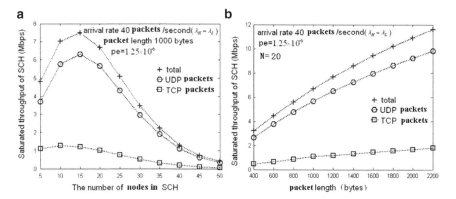

Fig. 132.1 Analysis results of saturated throughput in SCH. (**a**) Saturated throughput in terms of the number of nodes. (**b**) Saturated throughput in terms of packet length

1. Saturated throughput of SCH increases with the number of nodes, but when N is larger than 15, saturated throughput begins to decrease. This is because VANET has little data traffic when N is small. However, when N is large, collision probability increases; it makes nodes having lesser chance to compete channel successfully.
2. Saturated throughput of SCH increases with the increase of packet length.
3. Saturated throughput of UDP packets is larger than TCP packets; it means QoS of high-priority services is ensured adequately.

132.4 Conclusion

In this paper, we present a service channel assignment scheme to improve SCH transmitting performance in VANET. SCH is classified into three ACs to support QoS requirements of two priorities (UDP packets and TCP packets). During SCH interval, instead of static queue management in EDCA, our scheme achieves delivering data packets into optimal AC queue dynamically according to packet type, transmission delay, and network scale. During CCH interval, the scheme achieves adaptive SCH reservation according to SCH transmission quality and payload. Performance of our scheme is analyzed based on theoretical model; the result shows that our scheme is an efficient way to solve bottleneck of high-priority services and improve transmission performance of VANET.

However, with the increase of network scale or traffic load, performance of the scheme becomes worse, so the novel channel assignment scheme based on contention-free mechanism must be presented for high-density VANET. How to solve this problem is our further research work. In addition, we will extend the research to building simulation model and multi-hop wireless environment with routing protocols based on node location in VANET.

Acknowledgements This work is supported by National Natural Science Foundation of China (No. 61174175) and Natural Science Foundation of Shandong Province of China (No. ZR2010FM036).

References

1. IEEE Std.1609.4. (2010). *IEEE standard for wireless access in vehicular environments (WAVE)-multiple channel operation*. Intelligent Transportation Systems Committee of the IEEE Vehicular Technology Society, the Institute of Electrical and Electronics Engineers, Inc. New York, USA.
2. Marica, A., Claudia, C., & Antonella, M. (2012). Enhancing IEEE 802.11p/WAVE to provide infotainment applications in VANETs. *Ad-Hoc Networks, 10*(2), 253–269.
3. Wang, Q., Leng, S., Fu, H., & Zhang, Y. (2012). An IEEE802.11p-based multichannel MAC scheme with channel coordination for vehicular Ad-Hoc networks. *IEEE Transactions on Intelligent Transportation Systems, 7*(2), 449–458.
4. IEEE Std 802.11e-2005. (2005). *Wireless LAN medium access control (MAC) and physical layer (PHY) specifications amendment 8: MAC quality of service enhancements*. LAN/MAN Committee of the IEEE Computer Society, the Institute of Electrical and Electronics Engineers, Inc. New York, USA.
5. Gallardo, J. R., Makrakis, D., & Mouftah, H. T. (2010). Mathematical analysis of EDCA's performance on the control channel of an IEEE 802.11p WAVE vehicular networks. *EURASIP Journal on Wireless Communications and Networking, 2010*(1), 1–15.
6. Mao, J.-B., Mao, Y.-M., Leng, S., & Bai, X. (2010). Research of the QoS-supporting IEEE 802.11 EDCA performance. *Journal of Software of China, 21*(4), 750–770.
7. Xiao, Y. (2005). Performance analysis of priority schemes for IEEE 802.11 and IEEE 802.11e wireless LANs. *IEEE Transactions on Wireless Communications, 4*(4), 1506–1515.
8. Huang, C.-L., & Liao, W. (2007). Throughput and delay performance of IEEE 802.11e enhanced distributed channel access (EDCA) under saturation condition. *IEEE Transactions on Wireless Communications, 6*(1), 136–145.
9. Bianchi, G. (2000). Performance analysis of the IEEE 802.11 distributed coordination function. *IEEE Journal on Selected Areas in Communications, 18*(3), 535–547.

Chapter 133
An Exception Handling Framework for Web Service

Hua Guan, Shi Ying, and Caoqing Jiang

Abstract According to the problems of exception handling for service-oriented software, this paper presents a framework for Web service exception handling (EHF-S) based on policy driven. The EHF-S processes the response message of invoking Web service and produces a response message which is added exception information and exception handling message. We introduce the realization principle, the component, and the key technology for EHF-S. This framework can support the development and integration of exception handling logic for Web service process, improve the exception handling capability, and simplify the exception handling process for Web service.

133.1 Introduction

Web services are rapidly becoming a fundamental program paradigm for the development of complex Web applications. Because of the dynamics and uncertainty during runtime as well as the autonomy and loose coupling in the service resources, there appears the diversity and complexity of exception. This paper focuses on the exception of receiving response message and real-time checks whether an exception has occurred during execution of Web services through a response message returned by the service node.

There exist the following problems in exception handling of Web service: They lack system approach and enough processing capacity to support quickly and efficiently the exception handling logic development. They can't resolve all kinds of

H. Guan (✉)
The State Key Lab of Software Engineering, Wuhan University, Wuhan 430072, China

Network Center, Wuhan Polytechnic University, Wuhan 430023, China
e-mail: gh@whpu.edu.cn

S. Ying • C. Jiang
The State Key Lab of Software Engineering, Wuhan University, Wuhan 430072, China

exception, and exception handling cannot be reused. They lack the necessary exception handling set of resources. In response to these shortcomings, this paper proposes a framework of exception handling for Web services, referred to as the EHF-S, realized adding exception handling ingredients to the original Web service response message, and forming Web service response message with exception handling capabilities.

133.2 The Principle of EHF-S

The main function of EHF-S is to listen to response message of invoking Web service and identify exception of the Web service, using the exception handling logic provided by the framework to process the exception, resulting in response message of invoking Web service with exception handling capabilities. Response message with exception handling capabilities may still invoke the original service, but enrich exception description; it may invoke the new alternative Web services; it may also invoke the series of exception handling services, such as logging service and notification service. In order to use the logging service and exception notification service, we must produce exception information in detail, so we built the exception information and add exception handling information (message routing). Therefore, we specify the destination address and invoking information of the new Web services in the head tag of the SOAP message using WS-Addressing routing protocol. WS-Addressing provides a standard mechanism to identify Web services and Web services messages regardless of the transport protocol that is used. The SOAP protocol defines syntax specification of exception message, so we focus on SOAP message with the exception-associated fault tags, perfecting exception information (called SOAPFault), and extending the SOAP message v adding exception handling (WS-Addressing). Exception information of SOAP message is extracted at the process of exception handling for Web services, and based on the WS-Addressing we process the exception.

EHF-S detects exception information, collects the associated contents information in the exception listening and identifying module, and generates WS-Address and SOAPFault relevant content information in the exception handling module. We need to design a corresponding exception handling policy according to the type of the service layer exception. The implementation of exception handling policy will produce a corresponding action mode for exception handling (such as retry, replace, and skip); each action mode for exception handling will enrich the SOAPFault and WS-Address tag of SOAP message with exception handling capabilities. According to the corresponding WS Addressing format, the action mode adds information about the destination address. Finally, in the message processing module of EHF-S, the SOAPFault and WS-Addressing tag are added to the original SOAP message of invoking Web service, which generate a SOAP response message with exception handling capabilities, and the message is routed by EHF-S message processing module and then sent to the framework users or the appropriate service invoking.

133.3 The EHF-S Framework

EHF-S provides extension mechanism, supports developing exception handling for specific application exception, and registers in the framework as a new component for the exception handling resource sets. The input information of EHF-S is the response message of Web service. The output information of EHF-S is the fault-tolerant response message of Web service. At the same time, EHF-S provides logging and notification service; when an exception occurs, EHF-S will log the exception information and notify the framework's user to manually process; extension mechanism is also provided to support developers to develop exception handling services for particular application exception, and these services can be registered in the EHF-S as a new part of the resource set of exception handling.

133.3.1 The Implementation Principle of EHF-S

In the following, we give an introduction of the work principle of EHF-S, as shown in Fig. 133.1, and we describe the detailed working process of the EHF-S:

1. Exception listening service intercepts request and response messages, real-time detects service invoking execution, captures message's exception of service layer, and checks whether exception information exists on output message of service. If it successfully detects exception, it produces a series exception handling characteristic information and exports to exception identifying service.
2. Exception identifying service receives exception information from exception listening service. Exception identifying service matches the exception characteristic information according to the exception identifying resource set and judges the exception type. The process may use exception transform service which can transform the current exception information to the specific exception type and transmit the exception category information to exception policy management service.
3. The service of exception policy management matches the exception handling resource according to the information of identifying exception type. Finding corresponding exception handling policy, it will invoke related exception handling action.
4. If execution service of exception handling action can't tackle the exception event that occurred, it will invoke the exception notification service and send the exception information to the service provider and service requestor. Meanwhile, it will invoke exception logging service and save the exception information to logging database.
5. If there exist untreated exception and newly throwing exception in the execution of exception action service, then it will return to step two and begin the exception logic handling again.

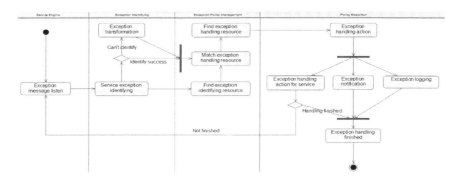

Fig. 133.1 The activity diagram of EHF-S

6. Service provider perfects exception log information exception handling policy repository, when the reoccurrence of such an exception, exception management processing.

133.3.2 The Architecture of EHF-S

As shown in Fig. 133.2, EHF-S is mainly composed of the following key components: service engine, message listener management, AOP message processing, message router management, exception identifying management, exception transformation management, exception notification service, exception logging service, exception policy management, execution service of exception action, exception identifying resource set, exception handling resource set, etc.

The foundation of EHF-S is service engine; all the exception handling is published and registered through Web service. The service engine is in charge of the execution of service component, service release and registration, service discovery, service invoking, etc. In the service engine, we can find and invoke the exception handling services, including the services of exception log and exception notification service.

The message listener management listens the response message of invoking service and determines whether there exists exception information in response message. If there exists Web service execution exception, check whether it contains custom exception in the specific parameters; if it finds, then extract the relevant exception information. The AOP message processor management extracts and analyzes the message that is accepted, identifies the format and content, and extracts the key values of exception information. It can modify the structure and add the exception handling content of the sending message, then composites to message format that fits to SOAP specification. In the message extracting function, it can support dividing the large message into small blocks and compositing to an integrated message at the destination.

133 An Exception Handling Framework for Web Service 1177

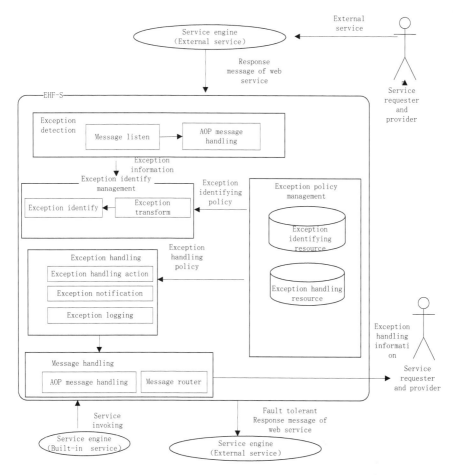

Fig. 133.2 The architecture of EHF-S

Exception transformation management realizes the conversion of exception types and transforms currently exception type into specific exception type. Exception identifying service captures the exception information of message, then matches the exception information and according to the exception identifying resource, judges the exception type of Web service. Exception identifying resource set saves all kinds of exception, and its characteristic of services provides related knowledge database for exception identifying service. Exception handling resource set provides all kinds of reusable exception handling resources. Such as ignore action, retry action for exception handling. At the same time, it has extensibility mechanism, allows adding the custom exception handling service made by developer into resource set.

Exception policy management realizes the integrated management of policy and publishes exception handling policy as Web service. It deploys and configures exception handling logic in runtime environment and deploys exception handling policy of exception handling logic into policy database; meanwhile, it adds the customizing exception handling made by developer into exception handling resource set. Exception action execution management executes the action that meets the exception handling rules; exception notification service sends the exception information to related EHF-S's users and notifies them to manual handling when the exception can't be processed. Exception logging service provides universal exception logging service.

133.3.3 The Key Implementation Technology of EHF-S

The implementation of service engine is based on Apache CXF which is an open source service-oriented framework; Apache CXF is based on a series of packaging and extended to make it able to support the deployment of service components, uninstall, start, stop, and configure. Message listening service is based on Apache CXF too; CXF provides message interception mechanism which builds on a general message layer. The implementation of message router is based on Apache Camel; message splitting divides message into multiple parts that is of fixed length. Through the aggregator component of Camel, the message composition aggregates message block based on related message ID. AOP message handling is based on AspectJ Development Tools (AJDT) of Eclipse Foundation; it can develop AOP programs. AOP message handling service utilizes relevant components of Camel through the splitter component of Camel.

The implementation of exception notification service is based on WS-Notification (WSN) criterion. The exception handling policy set uses WS-Policy criterion. We utilize the Apache Neethi implementation of the WS-Policy framework for editing and storing policies and convert the rules of policy set into rule engine's files. The implementation of rule engines is based on JBOSS Drools. We reference the rule file of WS-Policy files through WS-PolicyAttachment and use RuleFlow module of JBOSS Drools to orchestrate exception handling rules; it executes the exception handling action through Drools Fusion module. Policy management uses Guvnor module of Drools to edit and manage the policy files (or rules). Guvnor uses Web-based business rules management system, implements rules management and dynamic updates, provides a knowledge base of rules management, and enables developers and system administrators to manage the business rules online.

133.4 Related Research

Many scholars present exception handling method and framework for Web service. Sheng Quan et al. present SELF-SERV platform which uses configurable exception handling strategies, and based on the predefined exception handling strategies deal with the runtime exception of Web services with peer-to-peer network environment [1]. Giuliana Teixeira Santos et al. propose a Web service fault-tolerant infrastructure. The facility provides an agent that can be used on the interaction between client and server, and the agent adopts active fault-tolerant technology to achieve transparent fault-tolerant of client [2]. Hai L proposes a method to capture an exception of the outer layer in the process of Web services session and introduces enhanced Web service session context (CeWSC) mechanism to obtain the external context of the participants using the SOAP header and the confirmation message. An event-driven mechanism is presented to merge the context of an exception to the exception handling of composite Web services [3]. Gerald Kotonya et al. describe a differentiation-aware fault-tolerant framework for Web services; it supports for fault-tolerant framework of service-oriented computing, the framework uses asynchronous messaging agent LAMB to provide a news environment for different fault-tolerant protocol, it provides a plug-in way to express fault-tolerant protocol for a processing model, and fault-tolerant services container (sandbox) makes the Web service to be exposed and be found [4]. Liu Chen proposed a uniform rule-based exception handling of service-oriented software and the corresponding exception handling framework [5].

Most researches of exception handling mechanism focus on specific exception handling policy; there is little research on configurable and extendable exception handling framework. Our framework can significantly improve the ability of these aspects.

133.5 Conclusion

We propose a policy-based Web service's exception handling framework to handle exceptions in business processes, the framework realizes adding exception handling ingredients to the original Web service response message, thus forming a new Web service response message with exception handling capabilities. The framework provides exception characteristics resource set of exception type and exception handling resource set for each exception type and support developing exception handling service for particular application exception. Our framework simplifies the development and maintenance of business processes. Therefore, the developers could fast reuse the existing exception handling model.

Acknowledgements This work has been supported by the National 863 Program of China under Grant No. 2013AA102302 and the National Natural Science Foundation of China under Grant Nos. 61070012, 61272108, 61272113, and 61170022.

References

1. Sheng Q. Z., Benatallah, B., Dumas, M., & Oi-Yan Mak, E. (2002). SELF-SERV: A platform for rapid composition of web services in a peer-to-peer environment. In *Proceedings of the 28th VLDB conference, VLDB endowment* (pp. 1051–1054), United States.
2. Santos, G. T., Cheuk Lung, L., & Montez, C. (2005). FTWeb: A fault tolerant infrastructure for web services. In *Proceeding of ninth IEEE international enterprise computing conference (EDOC 2005)* (pp. 95–105), Enschede, The Netherlands.
3. Liu, H., Li, Q., & Chiu, D. K. W. (2007). Enhancing web services conversation with exception contexts for handling exceptions of composite services. In *The fourth IEEE international conference on enterprise computing, E-commerce, and E-services* (pp. 39–46). Piscataway: IEEE Computer Society Press.
4. Kotonya, G., & Hall, S. (2010). A differentiation-aware fault-tolerant framework for web services. In *International conference on service oriented computing* (pp. 137–151). German: Springer.
5. Liu, C., Xu, Y., Deng, F., et al. (2010) A rule-based exception handling approach in SOA. In *2010 international conference on computer application and system modeling (ICCASM 2010)* (pp. 137–141). San Antonio, TX: IEEE CPS.

Chapter 134
Resource Congestion Based on SDH Network Static Resource Allocation

Fuyong Liu, Jianghe Yao, Gang Wu, and Huanhuan Wu

Abstract In order to reduce the operation blocking rate of static resource allocation in SDH Mesh network effectively, balance network traffic, optimize the allocation of network resources, enhance the success rate of multiline information routes, and improve the overall performance of the network. This chapter introduced resource congestion avoidance algorithm (RCAA) based on the adjustment, which can effectively solve the resource congestion in the static resource allocation. In order to prove the feasibility of this RCAA, three simulation examples of resource allocation theory were adopted. Through analysis validation of these three examples, this article proved that RCAA based on the adjustment proposed in this paper can effectively reduce the blocking rate and improve the overall performance of SDH network. RCAA based on adjustment is more superior to ANM. RCAA can avoid resource congestion problems caused by the allocation of resources effectively.

134.1 The Concept and Background of SDH Network

SDH (Synchronous Digital Hierarchy) [1] network consists of some basic network elements (NE), fuses the functions of multiple connection, line transmission, and exchange, and is a summarized information-transferring network operated by unified network management system. Thus, it can be a general technology system which is both suitable for optical fiber, microwave, and satellite transmission. It has the functions of effective management, real-time operation monitor, dynamic network maintenance, and interoperability of different manufacturers' equipments

---

F. Liu (✉) • J. Yao • G. Wu • H. Wu
College of Information Engineering, Tarim University, A'er'la 843300, Xingjiang, China
e-mail: feng_yong2122@163.com; 417416506@qq.com1

so as to greatly improve the utilization of network resource, reduce the cost in management and maintenance, and achieve flexible and reliable as well as efficient operation and maintenance. So it has been the hot issue for development and application in transmission technology and has attracted widespread attention.

PDH (Plesiochronous Digital Hierarchy) has been generally applied in transmission network before SDH. With the rapid development of information technology and the tremendous increase of user, users want to be provided with all kinds of circuits quickly, economically, and effectively by transmitting network. The inherent defects on PDH have dissatisfied the requirement of the modern information network transmission. In order to adapt to the rapid development of modern information society, the experts in American BELL Communications Research institute put forward synchronous optical network (SONET) and the corresponding standards which have been established as the new standards for the digital system in 1986. The Consultative Committee of International Telegraph and Telephone (CCITT) decided to make the modest modifications to SONET and renamed it as Synchronous Digital Hierarchy (SDH) system in 1988 [1, 2].

The allocation problem of communication network resources can be divided into four aspects: resource scarcity relatively, service diversification, decentralization of the resource distribution, and commercialization of the application. Increasing users made the network resources scarce relatively. Due to improper management of resource allocation, the entire network system cannot yet achieve good performance, and the users' investment also cannot get the corresponding returns. Such results may be caused by link fault, virus, or server trouble light reason. However, in fact, one of the main reasons is unreasonable allocation of resources which causes flow bandwidth contention.

Now we have a variety of SDH network routing algorithms. Key Link Routing algorithm is based on the static network resource allocation algorithm, and heuristic algorithm is based on the shared sets. Because these two algorithms only consider the current operation and network status rather than the future operation, they hold some limitations on improving the overall performance. Though dynamic resource allocation algorithms, such as minimum hop routing algorithm (MHA), the shortest path algorithm (MSP), and minimum interference routing algorithm (MIRA), can solve some certain problems, it does not take into account the influence on other panel points between in-and-out panel points and in the flow between in-and-out panel points or links so as to appear network bottleneck effects, cause congestion, and lead to low utilization. In this paper, the main research direction is the optimization of static network resources. And based on the predecessors' study of various static resource allocation algorithms, this paper proposed an improved static resource allocation algorithm [3, 4].

134.2 The Presence of the Resource Congestion Avoiding Algorithm

In any process of network communication, we have to face to network bandwidth resource allocation, while the static resource allocation in SDH network is just a branch of communication network resource allocation which refers to reserving required resources before transferring information. Static resource allocation of SDH Mesh network is a research hotspot currently. Planning network resource can effectively reduce blocking rate and improve the overall performance of the network.

When allocating the static resource, we count information by routes item by item. So we only consider the current information transmission and network status without considering the impact on the future information transmission during this process. So the earlier information is likely to occupy some resources, which results in the failed transmission of the later information. This is called resources allocation congestion [5]. In order to achieve the real equilibrium of network traffic, we should optimize the allocation of network resources to solve the problem of resource blocks.

Figure 134.1 is a simulative SDH network with seven links and six node points. Assume that the remaining available capacity of each link is 1; two users deliver information at the same time. They are respectively from node E to node F and node D to node B. The requesting bandwidth is 1. When calculating the first path, we get the shortest path B-E-C-F based on the current topology and link costs. Now, we figure out the second path. Because of the occupation of B-E and C-F, we cannot allocate the resources. Therefore, the second path cannot transfer or cause delay of information transmission.

It is a complex problem to make multichannel route information successfully. The resource congestion avoiding algorithm (RCAA) based on the adjustment put forward in this paper can effectively solve such problems. RCAA can adjust the information transmission path for the first user, make the business not take up B-C link, and get C-D-E-F link. Then, we need to calculate the second user's information to get a D-C-B link so as to achieve the equilibrium of the network traffic. The key point of the algorithm lies in adjusting the unreasonable take-up information resources to make the following delivery information calculate the path smoothly.

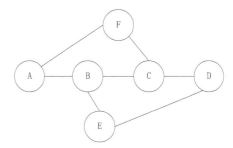

Fig. 134.1 Topology

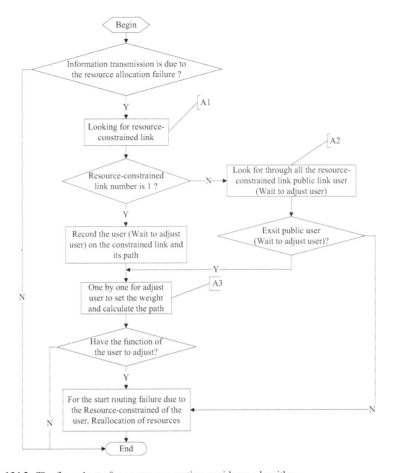

Fig. 134.2 The flow chart of resource congestion avoidance algorithm

134.2.1 Resource Congestion Avoidance Algorithm

The basic procedure of path calculation in simulative SDH network is as follows: weigh the design, calculate the path, and allocate the time slot. There may be some special cases in this process. If user A cannot route because the available resources of the link L is not enough, resource congestion may cause the failure of routing. If we move user B occupying in link L, the available resources on the link are greater than or equal to the bandwidth of the user As. Thus, user B will be able to succeed in allocating resources. This is the core idea of resource congestion avoidance algorithm—RCAA [6]. The procedure is as follows (Fig. 134.2).

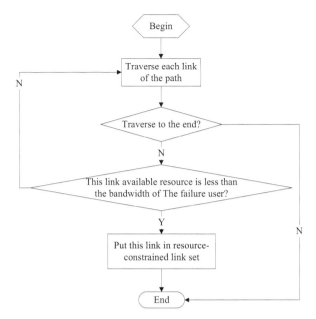

Fig. 134.3 The flow chart of looking for resource-constrained link

134.2.2 Algorithm Analysis of the Key Procedure of RCAA

1. A1: Search the resource-constrained link of the failed user in resource allocation. From the information of the static resource allocation algorithm, we can see that some users may meet with resource allocation failure after successful path calculation. For this kind of business, we traverse each link and judge whether the available capacity of each link is less than the bandwidth of the failed business. If it is, the link is resource-constrained; otherwise, the link is not resource-constrained. So we can get all the resource-constrained links. Specific process is shown in Fig. 134.3.
2. A2: When the resource-constrained link is greater than 1, search all users through the resource-constrained link.
   First, remove the users who go through the first two resource-constrained links respectively, and set out the intersection in this two. And then, remove the users who go through the third resource-constrained link, and set out the intersection with user collection. It doesn't stop by this method until all resource-constrained routes are went through.
3. A3: Weigh and calculate the path for the pending adjust users one by one.
   Weigh the path for the pending adjust users. Adopt the static design weighing method [4] in resources allocation, stop all resource-constrained links, and forbid the pending adjust users in going through those links.

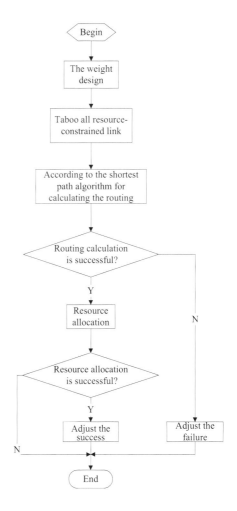

Fig. 134.4 The flow chart of link adjustment

Apply the shortest path algorithm—Dijkstra Algorithm [7]—to calculate the path of the user. If the path calculation succeeds, the resources allocation and adjustment succeed. If not, the user will go back to the original path, so the adjustment fails. Specific process is shown in Fig. 134.4.

134.3 Theoretical Simulation and Result Analysis of RCAA

Several examples will be provided to verify the performance of the RCAA, and theoretical simulation comparison will be made to the method by business adjustment according to the weight calculation. To describe easily, we call this ANM (algorithm with no modulation) which does not consider business adjustment.

Fig. 134.5 Topology of panel point

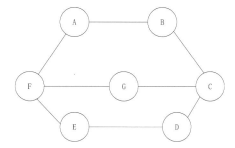

Table 134.1 Routing calculation comparison

Algorithm	Calculation results
ANM (without adjustment)	User 1: Work way route: F→G→C
	Protect the road route: F→A→B→C
	User 2: Calculate failure
RCAA	User 1: Work way route: F→A→B→C
	Protect the road route: F→E→D→C
	User 2: Work way route: F→G→C

Fig. 134.6 Topology of panel point

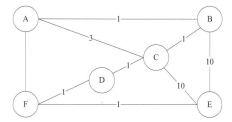

There are two service requests (shown in Fig. 134.5). The first service requests is user 1 from node F to node C with one VC4 bandwidth and without a hop count limit. The second service requests is user 2 from node F to node C with one VC4 bandwidth and without two hop count limits. Each link only has one VC4 bandwidth to be available. The optimization goal of the network is the smallest hop count limit.

Table 134.1 shows the simulation results for routing computation of the business. Seeing from this table, although ANM can guarantee two successful separate ways to user 1, at the same time, it takes the resources of path F→G→C, and only path F→E→D→C is free. But the path is not in conformity with the hop constraints of user 2. So the resource allocation calculation for user 2 fails. RCAA adjusts the links F→G and G→C and recalculates a work path for user 1, allowing user 2 to route successfully.

As shown in Fig. 134.6, the numbers on the link stand for the weights of the links. There is a user who has a request. He or she needs to compute a special

Table 134.2 Routing calculation comparison

Algorithm	Calculation results
ANM (without adjustment)	Calculate failure
RCAA	Work way route: A→C→D→F→E
	Protect the road route: A→B→E

Table 134.3 Routing calculation comparison

Algorithm	Calculation results
ANM (without adjustment)	Calculate failure
RCAA	Work way route: A→B→E→D
	Protect the road route: A→F→C→D

protection user from node A to E. He requests separated link with a VC4 bandwidth. The optimization goal of network is minimum weight and maximum business rate, and the network link capacity is 16 VC4.

Table 134.2 shows simulation results of the routing calculation. As we can see from the results, ANM cannot calculate the path to business successfully according to the weight calculation routing algorithm directly without adjustment, but it can successfully calculate by using RCAA with considerable adjustment. Mainly because after the ANM according to Dijkstra Algorithm calculates the first work path A→B→C→D→F→E, link A→B and link F→E have no resources available, and the links cannot separate. This leads to the failure of protection route computation. And RCAA adjusts the link A→B to ensure the calculation of protection route successfully.

As shown in Fig. 134.2, assume that we need to compute a special protection business from node A to node D, and request link separation and one VC4 bandwidth. The link capacity of network is 16 VC4. The optimization goal of network is the minimum hop count.

Table 134.3 shows the path calculation result. We can see that ANM is a failure for path calculation business, while RCAA can calculate both paths. Mainly because after the ANM according to Dijkstra Algorithm calculates the first work path A→B→C→D, the second path for separation cannot calculate. So it fails. However, RCAA adjusts the links B→C and C→D so as to ensure the successful calculation of the protection route.

From the above three theoretical simulation experiments, we can conclude that RCAA based on adjustment is more superior to ANM. RCAA can avoid resource congestion problems caused by the allocation of resources, so as to improve the success rate of resource allocation in the SDH network planning.

134.4 Conclusion

This paper studied the resource congestion problem in static resource allocation of SDH network and then proposed an improved RCAA (resource congestion avoidance algorithm). RCAA can adjust resource-constrained links' path and avoid these links as much as possible, so the follow-up business would be able to collocate. These three examples with the theoretical analysis demonstrated that RCAA proposed in the paper can effectively reduce the blocking rate of network service routing as well as improve the overall performance of SDH network.

Acknowledgement *Fund project*: Date Sharing Platform Construction of Biological Science in Tarim University (TDZKPT201201).

References

1. Sun, X., & Mao, M. (2009). *SDH Technology* (pp. 56–93). Beijing: Post &Telecom Press.
2. Xiao, P., & Wu, J. (2008). *Principle and application of SDH* (pp. 25–78). Beijing: Post &Telecom Press.
3. Guerin, R., Orda, A., & Williams, D. (1997). Qos routing mechanisms and OSPF extensions. *Proc. of IEEE GLOBECOM'97 IEEE, 3*, 1903–1908.
4. Kodialam, M., & Lakshman, T. V. (2000). Minimum interference routing with applications to MPLS traffic engineering. *Proc of IEEE INFOCOM, IEEE., 2*, 884–893.
5. Gu, S. (2009). *Equipment principle and application of SDH* (pp. 61–179). Beijing: Beijing University of Posts and Telecommunications Press.
6. Wang, Y., & Li, L. (2001). Considering the link load balancing and capacity limit protection design of WDM Optical Transport Network. *Chinese Journal of Electronics., 29*(10), 1319–1322.
7. Chen, B. (2005). *Optimization theory and algorithm* (pp. 101–203). Beijing: Tsinghua University press.

Chapter 135
Multilayered Reinforcement Learning Approach for Radio Resource Management

Kevin Collados, Juan-Luis Gorricho, Joan Serrat, and Hu Zheng

Abstract In this paper we face the challenge of designing self-tuning systems governing the working parameters of base stations on a mobile network system to optimize the quality of service and the economic benefit of the operator. In order to accomplish this double objective, we propose the combined use of fuzzy logic and reinforcement learning to implement a self-tuning system using a novel approach based on a two-agent system. Different combinations of reinforcement learning techniques, on both agents, have been tested to deduce the optimal approach. The best results have been obtained applying the Q-learning technique on both agents, clearly outperforming the alternative of using non-learning algorithms.

135.1 Introduction

The management of resources made on the radio interface for mobile access networks has traditionally followed a static approach [1, 2]. Any mobile operator, on pursuing a satisfactory quality of service, determines the amount of resources to be deployed on each base station, including the split in between those resources devoted to handovers and the remaining resources available to set up new connections [3]. Nevertheless, this working strategy seems to be too short-sighted for what will be necessary in the near future when upcoming optimization challenges will come into play. Key issues like minimizing the energy consumption, sharing the infrastructure among different operators on deploying the 4G mobile systems, or even borrowing radio resources among them are becoming desirable targets for the future mobile communication systems.

K. Collados • J.-L. Gorricho (✉) • J. Serrat
Telematics department, Polytechnic University of Catalonia (UPC), Barcelona, Spain
e-mail: juanluis@entel.upc.edu

H. Zheng
Key Laboratory of Universal Wireless Communication (BUPT), Ministry of Education,
Beijing University of Posts and Telecommunications (BUPT), Beijing, China

In this global scenario, one of the most promising approaches for an intelligent management of available resources comes from the use of machine learning techniques, learning from the system behavior to deduce suitable policies for managing those available resources [4], and policies pursuing different goals, ranging from an optimal quality of service for an individual operator to a global inter-domain efficient system. Ideally, the ultimate goal will be the implementation of self-tuning systems due to applying, for example, fuzzy neural methodologies for the radio resource management [5], similar to our present study, although our innovative contribution comes from using a two-agent approach.

135.2 Working Scenario

In our study case, we envisage a working scenario where several mobile operators are providing service in a completely overlapping fashion, as opposed to the approaches focusing on a single provider owning all the infrastructure [6]. More than that, as already pointed out for the future 4G mobile systems, we assume that all operators share the same base station infrastructure. This way we take a step forward from other approaches dealing with more than one access technology on any base station but still focusing on a single operator [7]. As usual, the geographical area is divided into many cells, a different base station provides service for each cell, and there is some overlapping on the coverage area provided by neighboring base stations.

Our aim is to allocate radio resources on the air interface for each base station to satisfy some given quality of service with a self-tuning system for the parameters governing the base stations' operation. To this end our strategy combines the use of fuzzy logic and reinforcement learning [4, 5], but, in our case, on each base station, two different agents will work together in order to manage the corresponding resources. It has been done in this way to separate two different goals, the quality of service and the operator's economic benefit. Both agents try to maximize their corresponding goals, although only one of the agents takes actions to modify the operating parameters on the base station, as shown in Fig. 135.1. As we can see, there is in practice a closed-loop control as the QoS layer influences the Profit layer jointly with the status information from the environment, and the Profit layer takes an action, which will influence the QoS layer on the next cycle. Hence, there is a mutual interaction between both agents, although not carried out directly from the Profit layer to the QoS layer due to the way we have implemented our approach.

In this scenario the operational parameters to work with are the following:

1. The coverage area per base station, configurable by tuning the power control mechanism.
2. The distribution of channels per base station, configurable by splitting up the channels in different categories on dealing with different types of services or establishing dedicated channels for handover.

Fig. 135.1 Structural model

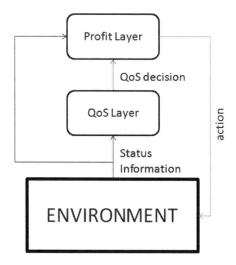

3. The total amount of channels per operator at each base station. In this case a trading mechanism is considered, so different operators exchange channels on their own benefit due to the irregular demand from their respective end users.

Using fuzzy logic we can easily cope with a continuous input space defining the possible system states and actions to be taken. The alternative would be the use of reinforcement learning alone; but, in this case, the disadvantage comes from working with a discrete number of states and actions [6, 7], producing an approach with a worse performance.

Two alternatives on implementing the reinforcement learning mechanism have been tested, the actor-critic and the Q-learning techniques, as can be shown in Figs. 135.2 and 135.3. For both algorithms, each action takes a different fuzzy logic weight α_{R_n}, which is the output due to applying the fuzzy rule, but also an additional weight w_t to be learned [4].

For comparative reasons, a non-learning system has also been implemented using only the fuzzy logic technique. Regarding the time domain, all simulations work with the same sliding window to obtain the input for any approach. In the following sections, we describe in details how both agents of our architectural model work.

135.3 QoS Agent Design

The purpose of this agent is to achieve a given quality of service (QoS) regarding two basic parameters: the blocking rate (measuring the unavailability to set up a new service connection) and the dropping rate (measuring the unavailability to hand over a connection between two base stations due to the user's travelling

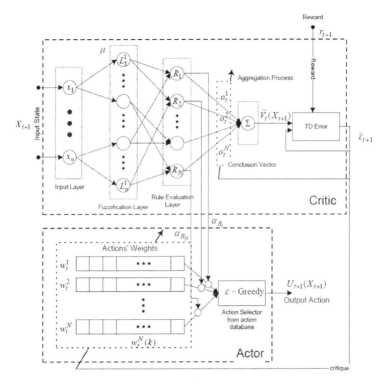

Fig. 135.2 Actor-critic technique

trajectory). Based on these two parameters, the reward expression needed to implement the reinforcement learning mechanism is formulated as

$$Reward = (T_B - B) + \beta(T_D - D)$$

where $T_B - B$ is the difference between the actual blocking rate and a given target value and $T_D - D$ is the difference between the actual dropping rate and a given target value. Usually the dropping rate is much more critical than the blocking rate on measuring the quality of service; consequently, a β-factor is added to emphasize this parameter in front of the other. The two inputs, the blocking and dropping rates, are labeled according to six fuzzification categories; for each category we obtain a membership degree of the input through a fuzzification stage as shown in Fig. 135.4.

The fuzzification-stage outputs are the inputs to a rule matrix (2 dimensions) as shown in Fig. 135.5; the rule matrix defines 5 rules (fuzzy rules) to produce the corresponding decision in a simple manner.

The rule-matrix output will be processed by the RL algorithm according to the reward definition, producing a decision due to the following procedure:

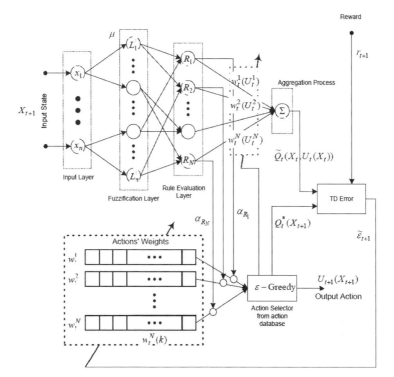

Fig. 135.3 Q-learning technique

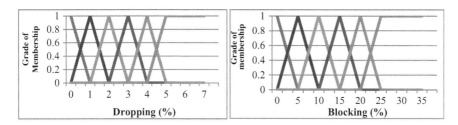

Fig. 135.4 Dropping and blocking rate labeling

Fig. 135.5 Rule matrix

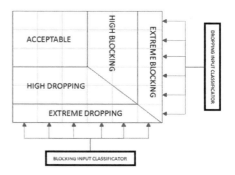

1. The rule weight $\alpha^i_{R_n}$ and action weight w^i_t are combined to produce a s^i_t selection weight; this way a compromise is acquired between the present needs and the learned behavior up until now.
2. Each weight s^i_t is evaluated following some predefined criteria; the criteria will be the requests of the agent for each rule; these are modification of the coverage area, modification of the distribution of channels in between handover and new service channels, and finally the request for extra channels; so for each s^i_t we will have three associated components.
3. The final decision is obtained by a combination of all s^i_t, deducing the definitive three-component request for the present input. This final decision is sent to the next agent, becoming one of its inputs.

135.4 Economical Benefit Agent Design

The objective of this agent is to obtain the maximum economical benefit for the operator; to accomplish this objective, the reward function used by the reinforcement learning mechanism is defined as

$$Reward = \frac{Load * Price * Num\_ChannelsOwned}{Call\_Duration} - Cost * Num\_ChannelsOwned$$

It is noteworthy that the QoS is not included in the reward function; this is because the QoS agent has already taken that into account, and it provides its input to the economical benefit agent. According to this, for the second agent, the inputs are:

- The system load
- The requests from the QoS agent
  - Modification of the coverage area
  - Modification of the distribution of channels
  - Need of extra channels

To avoid learning actions that are not feasible, some additional considerations must be applied invalidating the action. These considerations are the following: selling channels is forbidden if the minimum amount of channels that the operator must maintain is reached, reducing the coverage area is forbidden if the minimum cell radius that must be kept to assure some overlapping is reached, and increasing the coverage area is forbidden if the maximum transmitted power has been reached.

The system load is labeled according to three different categories, as shown in Fig. 135.6, low, medium, and high. When the system detects a low-load state, the weight for buying new channels is reduced and the weight for selling is increased and vice versa. This way the system tries to be led to a medium-load state.

Fig. 135.6 Cell load labeling

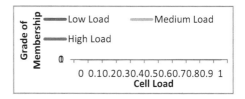

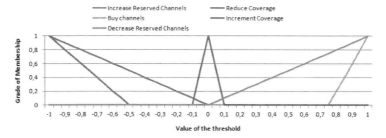

Fig. 135.7 Request of the QoS agent: modification of channel distribution

For labeling the requests of the QoS agent, we need to classify the modification of the channel distribution; in Fig. 135.7 we show the classification of this input; for this study the distribution of channels is split in between channels reserved for handover and channels available for new services. A negative value of this parameter implies an increase on the number of channels reserved for handover, but if the demand is high, the problem cannot be solved by only increasing the channels reservation; another solution is needed; in this case a reduction of the coverage area is applied, trying to delay the time for the handover execution and also to reduce the amount of handovers managed by the base station. A positive value of the parameter allows us to free channels for new connections. As it happened before, if the demand is high, the problem cannot be solved only by this approach; in this case the alternative will be to buy new channels if available. Finally, if the system performs properly in terms of QoS, it tries to increase the coverage area to benefit from more incoming calls.

The other requests of the QoS agent are evaluated directly without being classified, because there is no alternative to be applied. To avoid overacting on the system, a given time between actions is imposed, time defined by the skilled technician; this way the system can better learn the optimal action before acting again.

135.5 Simulation Results

To test the correctness of the algorithms by themselves without being affected by the behavior of another learning technique, each learning technique has been tested individually in the same scenario and under similar circumstances.

Fig. 135.8 Simulation: dropping and blocking rate results

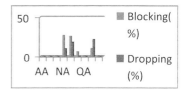

Fig. 135.9 Simulation: income and cost results

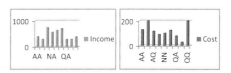

To simplify the notation used in the following figures, any learning system will be labeled with two consecutive letters—the first one applies for the economical benefit agent and the second for the QoS agent. Also the names of the learning algorithms are labeled by their initials; hereafter A means the agent using actor-critic technique, Q means the agent using Q-learning technique, and N means the agent using a non-learning algorithm. Independently of the combination of learning techniques, we have run the simulations to assure that the overall system remains in a medium-load state, which is one of the inputs of the economic benefit agent.

As shown in Fig. 135.8, the average degradation in terms of QoS for the same client demand is higher when the non-learning technique is used on any or both agents. The QoS for the remaining learning combinations results in a dropping and blocking rate even below their target values, 1 % for dropping and 5 % for blocking.

On the other hand, the operator's profit is directly proportional to the amount of established calls, as in all simulations the price for call is kept constant; besides, the cost of maintenance is proportional to the amount of channels managed by the operator; this way more channels turn into more established calls, but the cost of maintenance is also increased. In Fig. 135.9 the cost and the income for each combination of learning techniques are shown.

Those systems having reached an inferior QoS usually have a better economic benefit; one of the reasons is that only the economic benefit agent is able to act on the environment.

One of the drawbacks on using a non-learning algorithm is its dependency with respect to the size of the sliding window used to obtain the sequence of inputs; if the window size is too small, it overacts on sudden peaks of clients' demand, producing unnecessary changes on the system parameters to cope with the incoming demand; on the other hand, if the size is too big, it works with an unrealistic view of the environment behavior.

Figure 135.10 shows the alteration of the coverage area and the percentile of channels reserved for handover; usually those systems with a higher income also work with smaller radius of coverage per cell.

Fig. 135.10 Simulation: coverage and channel reservation results

135.6 Conclusion

In this paper we have evaluated the combined use of fuzzy logic and reinforcement learning to implement a self-tuning system governing the working parameters of the base stations on a mobile network system. Targeting a double goal to assure some given quality of service and to maximize the economic benefit, the simulation results have shown that the two-agent approach that we have considered is suitable to handle both goals simultaneously; nevertheless, depending on the learning technique used on both agents, the behavior is slightly different. In any case, the obtained results applying reinforcement learning outperform the alternative of using non-learning algorithms.

Acknowledgements This work has been done in the framework of the EVANS project (PIRSES-GA-2010-269323), and with the support of projects TEC2012-38574-C02-02 and TEC2012-32531 from Ministerio de Ciencia y Educacion, and grant 2009-SGR-1242 from Generalitat de Catalunya.

References

1. Tölli, A., Hakalin, P., & Holma, H. (2002). Performance Evaluation of Common Radio Resource (CRRM). In *IEEE International Conference on Communications* (pp. 3429–3433).
2. Pérez-Romero, J., Sallent, O., Agustí, R., Karlssot, P., Barbaresi, A., & Wang, L. (2005). Common radio resources management: Functional models and implementation requirements. In *IEEE Personal, Radio and Mobile Communications* (pp. 2067–2071).
3. Altman, Z., Dubreil, H., Nasri, R., Nawrocki, M. J., Dohler, M., & Hamid Aghvami, A. (2006). *Understanding UMTS radio network modelling, planning and automated optimisation*. Chichester: Wiley.
4. Naeeni, A. F. (2004). *Advanced multi-agent fuzzy reinforcement learning*. Master Dissertation, Computer Science Department, Dalarna University College, Sweden.
5. Giupponi, L., Agustí, R., Pérez-Romero, J., & Sallent, O. (2008). A novel approach for joint radio resource management based on fuzzy neural methodology. *IEEE Transaction on Vehicular Technology, 57*(3), 1789–1805.
6. Stefan, A. L., Ramkumar, M., Nielsen, R. H., & Prasad, N. R. (2011). A QoS aware reinforcement learning algorithm for macro-femto interference in dynamic environments. In *International Congress on Ultra Modern Telecommunications and Control Systems* (pp. 1–7).
7. Vucevic, N., Pérez-Romero, R., Sallent, O., & Agustí, R. (2011). Reinforcement learning for joint radio resource management in LTE-UMTS scenarios. *Computer Networks, 55*(7), 1487–1497.

Chapter 136
A Network Access Security Scheme for Virtual Machine

Mingkun Xu, Wenyuan Dong, and Cheng Shuo

Abstract Virtual machines have been widely adopted as servers nowadays. They have essential difference with physical machine. We can utilize the feature of virtual machine to let them be safer and resist an attack from Trojan and hackers. This paper introduces a kind of network access security scheme, which deploys the execution of security strategy outside virtual machine and monitors virtual machine's access to security-sensitive device. The measurements above can transfer the control for key hardware from upper Guest OS to host a platform. Even if Guest OS is affected by virus or Trojan, host can still effectively monitor the network communication of upper virtual machine. In this project, software running in Host OS is programmed to realize the scheme introduced above, it monitors the network communication of virtual machine according to the rules written in XML format. The software can prevent Guest OS or an application running on the virtual machine from communicating with designated domain or IP address successfully, which verifies the effectiveness of the proposed security scheme.

136.1 Introduction

Virtual machines have been widely adopted as ordinary servers or cloud servers nowadays [1, 2]. This paper researches and realizes a kind of security scheme for network access based on virtual machine, to provide a high-standard security solution of network access.

Let us introduce virtualization first. Virtualization allows several virtual machines to run on the same physical computer. Each virtual machine has a set

---

M. Xu (✉) • W. Dong
Beijing University of Posts and Telecommunications, Beijing 100876, China
e-mail: henry7120@hotmail.com; thisisapollo@163.com

C. Shuo
The University of Macau, Macau 999078, China
e-mail: xcsbruce@163.com

of hardware of its own, such as RAM, CPU, and NIC card, which supports the OS and applications to run. No matter what practical physical hardware is mounted, they are viewed as a set of standard hardware. Virtual machine is sealed in files, so that it can be quickly saved, copied, and deployed. The whole virtual system including application with complete configuration, OS, BIOS, and virtual hardware can be move from one server to another, to maintain service without interruption and keep continuous balance of computation load [3].

This paper arranges the execution of security scheme out of Guest OS, the scheme runs independent of Guest OS, and monitors the access of virtual machine to security-sensitive hardware device; this prevents virtual machine from external attack.

No matter whether virtual machine is attacked passively by outside Trojan and virus or virtual machine actively communicates with network, its communication packet is monitored by the network access security filtration software VFirewall introduced in this paper. Meanwhile the security measurements in original OS environment are still applicable [4]. Thus, any kind of communication between hacker and virus-affected Guest OS will be detected out.

136.2 General Design of a Network Access Security Scheme for Virtual Machine

136.2.1 The Construction of Virtual System

VMware or XEN is a well-known virtual machine software which is illustrated as VM monitor in Fig. 136.1. They can be installed on Linux to support Guest OS noted as virtual OS in Fig. 136.1. Configuration of VMware is relatively simple, here take XEN as an example, to illustrate the construction of virtual system [4].

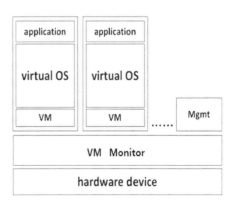

Fig. 136.1 The principle of virtual system

Fig. 136.2 The network access control scheme for virtual machine

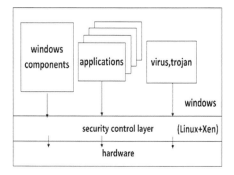

Here take CentOS, the open source version of Red Hat Linux as Host OS. Pay attention to that only if CPU chip supports VT (Intel) or AMD-V (AMD) functionality; Windows can be installed on XEN. Under Linux Gnome graphic interface, XEN Virtual Machine Manager should be installed first, and then Windows OS can be installed on XEN [5].

136.2.2 The Principle of Network Access Control for Virtual Machine

This paper takes Linux as Host OS, which directly runs on hardware, and takes VMware or XEN as virtual machine software. Network security control layer is realized as application of Linux [3]. The architecture of the network access control scheme for virtual machine is illustrated in Fig. 136.2.

136.3 The Realization of Network Access Control Scheme for Virtual Machine

136.3.1 The Architecture of Netfilter Software

VFirewall software introduced in this paper calls the functionality supplied by Netfilter software. Netfilter is a firewall framework designed by Rusty Russell in Linux kernel, which can realize many functionality of security process, such as data packet filtration, data packet process, address disguising, transparent proxy, dynamic Network Address Translation (NAT), Media Access Control, address filtering based on user identity, filtering based on status, and speed control of packet [6].

Netfilter has defined five hook functions for IP protocol. These hook functions may be called on five key locations of the protocol stack that data packets pass

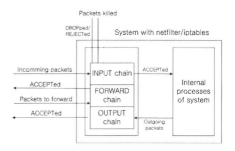

Fig. 136.3 The packet process model by Netfilter module in Linux kernel

through. In every key location of the five, data packet passed by will be captured and compared with corresponding rule chain in IP tables. Packet's destination depends on the comparison result: some packet is put back into IPv4/v6 protocol stack and transmitted continually; some packet is modified and put back into IPv4/v6 stream; some packet is dropped directly.

Figures 136.3 and 136.5 show the packet selection system in Netfilter: IP tables model. In the figure above, IP packet process procedure includes tracing connection, packet content modification, packet filtration, and so on. IP tables' model demonstrates the packet process functionality of Linux kernel, each of the functionality should be registered on corresponding hook in Netfilter, and the corresponding functionality is ultimately realized by the program.

Combination of above functionality flexibly can achieve comprehensive network security strategies. This paper mainly makes use of the packet-filtering ability of Netfilter.

136.3.2 The Architecture of VFirewall Software

In this paper, the network access control software VFirewall is developed in C++ [7]. It parses the network access control strategies described in XML file, checks and records the communication between the Guest OS and special IP segment or domain. VFirewall interacts with Netfilter in Linux kernel space through IP tables in Linux user space, to set network access rule and to capture network access event [8].

The data flow chart in VFirewall is showed below.

As shown in Fig. 136.4, the network security strategies are described in a XML file. The XML file, which is written in pure text mode, can be edited with any text editor. Network access rules described in XML file can be updated through human-machine interface of VFirewall.

The following access security strategy configuration in XML format can be recognized and accepted by VFirewall.

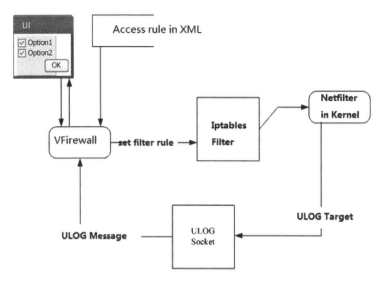

Fig. 136.4 The data flow chart in VFirewall software

- Single IP:
  <block IP="210.25.132.38"/>
- Single domain:
  <block NS="www.bupt.edu.cn"/>
- IP address segment:
  <block IPRANGE="210.23.132.0-210.25.132.255"/>

Having obtained network access control rules from human-machine interface or XML file, VFirewall writes corresponding rules into rule chain in Netfilter. Meanwhile, VFirewall registers a log target (ULOG). If Netfilter captures the IP data packet in line with the security access rule, it will send corresponding report message to VFirewall through ULOG. Then, VFirewall writes the message into log file.

136.3.3 Related Work

- The IP address of virtual machine can be hidden to external network; thus, a lot of attack based on IP address will not work [9].
- As shown in Fig. 136.5, the device driver of Linux may be modified to add data stream inspection mechanism in it, so that it can also monitor and filter data received and sent by upper Guest OS. In fact this work has already been done by the same author [10]. Since the functionality of program instruction set in driver

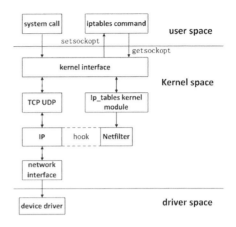

Fig. 136.5 Guest OS data stream can be inspected in device driver space

is less than that in user space, packet inspection in driver is incomplete compared with the VFirewall software proposed in this paper.

However, VFirewall is only suitable for virtual machine, while packet inspection in driver is practicable for both virtual machine and physical machine.

136.4 Conclusion

The running result of the software demonstrates that VFirewall successfully prevents the connection between Guest OS and designated domain or IP address without mounting any firewall in Guest OS, no matter whether unique Guest OS or several Guest OSs are installed in host platform; thus, the effectiveness of the network access security scheme is verified.

The communication efficiency, however, declines a little, because of the security check to access a network. We observed that the packet transmission speed may delay for about a few hundred milliseconds or seconds correspondent with the number of virtual machines launched on Linux.

Acknowledgements This paper is partially supported by the National High Technology Research and Development Program, No. 2011AA010704: the Key Technology and Verification of Network Security Based on IPv6 in Designated Scope.

References

1. Green, M. (2013). The threat in the cloud. *IEEE Security & Privacy, 11*(1), 86–89.
2. Popa, L., & Kumar, G. (2012). FairCloud: Sharing the network in cloud computing. *Computer Communication Review, 42*(4), 187–198.

3. Smith, J. E., & Nair, R. (2006). *Virtual machines-versatile platforms for systems and process* (pp. 10–35). Singapore: Elsevier Pte Ltd.
4. Qin, Z. Y. (2012). Survey on virtual system security. *Application Research of Computers, 29* (5), 1620–1622 (In Chinese).
5. Jang, M. (2011). *Security strategies in Linux platforms and applications* (pp. 62–65). Sudbury, MA: Jones & Bartlett Learning.
6. Russel, R. (2002). *Linux 2.4 Packet Flitering HOWTO*. Retrieved from http://www.netfilter.org/documentation/HOWTO//packet-filtering-HOWTO.html
7. Lippman, S. B. (2005). *C++ primer* (pp. 20–150). New York, NY: Pearson Education.
8. Cheng, S. L. (2009). *Network access control system model research based on virtualization*. Beijing, China: Beijing University of Posts and Telecommunications (In Chinese).
9. Nestler, V. J. (2006). *Computer security lab manual* (pp. 255–261). New York, NY: McGraw-Hill.
10. Xu, M. K. (2005). Encrypt data through streams module in kernel. *Computer Engineering and Design, 26*(7), 1710–1711 (In Chinese).

Chapter 137
Light Protocols in Chain Network

Ying Wang, Yifang Chen, and Lenan Wu

Abstract Aiming at some special applications, such as monitoring of high-speed rail and monitoring of large farm field, a wireless sensor network based on chain structure is proposed. Considering of simplicity and energy saving, two light protocols, which are based on time slot and competition, respectively, are applied in the above network. Finally, the two light protocols are compared with IEEE802.15.4 protocol by OPNET simulation, and the results show that the proposed light protocols have good reliability and low energy consumption.

137.1 Introduction

Wireless sensor network (WSN) has caught great attention and achieved great development in recent years. Theoretically, as long as the nodes are distributed densely enough, the WSN can implement communication by self-organization, relay, and multiple hops [1, 2]. But due to the limitations of volume, cost, and battery energy, the communication distance of WSN is greatly limited for some typical applications. When the sensor nodes in the WSN are sparsely distributed, for example, in the road subsidence monitoring in the highway or high-speed rail, in the stress monitoring of the super-large bridge, as well as in the hydrological monitoring of the rivers, the network topology is often a simple chain type, and the nodes are not randomly distributed. Therefore, the traditional protocols, which can support self-organization and are applied more often in the networks with dense nodes,

---

Y. Wang (✉)
School of Information Science and Engineering, Southeast University, Nanjing 210096, China

Hunan Post and Telecommunication College, Changsha 410015, China
e-mail: wangying_only@163.com

Y. Chen • L. Wu
School of Information Science and Engineering, Southeast University, Nanjing 210096, China

may not be optimum in sparse networks [3]. Therefore, this paper proposes a chain-type structure in WSN. By OPNET simulation, two light protocols in the MAC layer, based on time slot and competition, respectively, are analyzed. In this condition, "light" means simpler and more energy saving. Finally, we compared the proposed protocols with IEEE802.15.4, and the results show that the proposed protocols are superior to IEEE802.15.4 protocol in chain network with good reliability and low energy consumption.

137.2 The Light Protocol Based on Time Slot

As shown in Fig. 137.1, this is a chain network. All the nodes can produce and forward data frames, but each node could communicate with adjacent nodes, and the remote nodes could only communicate with the help of mid-nodes.

First, we propose the light protocol based on time slot for the above network, the "coordinator0" in the middle of the network is coordinator and the other nodes are ordinary nodes. The coordinator contains the information of time slot distribution in a beacon frame and regularly broadcasts it to the ordinary nodes in the network. By receiving the beacon frame and extracting the information, the ordinary nodes can know which timeslot belongs to them, so they can send the packets within the time slot. Therefore, in the light protocol based on time slot, conflicts can be avoided by unified beacon frame sent by the coordinator.

We design three kinds of frame formats in the light protocol based on time slot: beacon frame, data frame, and confirmation frame. Beacon frame is sent by the coordinator through broadcasting, which does not need to be confirmed by ACK. Confirmation frame is sent by nodes after receiving data frames. Confirmation frame itself does not need to be confirmed. Data frames are needed to be confirmed, and each data frame has a unique ID, which can be used to ensure the reliability.

The data flow and time slot allocation are shown in Fig. 137.2, the numbers in the figure represent that in which time slot will the nodes transmit data, and the arrows represent the direction of data transmission. Through this time slot allocation, conflict can be avoided, and the demand of real time can be guaranteed.

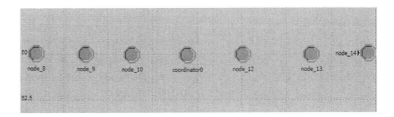

Fig. 137.1 Chain network

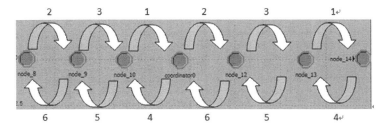

Fig. 137.2 Time slot allocation

| 0 | 1 | 2 | 3 | 4 | 5 | 6 | 1 | 2 | 3 | 4 | 5 | 6 | 1 | ...... |

Fig. 137.3 Optimized allocation of time slot

In this condition, a specific time slot no longer belongs to the packets produced by the node itself (also including the forwarding data produced by other nodes). Therefore, flexible allocation of time slot according to the amount of packets is not very suitable. So, in this protocol, the time slot belonging to each node is fixed. The coordinator sends a beacon frame, carrying only synchronization information.

In fact, in the chain network with a relatively stable environment, there is no such severe motion in the nodes that the WSN will lose synchronization in a very short period of time. Therefore, the protocol can be optimized: ordinary nodes will no longer receive a beacon frame each time after sending data frame; instead, the time slot can be allocated n times by one beacon frame (shown in Fig. 137.3). This can greatly reduce the number of beacon frames, which can lower the system energy consumption. According to simulation (shown in Fig. 137.4, the upper line in each figure represents the number of beacon frame a node received, while the lower one represents the number of data packets), when time slot allocates for 50 times in one superframe, the number of beacon frame can be reduced to only about 5 % of the original, so the energy cost on the beacon frame can save 95 %.

137.3 Light Protocol Based on Competition

In the light protocol based on time slot, whether the node has data to send or not, the coordinator always preserves certain time slots for it. When some nodes have large data amount, the others have small, it is difficult for the time slot protocol to coordinate. In order to guarantee the reliability of the protocol, the nodes need to intercept the channel continuously, which causes large energy consumption, especially in low-rate WSN. Therefore, this paper further proposes the light protocol based on compctition to solve the above problems.

Fig. 137.4 Results of optimized allocation of time slot. (**a**) Before optimization. (**b**) After optimization

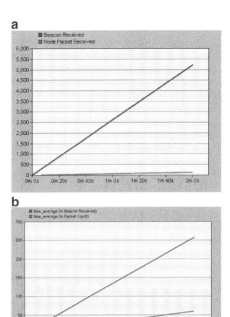

The light protocol based on competition relies on CSMA/CA algorithm [4]. In this chain-type network, synchronization is no longer need to be considered, so all the nodes are set as common nodes.

Figure 137.5 shows the flow chart of the light protocol based on competition. After initialization, if there are packets need to be sent, CSMA/CA algorithm will be performed. After successful access to the channel, the node occupies the channel alone to send the packets. Once the packet is lost, no matter whether the node has other data packets to send or not, the node is forced to give up the channel. Otherwise, if the packet has been successfully received and there are other data packets in the queue, the procedure repeats until the queue is empty. Then, the node releases the channel, and the other nodes can compete for the channel again.

In this flow chart, optimization has also been made to the traditional competitive protocol: a statement is added to judge whether there is a packet lost. Without this statement, the protocol can also work, but it is not good enough. Because when packet loses, the traditional protocol always removes the packet from the queue without trying more times. On the other hand, resending the packets may bring more dramatic conflict and cause more packet loss. Thereby, when a node hasn't received the confirmation frame after expectant time, it may suggest that this channel is not suitable for the node to transmit the packets right now. In this consideration, the node is forced to abdicate the channel. But, the lost packet is not deleted from the queue, which can be sent again after accessing the channel next time. Simulation shows that the optimization reduced the amount of lost packets significantly.

Fig. 137.5 Flow chart of light protocol based on competition

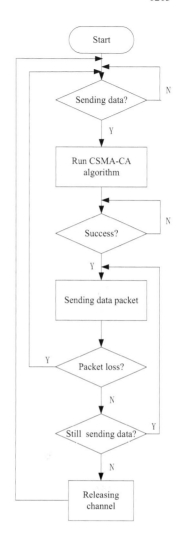

137.4 Comparison Between Light Protocols and the Standard Protocol

In this section, the light protocols are compared with IEEE802.15.4 [5]. First of all, IEEE802.15.4 is realized in OPNET [6]. The chain-type topology is shown in Fig. 137.6; the coordinator sends a beacon frame first, which is received by the neighboring node "node_9" and transmits to the next adjacent node; the procedure repeats until "node_4" receives the beacon. The nodes that have already received the beacon frame will immediately step into the CAP (contention access period). In this period, if they have the GTS (guaranteed time slot) data, the nodes will send

Fig. 137.6 Chain-type topology for IEEE802.15.4

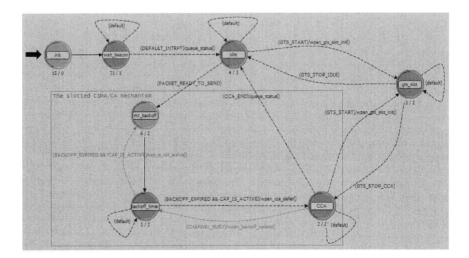

Fig. 137.7 State diagram of IEEE802.15.4 in OPNET

command frame to request the allocation of time slot. After receiving the command frame, the coordinator will contain the information of time slot allocation in a beacon frame and send it out. Then, the nodes can get their slot and send the packets during the time slot. If the GTS data needs to be forwarded, it will be forwarded as GTS data in the time slot of the transmitting node.

According to the analyses above, IEEE802.15.4 protocol is built in OPNET, and the state diagram is shown in Fig. 137.7.

Below is the comparison between IEEE802.15.4 protocol and the two kinds of light protocol in the chain network.

Simulation Settings (see Fig. 137.6): The packets are generated in "node_9" and the generated time interval of CAP packets is 1 s, the data load is 160 bits, the destination address is "node_4," the simulation time is 20 min, the code rate is 250 kbps, and the modulation mode is BPSK. The collected statistics are the number of packets successfully received and the energy consumption in the network.

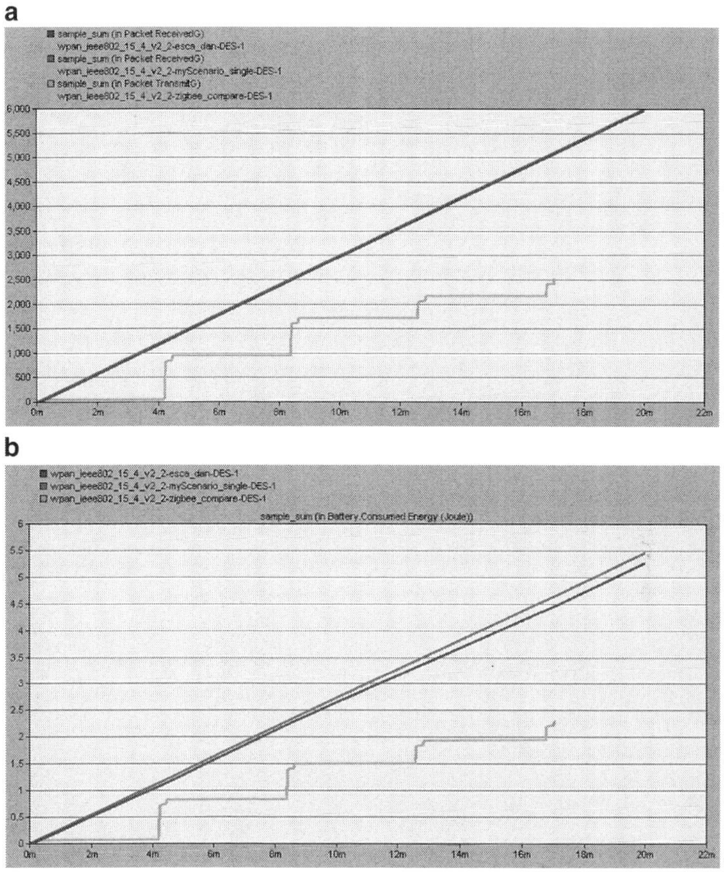

Fig. 137.8 Comparison results. (**a**) Packets received. (**b**) Energy consumption

The results are shown in Fig. 137.8; the curves of IEEE802.15.4, the light protocol based on competition, and the light protocol based on time slot are respectively represented from the bottom to the top in each figure.

By calculating, the energy consumption per packet by using the three protocols respectively is

IEEE802.15.4: $2.35/2500 = 9.4\text{e-}4$ (J)
Light protocol based on time slot: $5.5/6000 = 9.2\text{e-}4$ (J)
Light protocol based on competition: $5.25/6000 = 8.75\text{e-}4$ (J)

From the calculation, the energy consumption of the light protocol based on competition is the lowest. Two light protocols have sent the same number of packets with better real-time performance. IEEE802.15.4 protocol costs the most energy, and the number of packages received successfully is less than half of that generated.

This is partly due to packet loss; more importantly, more packets are failed to be sent out in time. According to the results, the two light protocols are more reliable and more energy saving than IEEE802.15.4 protocol in chain network.

137.5 Conclusion

For chain-type WSN in the actual engineering, there is no mature protocol. This paper designs two kinds of light protocols: protocol based on time slot and protocol based on competition, which are superior to IEEE802.15.4 protocol both in reliability and energy efficiency. The research and simulation in this paper have built the theoretical foundation for the engineering application of the two light protocols.

References

1. Cui, L., Hailing, J., Yong, M., Tianpu, L., Wei, L., & Ze, Z. H. A. O. (2005). Overview of wireless sensor network. *Journal of Computer and Development, 42*(1), 163–174.
2. Limin, S., Jianzhong, L., Chen, Y., & Hongsong, Z. (2012). *Wireless sensor network* (pp. 4–11). Beijing, China: Tsinghua University Press.
3. Weili, X., Mengna, T., & Baoguo, X. (2010). Research of MAC protocol based on OPNET in wireless sensor networks. *Chinese Journal of Sensors and Actuators, 23*(1), 139–143.
4. Yueping, W., Xiaoju, G., Chun-tao, L. I. U., & Jian-de, L. U. (2009). Low energy consumption research on MAC protocol of IEEE 802.15.4. *Computer Technology and Development, 19*(12), 139–142.
5. Yang, H., Yang, G., et al. (2005). Performance research of sensor network based on IEEE802.15.4. *Journal of Signal Processing, 21*(1), 444–447.
6. Chen, Y. (2013). *Research on light protocols using digital interphone as sparse nodes of wireless sensor network*. Nanjing, China: Master dissertation of Southeast University.

Chapter 138
Research and Implementation of a Peripheral Environment Simulation Tool with Domain-Specific Languages

Maodi Zhang, Zili Wang, Ping Xu, and Yi Li

Abstract The importance to build relevant peripheral environment in the testing process for complex embedded software is becoming higher. This paper discussed the current design method of simulation test environment for the embedded software and then presented a modelling method which is used to build peripheral simulation environment for the SUT (system under test) through ICD (interface control data) documents and the software requirement specification. Using this method, the peripheral environment simulation tool which consisted of relevant database and simulation model was set up with Ruby program language. This tool could provide necessary control commands and data support just like in a real running environment for the SUT. Furthermore, a DSL (domain-specific languages) design method for this domain was researched on the basis of the model. The experiment result has demonstrated that it's feasible to set up a peripheral environment for embedded system with our simulation tool.

138.1 Introduction

Most of complex embedded system consists of many subsystems, and there is a large amount of data interaction between each subsystem. This feature makes the software testing different from those in a simple embedded system. When we want to test the core software in the system, we must prepare a complete runtime environment. The cost to do this may be very huge, and in some cases, it's difficult or even impossible to establish a physical system for the testing. Therefore, using simulation method to establish the simulation environment is worth considering.

M. Zhang (✉) • P. Xu • Y. Li
Key Laboratory of Science and Technology on Reliability and Environment Engineering, School of Reliability and System Engineering, Beihang University, Beijing 100191, China
e-mail: canbi007@163.com

Z. Wang
Institute of Reliability Engineering, Beihang University, Beijing 100191, China

In our previous work, there is an embedded system emulator designed by our laboratory which can run operating system and has the function of fault injection. It also needs peripheral environment if we want to test the software running on the emulator. Some researchers have provided a real-time test script for automated test equipment [1]. They use the test script to drive the simulation test model and put forward a method to simulate the peripheral environment through ICD (interface control data) documents. But its focus is just on the design of the test script, not the establishment of the simulation models, and like most of the other simulation environment for embedded system, their application object isn't suitable to a simulation system [2]. So we also need to explore a method of simulation modelling.

The establishment of simulation model needs the analysis for the specific requirement. So the extensibility of the model is a question that worth considering. Domain-specific languages are designed for domain requirement, which abstract from domain entitles and operations [3]. Establishing the simulation environment through domain-specific languages can make the model have better extensibility in this domain and promote its application values. Otherwise, the DSL could make it easy for the professionals in this domain to set up simulation models.

Ruby is an object-oriented dynamic programming language, with simple and beautiful syntax. Ruby has a powerful meta-programming capability. It can support dynamic method, introspection, code block, etc., which can make Ruby suitable to design a DSL as mother language [4, 5].

The rest of this paper is organized as follows. Section 138.2 gives a design method of database and simulation model. The implementation of the simulator consists of database; simulation models are described in Sect. 138.3. Section 138.4 shows a preliminary experiment, and Sect. 138.5 concludes this paper and our future work.

138.2 Design Requirement of the Peripheral Environment Simulation

The peripheral environment simulation is mainly based on the ICD and the software requirement specifications, so that this simulation environment does not support the data processing ability just like in the real environment. It just provides necessary incentive information and the response data according to the requirements of the real operation. So we can conclude that the most important two elements to build peripheral simulation environment are the following:

1. Database: provides the necessary control and data information
2. Simulation model: helps to realize the basic functions and running process

The rest of this paper will give the detail method for database and simulation model.

138.2.1 Design Method for Data File

The establishment of the database is mainly based on the ICD document. In the ICD document, it provides the whole data which is used in the real running process, with the specification of the data format, size, type, and so on.

Now, through the analysis of the ICD document, there are two methods to build the database.

First, if all of the data has the feature of clear rules, simple structure, and single type, we can adopt the method of mathematical simulation and get the required data through some programmed algorithm. This method can help us to simplify the process of data processing.

Second, when the data doesn't have obvious characteristics as in the first condition, we can store the data which is needed to the test in the form of data files. In this process, we need to design a storage format according to the actual needs. For example, the *message type*, the *subtype*, the *size*, the *corresponding model*, and the data subject are useful to identify the uniqueness of the data in the database. We can also combine with the first method to simplify the data structure.

The implementation of the data files may be diverse. We can store them in a database like Oracle, MySQL, SQLite, or store them as the CSV format, or even in an array or Hash.

138.2.2 Design Method for Domain Meta-Model

Meta-model is a kind of model which has a higher abstraction degree and has the characteristics and commonalities of the general models. The relationship between meta-model and general model is similar to the relationship between class and object in the programming language.

The general theory of DSL shows that domain modelling is the first step to design a DSL [6]. In our condition, we can abstract the running process to be some small models which could present a specific task in the process. And these small models are the instances of a meta-model which has a higher abstract level. So the meta-model could be the basis of our DSL designed in the future.

To design the domain meta-model for peripheral environment, we parse the running process through the software requirements specification and the ICD documents and make the running process be some discrete events. All these discrete events were driven by their corresponding models, which could have their attributes and operation method. We can conclude the meta-model from the abstraction of these models. The detail process of domain modelling is shown in Fig. 138.1.

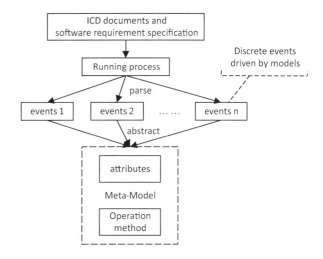

Fig. 138.1 Domain modelling process for peripheral environment simulation

138.3 Implementation of the Simulation Model with Ruby

Figure 138.2 shows our framework of the simulator; now we have 3 main parts in the peripheral environment simulator, Ruby interpreter, simulation models, and data transceiver. The simulator obtains the data from the database and, driven by the test script, finally sends the data to the SUT through pipe file.

138.3.1 Model Structure

In our simulation meta-model, we have designed four attributes: *name*, *state*, *type*, and *subtype*. The details for each attribute are showed in Table 138.1. These four attributes will be initialized in the process of the model instantiation.

Then we designed some basic method for the meta-model according to the requirement of the simulation:

1. Data transmission method: open or create a pipe file, send or receive data through the pipe file.
2. Data match method: obtain the data which is matched to the model attribute according to the initialized model.
3. Data modification method: according to different data type and structure, using different algorithms to modify the original data.

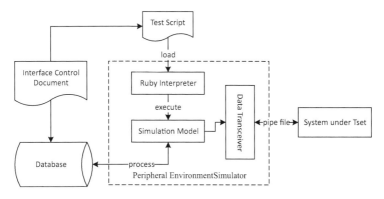

Fig. 138.2 Main framework of the peripheral environment simulator

Table 138.1 The main attributes in the meta-model

Attribute name	Description	Note
Name	Define the name of this model	Specific by keywords
State	Define the state of this model	Only to be "normal" or "fault"
Type	Define the type of data in this model	
Subtype	Branch of the data type	This is an optional attribute

138.3.2 Data Structure and Processing

In this section, we conclude the standard data format from the ICD document. Most of the data we need to process are the BIT information and some other control command (Fig. 138.3).

There are no commonalities between these data in addition to follow the above data structure. The data length may range from 3 to 700 bytes. Especially in the BIT answer data, it contains a lot of subtypes (Table 138.2).

From the above analysis, it's impossible to design an algorithm to generate data automatically or store all kinds of data in detail. So, we have only stored the necessary control information and the normal data of each type of the BIT information. And these data are stored together with their data ID and data type, which may be important parameters in the process of data matching and processing.

The process of the data matching and processing is mainly based on the data ID, subtypes, and their length. Figure 138.4 shows a flow diagram of the data processing.

Fig. 138.3 Format of the data information

| Data ID | Data Length | Data Details |

Table 138.2 Data information

Data ID	Data label	Description	Size (bytes)
0x01	BitAnsProc	BIT response	8–14
0x02	InfoReq	Request for information	3
0x1D	CanRtr	CAN node status	5
0x41	SysSw	Master–slave switch command	3
0x42	BitReq	BIT request	3
0x43	MSListAll	Module state synchronization	724
0x49	FunListAll	Function state synchronization	154
0x5A	MSListAck	Ack for module state update	3

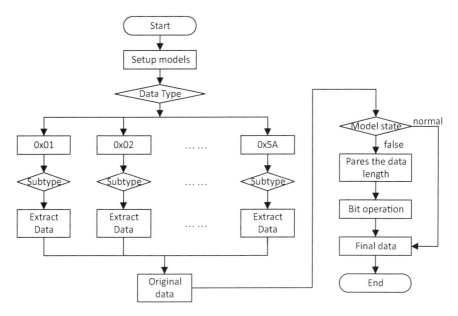

Fig. 138.4 Flow diagram of data processing

From the data structure, we can know that the low 16 bits of every data present the data ID and data length. So we can make bit operation to the high data bits of the original data, for example:

1. Control command of 3 bytes: final_data = original_data | 0x010000
2. BIT information of 8 bytes: final_data = original_data | 0xFFFFFFFF000000000

Then, the final data would be sent to the SUT through the data transceiver method.

138.4 Experiments and Analysis

In our current experimental situation, we make a simple data receiver for the test. This data receiver will create a new pipe file at the run time and will always be monitoring this pipe file.

In this experiment, we set up two models through the meta-model: ("SysSw," "normal," "0x41") and ("BITAnsProc," "fault," "0x01," "ZHXD1_M_JK1"). The first model presents a control model, which would send a normal excitation signal for the software under test. The second model presents BIT response model; it could send a fault BIT information of the subtype named "ZHXD1_M_JK1" to the SUT. The result of this experiment shows in Fig. 138.5.

From the running result, we can see that the "SysSw" model sends a normal control command "0x000341" to the data receiver. And the "BITAnsProc" model sends a fault BIT information "0xffffffffffffffffff0000320e01" of a subtype named "ZHXD1_M_JK1." The data receiver has received the control command and the BIT information.

We can conclude that our simulation tool has the following advantages compared with other simulation tools:

1. This is a full virtualized peripheral environment simulation tool; it could be an effective test method for this equipment which is difficult and high cost to be tested.
2. This simulation tool is designed for another virtual platform which can run an OS and their software; we haven't found a similar tool that can realize peripheral environment simulation functions for virtual platforms.
3. Our design is supported by DSL design ideal, which makes the communication between the programmer and domain experts, and this test script could be very easy to understand and write for the domain experts.
4. We have proposed the modelling and simulation method for peripheral environment, and our target system is mainly avionics systems. The meta-model is designed through the standard ICD documents in the avionics systems. We can

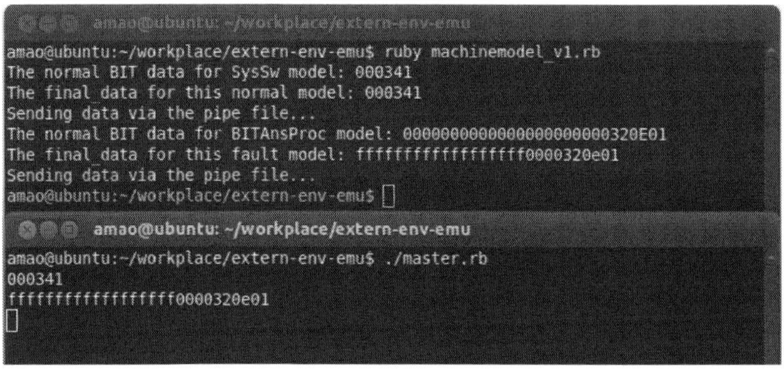

Fig. 138.5 The experiment result

change the meta-models or the model attributes easily through this method and make it suitable for other models and systems. So this tool has a good generality for avionics systems or other systems.

138.5 Conclusion and Further Work

In this paper, we summarized the current methods of running environment simulation and then we proposed a domain modelling method for peripheral environment simulation through ICD document and software requirement specification. The modelling and simulation methods have a good generality for the avionics systems and other systems. After all, we setup a small simulator with a data source and some models through this method. The feasible was confirmed in a small experiment.

In the future, we still have to optimize the model structure, and add more functions according to the actual requirement, make our simulator have the ability to send and receive data at the same time, and in the last communicate with the QEMU simulator.

Acknowledgements This research was supported by the Technological Foundation Project of China National Defence Science and Engineering Bureau under grant No. Z132012A004.

References

1. Jiang, C., Liu, B., Yin, Y., & Liu, C. (2009) Study on real-time test script in automated test equipment. *In 8th International Conference on Reliability, Maintainability and Safety* (pp. 738–742). Chengdu, China: IEEE.
2. Zainzinger, H.J., & Austria, S.A. (2002). Testing embedded systems by using a C++ script interpreter. *Proceedings of the 11th Asian Test Symposium* (pp. 380–385). Guam, USA: IEEE.
3. Gunther, S., Haupt, M., & Splieth, M. (2010). Agile engineering of internal domain-specific languages with dynamic programming languages. *In 5th International Conference on Software Engineering Advances* (pp. 162–168). Nice, France: IEEE.
4. Gunther, S. (2010). Multi-DSL application with Ruby. *IEEE Software, 27*(5), 25–30.
5. Cuadrado, J., & Molina, J. (2007). Build domain-specific languages for model-driven development. *IEEE Software, 24*(5), 48–55.
6. Dinkelaker, T., & Mezini, M. (2008). Dynamically linked domain-specific extensions for advice languages. *ACM Proceedings of the 2008 AOSD Workshop on Domain-Specific Aspect Languages, New York* (pp. 1–7).

Chapter 139
Probability Model for Information Dissemination on Complex Networks

Juan Li and Xueguang Zhou

Abstract In order to analyze the regulation of information dissemination on the complex network, SIR probability model has been built to represent the peoples' interaction during information dissemination on complex networks. By introducing and computing the state transiting probabilities of the net nodes, we can effectively analyze and update the nodes' states at each step in information dissemination. Accordingly, the evolution algorithm of information dissemination is designed and realized by simulation. Simulation experiments of information dissemination on ER network and BA network with different parameters reveal that the density of final awareness will not be affected by the total of nodes, but increase progressively following the increase of average degree until a certain value. Different degree distributions can also be effect on the density of final awareness. SIR probability model can accurately reflect the process of information dissemination on complex networks. It can be used for the description and analysis of information dissemination on complex networks.

139.1 Introduction

How many persons will know the message during information dissemination in the realistic society or Internet space? Which factors will influence or determine the results of information dissemination? Whether the different network structure will affect information dissemination? How will these factors affect information dissemination? These are all that interest us for research on the information dissemination on complex networks.

J. Li (✉) • X. Zhou
Naval University of Engineering, WuHan 430033, China
e-mail: lijuan770107@126.com

The information dissemination is a typical evolution process of complex system [1]. To understand the information dissemination in the complex network well, first we need a proper model to describe the people's interaction during information dissemination on complex networks. Basing on SIR (susceptible, infective, recovered) model, this paper presents SIR probability model by introducing the state transiting probability of the net node. First, we can compute the state transiting probabilities of all nodes in the complex network. Then we will analyze and update the states in the next moment accordingly. Thus, we can describe the process of information dissemination accurately.

Furthermore, to discover the regulation of information dissemination on the complex networks, experiments of information dissemination on different networks with different parameters are needed. In this paper, many emulation experiments have been done for the information dissemination on ER network and BA network with different parameters. Many parameters of the network can be effect on the result of information dissemination. We have analyzed the total of nodes, average degree, and topology of complex networks affecting the density of final awareness.

139.2 SIR Probability Model

139.2.1 The Basic SIR Model

SIR model [2] can be used for information dissemination. Here, "S" is the state for who have not known the information yet. "I" is the state for who have known the information and would transmit it. "R" is the state for who have known the information but will not transmit it. Many persons including Sudbury, Zanette, and Zonghua Liu had researched on this model. The dissemination rules they made can be expressed with the following formula:

$$\begin{cases} I(i) + S(j) \to I(i) + I(j) \\ I(i) + I(j) \to R(i) + I(j) \\ I(i) + R(j) \to R(i) + R(j) \end{cases} \quad (139.1)$$

Disseminator staying in "I" state would select a neighbor randomly to spread the information. If his selection is in "S" state, this neighbor will obtain the information and become a new disseminator. His state changes from "S" to "I". Otherwise, when the selection is in "I" or "R" state, the disseminator will lose the interest to spread the news, he will change his state to "R," and the neighbor he selected has no change.

139.2.2 State Changing Probability in SIR Model

The value kv is defined for the degree of the node v in the spreading network. It is the scalar of his neighbors. kvS, kvI, and kvR are defined respectively for the scalar of his "S," "I," and "R" state neighbors. By all appearances,

$$k_{vS} + k_{vI} + k_{vR} = k_v \tag{139.2}$$

If v is a node in "I" state, we define pvI as the probability for node v to keep "I" state and pvR as the probability for node v changing to "R" state next time. So, according to dissemination rules of SIR model, we can calculate pvI and pvR:

$$p_{vI} = \frac{k_{vS}}{k_v} \tag{139.3}$$

$$p_{vR} = 1 - p_{vI} = \frac{k_v - k_{vS}}{k_v} \tag{139.4}$$

If u is a node in "S" state, we define puS as the probability for node u to keep "S" state and puI as the probability for node u changing to "I" state next time. We can calculate puS and puI with the following expressions:

$$p_{uS} = \prod_{v=I} p_{vR} = \prod_{v=I} \frac{k_v - k_{vS}}{k_v} \tag{139.5}$$

$$p_{uI} = 1 - p_{uS} = 1 - \prod_{v=I} \frac{k_v - k_{vS}}{k_v} \tag{139.6}$$

In (139.5) and (139.6), v is the neighbor with "I" state of the node u.

In the information dissemination network, N is defined for the total of nodes as well as the numbers of crowds. Ct is defined for the numbers of crowds who known the information at t moment. Then, we can estimate it with the following expression:

$$C_t = N - \sum_{u=S} p_{uS} \tag{139.7}$$

Here, u is the node with "S" state in t-1 moment.
Unitarily, ct is defined for the density of the known:

$$c_t = \frac{C_t}{N} \tag{139.8}$$

139.2.3 Evolving of SIR Probability Model

In the SIR model, three states are existing: "S," "I," and "R." In the beginning, the number of nodes with "I" state is $C0$. Others are in "S" state. None is in "R" state. Node u is in "S" state. If there are some neighbors in "I" state, node u will change its state to "I" according to puI. The evolving algorithm is the following:

1. Initializing the states and information of all nodes in the complex network. There into, $C0$ is the initialization of dissemination nodes. The initialization of time t is 0. The terminate time is set with T.
2. Calculating probability value pvI and pvR for each node v with "I" state and calculating probability value puS and puI for each node u with "S" states.
3. Changing the next time state of the node v with "I" state to "R" according to pvR, as well as changing the next time state of the node u with "S" state to "I" according to puI.
4. $t = t + 1$. If $t = T$ or there has no node in "I" state, it is end. Otherwise, go to the step (2).

139.3 Emulation of SIR Probability Model in Complex Network

SIR probability model and its evolvement algorithm are independent of the structure of network. However, information spreading in the network is nearly correlative to the network topology [3]. No matter whether information spreads in social network or internet, all these networks are typical complex networks. They have complicated dynamics action. To study the effect of complex network structure on information dissemination, in this paper, the two mature complex network models, ER (Erdös–Rényi) stochastic network and BA (Barabási–Albert) scale-free network, are selected for emulating and analyzing.

139.3.1 Creating of Complex Network

1. ER Model
   The following is the algorithm for creating ER network:

   (a). Creating a network only including N nodes. N is the total of nodes.
   (b). For each pair of nodes, the edge is created following the probability value p. $p = <k>/(N-1)$, $<k>$ is the average degree of the nodes in the network.

Fig. 139.1 The relation between total of nodes and density of awareness

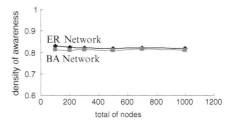

2. BA Model

Network like Internet, WWW, and so on, their degree distributions obey power law. They have the characteristic of scale-free. BA network is a typical scale-free network [4–6]. We can create BA network with the method of dynamic growth as the following:

1. Creating the initial network including *m0* nodes.
2. Adding new node *v*. *m* edges from the new node link to the existing nodes according to preference rule. The node with great degree has more probability to get new edge, i.e., 3. repeating 2. until the total of nodes is enough.

N is setting for total of nodes in network. $<k>$ is setting for the average degree. Then, we can set the initial network as a complete network with *m0* (m0 = $<k>$/2) nodes. $<k>/2$ (m = $<k>/2$) edges will be created for every new node. Accordingly, we can create BA network with N nodes.

139.3.2 Experiment and Analysis

Setting the initial value *C0* for density of known crowd, we can process the evolvement of SIR probability model in ER network and BA network. We can review the effects of various facts on information dissemination.

1. Total of Nodes Affecting Information Dissemination

Assuming only one node with "I" state in the first, emulation experiments have been processed for $<k> = 20, N = 100, 200, 300, 500, 700, 1,000$. Through these experiments, we can examine the effect degree of total of nodes on information dissemination. To assure the correctness of the result, each experiment has been done 100 times independently. The average of the results is shown in Fig. 139.1.

From Fig. 139.1, we can know that the total of nodes affects little on the density of awareness at end.

Fig. 139.2 The relation between average degree and density of awareness

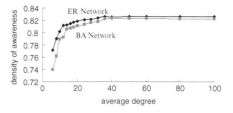

Fig. 139.3 Increasing peoples with "R" state

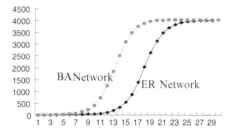

2. Average Degree Affecting Information Dissemination

Assuming only one node with "I" state in the first, emulation experiments have been processed for $N = 500$, $<k> = 12, 16, 20, 24, 28, 32, 36, 40, 50, 60$. Through these experiments, we can examine the effect degree of average degree on information dissemination. To assure the correctness of the result, each experiment has been done 100 times independently. The average of the results is shown in Fig. 139.2.

From Fig. 139.2, we can know that density of awareness at end will increase to a certain degree following augment of average degree. While $<k>$ reaches about 36, the density of awareness will keep tranquilization.

3. Network Structure Affecting Information Dissemination

The structures of ER network and BA network are different. From Figs. 139.1 and 139.2, we can realize that the density of awareness end in BA network is always less than that in ER network generally. To analyze the network structure affecting information dissemination more, emulation experiments have been processed for $N = 5,000$, $<k> = 10$, $C0 = 0.002$. During the processes of information dissemination in these two types of network, the totals of people in "R" state (who know the information but will not transmit it) are changing as shown in Fig. 139.3.

From Fig. 139.3, we can find that increasing speed of peoples with "R" state is rapid during the processes of information dissemination. Because there are more nodes with great degree in scale-free network than in stochastic network, the information is easy to be send to the nodes with great degree. When these nodes are in "I" or "R" state, the others will connect them with big probability and then change themselves in "R" state. So, information dissemination ends more quickly in BA network than in ER network. It leads that density of awareness at end in BA network is less than in ER network. Figs. 139.1, 139.2, and 139.3 all show the characteristic.

139.4 Conclusion

The paper uses probability method to develop the SIR model. We apply this probability SIR model in the ER network and BA network to study the information dissemination in complex networks. The results of emulation are same as the theoretic existed. It indicates the validity of the model we created.

References

1. ShiZhe, G., Zheming, L. (2012). *Complex network basic theory[M]*. Beijing: Science Press. 6, 143–146.
2. Guanrong Chen, Xiaofan Wang, & Xiang Li. (2012). *Introduction to complex networks: Models, structures and dynamics*[M]. Beijing: Higher Education Press. 5, 187–198.
3. Jie, Z., Zonghua, L., & Baowen, L. (2007). Influence of network structure on rumor propagation [J]. *Physical Letters A, 368*, 458–463.
4. Yi, C. X. (2012). Weighted BA scale-free random graph model [J]. *Advanced Materials Research., 1566*, 2780–2783.
5. Kaiying, D., Jingwei, D., & Yingxing, L. (2012). Research and application of improved BA networks evolving models[J]. *Computer Systems & Applications., 8*, 116–119.
6. Barabási, A.-L. (2003). Linked: The new science of networks[J]. *American Journal of Physics, 71*, 409–410.

Chapter 140
Verification of UML Sequence Diagrams in Coq

Liang Dou, Lunjin Lu, Ying Zuo, and Zongyuan Yang

Abstract The UML is a semiformal modeling language which only has syntax and static semantics precisely defined. The dynamic semantics for the UML is specified neither formally nor algorithmically. When using UML at the design phrase, there does not exist a systematic way that allows the model designer to specify its formal semantics and automatically verify correctness properties of the described model. The UML sequence diagrams are widely used to describe the behaviors of software. Reasoning about properties of sequence diagrams at the analysis and design phrase may reveal software faults before software implementation. We propose to use the theorem proof assistant—Coq to verify syntax and semantics constrains of sequence diagrams. The verification and proof process are useful for improving the correctness of sequence diagrams and hence increases the software quality.

140.1 Introduction

In practical software engineering, the UML (Unified Modeling Language) has become the de facto modeling language. The UML sequence diagrams [1] provide a graphical notation to describe dynamic aspects of software during the design phase and have been proven very useful for modeling system behavior in model-based development.

However, the UML is a semiformal modeling language which only has syntax and static semantics precisely defined. There should not be a systematic way that allows the model designer to specify formal semantics for UML so that their

---

L. Dou (✉) • Y. Zuo • Z. Yang
Department of Computer Science and Technology, East China Normal University, Shanghai 200241, China
e-mail: ldou@cs.ecnu.edu.cn

L. Lu
Department of Computer Science and Engineering, Oakland University, Rochester, MI 48309, USA

behavior can be precisely and unambiguously understood, and safety-critical properties of system specifications can be verified automatically. As a result, it is highly desirable to use formal methods to analyze UML models in the design phrase.

Coq [2] is an interactive theorem proof assistant based on the Calculus of Inductive Constructions and widely used in program verification and theorem proving. In this chapter, we explore the possibility of using Coq to formalize the semantics of sequence diagram and verify the desired properties. The verification and proof process will provide increased reliability guarantee in the design phrase, thereby improving software quality and reducing development cost. In our solution, the syntax of a sequence diagram is represented as inductive types and its denotational semantic is represented as a recursive function in Coq. Based on trace semantics, we could verify the syntax and semantics constraint on models. The desired properties of the semantics can be stated as lemmas and the proof ensures that the semantics satisfy their corresponding properties.

The chapter is organized as follows. In Sect. 140.2 the trace semantics of sequence diagrams is introduced formally, together with a brief introduction to Coq. In Sect. 140.3 we describe the representation for sequence diagrams in Coq. We then describe how to verify the syntax and semantic restrictions and prove desired properties in Coq in Sect. 140.4. Section 140.5 uses a case study to illustrate our approach. Section 140.6 discusses the related work and we conclude in Sect. 140.7. All the proofs and samples can be found at https://github.com/lisa-dou/VerifySD.

140.2 Background

140.2.1 Abstract Syntax

Based on a previous work [4], we give the abstract syntax of sequence diagrams. *Name* is a denumerable set of names. An event *e* in *Evt* has the following structure. An event sending a message with signal $S \in Name$, transmitter $T \in Name$, and receiver $R \in Name$ is written as (!,S,T,R), and its corresponding receiving event is written as (?,S,T,R). We abstract from details of guard conditions *c* in *Cnd* and require that the collection of guard conditions is closed under classical logical negation ($\neg$), conjunction ($\wedge$), and disjunction ($\vee$) operations. Let τ represent unobservable events. The abstract syntax for sequence diagrams in *D* is given below:

$$D := \tau|e|strict(D_1,D_2)|alt(c,D_1,D_2)|opt(c,D)|par(D_1,D_2)|seq(D_1,D_2)|loop(c,D)$$

140.2.2 Semantics

Let Σ be an alphabet. $\Sigma*$ denotes the set of all strings over Σ. The interleave of two strings is the set of strings obtained by interleaving the two strings in all possible ways. Let $x, y \in \Sigma$ and $u, v \in \Sigma*$. We definite of the interleave operator $\|$ as in [5]:

$$\varepsilon\|\mu = \mu\|\varepsilon = \mu$$

$$x\mu\|yv = \{x\}(\mu\|yv) \cup \{y\}(x\mu\|v)$$

where • is the language concatenation operator and ε represents the empty string. Let $\oplus$ be a binary operation on domain S. Then $\oplus^\#$ defined below is a binary operation on the power set $\varphi(S)$ of S:

$$X\oplus^\# Y = \{x \oplus y | x \in X \land y \in Y\}$$

Semantic Domain. A sequence diagram is a partial specification of required and prohibited behaviors of an application. Our work is concerned only with required behaviors. The semantics of sequence diagram is trace based. A trace is a sequence of tokens, each of which is either an event or a guard condition. The domains of tokens and traces are respectively:

$$Tk = Evt \cup Cnd$$

$$Tr = Tk^*$$

An obligation may contain more than one trace. Once an obligation is chosen, all traces in the obligation are required in that for each trace t in the obligation, there is an interaction that produces t. The domain of the obligation Ob and the semantic model Mo are respectively:

$$Ob = \{o \in \varphi(Tr)\}$$

$$Mo = \{m \in \varphi(Ob)\}$$

Semantic Function. The semantics of a sequence diagram D is denoted as $[\![D]\!]$ follows:

1. Unobservable and observable events:

$$[\![\tau]\!] = \{\{\varepsilon\}\}$$

$$[\![e]\!] = \{\{e\}\}$$

2. *Strict* fragments:

$$[\![strict(D_1, D_2)]\!] = [\![D_1]\!] \cdot {}^\#[\![D_2]\!]$$

3. *Alt* fragments: *Alt*, *opt*, and *loop* fragments introduce guards to traces. Define $c \triangleright \varepsilon = \varepsilon$, $c \triangleright \sigma = \sigma$ where $c \in Cnd$ and $\sigma \in Tr$ such that $\sigma \neq \varepsilon$. Let $c \triangleright {}^\# M = \{\{c \triangleright \sigma | \sigma \in O\} | O \in M\}$. We have:

$$[\![alt(c, D_1, D_2)]\!] = (c \triangleright {}^\#[\![D_1]\!]) \cup (\neg c \triangleright {}^\#[\![D_2]\!])$$

4. *Opt* fragments: The semantics of *opt* fragments is obtained similarly:

$$[\![opt(c, D)]\!] = (c \triangleright {}^\#[\![D]\!]) \cup {}^\# \{\{\varepsilon\}\}$$

5. *Par* fragments: Parallel interleaving produces a set of alternative obligations from O_1 and O_2. Define

$$O_1 \Uparrow O_2 = \{O | \forall \sigma_1 \in O_1, \forall \sigma_2 \in O_2 : \exists \sigma \in O.(\sigma \in \sigma_1 \| \sigma_2)\}.$$

The semantics of *par*(D_1, D_2) is defined:

$$[\![par(D_1, D_2)]\!] = \cup_{O_1 \in M_1, O_2 \in M_2} O_1 \Uparrow O_2$$

6. *Seq* fragments: The interaction operator *seq* combines traces from component sequence diagrams via weak sequencing. By $t_1 \frown t_2$ we denote the trace consisting of t_1 immediately followed by t_2. By $A \Theta t$ we denote the trace obtained from the trace t by removing all elements that are not in the set of elements A. Let $e \cdot l$ denote the set of events that may take place on the lifeline l; weak sequencing of trace sets is defined as follows [6]:

$$s_1 \succ s_2 = \{h \in H | \exists h_1 \in s_1, \exists h_2 \in s_2 : \forall l \in L, e \cdot l \Theta h = e \cdot l \Theta h_1 \frown e \cdot l \Theta h_2\}$$

The semantics of *seq* fragments is defined:

$$[\![seq(D_1, D_2)]\!] = \{t_1 \succ t_2 | t_1 \in [\![D_1]\!] \wedge t_2 \in [\![D_2]\!]\}$$

7. *Loop* fragments: We use a natural number n to indicate the counter in the *loop* fragment. The semantics of *loop* fragments is defined:

$$[\![loop(0, c, D)]\!] = \{\{\varepsilon\}\}$$
$$[\![loop(n+1, c, D)]\!] = c \triangleright {}^\# [\![seq(D, [\![loop(n, c, D)]\!])]\!]$$

Example Let m, n be signals, l_1, l_2 be lifelines, c be a condition, $f_1 = (!,m,l_1,l_2)$, $f_2 = (?,m,l_1,l_2)$, $f_3 = (!,n,l_1,l_2)$, and $f_4 = (?,n,l_1,l_2)$. Put $D_1 = strict(f_1,f_2)$ and $D_2 = strict(f_3,f_4)$. Then we have:

$$[\![D_1]\!] = \{\{f_1 f_2\}\}$$
$$[\![D_2]\!] = \{\{f_3 f_4\}\}$$
$$[\![strict(D_1,D_2)]\!] = \{\{f_1 f_2 f_3 f_4\}\}$$
$$[\![alt(c,D_1,D_2)]\!] = \{\{cf_1 f_2\},\{\neg c f_3 f_4\}\}$$
$$[\![opt(c,D_1)]\!] = \{\{cf_1 f_2\},\{\varepsilon\}\}$$
$$[\![par(D_1,D_2)]\!] = \{\{f_1 f_2 f_3 f_4\},\{f_1 f_3 f_2 f_4\},\{f_1 f_3 f_4 f_2\},\{f_3 f_4 f_1 f_2\},\{f_3 f_1 f_4 f_2\},\{f_3 f_1 f_2 f_4\}\}$$
$$[\![seq(D_1,D_2)]\!] = \{\{f_1 f_2 f_3 f_4\},\{f_1 f_3 f_2 f_4\}\}$$
$$[\![loop(2,c,D_1)]\!] = \{\{ccf_1 f_2 f_1 f_2\},\{ccf_1 f_1 f_2 f_2\},\{cf_1 cf_2 f_1 f_2\},\{cf_1 cf_1 f_2 f_2\},\{cf_1 f_2 cf_1 f_2\}\}$$

140.3 Theorem Proof Assistant: Coq

Coq is an interactive theorem proof assistant [3]. All data is represented with inductive data types in Coq. The Coq library contains elementary data types, including natural numbers, strings, and lists, among others. They are "predefined" types and equipped with many useful functions. As an example, Coq defines natural numbers and list of (parametric) type *A* using the following inductive definitions:

```
Inductive nat: =| O: nat | S: nat -> nat.
Inductive list A: =
| nil: list A
| cons: A -> list A -> list A.
```

Here, *O* and *S* are the *constructors* of type *nat*. *nil* and *cons* are the *constructors* of type list *A* for any type *A*. Type *nat* and *list* are inductive types, which means their elements are obtained as finite combinations of the constructors. For example, there is a number O (zero), and for every number *n*, there is another number *S n* (the successor of *n*). So we could write number 3 as *S (S (S O))*. In Coq, the empty list is denoted by *nil* while *a::b* is a notation for *cons a b*, so the list 1, 2, 3 can be written as *1::2::3::nil*. Inductive types are at the core of powerful programming and reasoning techniques.

The keyword *Definition* is used to define new types, for example:

```
Definition pair string: Set: = string * string
```

The *pair string* is the type of pairs of strings. The * is the pair type constructor.

We can use the keyword *Fixpoint* to define well-founded recursion for inductive types. For example:

```
Fixpoint fact (n: nat) {struct n} := match n with
| 0 => 1
| S p => n * fact p
end.
```

The *fact* is a simple function that calculates a number's factorial using recursion. The *struct n* annotation states that it is structurally recursive on its *n* parameter, and therefore guaranteed to terminate. The *match with* construct represents pattern-matching on the shape of the inductive type and handle all possible cases.

The keyword *Lemma* can be used to signify proofs. Coq enters interactive proving mode to assist us in building a proof. When proving properties of the inductive type, we could do an induction on the structure of the type.

140.4 Formalize the Trace Semantics in Coq

In this section, we show how to formalize the trace semantics of sequence diagrams in Coq. Due to space limitations, only important definitions and functions are shown.

The first step is to map sequence diagrams to the formal semantic models. We begin by defining the semantic model in Coq as follows:

```
Inductive kd := | Send : kd | Receive : kd. (* Define the kind of event *)
Definition sg := string. (*Define the signal *)
Definition lf := string. (*Define the lifeline *)
Definition evt := kd * sg * lf * lf. (*Define the event *)
(*Define the syntax of condition *)
Inductive id := Id : nat -> id.
Inductive cnd := | Bvar : id -> cnd | Btrue : cnd | Bfalse : cnd
| Bnot : cnd -> cnd
| Band : cnd -> cnd -> cnd | Bor : cnd -> cnd -> cnd | Bimp : cnd
-> cnd -> cnd.
Inductive tk := | ev : evt -> tk | cd : cnd -> tk. (* Inductive type for token *)
Definition tr := list tk. (* A trace is a list of tokens *)
Definition ob := set tr. (* An obligation is a set of traces *)
Definition mo := set ob. (* A model is a set of obligations *)
Definition state := id -> bool. (* States associate values to variables *)
```

Then we define the sequence diagrams as an inductive type:

```
Inductive sd := | Dtau : sd | De : evt -> sd | Dstrict : sd ->
sd -> sd
| Dopt : cnd -> sd -> sd | Dalt : cnd -> sd -> sd -> sd | Dpar :
sd -> sd -> sd
| DSeq : sd -> sd -> sd | Dloop : cnd -> nat -> sd -> sd.
```

We can now write the denotational functions for each inductive constructor of the sequence diagram *sd*. Because the semantic model is the power set of obligations, and the obligation is the power set of traces, the denotational function needs to compute on the three-tier model.

When defining the denotational functional for the par operator, the interleave operator ‖ fulfills the foundational computation. We implement the interleave operator as a recursive function *intlev*:

```
Fixpoint intlev (t1 : tr) {struct t1}: tr -> set tr :=match
t1 with
| nil => fun t2 => (t2 :: nil)
| x :: u => fix aux (t2 : tr) {struct t2} : (set tr) :=
  match t2 with
| nil => (t1 :: nil)
| y :: v => set_union tr_dec (addEvtOb x (intlev u t2))
(addEvtOb y (aux v))
end
end.
```

Here the *addEvtOb* is a function that adds a particular event before each trace of an obligation. The *aux* is a local recursive function.

When implementing the denotational function for the *seq* operator, the ordering of events on each lifelines and the ordering between transmission and receipt are all preserved, while all other ordering of events are arbitrary. The following two functions *filter* and *isWeak* are the most important ones:

```
Fixpoint filter (evs : set evt) (t : tr) {struct t} : tr :=
match t with
|nil => nil
|ev e :: tail => if (evtMem e evs) then (ev e) :: (filter evs
tail) else (filter evs tail)
|cd c :: tail => filter evs tail
end.
Fixpoint isWeak (st : state)(es : set evt)(h1 h2 t: tr)(ls :
set lf) : bool :=match ls with
|nil => true
|l :: tail => (andb (beq_tr st (filter (projLf l es) t) ((fil-
ter (projLf l es) h1) ++
(filter (projLf l es) h2))) (isWek st es h1 h2 t tail))
end.
```

The *evtMem* returns a Boolean value, checking whether an event belongs a set of events. The filtering function *filter* is used to filter away elements from a trace, corresponding to operator Θ. The function *projLf* is corresponding to operator $e \cdot l$ while the *isWeak* checks weather trace t is a member of weak sequencing of trace $h1$ and $h2$.

Finally, the denotational semantics is presented as a recursive function *interp* that associates a sequence diagram to the semantic model *mo* in the state *st*. We use the corresponding function to match the shape of the sequence diagram d for each case. The *interp* works as an interpreter, which can evaluate a sequence diagram in a known environment:

```
Fixpoint interp (st : state)(d : sd){struct d} : mo := match
d with
| Dtau => ((nil) :: nil) :: nil | De e => ((ev e :: nil) :: nil)
:: nil
| Dstrict d1 d2 => strMo (interp st d1) (interp st d2)
| Dopt c d => optMo st c (interp st d)
| Dalt c d1 d2 => altMo st c (interp st d1) (interp st d2)
| Dpar d1 d2 => parMo (interp st d1) (interp st d2)
| DSeq d1 d2 => seqMo st (getEvts (DSeq d1 d2))(getLfs (DSeq
d1 d2))(interp st d1) (interp st d2)
| Dloop c n d => loopMo st c n (interp st d)
end.
```

140.5 Verification of Sequence Diagrams

Based on the trace semantics presented in previous section, we could impose some restrictions on the syntactically correct sequence diagrams and verify them directly in Coq. As an example, if both the transmitter and the receiver lifelines of a signal are present in a diagram, then the corresponding receive event of any transmit event must be in the diagram, and vice versa. The function *checkEvt* in Coq is used to check whether an event e satisfies this restriction:

```
Definition chkEvt (d : sd)(e : evt) :=
if (andb (set_mem lf_dec (getTrLf e) (getLfs d)) (set_mem
lf_dec (getReLf e) (getLfs d))) then set_mem evt_dec
(revEvt e) (getEvts d) else true.
```

Here, the function *getTrLf* and *getReLf* are used to get the lifeline of transmitter and receiver from event, respectively. The function *revEvt* is used to get the corresponding receive event of a transmit event or the transmit event of a receive event.

Furthermore, we could verify the semantic constraint on traces. As an example, regarding all traces, if at a point in a trace we have a receive event of a signal, then up to that point we must have had at least as many transmits of that message as receives. We write the following function *chkTr* to check whether trace *t1* satisfies this restriction, in which the *countKd* is used to count the number of kind *k* in trace *t*:

```
Fixpoint chkTr (t1 : tr) : bool :=match t1 with
|nil => true
|ev (?, s, t, r) :: tail => if (ble_nat (countKd Receive
(filter ((!,s,t,r)::(?,s,t,r)::nil) t1))
(countKd Send (filter ((!,s,t,r)::(?,s,t,r)::nil) t1)))
then chkTr tail else false
|_ :: tail => chkTr tail
end.
```

Finally, the denotation function *interp* can also be used to prove properties of the semantics. For example:

```
Lemma eventNil : forall (st : state)(d : sd), getEvts d =
nil-> interp st d =(nil::nil)::nil.
```

Our definition of sequence diagrams as an inductive type enables the proof of this lemma to be straightforward. The proof of this lemma is a simple induction on the structure of *d* and use of the inversion tactic of Coq for each case.

140.6   Case Study

Figure 140.1 shows a UML sequence diagram *LoginDiag* that describes a common login interaction. A client sends its id and password to the server and gets a login reply (*loginsucc* or *loginfail*). If the client successfully login, it can send a command to the server. The diagram *LoginDiag* can be transformed to Coq representations as follows:

Definition LoginDiag := Dstrict (Dstrict (Dstrict (De sid)(De rid)) (Dstrict (De spwd)(De rpwd))) (Dalt (Bvar flag1) (Dstrict (Dstrict (De sloginsucc) (De rloginsucc)) (Dopt (Bvar flag2) (Dstrict (De scmd) (De rcmd)))) (Dstrict (De sloginfail) (De rloginfail))).

By executing the denotational function *interp* mentioned before, Coq prints the computation result which is the right interactive traces between object: *((sid::rid:: spwd::rpwd::flag1::sloginsucc::rloginsucc::flag2::scmd::rcmd::nil)::(sid::rid:: spwd::rpwd::(notflag1)::sloginfail::rloginfail::nil)::nil)::((sid::rid::spwd::rpwd:: flag1::sloginsucc::rloginsucc)::(sid::rid::spwd::rpwd::(notflag1)::sloginfail:: rloginfail::nil)::nil)::nil*.

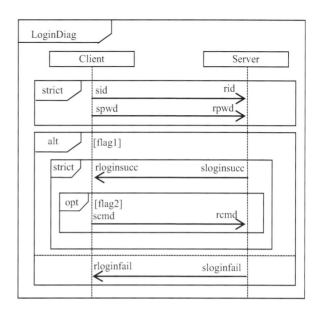

Fig. 140.1 An example of UML sequence diagram

Moreover, we could verify the syntax and semantic constraint by executing the verification functions in Coq. For example, when executing *Compute checkSd (LoginDiag)*, Coq returns *true*, which represents all the interactive traces from the diagram meet the restriction of *chkTr* defined before.

140.7 Related Work

The researchers survey the representative formal semantics proposed for UML sequence diagrams [7, 8]. However, the existing approaches mainly focus on getting a formal representation of sequence diagrams. The area is lack of a systematic methodology to analyze and verify the important properties.

The sequence diagrams have been translated to existing formalisms. The researchers use the formal specification language Z to present the semantics of sequence diagrams, well-formed rules, and consistent constraints and demonstrate that the work supports the validation [9]. The ASM semantic model for sequence diagram is proposed to describe the model's characteristics and improve the testing process of large systems [10]. A sequence diagram has been translated to a Petri net [11, 12]. These approaches can take advantage of the existing well-established tools to reason after translation. However, unlike the approach presented here, these techniques are not intuitive, and the interactive traces between objects are not clear.

There has been some work on formalize UML using a higher-order proof assistant. The HOL-OCL system is an interactive proof environment for UML (mainly focus on the class diagrams) and OCL specifications, supporting object-oriented modeling and reasoning [13]. A corrected version for the UML 2.3 metamodel relating to state machine is provided with the help of the Coq [14]. In contrast to these approaches, here we apply Coq to support model-based development by offering the capability of analyzing the relations on models or specifications.

140.8 Conclusion and Future Work

We propose to use the theorem proof assistant—Coq to formalize the syntax and semantics of UML sequence diagrams and verify the syntax and semantics constraint on models. The desired properties can be stated as lemmas and the proof of the lemma ensures that the semantics satisfy their corresponding properties. The semantics is well constructed and takes into account the most popular combined fragments. This work will provide increased reliability guarantee for the future design and implementation in model-based development. A tool which can transform sequence diagram models into the Coq representations has been developed.

In the future, we will extend the syntax and semantics to cover a larger subset of UML sequence diagrams. Another topic is to conduct and evaluate larger case studies.

Acknowledgement This work is supported by National Natural Science Foundation of China (No.61070226 and No.61003181).

References

1. OMG. (2005). Unified modeling language: superstructure. Version 2.0.
2. Coq, [EB/OJ]. Retrieved from http://coq.inria.fr/.
3. Yves, B., & Castéran, P. (2004). *Interactive theorem proving and program development—Coq'Art: The calculus of inductive constructions (pp. 1–496)*. Berlin: Springer.
4. Lu, L., & Kim, D. (2011). Required behavior of sequence diagrams: Semantics and refinement. In: *Proceedings of the 2011 16th IEEE international conference on engineering of complex computer systems (ICECCS)* (pp. 127–136). Washington, DC: IEEE Computer Society.
5. Störrle, H. (2003). Semantics of interactions in UML 2.0. In: *Proceedings of the 2003 I.E. symposium on human centric computing languages and environments* (pp. 129–136). Washington, DC: IEEE Computer Society.
6. Lund, M. S. (2008). *Operational analysis of sequence diagram specifications*. Ph.D. Thesis, Faculty of Mathematics and Natural Sciences, University of Oslo.
7. Micskei, Z., & Waeselynck, H. (2011). The many meanings of UML 2 sequence diagrams: A survey. *Software and Systems Modeling., 10*(4), 489–514.
8. Lund, M. S., Refsdal, A., & Stølen, K. (2010). Semantics of UML models for dynamic behavior: A survey of different approaches. In *Proceedings of the 2007 international Dagstuhl conference on model-based engineering of embedded real-time systems* (pp. 77–103). Berlin: Springer.

9. Jingfeng, L., Yan, L., & Ping, C. (2003). The Z specification-based method for the semantic analysis of UML sequence diagrams. *Journal of Xidian University (Natural Science), 30*(4), 519–524.
10. Xiang, Z., & Zhi-qing, S. (2009). ASM semantic modeling and checking for sequence diagram. In *Proceedings of the fifth international conference on natural computation* (pp. 527–530). Washington, DC: IEEE Computer Society.
11. Eichner, C., Fleischhack, H., Schrimpf, U., & Stehno, C. (2005). Compositional semantics for UML 2.0 sequence diagrams using Petri Nets. In *Proceedings of the 12th international conference on model driven* (pp. 133–148). Berlin: Springer.
12. Fernandes, J. M., Tjell, S., Jorgensen, J. B., & Ribeiro, O. (2007). Designing tool support for translating use cases and UML 2.0 sequence diagrams into a Coloured Petri Net. In *Proceedings of the sixth international workshop on scenarios and state machines* (p. 2). Washington, DC: IEEE Computer Society.
13. Brucker, A. D., & Wolff, B. (2008). HOL-OCL—a formal proof environment for UML/OCL. In *Proceedings of the 11th international conference on fundamental approaches to software engineering* (pp. 97–100). Berlin: Springer.
14. Barbier, F., & Ballagny, C. (2010). Proved metamodels as backbone for software adaptation. In *Proceedings of the 12th international symposium on high-assurance systems engineering* (pp. 114–121). Washington, DC: IEEE Computer Society.

Chapter 141
Quantitative Verification of the Bounded Retransmission Protocol

Xu Guo, Ming Xu, and Zongyuan Yang

Abstract In order to verify the reliability of the bounded retransmission protocol, probabilistic model checking technology is used in this paper. The integer semantics approach is introduced, which allows working directly at the level of the original probabilistic timed automaton (PTA). In such a method, clocks are viewed as counters storing nonnegative integer values, which increase as time passes. The PTA modeling the system can then be seen as a discrete-time Markov chain. Based on this fact, the protocol is modeled directly with DTMC. Properties are described in probabilistic computation tree logic. By making an analysis of the quantitative properties of the protocol, a threshold is obtained. Experimental result shows that no matter how many chunks to be transmitted, if the maximum retransmitted time is greater than or equal to 3, the protocol can be considered reliable. Method in this paper can not only verify the correctness of a system but also make analysis of nonfunctional indices of a system such as reliability or performance.

141.1 Introduction

The bounded retransmission protocol (BRP) is a data link layer protocol. It has been designed to transmit messages which are divided into small frames over unreliable channels. The protocol does not rely on fairness of data transmission channels, i.e., repeated transmission of a frame does not guarantee its eventual arrival. The number of retransmission attempts is also limited. So the pressure for reliability of the protocol involved poses an important challenge to verification techniques.

X. Guo (✉)
East China Normal University, Shanghai 200241, China

Shanghai Dianji University, Shanghai 200092, China
e-mail: neuguox@126.com

M. Xu • Z. Yang
East China Normal University, Shanghai 200241, China

This paper applies probabilistic model checking techniques to the quantitative verification of the protocol. Following D'Argenio's work [1, 2], we model the protocol in the framework of discrete-time Markov chain (DTMC) and then use PRISM [3, 4], a probabilistic model checking tool, to analyze the resulting model of the BRP. We establish quantitative properties of the protocol in order to verify its reliability, such as the minimal correctness of the protocol and the minimum number of retransmissions that satisfies our probabilistic requirements.

Related work. D'Argenio investigates what extent real-time aspects are important to guarantee the protocol correctness by using model checker SPIN and UPPAAL [2]. Ravn uses labeled transition systems to specify behavior of BRP and compositional reachability of the protocol based on its software architecture [5]. Forejt presents a framework for analyzing multiple quantitative objectives of system that exhibit both nondeterministic and stochastic behavior [6]. Kwiatkowska captures some quantitative properties of zero-conf protocol using probabilistic computation tree logic (PCTL) [7].

Organization of the paper. After a brief description of the BRP and related model technology (Sect. 141.2), the paper proceeds by giving the modeling of BRP in the framework of DTMC (Sect. 141.3). In the rest of the paper (Sects. 141.4 and 141.5), we present the PRISM tool dedicated to the model checking of probabilistic systems with a brief description of its theoretical framework and the results of experiments.

141.2 Preliminaries

141.2.1 Sketch of the Protocol

The protocol control procedures will be described by means of a sender S, a receiver R, and two lossy communication channels K and L. The sender sends elements of a file one by one over K to the receiver. After sending the frame, the sender waits for an acknowledgement or for a timeout. In case of acknowledgement, if it corresponds to the last chunk, the sending client is informed of correct transmission (signal OK); otherwise the next element of the file is sent. If a timeout occurs, the frame is resent, or the transmission of the file is broken off.

The receiver waits for a frame to arrive. This frame is delivered at the receiving client informing whether it is the first (FST), an intermediate (INC), or the last one (OK). Afterwards, an acknowledgement is sent over L to the sender. Then the receiver simply waits for more frames to arrive.

141.2.2 Basic Models

We introduce two basic models that will be used later in the paper.

Definition 1. A discrete-time Markov chain (DTMC) is a tuple D = (S, s, P, L) where S is a finite set of states, s is the initial state, and P:S×S→[0,1] is the transition probability matrix; this gives the probability P(s; s') that a transition will take place from state s to state s'. And L : S →2^{AP} is function labeling states with atomic propositions.

Classically, analysis of DTMCs often focuses on transient or steady-state behavior, i.e., the probability of being in each state of the chain at a particular instant in time or in the long run, respectively. Probabilistic model checking adds to this the ability to reason about path-based properties, which can be used to specify constraints on the probability that certain desired behaviors are observed. Properties are then expressed using temporal logic. For DTMCs, specifications can be written in PCTL, a probabilistic extension of the temporal logic CTL [7].

Definition 2. The syntax of PCTL is given by

$\Phi ::= $ true|a |$\Phi \wedge \Phi$|$\neg \Phi$|$P_{\sim p}[\Psi]$
$\Psi ::= X\Phi|\Phi \cup^{\leq k}\Phi|\Phi \cup \Phi$

where a is an atomic proposition, p∈{0,1} is a probability bound, $\sim \in \{<,>,\leq,\geq\}$, k∈N.

The key operator in PCTL is $P_{\sim p}[\Phi]$ which means that the probability of a path formula Φ being true in a state satisfies the bound ~ p.

141.2.3 Probabilistic Model Checking and PRISM

Probabilistic model checking is based on the construction of a probabilistic model from a precise, high-level description of a system's behavior. The model is then analyzed against one or more formally specified quantitative properties, usually expressed in temporal logic. These properties not only contain the correctness of the system but also a wide range of indices such as reliability or performance [8].

PRISM is a probabilistic model checker which provides support for analysis of DTMC and performs verification of PCTL formulae for DTMC.

141.3 Modeling

We focus on a restricted set of reachable properties. For verifying probabilistic reachable properties, we must derive an equivalent finite-state (probabilistic) system. In the non-probabilistic framework, possible methods of reduction are region

Fig. 141.1 Schematic view of the BRP

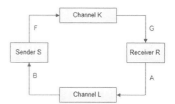

Fig. 141.2 Model of sender

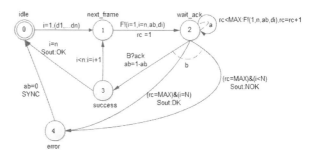

equivalence [9], forward exploration [10], and integer semantics [11]. These methods have been extended to the probabilistic framework. The first two methods require in practice the preliminary construction of an abstraction of the original probabilistic timed automaton [3]. For the sake of simplicity, we choose the third approach (integer semantics), which allows us to work directly at the level of the original probabilistic timed automaton. In such a method, clocks are viewed as counters storing nonnegative integer values, which increase as time passes. The PTA modeling the system can then be seen as a finite-state DTMC.

The model of the BRP protocol consists of four components operating in parallel, namely, sender, receiver, channel K, and channel L (see Fig. 141.1).

141.3.1 Sender

The sender S (see Fig. 141.2) has three system variables: $ab \in \{0,1\}$, indicating the alternating bit that accompanies the next chunk to be sent; $i, 0 \leq i \leq n$, indicating the subscript of the chunk currently being processed by S; and rc, $0 \leq rc \leq MAX$, indicating the number of attempts undertaken by S to transmit a certain chunk.

On receipt of a new file, S sets i to 1. Going from state n_frame to wait_ack, chunk di is transmitted with corresponding information and rc is reset. In state wait_ack, there are several probabilities: in case the maximum number of retransmissions has been reached, S moves to an error state while resetting x and emitting an DK or NOK indication depending on whether di is the last chunk or not.

Fig. 141.3 Model of receiver

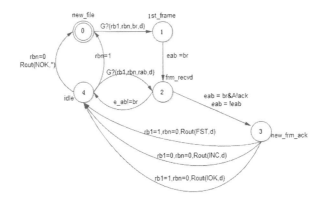

If rc<MAX, either an ACK is received and S moves to the success state while alternating ab or a retransmission is initiated. If the last chunk has been acknowledged, S moves from state success to state idle indicating the successful transmission of the file.

141.3.2 Receiver

The receiver R (see Fig. 141.3) has one system variable: $e\_ab \in \{0,1\}$ indicating the alternating bit. In the state new_file, R is waiting for the first chunk of the new file to arrive. On receiving the chunk, e_ab is set to the just received alternating bit and R enters the state frame_received. If e_ab agrees with the just received alternating bit (which, due to the former assignment to ab, is always the case for the first chunk), then an ACK is sent via A and e_ab is toggled. R is now in the state idle and waits for the next frame to arrive. It moves to the states frame_received if such a frame arrives in time. And the above described procedure repeats. If timeouts occur, in case R has not just received the last chunk of a file, then an indication NOK is sent, and in case R has just received the last chunk, no failure is reported.

141.3.3 Channels

The channels K and L are lossy channels. In this model we assume that a frame is lost with probability 0.02, and acknowledgement is lost with probability 0.01.

141.4 Experiments

The experiment we perform is to try to find the minimum number of retransmissions that satisfies our probabilistic requirements for these properties when the transmitted file length N is an input parameter.

Properties 1–3 are concerned with transmissions that the sender does not consider successful, while property 4 considers an attempt for transmission with no reaction at the receiver side. The four main properties we have verified are as follows:

Property 1 represents the sender does not report a successful transmission. This is written (in PCTL): P=? [true U s=4].

Figure 141.4 shows the probability that satisfies property 1 for different values of the transmitted file length (N) and the maximal retransmission times (MAX). The value of parameter N equals to 16, 32, 48, 64 and that of MAX equals to 2, 3, 4, 5, respectively. When MAX is greater than or equal to 3, no matter how many chunks to be transmitted, a failed transmission is rather slim, for the probability is less than 0.00025. When MAX is 2, as N increases, the probability of a failed transmission increases fast. So if MAX is greater or equal to 3, we can consider that the transmission is reliable.

Property 2 represents the sender that reports an uncertainty on the success of the transmission. The property written in PCTL is P=? [true U s=4 & srep=2].

Figure 141.5 shows the probability that satisfies property 2 for different values of N and MAX. When MAX is greater than or equal to 3, we can ignore this phenomenon that the sender receives a DK instead of OK acknowledgement after a success transmission for the probability nearly equals to 0. So if MAX is greater than or equal to 3, we can consider that the transmission is reliable.

Property 3 represents the sender that reports a certain unsuccessful transmission after transmitting half a file. Written in PCTL is P=? [true U s=4 & srep=1 & i>N/2].

Fig. 141.4 PRISM measurement for property 1

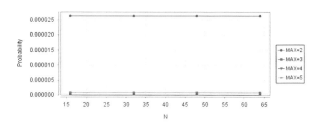

Fig. 141.5 PRISM measurement for property 2

Property 4 represents the receiver does not receive any chunk of a file. Written in PCTL is P=? [true U !(srep=0) & !recv].

From the result, we can see that if the value of MAX is greater than or equal to 3, the probability of the sender reports a certain unsuccessful transmission after transmitting half a file is less than 0.00025. And the probability that the receiver does not receive any chunk of a file is less than 0.00001. Due to space constraints, experiment results of properties 3 and 4 are emitted.

So we can conclude that the protocol is reliable no matter how many chunks to be transmitted; if we set a proper value of maximal retransmission time, the threshold is 3.

141.5 Conclusion

In this paper, we applied probabilistic model checking technology to the verification of quantitative properties of BRP.

A crucial problem in our work is the assumption of 0 delay in the transmission channels K and L. This reduces the size of region space significantly. As a result of these zero delays, the receiver may detect a transmission failure, while the sender has not aborted the transmission.

Our work provides a fully automatic verification of the parameterized version of the BRP. With transmitted file length N and the maximal retransmission counter MAX as parameters, PRISM can measure the properties. We can extend the reachable properties of the protocol with probability. Such properties are expressed in terms of a probability bound, for example, the property the sender reports an uncertainty on the success of the transmission can be expressed as a probability bound.

Future work can address the application of PCTL* to describe the properties of one model. As PCTL* subsumes PCTL and LTL, it is a more expressive logic. Application of PRISM to model and analysis mobile ad hoc networks is underway.

Acknowledgements This work is supported by National Natural Science Foundation of China (No. 61070226).

References

1. D'Argenio, P. R., Jeannet, B., Jensen, H. E., & Larsen, K. G. (2001). *Reachability analysis of probabilistic systems by successive refinements. Process algebra and probabilistic methods. Performance modelling and verification* (pp. 39–56). Berlin: Springer.
2. D'Argenio, P. R., Katoen, J. P., Ruys, T. C., & Tretmans, G. J. (1997). *The bounded retransmission protocol must be on time!* (pp. 416–431). Berlin: Springer.
3. Kwiatkowska, M., Norman, G., & Parker, D. (2011). *PRISM 4.0: Verification of probabilistic real-time systems. Computer aided verification* (pp. 585–591). Berlin: Springer.

4. Hinton, A., Kwiatkowska, M., Norman, G., & Parker, D. (2006). *PRISM: A tool for automatic verification of probabilistic systems. Tools and algorithms for the construction and analysis of systems* (pp. 441–444). Berlin: Springer.
5. Ravn, A. P., Srba, J., & Vighio, S. (2011). *Modelling and verification of web services business activity protocol. Tools and algorithms for the construction and analysis of systems* (pp. 357–371). Berlin: Springer.
6. Forejt, V., Kwiatkowska, M., Norman, G., Parker, D., & Qu, H. (2011). *Quantitative multi-objective verification for probabilistic systems. Tools and algorithms for the construction and analysis of systems* (pp. 112–127). Berlin: Springer.
7. Kwiatkowska, M., Norman, G., & Parker, D. (2010). Advances and challenges of probabilistic model checking. In *48th Annual Allerton Conference on Communication, Control, and Computing (Allerton), IEEE*, Los Alamitos (pp. 1691–1698).
8. Filieri, A., Ghezzi, C., & Tamburrelli, G. (2011). Run-time efficient probabilistic model checking. In *Proceedings of the 33rd International Conference on Software Engineering, ACM*, New York (pp. 341–350).
9. Aceto, L., Ingólfsdóttir, A., & Larsen, K. G. (2007). *Reactive systems: Modelling, specification and verification* (pp. 121–136). Cambridge: Cambridge University Press.
10. David, A., Illum, J., Larsen, K. G., & Skou, A. (2010). Model-based framework for schedulability analysis using uppaal 4.1. *Model-Based Design for Embedded Systems, 7*(4), 93–119.
11. Hartmanns, A., & Hermanns, H. (2009). A Modest approach to checking probabilistic timed automata. In *Sixth International Conference on the Quantitative Evaluation of Systems, 2009. QEST'09, IEEE*, Los Alamitos (pp. 187–196).

Chapter 142
A Cluster-Based and Range-Free Multidimensional Scaling-MAP Localization Scheme in WSN

Ke Xu, Yuhua Liu, Cui Xu, and Kaihua Xu

Abstract As using traditional MDS-MAP algorithm to locate nodes' position in irregular WSN leads to low positional accuracy, based on this fact, this chapter presents an improved algorithm named MDS-MAP(C, RF). The algorithm can effectively divide a WSN into several clusters, and each cluster locates all nodes' position in it and forms a local position map. After all clusters get local position maps, the algorithm merges all the local position maps together using the information of inter-cluster nodes. Simulations demonstrate the proposed algorithm yields smaller accuracy error in irregular WSN.

142.1 Introduction

A sensor network is composed of a large number of sensor nodes, which are densely deployed either inside the phenomenon or very close to it. The goal of a sensor network is to perceive, collect, and process the information of specific objects and send the information to observers [1]. The localization refers to computing all nodes' position based on a few nodes' position information in a sensor network. Localization plays a key role within the application of WSN. Localization is a hot research topic [2–4]. Localization techniques can be classified in a different way [4]. MDS-MAP algorithm is first proposed in the year of 2003 [5]; the noted advantage of MDS-MAP is that with a few anchor node information, it can compute

K. Xu • Y. Liu (✉) • C. Xu
School of Computer, Central China Normal University, Wuhan 430079, China
e-mail: yhliu@mail.ccnu.edu.cn

K. Xu
College of Physical Science and Technology, Central China Normal University, Wuhan 430079, China

the position of each node, and its localization accuracy is higher than most known localization technology. The main disadvantage of MDS-MAP is that it does not work well in irregular sensor networks, and the position error is high.

This chapter proposes the MDS-MAP(C, RF) algorithm. The core idea is dividing a given network into several clusters; each cluster computes its local position map through MDS-MAP algorithm, respectively, then merges all the local position maps together to get a global map.

142.2 MDS-MAP

142.2.1 MDS

MDS (Multidimensional Scaling) is a set of data analysis technology which can transform the given data into geometry model, thus problems can be visually solved. Torgerson had firstly given the terminology MDS based on the work of Richardson and proposed the first MDS method [6].

142.2.2 Procedure of MDS-MAP

MDS-MAP consists of three steps [5]:

Step 1: Calculate the shortest distances between nodes in WSN. The time complexity of this step is $O(n^3)$, where n is the number of nodes.

Step 2: Calculate the first r maximum eigenvalues of r-dimensional space to construct the relative location map of nodes. The time complexity of this step is $O(n^3)$, where n is the number of nodes.

Step 3: Based on the location information of anchor nodes (it needs at least r + 1 nodes in r-dimensional space), the algorithm transforms the relative location map to absolute location map. The time complexity of this step is $O(m^3)$, where m is the number of anchor nodes.

142.3 Improved MDS-MAP

Traditional MDS-MAP could not work well in irregular WSN because the location error is quite large.

The procedure of MDS-MAP(C, RF) is shown below:

Step 1: Divide a wireless sensor network into several clusters; the method of dividing the cluster is k-hop clustering.

Step 2: Use traditional MDS-MAP algorithm to build the location map of each cluster which is produced in step 1.

Step 3: Merge the location map of each cluster together to form a global location map.

142.3.1 Clustering

142.3.1.1 Selection of Cluster Heads

At the initial stage, all sensor nodes are randomly placed in a large sensor field and broadcast Hello messages. Then each node discovers its node degree (the number of nodes it can reach by one hop), the IDs of all neighboring nodes, and distances to these neighboring nodes. A node with the lowest ID becomes an initiator of cluster generation step and starts a timer $T_1(i)$; formula (142.1) is the computation of $T_1(i)$:

$$T_1(i) = \alpha_{deg} \max\left(0, 1 - \frac{\deg(i)}{\theta_{deg}(i)}\right) + t_{BCD} \qquad (142.1)$$

where t_{BCD} is a small broadcasting random delay, $\deg(i)$ is the connectivity degree of node i, $\theta_{deg}(i)$ denotes the largest degree of neighbor nodes, and α_{deg} is a weight factor for the degree.

The initiator, node i, becomes a cluster head node when $T_1(i)$ is time out; name the cluster as Cluster-i. However, if it receives a cluster head declaration message from other nodes before $T_1(i)$ expires, the node will stop the timer and become a member of the cluster.

The cluster head floods declaration messages to nodes within 2k hops. Any node within k hops to Cluster-i head nodes becomes a member of Cluster-i; any node located between (k + 1) hops and 2k hops to node i becomes a candidate of a new cluster head node. Each candidate starts a timer $T_2(j)$ (j is the candidate node's ID):

$$T_2(j)^2 = \alpha_{deg} \max\left(0, 1 - \frac{\deg(j)}{\theta_{deg}(j)}\right) + \alpha_{dist}\left(1 - \frac{chdist(j)}{2k}\right) + t_{BCD} \qquad (142.2)$$

where α_{dist} is a weight factor for the distance and $chdist(j)$ denotes the number of hops between current node i and the cluster head node($k + 1 \leq chdist(j) \leq 2k$). A node with an expired timer $T_2(j)$ becomes a new cluster head node and floods cluster head declaration messages declare-j, continuing the cluster generation process. Using random delay values t_{BCD} prevents a broadcast storm happen.

Figures 142.1 and 142.2 give an example of k-hop clustering (k = 2). In Fig. 142.2, nodes 2, 3, 4, 5, 8, 12, and 13 are members of Cluster = 1 (node 1 is its cluster head). Nodes 5, 6, 7, 9, 10, 11, 13, 14, 15, and 16 are members of Cluster-11, where node 11 is its cluster head.

Fig. 142.1 Selection of cluster heads

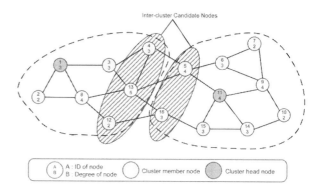

Fig. 142.2 Result of k-hop clustering

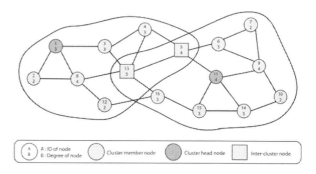

142.3.1.2 Inter-cluster Nodes

There exists a special kind of nodes between two adjacent clusters: they receive more than one declares messages. As shown in Fig. 142.2, nodes 4, 12, 13, 5, and 16 receive declare from Cluster-1 and Cluster-11. These nodes are called inter-cluster candidate nodes. Nodes 4, 12, and 13 are inter-cluster candidate nodes of Cluster-1, and nodes 5 and 16 are inter-cluster candidate nodes of Cluster-11, respectively. Each cluster head node randomly selects one of the inter-cluster candidate nodes as an inter-cluster node to the neighboring clusters. These nodes will be used in the phase of merging local map together to construct a global map. As shown in Fig. 142.2, nodes 5, 13, and 16 are the inter-cluster nodes of Cluster-1 and Cluster-11.

142.3.2 Building Cluster Location Map

The step of this process is shown below:

Step 1: After k-hop clustering is completed, each cluster head node calculates distance based on RSSI and the IDs of neighboring nodes [7]. Using the distance information which is expressed in a distance matrix and the shortest path algorithm, Dijkstra or Floyd, the cluster head node constructs a shortest distance matrix **D**.

Step 2: Using this shortest distance matrix **D**, the MDS-MAP algorithm produces a relative location map within the cluster.

Step 3: Refine cluster location map.

Step 1 and step 2 are clearly described [8]. This chapter only gives the detail of step 3.

Define $NB(i)$ as a set of nodes in Cluster-i; the cluster head node is node i, $v_r, v_s \in NB(i)$, $v_r(ID) \leq v_s(ID)$, and $\varepsilon(> 0)$ is a threshold which has a small value for the purpose of high accuracy. The process of refinement is illustrated below:

1. Define $\vec{X}_{ir} = (x_{r1}, x_{r2}, \ldots, x_{rm})$ to indicate row vector of node r within Cluster-i in m-dimensional space; set any values to nodes' initial coordinates in Cluster-i.
2. Calculate Euclidean distance between nodes by formula (142.3):

$$d_{rs} = \sqrt{\left(\vec{X}_r - \vec{X}_s\right)\left(\vec{X}_r - \vec{X}_s\right)^T}, V_r(ID) \leq V_s(ID) \quad (142.3)$$

3. Use PAV technique to compute the difference $\hat{d}_{rs}$ of d_{rs}.
4. Compute Stress 1:

$$Stress1 = \sqrt{\frac{\sum (d_{rs} - \hat{d}_{rs})^2}{\sum d_{rs}^2}} \quad (142.4)$$

If $Stress1 < \varepsilon$, end the process, and $\mathbf{X}_{cluster-i} = \left[\vec{X}_{i1}, \vec{X}_{i2}, \cdots, \vec{X}_{in}\right]^T$ is the coordinate matrix of Cluster-i, else go to step 5.

5. Compute the new coordinate of each node using formula (5):

$$\vec{X}_r^* = \vec{X}_r + \frac{\alpha}{M-1} \sum_{V_s \in NB(i)} \left(1 - \hat{d}_{rs}/d_{rs}\right)\left(\vec{X}_s - \vec{X}_r\right) \quad (142.5)$$

where α is step factor and M depicts the number of $NB(i)$.
6. Update **X**; go to step 2.

142.3.3 Merging Cluster Location Map

If $NB(i) \cap NB(j)$ have more than three nodes, then using formula (6) can transform the coordinates of nodes in Cluster-j to coordinate in Cluster-i:

$$\mathbf{X}_{cluster-i} = sR(\mathbf{X}_{cluster-j}) + \mathbf{X}_0 \qquad (142.6)$$

where s is zoom factor, $R(\cdot)$ depicts rotation transformation, $\mathbf{X}_0$ indicates translation transformation; the technique of compute s, $R(\cdot)$, $\mathbf{X}_0$ is shown below.

Suppose $\mathbf{S}_{cluster-i}$, $\mathbf{S}_{cluster-j}$ are coordinate matrices of nodes of $NB(i) \cap NB(j)$ in Cluster-i and Cluster-j. The row vector of $\mathbf{S}_{cluster-i}$, $\mathbf{S}_{cluster-j}$ is depicted by $\vec{S}_{cluster-i,i}$, $\vec{S}_{cluster-j,i}$. Matrix $\mathbf{D}$ is the nodes' coordinate matrix of $NB(j) - NB(i) \cap NB(j)$ in Cluster-i, $|NB(i) \cap NB(j)| = n$, $(n \geq 3)$. Expand the nodes' coordinate in $\mathbf{S}_{cluster-i}$ and $\mathbf{S}_{cluter-j}$ to 3-dimensional space, and set 1 to the third coordinate value of each node. The central points of $\overline{S}_{cluster-i}$, $\overline{S}_{cluster-j}$ are computed through formula (142.7):

$$\overline{S}_{cluster-i} = \frac{1}{n}\sum_{k=1}^{n} \vec{X}_{cluster-i,k}, \quad \overline{S}_{cluster-j} = \frac{1}{n}\sum_{k=1}^{n} \vec{X}_{cluster-j,k} \qquad (142.7)$$

Assume $\vec{S}'_{cluster-i,k} = \vec{S}_{cluster-i,k} - \overline{S}_{cluster-i}$, $\vec{S}'_{cluster-j,k} = \vec{S}_{cluster-j,k} - \overline{S}_{cluster-j}$; the zoom factor s in formula (142.6) is calculated through formula (142.8):

$$s = \sqrt{\sum_{k=1}^{n} \left\| \vec{S}'_{cluster-i,k} \right\|^2 \Big/ \sum_{k=1}^{n} \left\| \vec{S}'_{cluster-j,k} \right\|^2} \qquad (142.8)$$

The covariance matrix $\mathbf{C}$ of $\vec{S}'_{cluster-j}$, $\vec{S}'^T_{cluster-i}$ can be computed by using formula (142.9).

$$\mathbf{C} = \sum_{k=1}^{n} \vec{S}'_{cluster-j,k} \vec{S}'^T_{cluster-i,k} = \begin{bmatrix} C_{xx} & C_{xy} & C_{xz} \\ C_{yx} & C_{yy} & C_{yz} \\ C_{zx} & C_{zy} & C_{zz} \end{bmatrix} \qquad (142.9)$$

Define matrix $\mathbf{U}$ as formula (142.10):

$$\mathbf{U} = \begin{bmatrix} C_{xx}+C_{yy}+C_{zz} & C_{yz}-C_{zy} & C_{zx}-C_{xz} & C_{xy}-C_{yx} \\ C_{yz}-C_{zy} & C_{xx}-C_{yy}-C_{zz} & C_{xy}+C_{yx} & C_{zx}+C_{xz} \\ C_{zx}-C_{xz} & C_{xy}+C_{yx} & -C_{xx}+C_{yy}-C_{zz} & C_{yz}+C_{zy} \\ C_{xy}-C_{yx} & C_{zx}+C_{xz} & C_{yz}+C_{zy} & -C_{xx}-C_{yy}+C_{zz} \end{bmatrix}$$

$$(142.10)$$

Compute $\vec{\lambda}_m$ which is composed by eigenvalue of matrix $\mathbf{U}$: $\vec{\lambda}_m = (\lambda_0, \lambda_1, \lambda_2, \lambda_3)$
Compute $R(\cdot)$ of formula (142.6) using formula (142.11):

$$\mathbf{R} = \begin{bmatrix} \lambda_0^2 + \lambda_1^2 - \lambda_2^2 - \lambda_3^2 & 2(\lambda_1\lambda_2 - \lambda_0\lambda_3) & 2(\lambda_1\lambda_3 + \lambda_0\lambda_2) \\ 2(\lambda_2\lambda_1 + \lambda_0\lambda_3) & \lambda_0^2 - \lambda_1^2 + \lambda_2^2 - \lambda_3^2 & 2(\lambda_2\lambda_3 - \lambda_0\lambda_1) \\ 2(\lambda_3\lambda_1 - \lambda_0\lambda_2) & 2(\lambda_3\lambda_2 + \lambda_0\lambda_1) & \lambda_0^2 - \lambda_1^2 - \lambda_2^2 + \lambda_3^2 \end{bmatrix} \quad (142.11)$$

Compute X_0 through formula (142.12):

$$\vec{X}_0 = \overline{S}_{cluster-i} - \overline{S}_{cluster-j}\mathbf{R}^T \quad (142.12)$$

Using formula (142.6), (142.8), (142.11), and (142.12), the matrix $\mathbf{D}$ (the nodes' coordinate matrix of $NB(j) - NB(i) \cap NB(j)$) can be transformed to the coordinate system of Cluster-i; the technique is shown as formula (142.13).

$$\mathbf{F} = \mathbf{D}\mathbf{R}^T + \mathbf{I}X_0 \quad (142.13)$$

where $\mathbf{I}$ is an all-1 matrix which has an equal row number as matrix $\mathbf{D}$.

Using this technique, all clusters' location map can be merged to one location map.

142.4 Experiments and Simulations

142.4.1 Simulations of Two Types of WSN

Suppose the transmission radius of sensor node is 1.5r, all nodes are deployed in a 10r × 10r area. The simulations have two types. Type 1, 170 nodes are randomly distributed within a sensing field that is C shaped in the middle. Type 2, 170 nodes are randomly distributed in an environment with a horseshoe-shaped hole in the middle of the sensing field.

Figures 142.3 and 142.4 show the results of two types of simulations. Figure 142.3 depicts localization results where a sensor topology is configured in a C shape. Figure 142.3a indicates large errors are produced by conventional MDS-MAP in such environments. The average error of conventional MDS-MAP is 2.4r. Figure 142.3b is the localization result of MDS-MAP(C,RF); its average error is quite small compared to (a), it is 0.35r. MDS-MAP(C,RF) improves localization accuracy up to 585 % in a C-shaped sensor topology compared to the conventional MDS-MAP.

Figure 142.4 shows localization results in a sensor topology with a horseshoe-shaped hole. The localization errors are 1.79r for conventional MDS-MAP and

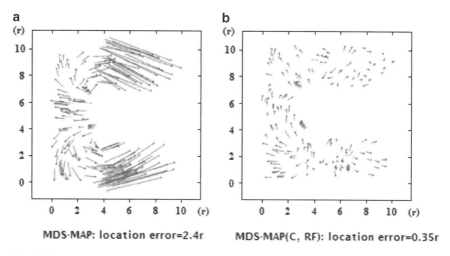

Fig. 142.3 Localization results where a sensor topology is configured in a C shape

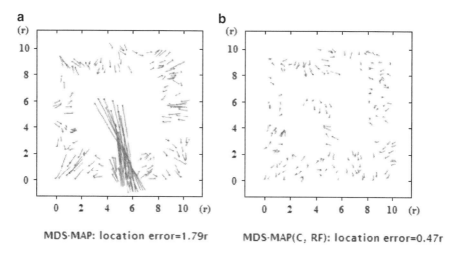

Fig. 142.4 Localization results in a sensor topology with a horseshoe-shaped hole

0.47r for MDS-MAP(C,RF). MDS-MAP(C,RF) improves localization accuracy up to 280 % in a horseshoe-shaped sensor topology compared to the conventional MDS-MAP.

From the simulation results, we conclude that the MDS-MAP(C,RF) produces better result than conventional MDS-MAP in irregular WSNs.

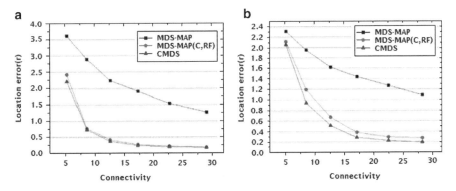

Fig. 142.5 Localization error with connectivity in C-shaped WSNs (**a**), horseshoe-shaped WSNs (**b**)

142.4.2 Localization Error with Connectivity

Figure 142.5 shows the localization errors with connectivity in two types of environments. Figure 142.5a shows that the localization errors of the HMDS and MDS-MAP(C,RF) are quite smaller than the conventional MDS-MAP under the different connectivity in a sensor network within a C-shaped hole. In sensor networks where a horseshoe-shaped hole is in the middle, as is shown in Fig. 142.5b, the localization errors of HMDS and MDS-MAP(C,RF) are not hugely different, but conventional MDS-MAP's localization error is quite larger than the former ones'. Figure 142.5b illustrates that MDS-MAP(C, RF) is better than conventional MDS-MAP. MDS-MAP(C, RF) improves a bit compared to HMDS.

It can be concluded from the simulation results that MDS-MAP(C, RF) works better in irregular WSNs and can gain a high precision localization result.

142.5 Conclusion

This chapter proposes MDS-MAP(C, RF) algorithm based on conventional MDS-MAP. Simulation results indicate that MDS-MAP(C, RF) is better than MDS-MAP for it can improve localization accuracy. Since MDS-MAP(C, RF) is a cluster-based technique which can reduce the burden of center nodes, it can extend the life span of WSN. The next stage work is focused on the effect to WSN's life span of MDS-MAP(C, RF).

References

1. Mao, G., Fidan, B., & Anderson, B. (2007). Wireless sensor network localization techniques. *Computer Networks., 51*(10), 2529–2553.
2. Pal, A. (2010). Localization algorithm in wireless sensor networks: Current approaches and future challenges. *Network Protocols and Algorithms., 2*(1), 45–73.
3. Stefano, G. D., & Petricola, A. (2008). A distributed AOA based localization algorithm for wireless sensor networks. *Journal of Computers., 3*(4), 1–8.
4. Niewiadomska-szynkiewicz, E. (2012). Localization in wireless sensor networks: Classification and evaluation of techniques. *International Journal of Applied Mathematics and Computer Science., 22*(2), 281–297.
5. Shang, Y., Ruml, W., Zhang, Y., & Fromherz, M. (2003). Localization from mere connectivity. In Proceeding of the fourth ACM international symposium on Mobile ad hoc networking & computing (pp. 201–212). New York: ACM.
6. Torgeson, W. S. (1965). Multidimensional scaling of similarity. *Psychometrika., 30*(4), 379–393.
7. Xu, K., Wang, Y., & Liu, Y. (2007). A clustering algorithm based on power for WSNs (Vol. 4489, pp. 153–156). Lecture Notes in Computer Science, Springer.
8. Shang, Y. &, Ruml, W. (2004). Improved MDS-based localization. In INFOCOM 2004, twenty-third annual joint conference of the IEEE computer and communications societies (pp. 2640–2651), Hong Kong, China.

Chapter 143
A Resource Information Organization Method Based on Node Encoding for Resource Discovering

Zhuang Miao, Qianqian Zhang, Songqing Wang, Yang Li, Weiguang Xu, and Jiang Xiao

Abstract In order to discover a variety of network resources of structured P2P, resource information organization methods are required, which should have scalability and robustness. However, structured P2P has bad performance because of churn, so it cannot be widely used currently. To solve the problem, a resource information organization method based on node encoding is provided in this chapter. A node group-based resource information organization and resource distribution-based node encoding algorithm are presented. Redundancy tables are established based on the overlay of the node. The proposed algorithm can decrease the burst of transmission and reduce the traffic load of transited information. The experiment results show that the presented method is tolerant to churn.

143.1 Introduction

The structured P2P-based resource discovery protocol is the best choice for building a resource discovery system [1]. It has the best performance in scalability and robustness. The route algorithm that the protocol adopts can keep high performance and low cost. However, it cannot be widely used in the present world with bad performance for churn.

Traditional resource information organization modes of structured P2P need to transfer a great deal of information in a short period under churn. The high burst of transmission and big amount of information will dispend all the bandwidth for a long time. There will be no bandwidth to deal the users' requests. The route algorithm which the structured P2P-based resource discovery protocol adopts has low success rate of route with churn.

Z. Miao (✉) • Q. Zhang • S. Wang • Y. Li • W. Xu • J. Xiao
Institute of Command Information System, PLA University of Science and Technology, Nanjing 210000, China
e-mail: emiao_beyond@163.com

In this chapter, a resource information organization method based on node encoding is presented. The method can solve the churn problem of structured P2P. In the method, resource information is organized to be distributed according to geographic position. Nodes of the structured P2P are encoded and grouped according to the overlay of the node. The new information organization method can decrease the burst of transmission and reduce the amount of transited information.

143.2 Related Works

To reduce the traffic load of transportation, data redundancy strategy [2] and node selection strategy [3] are used. Multiple backups of the resource information are stored in some different nodes in data redundancy strategy. The node selection strategy uses geographic position and node load and so as the standard to select node as the neighbor. The success rate of route can be improved by mending the route table maintain strategy [4] and route configuration strategy [5]. However, these strategies lack pertinence for structured P2P-based resource discovery protocol. Ou presented a performance evaluation of a structured communication-oriented P2P system in the presence of churn [6].

To eliminate churn, a general solution is proposed which makes a P2P network need not pay much attention to churn problem by introducing a logic layer named Dechurn [7]. Adding the churn resilience, which is achieved by employing the properties of threshold secret sharing schemes, an existing anonymous peer-to-peer network design is improved [8]. The impact of churn on object management policies in CDN-P2P systems is studied and the effectiveness of buffer enlargement and replicating control in reducing the effects of churn on these policies is analyzed [9]. A peer-churn resistant video multicast system over P2P networks, which takes into account both the link delay and the peer stability in order to achieve a seamless video streaming multicast service with a low delay, is presented [10]. There is no overall design for churn-tolerant resource protocol in existing strategies. So to solve churn is an open-ended problem [11].

143.3 Node Group-Based Resource Information Organization

Resource information organization consists of overlay structure, resource information storing, and transporting rules. The structured P2P resource information organization has good performance in scalability and robustness. It should be used widely, but it cannot adapt to churn. There will be a great deal of information needed to transfer in a short period. This makes cost out of existing bandwidth. There will be no more bandwidth to deal users' requests.

The main reason of the problem is that the node encoding is unique and random. The uniqueness makes that there is only one node on one position in code space. So if the node leaves, the information stored in one must be moved to other nodes immediately. The randomness makes that the resource information cannot be stored in local node. These information have different actions with the node that stores them, so they need to be moved frequently.

To solve the problems, we present a resource distribution-based node encoding algorithm. It breaks the limitation of the uniqueness and randomness of node encoding. Based on this algorithm and FreePastry [12], a redundancy table is provided and the redundancy table-based overlay is designed to decrease the burst of transmission of route information. To reduce the amount of resource information, a new resource information storing rule is given.

143.3.1 Resource Distribution-Based Node Encoding Algorithm

The resource distribution-based node encoding algorithm encodes node depending on the node's resource distribution and the local bandwidth and breaks the limitation of the uniqueness and randomness of traditional ones. The node codes will have relationship with the local resource distribution according to the new algorithm.

The main idea of the node encoding algorithm is as follows: *Limen* is called as threshold of encoding and equal to the local bandwidth; if the amount of one local resource information exceeds *Limen*, the node will have the code of this resource; if there is no resource information out of threshold of encoding in one node, this node will get a code randomly. The pseudo code of the node encoding algorithm is shown in Table 143.1.

After encoding there will be some difference from traditional algorithm: (1) one node may have more than one code, and (2) one code may be mapped to more than one node. The codes of one node will be adjusted with the change of amount of resource information.

143.3.2 Redundancy Table-Based Overlay

The resource information organization needs to maintain three tables, leaf table, route table, and neighbor set, because it is based on FreePastry. However, there may be more than one node on one position in code space, so a redundancy table is designed. Some or all table entries of one table can be mapped to multiple appropriate nodes with strategy of redundancy table. Then the overlay is provided based on the redundancy table, and Fig. 143.1 shows the structure of it.

Table 143.1 Pseudo code of resource distribution-based node encoding algorithm

Algorithm: Resource distribution-based node encoding algorithm

Input: encoding the *node*
R: resource set of *node*
c_k: amount of information of k-th resource of R
W: bandwidth of *node*
Output: codes of *node*
CODE: code set of *node*
Algorithm:
1 Allocate_Id(*node*){
2      $CODE \leftarrow \emptyset, k = 0, Limen = W$;
3   while($k < |R|$){
4   if($c_k > Limen$){
5              $CODE = CODE \leftarrow code_k$; //$code_k$: code of the k-th resource
6   }
7   k++;
8   }
9   if($|CODE| = 0$){
10         $CODE = CODE \leftarrow Random()$; // $Random()$: producing node code randomly
11   }
12   return CODE;
13 }

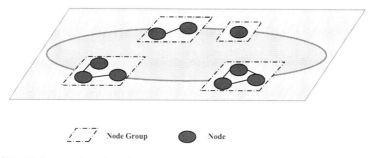

Fig. 143.1 Node group-based overlay

There may be more than one node on one position, so the construction rule of overlay is that all the nodes which have the same code are organized into a group, called node group. Nodes are constructed according to unstructured P2P in one group, called group intra-network. And one node group is connected to other groups according to the three redundancy tables, called group inter-network.

Because of multiple codes for one node, it can belong to more than one group. Node has independent action in each group which it belongs to. When constructing the group inter-network, we should select the node that is not itself.

There are multiple links between one node and one corresponding group. If one of the links is destroyed, protocol will not reconstruct it immediately. This work will be dispersed in a long time. The burstiness of transmission is abased.

143.3.3 Resource Information Storing Rule

Resource information storing rule defines the strategies on how to store, transfer, and renew the information. In the present protocol, the resource information cannot be stored in local node, unless the local node regards code of this resource as its code. When the amount of resource information exceeds the threshold, the resource information should be stored in local node. So the resource information of large amount is not transferred to far end and the information stored in local node will not be moved. The amount of information needed to be transferred under churn is reduced.

Information which doesn't exceed *Limen* will be transferred to other node group and is backed up by redundancy. The process of transferring is as follows:

Step1: Original node selects one node from the corresponding group randomly, and the information is transferred to it; go to **Step2**.
Step2: The node which stores information checks whether there are other nodes in the corresponding group; if so go to **Step3**, or else go to **Step5**.
Step3: The node selects a part of nodes which are in the same group to back up the information randomly; go to **Step4**.
Step4: The node in the same group with the last node checks whether the information is existing in local; if so go to **Step5** immediately; otherwise, store it and go to **Step5**.
Step5: End.

With the process, the information that is stored in far end will have multiple backups. When one node storing the information leaves, it is not necessary to transfer new information to other nodes. Further, the traffic load is reduced.

Resource distribution-based node encoding algorithm, redundancy table-based overlay, and resource information storing rule compose the node group-based resource information organization. In theory, the presented resource information organization can decrease the burst of transmission and reduce the load of information. The experiments will be given in Sect. 143.4.

143.4 Experiments

We verify the theoretical results by measuring the performance. The method is simulated by PeerSim [13] which is a simulator that can be extended easily. The simulator adopts the event-based simulation of PeerSim. The traffic load of the resource information organization is compared with FreePastry to prove that the burst and traffic load of transmission are both reduced under churn. Then, the simulator is presented to show that the route algorithm has good performance in route success rate with high efficiency and low cost as FreePastry.

Table 143.2 Main parameters of simulator

Parameter	Value
Number of node	1,000
Number of kind of resource for each node	$U(1, 64)$
Number of one resource for each node	$U(10, 100)$
Amount of each information	10
Bandwidth of each node	$U(10, 500)$
Length of code	128

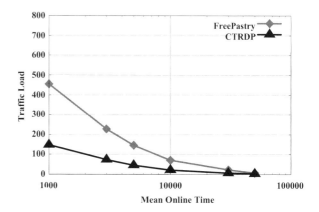

Fig. 143.2 Traffic load under churn

The churn model of our experiment is exponential distribution [14]. The churn level is adjusted by changing average online time of nodes in simulator. The longer the time is, the lower the churn level is.

Some main parameters of simulator are shown in Table 143.2, where U is the uniform distribution on the unit interval of value in the bracket. There are 1,000 nodes in the simulator. When one node leaves, a new node must join immediately. The size of each information is 10, including resource information and route table information. The code of resource and node is 128-bit BigInteger of Java.

Because our method is extended on FreePastry, the parameter of FreePastry is as follows: $b = 4, M = 32, L = 16$.

The experiments of transmission give comparison between our method and FreePastry on burst and traffic load under churn.

Traffic Load. The values of mean online time are as follows: 1,000, 3,000, 5,000, 10,000, 30,000, and 50,000. The comparison between our method and FreePastry on traffic load under churn is shown in Fig. 143.2.

The result shows that the traffic load of our method is less than FreePastry under churn. The higher the churn level, the more obvious is the advantage. The reason of the phenomenon is that a great deal of resource information is stored in local nodes. These resource information need not to be moved with node's leaving or joining. Although there are multiple backups for some resource information that is stored

Fig. 143.3 Traffic load of route information under churn

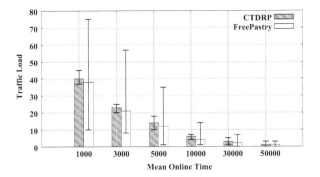

in far end, the amount of information doesn't increase dramatically. At the same time, the redundancy table doesn't make the traffic load increasing obviously. So the traffic load of our method is lower than FreePastry.

Route Information. This experiment compares our method and FreePastry on route information under churn. The same mean online times as the last one are adopted. The result can be found in Fig. 143.3.

In the figure, the bins represent the average traffic load of route information in 10,000 cycles for each node. The upper extreme points are the max value and lower extreme points are the min value in the period. The result shows that the average traffic load of our method is little more than FreePastry's. However, the stability of our method's traffic load is better than the old one's obviously. So we can get the conclusion that our method decreases the burst of transmission with a little more traffic load.

Summing up the above, node group-based resource information organization that our method adopts can reduce the traffic load and decrease the burst of transmission.

143.5 Conclusion

In this chapter, a resource information organization mode based on node encoding is provided to solve the churn problem. The encoding algorithm breaks the limitation of uniqueness and randomness of traditional ones. After encoding, a code may be mapped to more than one node. The redundancy table is designed on this character of encoding. A table entry can be mapped to multiple nodes in redundancy table. The overlay is constructed according to the redundancy table. The group intra-network is based on unstructured P2P and the group inter-network on structured P2P, FreePastry. The redundancy table-based overlay can abase the burstiness of transmission. The resource information storing rule is given, which makes a great deal of resource information stay in local node. So the traffic load of our method is lower than FreePastry.

Acknowledgements The authors are supported by the China Postdoctoral Science Foundation 2012T50844, by Provincial Nature Science Foundation of Jiangsu China BK2012512, and by the Advanced Research Foundation of PLA University of Science and Technology KYZYZLXY1205.

References

1. Ranjan, R., Harwood, A., & Buyya, R. (2008). Peer-to-peer based resource discovery in global grids: A tutorial. *IEEE Communications Surveys and Tutorials., 10*(2), 6–33.
2. Tian, J., & Dai, Y. F. (2007). Study on durable peer-to-peer storage techniques. *Journal of Software, 18*(6), 1379–1399.
3. Li, J. Y., Stribling, J., Morris, R., & Kaashoek, M. F. (2005). Bandwidth efficient management of DHT routing tables. In *Proceedings of second conference on symposium on networked systems design & implementation* (pp. 99–114), Berkeley, CA, USA.
4. Lam, S., & Liu, H. (2006). Failure recovery for structured P2P networks: Protocol design and performance under churn. *Computer Networks., 50*(16), 3083–3104.
5. Stutzbach, D., & Rejaie, R. (2006). Improving lookup performance over a widely-deployed DHT. In *Proceedings of 25th IEEE international conference on computer communications* (pp. 1–12). Piscataway, NJ: IEEE Press.
6. Ou, Z. H., Harjula, E., Kassinen, O., & Ylianttila, M. (2010). Performance evaluation of a Kademlia-based communication-oriented P2P system under churn. *Computer Networks., 54*(5), 689–705.
7. Meng, X. F., Chen, X. L., & Ding, Y. L. (2013). Using the complementary nature of node joining and leaving to handle churn problem in P2P networks. *Computers & Electrical Engineering., 39*(2), 326–337.
8. Alexandrova, T., Huzsak, G., & Morita, H. (2012). Churn resilience in network coding-based anonymous P2P system. In *Proceedings of 2012 international symposium on information theory and its Applications (ISITA)* (pp. 270–274). Piscataway, NJ: IEEE Press.
9. Melo, C. A. V., Vieira, D., & Liborio, J. M. (2012). Impact of churn on object management policies deployed in CDN-P2P systems. *IEEE (Revista IEEE America Latina) Latin America Transactions, 10*(3), 1811–1816.
10. Kwon, O. C., Song, H. J. (2012). A peer-churn resistant video multicast system over P2P networks. In *Proceedings of 2012 international symposium on communications and information technologies (ISCIT)* (pp. 1003–1008). Piscataway, NJ: IEEE Press.
11. Zhang, Y. X., Yang, D., & Zhang, H. K. (2009). Research on churn problem in P2P networks. *Journal of Software., 20*(5), 1362–1376 (In Chinese).
12. FreePastry protocol v2.1. (2012). Retrieved from http://www.freepastry.org/FreePastry.
13. PeerSim. (2012). Retrieved from http://wenku.baidu.com/view/.
14. Ou, Z. H., Harjula, E., & Ylianttila, M. (2009). Effects of different churn models on the performance of structured peer-to-peer networks. In *Proceedings of IEEE 20th international symposium on personal, indoor and mobile radio communications* (pp. 2856–2860). Piscataway, NJ: IEEE Press.

Chapter 144
The Implementation of Electronic Product Code System Based on Internet of Things Applications for Trade Enterprises

Huiqun Zhao and Biao Shi

Abstract In order to solve the EPC codec problems based on Internet of Things (IOT) applications for trade enterprises, in this chapter an EPC codec system is designed for enterprise applications. According to the "Tag Data Standards," we design the encoding and decoding algorithm/schema of SGTIN-96. On the basis of the algorithm/schema, the system has improved its coding and decoding algorithm, making the coding algorithm more simple and practical and improving the efficiency. Besides, it also has realized the transformation between SGTIN-96 and GTIN-14, which makes the final printed electronic tag contain both a bar code and the EPC code, and realizes the compatibility of bar code and EPC code. The coding and decoding system can code and decode well for the products of the trade enterprise. Through the system, SGTIN-96 labels can be generated and printed. And the content of SGTIN-96 labels can be decoded. Finally, we test the codec system to prove that it can achieve our established requirements.

144.1 Introduction

IOT is based on the Internet and uses the radio frequency identification (RFID), wireless data communications, and computer technology to construct an Internet that covers everything in the world [1, 2]. IOT EPC aims to build an open global network, where anything and its position in the logics chain can be identified [3]. IOT can apperceive the EPC code within the electronic tag on goods. When the perception of information is transmitted to the high-level, it will be identified, transferred and integrated by using the information processing software in the high-level. At last, the processing power of the Internet is taken advantage of to

---

H. Zhao • B. Shi (✉)
Department of Computer Science, North China University of Technology,
Beijing 100041, China
e-mail: zhaohq6625@sina.com; shibiao462@163.com

integrate the perception of information, forming the effective management and control capability to meet user demand for a variety of applications.

To achieve IOT, EPC encoding within the electronic tag on goods is vitally crucial. Moreover, IOT will be widely used firstly in the trade logistics management. In this chapter, the EPC codec system based on IOT applications for trade enterprise has been designed and implemented.

The chapter is organized as follows: Sect. 144.2 introduces EAN.UCC standards of GID (General Identifier) and its derivatives GTIN and SGTIN [4, 5] and proposes an encoding and decoding scheme for implementing the codec system, and the converting algorithm for mapping barcode into EPC is discussed. In Sect. 144.3, in order to support our technique, a system is designed and implemented. In Sect. 144.4, we test the encoding and decoding system. Finally as a key technique we mainly introduce related work and conclude our research work in Sect. 144.5.

144.2 The Encoding and Decoding Schema

In this part, we design the proposals. It includes some algorithms, such as the transformation algorithm from SGTIN-96 to GTIN-14, SGTIN-96 encoding algorithm, and so on. The purpose is to provide a general method for constructing the EPC codec system based on IOT applications for trade enterprises.

144.2.1 Introduction of the GTIN and the SGTIN

The Serialized Global Trade Item Number (SGTIN) is a new identity type based on the EAN.UCC Global Trade Item Number (GTIN) code defined in the General EAN.UCC Specifications [6]. A GTIN by itself does not fit the definition of an EPC pure identity, because it does not uniquely identify a single physical object.

The SGTIN is an EPC encoding scheme that permits the direct embedding of EAN.UCC system standard GTIN and serial number codes on EPC tags [7]. There are two launched encoding schemes: SGTIN-64 (64 bits) and SGTIN-96 (96 bits) now.

For instance, in the SGTIN-96 encoding, the limited number of bits prohibits a literal embedding of the GTIN. As a partial solution, a Company Prefix *Index* is used. This Index is assigned to companies that need to use the 96-bit tags, in addition to their existing EAN.UCC Company Prefixes [8]. The Index is encoded on the tag instead of the Company Prefix and is subsequently translated to the Company Prefix at low levels of the EPC system components (i.e., the Reader). While this means that only a limited number of Company Prefixes can be represented in the 96-bit tag, this is a transitional step to full accommodation in 96-bit and additional encoding schemes (Table 144.1).

The relationship between GTIN and SGTIN is shown in Fig. 144.1.

Table 144.1 The structure of SGTIN-96

Header	Filter value	Partition	Company Prefix *Index*	Item reference	Serial number
8	3	3	20–40	24–4	38

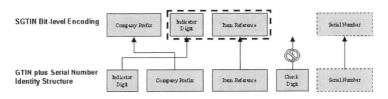

Fig. 144.1 The encoding of GTIN and SGTIN

The SGTIN-96 consists of the following information elements:

- The *Company Prefix*, assigned by GS1 to a managing entity. The Company Prefix is the same as the Company Prefix digits within an EAN.UCC GTIN decimal code.
- The *Item Reference*, assigned by the managing entity to a particular object class. The Item Reference for the purposes of EPC Tag Encoding is derived from the GTIN by concatenating the Indicator Digit of the GTIN and the Item Reference digits and treating the result as a single integer.
- The *Serial Number*, assigned by the managing entity to an individual object. The serial number is not part of the GTIN code, but is formally a part of the SGTIN.

144.2.2 *Transformation of GTIN-14 and SGTIN-96*

In Fig. 144.2, the process of the transformation from GTIN-14 to SGTIN-96 is demonstrated, and the converse process is the transformation from SGTIN-96 to GTIN-14. The following procedure creates an SGTIN-96 encoding from GTIN-14:

Given:

1. A GS1 GTIN-14 consisting of digits $d_1 d_2 \ldots d_{14}$
2. The Length L of the Company Prefix portion of the GTIN
3. A Serial Number S where $0 <= S < 2^{38}$, or a GS1-128 Application Identifier 21 consisting of character $S_1 S_2 \ldots S_K$
4. A Filter Value F where $0 <= F < 8$

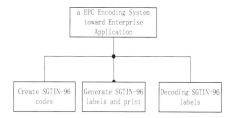

Fig. 144.2 The structure of the system

Yields:

5. An SGTIN-96 as a 96-bit string

Procedure:

Look up the Length L of the Company Prefix in the "Company Prefix Digits" column of the Partition Table;

If (the Length L exist){
  determinate the Partition Value, P;
  the number of bits M in the Company Prefix field;
  the number of bits N in the Item Reference and Indicator Digit field.;
  Construct the Company Prefix by concatenating digits $d_2 d_3 \ldots d_{(L+1)}$ and considering the result to be a decimal integer, C;
  Construct the Indicator Digit and Item Reference by concatenating digits $d_1 d_{(L+2)} d_{(L+3)} \ldots d_{13}$ and considering the result to be a decimal integer, I;
  If(Serial Number is provided&&an integer S where $0 <= S < 2^{38}$){
    Construct the final encoding =
    Header 00110000 (8 bits)+Filter Value F (3 bits)+Partition Value P (3 bits)+Company Prefix C (M bits)+Item Reference (N bits)+Serial Number S (38 bits);
  }else if(the Serial Number is provided as a GS1-128,$s_1 s_2 \ldots s_K$){
    If($s_1 s_2 \ldots s_K$ not a digit)
      Return;
    If(K > 1 and s1 = 0)
      Return;
    consider the result to be a decimal integer, S;
    If(S >= 2^{38}){
      this Serial Number cannot be encoded in the SGTIN-96 encoding;
      Return;
    }else{
      Construct the final encoding =
      Header 00110000 (8 bits)+Filter Value F (3 bits)+Partition Value P (3 bits)+Company Prefix C (M bits)+Item Reference (N bits) +Serial Number S (38 bits);
    }

          }
    }else{
        this GTIN cannot be encoded in an SGTIN-96;
        return;
    }

144.2.3 Algorithm of SGTIN-96 Encoding and Decoding

The SGTIN-96 encoding algorithm is to form an EPC code, according to the given and known information, such as the filter value, manufacturer identification code, packing instructions, and commodity types of product information. It includes obtaining the bar code by SGTIN-96 and GTIN-14. The following procedure creates an SGTIN-96 encoding:

Given:

6. Filter value
7. Company Prefix
8. Indicator Digit
9. All kinds of information of products
10. Information and the number to generate code for product

Yields:
All kinds of products and the specified number of SGTIN-96 code and GTIN-14 code

Procedure:

1. According to the company prefix, we look up the partition table to find out the partition value and obtain figures of item reference. If there are no corresponding values in the table, the digit company prefix is not valid and stops with a prompt error message.
2. According to the given filter value, we query the SGTIN-96 filter table to see if the selected filter value is in practical sense, i.e., not retention value. If it is reserved, the filter value is not valid and stops with a prompt error message.
3. Also we can query that the indicator digit is legitimate. If not legal, we prompt the error message and stop.
4. According to the number of commodity product information, the item reference field is assigned to the different parts, which is the kind of distribution of several fields. It means that the field assigned preserves the different information of commodities.
5. According to the number of kinds of goods code given, we encode for the commodities. If the number is not lengthy enough, we add leading zeros to achieve a specified number of bits. Then the code is stored in the database.

6. According to the generated code types, the number of product information, and the SGTIN-96 tag field format, we splice each field to generate EPC tags. According to the EPC label and the formation of GTIN-14 bar code, ITF-14 is structured (the ITF-14 is a kind of bar code of GTIN-14).
7. By means of the replacement encryption on the formation of the SGTIN-96 tag, we form 24 bits 16 hexadecimal code.
8. The labels are printed by a special printer, forming effective labels that contain IOT EPC and bar code.

144.3 Design and Implement

We design an EPC codec system according to the analysis of the proposal. It is mainly to form the proposal of SGTIN-96 encoding, based on the specific application coding information provided by company. At the same time, it can transform SGTIN-96 encoding to GTIN-14 encoding. After the label is formed, we can use the label printer to print. We use BarTender to complete the mission. Then, we send the label data generated by the system, including SGTIN-96 and ITF, to the driver of the BarTender, and the driver drives the printer to print the label. Besides, the system also realizes the decoding of specified label, that is, obtains the information contained in the code by the SGTIN-96. The structure diagram of the system is shown in Fig. 144.2.

144.3.1 Encoding Module of IOT Labels

Encoding module of the EPC label is to provide the unique assigned company prefix, the trade type of the product, the indicator value of the project reference field, and the information of the product to the system, according to the distribution of the SGTIN-96 field. The system will form the SGTIN-96 label based on this basic information and store the obtained data in the database.

144.3.2 Printing Module of IOT Labels

The printing module of IOT label, based on tag encoding, will need to print out the tags. When in printing, all kinds of information need to choose to print labels, as well as the number of print labels, in order to produce continuous tags and labels. Figure 144.3 shows the printing module interface of the system.

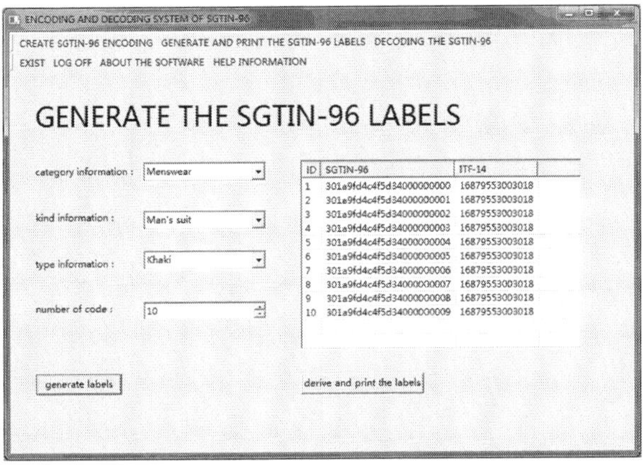

Fig. 144.3 The generating label interface of system

144.3.3 Decoding Module of IOT Labels

For the decoding module of IOT label, according to some parameters of each field in an SGTIN-96 distribution and encoding algorithm, we can respectively get the fields of the IOT label from SGTIN-96 such as the company prefix, item reference, indicator, and kinds of products field. Finally, we analyze the practical significance which the fields represent.

144.4 Testing the System

The encoding module can generate the same encoding results with the respected by inputing all kinds of the related testing data in the encoding module of the system. It also can export and print EPC tag efficiently. The decoding module can analyze the input label well. The system can complete the function well, put out the correct result, and have good operability and better fault tolerance. Besides, the system has good expansibility which is good for the updating and maintenance of the system.

Taking clothing industry as an example, we input the information needed about the clothing and formulate the information about the labels in the database. The printing module will form tags that contain both SGTIN-96 and ITF. Enter a valid label information in the decoding module; the information about the product will be resolved within the label. Finally, Fig. 144.4 depicts the printed label, which contains both SGTIN-96 in the inner chip and GTIN-14 on the surface of the label.

Given that the encoding and decoding system is the basis of the IOT and will be planted into other system, this system has good portability. However, because of being developed in the Windows 7 and uses SWT/JFace package, the system has a higher operational efficiency in the Windows 7 system than in the Linux system. According to the principle of the EPC encoding, EPC tag should have the

Fig. 144.4 The printing labels

confidentiality and security, but it is not considered much in the system. The system just does a simple substitution encryption to the generated EPC encoding, so the security of encoding needs to be strengthened.

The improved proposal: Encoding and decoding of the EPC label should be published as a service, so the system will have better cross-language, cross-platform, expansibility, support, and so on. Besides, the system will have higher confidentiality and security by combining the security and encryption technology.

144.5 Related Works and Conclusion

EPC and its application have become a popular research topic in the last decade, so some implementations of this EPC system have been applied in the business world. GS1, a management and development body for EPC standards, has been developing universal standards to regulate its encode and decode. However, some changes should be made to satisfy the specific requirements for trade enterprises. Thus we put forward our own algorithms/schemes for the codec system based on the customary EPC system. Because of the improvement of coding and decoding algorithm, the system can be used easily and the efficiency has been improved. Combined with practical work, an EPC codec system is designed. Our system can provide a unique physical marking for retail or trade items industry. The efficient implementation of the design and printing label provides the basis for the popularization of IOT.

References

1. Staake, T., Thiesse, F., & Fleisch, E. (2005). Extending the epc network—the potential of RFID in anti-counterfeiting. *Symposium on Applied Computing—SAC* (pp. 1607–1612).
2. Price, J., Jones, E., Kapustein, H., Pappu, R., Pinson, D., Swan, R., & Traub, K. (2003). *Auto-ID reader protocol 1.0* (pp. 29–33).
3. Hoag, J. E., & Thompson, C. W. (2006). Architecting RFID middleware. *IEEE Internet Computing, 10*(5), 88–92.
4. Harrison, M. (2004). EPC information service (EPCIS). *Auto-ID Labs Research Workshop* (pp. 29–30).

5. Ding, Z. H., Li, J. T., Zheng, W. M. & Feng, B. (2007). A filter design of RFID middleware in the progress of updating barcode to RFID. *1st Annual RFID Eurasia,* 18(2), 56–66 (In Chinese).
6. Rivest, R. L., Shamir, A., & Adleman, L. (1978). A method for obtaining digital signatures and public-key cryptosystems. *Communications of the ACM, 21*(2), 120–126.
7. Juels, A., Rivest, R. L., & Szydlo, M. (2003). The blocker tag: selective blocking of RFID tags for consumer privacy. *Proceedings of 10th ACM Conference on Computer and Communications Security, CCS 2003* (pp. 103–111).
8. Ham, Y. H., & Kim, N. S. (2005). A study on establishment of secure RFID network using DNS security extension. *2005 Asia-Pacific Conference on Communications, Perth, Western Australia* (pp. 3–5).

Chapter 145
The Characteristic and Verification of Length of Vertex-Degree Sequence in Scale-Free Network

Yanxia Liu, Wenjun Xiao, and Jianqing Xi

Abstract Many natural large-sized complex networks exhibit a scale-free, power-law distribution of vertex degree. To better understand the formation mechanism of power law in the real network, we analyze the general nature in scale-free network based on the vertex-degree sequence. We show that when the power exponent of scale-free network is greater than 1, the number of degree-k_1 vertices, when nonzero, is divisible by the least common multiple of 1, k_2^γ/k_1^γ, ..., k_i^γ/k_1^γ, and the length of vertex-degree sequence l is of order log N, where $1 \leq k_1 < k_2 < \ldots < k_l$ is the vertex-degree sequence of the network and N is the size of the network. We verify the conclusion by the coauthorship network DBLP and many other real networks in diverse domains.

145.1 Introduction

Complex networks are widespread in nature and human society, such as the Internet, the World Wide Web, protein networks, scientific collaboration networks, and transportation networks. As a key technology of depicting and studying the topology and behaviors of complex systems, complex network has become a hot spot of the multidisciplinary research of common concern in recent years.

Y. Liu (✉)
School of Computer Science and Engineering, South China University of Technology, Guangzhou 510006, China

School of Software Engineering, South China University of Technology, Guangzhou 510006, China
e-mail: cslyx@scut.edu.cn

W. Xiao • J. Xi
School of Software Engineering, South China University of Technology, Guangzhou 510006, China

Table 145.1 Scale-free networks with $\gamma < 2$ (N, M, γ)

Network	N	M	γ
E-mails [7]	56,969	84,314	1.81
Gnutella [8]	1,026	3,752	1.4
Word Web [9]	478,773	1.8×10^7	1.5
Software package [4]	1,439	1,723	1.6/1.4
Coauthorship in HEP [10]	56,627	9,796,471	1.2

The basic theory of complex network research is small-world network model [1] and scale-free network model [2]. In 1998, Watts and Strogatz proposed WS small-world network model by randomly reconnecting a small number of the edge of regular network [1]. As the transition from completely regular network to random network, the small-world model features localized clusters connected by sparse long-range edges, leading to a short average distance between vertices that grow logarithmically with the network size. Scale-free networks, on the other hand, show heterogeneous vertex connectivity, in which a fraction of the vertices is highly connected. In 1999, Barabási and Albert proposed a scale-free network model BA [2]. It demonstrated that the scale-free power-law distribution of vertex degrees in many large-sized networks is a direct consequence of two generic mechanisms that govern the network formation: (1) network expansion over time through addition of new vertices and (2) preferential attachment of new vertex to those existing ones that are already highly connected.

In this chapter, we focus on scale-free networks. Many real networks belong to scale-free networks which obey an approximate power-law distribution, i.e., $P(k) \propto k^{-\gamma}$. Usually the scale-free networks need to satisfy power exponent $\gamma \geq 1$; the extra $\gamma \geq 1$ requirement ensures that P(k) can be normalized. Furthermore, a large number of empirical research have shown that the exponent γ in real network is between 2 and 3. Consequently many of the scale-free network research are on the premise of $\gamma \geq 2$ by default [3–5], thus $\gamma \geq 2$ have been a precondition to study the general nature and dynamic behaviors of scale-free networks. In other words, $1 < \gamma < 2$ network has been ignored and few research pay attention to such networks. In fact, such networks are also widely used in the real world, such as Gnutella P2P network and HEP collaborators network. Table 145.1 lists the parameters for several real scale-free networks with $\gamma < 2$.

It's very meaningful to study the similarity and difference between such network and network with $\gamma \geq 2$, which can help us better understand the laws of the generation of different power law in the real network.

In the previous works [5, 6], we have presented necessary conditions for scale freedom in complex networks, which is based on the assumption $\gamma \geq 2$. Here, we extend the research results to the case $\gamma < 2$ and pay our attention to the general characteristic in scale-free networks with the exponent $\gamma > 1$. We proved that the length of vertex-degree sequence l of scale-free networks with the exponent $\gamma > 1$ is of order log N, where N is the size of the network. We further verify the conclusion and some applications are given.

145.2 Scale-Free Networks

Complex networks can be abstracted as undirected or directed graph G (V, E), where V is the set of vertices or nodes and E is the set of edges or links. The parameters that are of interest in this chapter appear in the following list:

M Number of edges; $M = |E|$
N Number of vertices; $N = |V|$
$d(v)$ Degree of the vertex; $v \in V$
$\bar{d}$ Average vertex degree of the network; $\bar{d} = 2M/N = \sum_{v \in V} d(v)/N$
n_k Number of degree-k vertices; $n_k = \{v | d(v) = k\}$
$P(k)$ Degree distribution or fraction of vertices that are of degree k, $P(k) = n_k/N$
l Length of vertex-degree sequence; $\{k_1, k_2 \ldots, k_l\}$ is the vertex-degree sequence of the network where $1 \leq k_1 < k_2 \ldots < k_l$

Most real complex networks are scale-free network. The degree distribution of scale-free networks is not like random network which is presented in the form of the Poisson distribution. Scale-free networks comparatively have the nature of the power-law distribution, which is presented in the following form:

$$P(k) = ck^{-\gamma}, \gamma > 1$$

where γ is power exponent, also known as degree distribution exponent or scaling exponent. The extra $\gamma > 1$ requirement ensures that $P(k)$ can be normalized. Power laws with $\gamma < 1$ rarely occur in nature. The term c is normalizing constant, defined as $c = (\sum_{k \in K} k^{-\gamma})^{-1}$, and K is the set of all node degrees occurring in the network.

Scale-free network aroused great interest of researchers. Despite extensive research on scale-free network within multiple scientific disciplines over the past decade, there are still gaping holes in our understanding of such networks. New results that shed light on the static structure and dynamic properties of different classes of large-scale networks are needed to facilitate further progress. In the previous works [5, 6], we have exposed some characteristics of scale-free network with the exponent $\gamma \geq 2$. We proved that when the vertex degree of a large-sized network follows a scale-free power-law distribution with exponent $\gamma \geq 2$, the number of degree-1 vertices, when nonzero, tends to be of the same order as the network size N and that the average degree is of order lower than log N. Our method provides an analytical tool that helps to answer the question of whether a network is scale-free because it relies on conditions that are static and easily verified for any network. Furthermore, we showed that the number of degree-1 vertices is divisible by the least common multiple $k_1^\gamma, k_2^\gamma \ldots k_l^\gamma$, where $k_1 < k_2 < \cdots < k_l$ is the vertex-degree sequence of the network, which leads a remodeling method to equip a scale-free network with small-world features.

Next, based on our previous research [5, 6], we will further investigate the general characteristic of scale-free networks by extending the exponent condition. We mainly focus on the characteristic of vertex-degree sequence in scale-free networks with the exponent $\gamma > 1$.

145.3 New Characteristic of Scale-Free Network and Its Derivation

In this section, we present the new characteristic of scale-free network and its mathematical derivation. Supposing that the vertex-degree sequence of network can be represented as $1 \leq k_1 < k_2 \ldots < k_l$, according to the definition of strict scale-free networks, we have

$$P(k_i) = \frac{n_{k_i}}{N} = ck_i^{-\gamma} \tag{145.1}$$

Here, c is a constant. When i = 1, we have

$$P(k_1) = \frac{n_{k_1}}{N} = ck_1^{-\gamma} \tag{145.2}$$

Thus, we can get the value of c

$$c = \frac{n_{k_1} k_1^{\gamma}}{N} \tag{145.3}$$

By substituting Eq. 145.3 into Eq. 145.1, we have

$$n_{k_i} = n_{k_1} \left(\frac{k_1}{k_i}\right)^{\gamma} \tag{145.4}$$

Recall that n_k denotes the number of vertices of degree k, apparently, we have

$$N = \sum_i n_{k_i} \tag{145.5}$$

The following identities follow from Eq. 145.4 and Eq. 145.5

$$N = \sum_i n_{k_i} = n_{k_1} k_1^{\gamma} \sum_i \frac{1}{k_i^{\gamma}} \tag{145.6}$$

Therefore, assuming $\gamma > 1$, according to Riemann-ζ function, we know that $\sum_i \frac{1}{k_i^{\gamma}}$ in Eq. 145.6 is convergent, which satisfied

$$\sum_i \frac{1}{k_i^\gamma} \leq \sum_{k=1}^{\infty} \frac{1}{k^\gamma} < +\infty \tag{145.7}$$

Supposing that $\sum_i \frac{1}{k_i^\gamma}$ converge to some constant a, i.e., $a = \sum_i \frac{1}{k_i^\gamma}$, clearly we have the following conclusion:

$$N = n_{k_1} k_1^\gamma a \tag{145.8}$$

Thus, we have shown that for scale-free networks with power-law exponent $\gamma > 1$, the size of scale-free network is related to some constant a. Furthermore, if $k_1=1$, then $N = n_{k_1} a$. It means that when γ gets smaller, constant a gets greater, thereby the ratio of the number of degree-1 vertices and the size of network gets smaller. Note that $k_1=1$ is not the necessary condition for the scale-free networks.

Following from the above analysis, we can obtain interesting results on the degree sequence of scale-free networks. In the preceding derivation, we have shown that Eq. 145.4 holds which leads to $n_{k_1}/n_{k_i} = k_i^\gamma/k_1^\gamma$. We may assume that γ is a rational number since any real number can be approximated infinitely by rational number sequence. Taking this equality to be exact and noting that the left-hand side is a rational number, we can easily prove that either k_i^γ/k_1^γ is an integer with $(k_i^\gamma/k_1^\gamma) | n_{k_1}$ or k_i^γ is an integer with $k_i^\gamma | n_{k_1} k_1^{\gamma 1}$. Thus, recall that $1 \leq k_1 < k_2 \ldots < k_l$ is the degree sequence of the network, we can obtain that either n_{k_1} must be divisible by the least common multiple of $1, k_2^\gamma/k_1^\gamma, \ldots, k_l^\gamma/k_1^\gamma$, denoted as $n_{k_1} = c[1, k_2^\gamma/k_1^\gamma, \ldots, k_l^\gamma/k_1^\gamma]$, or $n_{k_1} k_1^\gamma = c[k_1^\gamma, k_2^\gamma, \ldots, k_l^\gamma] \neq 0$ holds for some constant c. Using the method as above, in the case of $\gamma > 1$, we have the following estimate:

$$n_{k_1} k_1^\gamma \geq 2 \times 2 \times \cdots \times 2 \geq 2^{(l-2)} \tag{145.9}$$

Then taking all logarithms to be in base 2, Eq. 145.9 yields

$$O(\log N) > \log n_{k_1} k_1^\gamma \geq l - 2 \tag{145.10}$$

Therefore, we obtain our result that for scale-free network with $\gamma > 1$, the length of degree sequence is of order log N, i.e., $l \leq O(\log N)$.

This conclusion is very meaningful, which means the characteristic of the length of degree sequence is the general nature in scale-free network. All scale-free networks have a very small degree sequence compared with the size of the network. Using the conclusion, we can reconstruct the scale-free network presenting apparent small-world feature and further improve the current maximal degree search algorithm.

In addition, we must also stress that the above conclusion holds based on the premise that the network obeys strict power-law distribution. Actually many real scale-free networks are not exact which only have approximate scale-free characteristic, so there are subtle differences between the real network and the theoretical estimate. We further pointed out that for many real networks the length of degree sequence is also of order log ^bN at the very most, namely, $l \leq O(\log{^bN})$, where b is a very small constant, which means l is very small compared with the size of network. Next, we will verify our conclusion in some applications.

145.4 Analysis and Verification in DBLP and Many Other Real Networks

We start our analysis by investigating coauthorship network of scientists using bibliographic data drawn from DBLP. DBLP is a well-known computer science bibliography website hosted at Universität Trier in Germany which provides bibliographic information on major computer science journals and proceedings. Until now DBLP has indexed more than 2.1 million articles and DBLP data is available from the website which is stored as xml records.

For our research, we downloaded the latest dblp.xml file and parsed it using SAX parser in Java. Then we constructed the network of scientists in which a link between two scientists is established by their coauthorship of one or more scientific papers. Due to the large amount of data, the coauthorship network has to be constructed by year. We produced several separate networks by respectively extracting the articles generated in 2009, 2010, and 2012. Among them, the data in year 2012 only contains the articles generated before the March. Subsequently, we computed the parameters of each network which are listed in the following table.

From Table 145.2, we can see that the three networks have consistent statistical values which reflect the general characteristic in the whole DBLP data to a certain extent. First, the ratio of degree-1 vertex and the size of network approximately equal to 0.1, which means that about 10 % of scientists have only one coauthor when they wrote papers during 1-year period. Second, the average degree of the networks is roughly 5, which means that, on average, scientists wrote articles with five other people each year. Third, most importantly, we can see that $l \leq \log{^2N}$ holds which means compared with the size of network, the range of the possible numbers of collaborators per author is extremely small. Here, the reason why the length of degree sequence is not of order log N is that DBLP network doesn't strictly obey power-law distribution.

Table 145.2 The parameters in DBLP networks

Network	N	M	n_1	$\overline{d}$	l	$\log N$	b
DBLP_2009	57,623	160,456	5,266	5.569	83	15.814	2
DBLP_2010	74,899	197,849	7,048	5.283	88	16.193	2
DBLP_2012	17,846	49,443	1,820	5.541	56	14.123	2

Notes: For DBLP data, see http://dblp.uni-trier.de/

Table 145.3 The length of degree sequence in some real networks

Network	Type	N	M	$\overline{d}$	l	log N	b
TG city	TP	18,263	23,797	2.606	7	14.157	1
OL city	TP	6,105	7,029	2.303	5	12.576	1
US Air	TP	332	2,126	12.807	58	8.375	2
Linux	SW	5,285	11,352	4.296	51	12.368	2
Mysql2	SW	1,480	4,190	5.662	43	10.531	2
Helico	Bio	710	1,396	3.932	31	9.472	2
Elegans	Bio	314	363	2.312	17	8.295	2
Ncstrlwg2	SC	6,396	15,872	4.963	42	12.643	2

Notes: For data on transportation networks, see http://www.cs.fsu.edu/~lifeifei/SpatialDataset.htm; for software networks, see www.tc.cornell.edu/~myers/Data/SoftwareGraphs/index.html; for biological networks, see www.cosin.org/extra/data; the data of Ncstrlwg2 is provided by M.E.J Newman [11]

Besides DBLP network which belongs to scientific collaboration networks, we have found that various types of real network without exception exhibit the same characteristic, i.e., $l \leq O(\log N)$ or $l \leq O(\log^b N)$, where b is a very small constant. So far, we have done a lot of data verification and observed that the length of the degree sequence in the real network is less than the $\log^2 N$, i.e., b value is 2. Table 145.3 shows some real examples from the scale-free networks in diverse domains, including scientific collaboration networks, transportation networks, software packages networks, and biological networks.

145.5 Conclusion

In this chapter we show that when the vertex degree of a large-sized network follows a scale-free power-law distribution with exponent $\gamma > 1$, the number of degree-k_1 vertices is divisible by the least common multiple of 1, k_2^γ/k_1^γ, ..., k_i^γ/k_1^γ and the length of vertex-degree sequence l is of order log N, where $k_1 < k_2 < \cdots < k_l$ is the vertex-degree sequence of the network. Furthermore, we generalized the conclusion taking into account that many real networks are approximately scale-free. We pointed out that for many real networks the length of degree sequence is also of order $\log^b N$ at the very most, where b is a very small constant. We verify the conclusion by the coauthorship network DBLP and other real networks in different domains. Next, we will further research on searching problem based on the vertex-degree sequence in scale-free networks.

References

1. Watts, D. J., & Strogatz, S. H. (1998). Collective dynamics of 'small-world' networks. *Nature, 393*(6684), 440–442.
2. Barabási, A. L., & Albert, R. (1999). Emergence of scaling in random networks. *Science, 286* (5439), 509–512.
3. Albert, R., & Barabási, A. L. (2006). Statistical mechanics of complex networks. *Reviews of Modern Physics, 74*(1), 47–91.
4. Newman, M. E. J. (2003). The structure and function of complex networks. *SIAM Review, 45* (2), 167–256.
5. Xiao, W. J., Chen, W. D., & Parhami, B. (2011). On necessary conditions for scale-freedom in complex networks with applications to computer communication systems. *International Journal of Systems Science, 42*(6), 951–995.
6. Xiao, W. J., Jiang, S. Z., & Chen, G. R. (2011). A small-world model of scale-free networks: Features and verifications. *Applied Mechanics and Materials, 50–51*(2011), 166–170.
7. Ebel, H., Mielsch, L. I., & Bornholdt, S. (2002). Scale-free topology of e-mail networks. *Physical Review E, 66*(3), 1–4.
8. Jovanovic, M.A., Annexstein, F.S., & Berman, K.A. (2001). *Scalability issues in large peer-to-peer networks—A case study of gnutella* (Technical Report). Cincinnati: University of Cincinnati.
9. Cancho, R. F. I., & Solé, R. V. (2001). The small world of human language. *Proceedings of the Royal Society of London B, 268*(1482), 2261–2265.
10. Newman, M. E. J. (2001). Scientific collaboration networks. II. Shortest paths, weighted networks, and centrality. *Physical Review E, 64*(1), 016132.
11. Newman, M. E. J. (2001). Scientific collaboration networks. I. Network construction and fundamental results. *Physical Review E, 64*(1), 016131.

Chapter 146
A Preemptive Model for Asynchronous Persistent Carrier Sense Multiple Access

Lin Gao and Zhijun Wu

Abstract In order to analyze the problem of packet collision in the asynchronous mode of persistent carrier sense multiple access (p-CSMA), in which there is no time slot different from synchronous mode and propagation delay have a heavy effect on the probability of packet collision, a preemptive asynchronous p-CSMA probability model is established for the first time in the chapter. From sub-cycle conditional probability, the model gives the expectations of an idle and busy period. On the basis, performance targets, e.g., throughput/delay/success rate and channel efficiency, are gotten. For illustration, VDL2 (a typical asynchronous p-CSMA network) simulation model is set up on OPNET platform and experiments are also carried out to verify the correctness of this model in diverse scenarios. Through simulation, the results of fixed position distribution have the good consistency with the preemptive probability model. Finally, the conclusion is achieved that packet collisions will aggravate with the stations distribution becoming more uneven.

L. Gao (✉)
School of Electrical and Information, Tianjin University, Tianjin 300072, China

School of Science and Technology, Tianjin Economic and Financial University, Tianjin 300000, China
e-mail: gavingao71@sohu.com

Z. Wu
School of Electronics & Information Engineering, Civil Aviation University of China, Tianjin 300000, China
e-mail: zjwu@cauc.edu.cn

146.1 Introduction

146.1.1 p-CSMA Protocol

As a widely used channel access method, p-CSMA is evolved by ALOHA/CSMA. In the early random media access modes, such as ALOHA, each user is independent to send packet. The conflict probability increases exponentially with the payload increasing, which leads to low channel efficiency and throughput. In contrast to ALOHA, CSMA can sense the channel and judge if it is busy or not before transmitting packet. By this means, the user only has the opportunity to transmit when a channel is idle. As a result, the system performance is improved obviously.

With the function of probability decision, the basic principles of p-CSMA are as follows:

1. In the case of sensing that a channel is busy, the transmitting station keeps monitoring until it senses the channel to be free. Then, it will send packet with probability p or will not send with the probability $1 - p$.
2. In the state of channel being free, if there is at least one packet waiting to send and the packet is not sent out due to out of coverage of p-value, the station will not sense for an instant and will re-sense the channel after 2τ (in the chapter, τ represents the delay for radio propagation between stations). Then, it will continue to monitor, repeating the principle 1.

Furthermore, p-CSMA can be divided into synchronous mode and asynchronous mode according to whether or not time slot is applied. In synchronous mode, a channel is divided into time slots (all the nodes are synchronized by master clock) and a message can only be sent at the beginning moment of a slot. In contrast, asynchronous mode has no function both of time slot and of synchronization.

146.1.2 Related Research

The theoretical research into CSMA series has a long history. Early in the middle of the 1970s, Kleinrock, one of the Internet founders, published his landmark papers on packet switching by radio. As the theoretical basis, throughput–delay characteristics are given for ALOHA, CSMA, and p-CSMA and the large advantage which CSMA provides is shown as compared to the random ALOHA [1]. Furthermore, Kleinrock successfully discussed the problem of system stability [2]. The classic literatures supply us the basic principle and analytic methods for CSMA, e.g., the definition of system stability and the data stream modeling by means of Poisson distribution, which are applied in the chapter as the precondition of analysis.

In the premise of saturation, Bianchi model is established at the beginning of this century. The model is based on the two-dimensional Markov chain in analyzing the

DCF (CDMA/CA) performance [3]. In some sense, the model can be regarded as a specializing and deepening one of the Kleinrock model.

In recent years, Richard MacKenzie expounds how the throughput and delay for the traffic flows are affected by the relative p-persistence [4]. Taking the capture effect into account, Salim Abukharis discussed the analysis method of throughput and packet delay in the Rayleigh fading channel [5]. In Yayu Gao's paper, a view is pointed out: with the small transmitting probability, throughput has nothing to do with retransmission strategy, but it is only relevant to propagation delay [6]. Deqing Wang et al. discussed the long-delay hidden terminal problem which is decided by propagation delays between finite nodes [7]. Recently, the advantage of asynchronous p-CSMA is paid more attention: it saves the time of waiting in time slot and facilitates the realization. As a result, the access mode is applied in distributed network, especially in the next generation of aeronautical data link-VDL 2 [8].

146.2 Theory and Methodology

Different from others, this chapter establishes an asynchronous preemptive p-CSMA model, with full consideration of the propagation delay.

146.2.1 Basic Conception

For the preemptive multiplexing, the probability of collision plays a vital role: in the ideal channel, no collision means that a packet transmits successfully in a cycle. Otherwise, all of the transmissions fail in the cycle. Although p-CSMA have the ability of channel detection and probability decision, packet collision is still inevitable. The reasons why packet collision is destined to happen are as follows:

Reason 1: When more than one station both have packets to send and have detected the channel to be free within the coverage of p decision, they will send messages at the same time, which will cause packets to collide.

Reason 2: Due to the existence of propagation between the stations, a sent packet should only be sensed by a station after the propagation delay. Therefore, in the period of propagation delay, the station will think the channel to be idle and may send message, which will lead to packet collision.

It is obvious that collision probability of reason 2 is far greater than that of reason 1. Therefore, in the following analysis, collision from reason 1 is neglected and only collision from reason 2 is taken into consideration.

146.2.2 Assumptions

To facilitate the discussions and statements, we give two assumptions as follows:

1. With the same as Kleinrock model [1], we assume that the new-generated packets comply with Poisson distribution and a new-generated packet will be rescheduled to send in a future time if it can't be sent immediately. If the interval to retransmit is long enough, according to Kleinrock's assumption, the two kinds of packets can combine into a Poisson flow with a new intensity. Here, the combined Poisson intensity of station i is marked as λ_i.
2. In the chapter, we assume that packet length (l) complies with uniform distribution and we have $l \sim U(l_a, l_b)$, where l_a and l_b, respectively, represent the lower and upper bound of the distribution.

146.2.3 Modeling

In the chapter, idle period (T_0) is defined as the duration from the moment of channel being idle to t_0. In contrast, busy period (T_1) is defined as the duration from t_0 to the moment that all the M senders have just finished sending message. An idle period is followed by a busy period, vice versa. For S_1, we have:

$$P(S_i = S_1) = p_i \sum_{L=0}^{N-1} P(L) \prod_{j \neq i}^{L-1} P\left[\left(\frac{d_{0i}(0)}{C} + \tau_i\right) < \left(\frac{d_{0j}(0)}{C} + \tau_j\right)\right] \quad (146.1)$$

Supposing $a_{ij} = \frac{d_{0j}(0) - d_{0i}(0)}{C}$, the result is as shown in Eq. (146.2):

$$P\left[\left(\frac{d_{0i}(0)}{C} + \tau_i\right) < \left(\frac{d_{0j}(0)}{C} + \tau_j\right)\right] = \iint_{\tau_i - \tau_j < a_{ij}} \lambda_i e^{-\lambda_i \tau_i} \lambda_j e^{-\lambda_j \tau_j} d\tau_i d\tau_j$$

$$= \frac{\lambda_i}{\lambda_i + \lambda_j} e^{a_{ij} \lambda_j} \quad (146.2)$$

$$P(S_i = S_1) = p_i \sum_{L=0}^{N-1} \left(P(L) \prod_{j \neq i}^{L-1} \frac{\lambda_i}{\lambda_i + \lambda_j} e^{a_{ij} \lambda_j}\right) \quad (146.3)$$

Taking $T_0 = t_0$ into account, we have:

$$E[t_0 | S_i = S_1] = E\left[\frac{d_{0i}(0)}{C} + \tau_i\right] = \frac{d_{0i}(0)}{C} + E[\tau_i] = \frac{d_{0i}(0)}{C} + \frac{1}{\lambda_i} \quad (146.4)$$

$$E[T_0] = \sum_N E[t_0|S_i = S_1]P(S_i = S_1)$$

$$= \sum_{i=0}^{N-1} p_i \left[\frac{d_{0i}(0)}{C} + \frac{1}{\lambda_i}\right] \left[\sum_{N-1}^{L-1} P(L) \prod_{j \neq i} \frac{\lambda_i}{\lambda_i + \lambda_j} e^{a_{ij}\lambda_j}\right] \quad (146.5)$$

Similarly, if we suppose $b_i = \frac{d_{1i}(t_0) - d_{0i}(0)}{C}$, we have For S_2:

$$P(S_i = S_2) = \sum_N P(S_i = S_2|L)P(L)$$

$$= p_i \left[e^{-\lambda_i \left(t_0 - \frac{d_{0i}(0)}{C}\right)} - e^{-\lambda_i(b_i + t_0)}\right] \sum_{N-1}^{L-2} P(L) \prod_{j \neq i, j \neq 1} \frac{\lambda_i}{\lambda_i + \lambda_j} e^{a_{ij}\lambda_j} \quad (146.6)$$

$$E[t_1] = \sum_N E[t_1|S_i = S_2]P(S_i = S_2) = \sum_{N-1} p_i \left[\frac{d_{0i}(0)}{C} + \frac{1}{\lambda_i} - E[T_0]\right]$$

$$\left[e^{-\lambda_i \left(E[T_0] - \frac{d_{0i}(0)}{C}\right)} - e^{-\lambda_i(b_i + E[T_0])}\right] \left[\sum_{N-2}^{L-2} P(L) \prod_{j \neq i, j \neq 1} \frac{\lambda_i}{\lambda_i + \lambda_j} e^{a_{ij}\lambda_j}\right] \quad (146.7)$$

Due to $E[t_{N-1}] \ll \cdots \ll E[t_2] \ll E[t_1]$, we have:

$$E[T_1] = E\left[l + \sum_{i=0}^{N-1} t_i\right] \approx E[t_1] + E[l] \quad (146.8)$$

P-CSMA performance can be valued by the targets as follows:

$$P_s = \prod_{i \neq 1} [1 - P(S_i = S_2)] \quad (146.9)$$

$$\bar{S} = E[S] = E\left[\frac{P_s}{v} \lim_{n \to \infty} \frac{\sum_n l_i}{\sum_n T_i}\right] \approx \frac{P_s}{v} \lim_{n \to \infty} \frac{E\left[\sum_n l_i\right]}{E\left[\sum_n T_i\right]}$$

$$= \frac{P_s}{v} \frac{E[l]}{E[T_0] + E[T_1]} \quad (146.10)$$

$$\bar{\tau}_s = E[T_0] + (1 - P_s)E[T_1] \quad (146.11)$$

$$\eta \approx \frac{E[T_1]}{E[T_0] + E[T_1]} \quad (146.12)$$

In Eq. (146.10), v represents channel capacity, which is the maximum bit rate of a communication system. With limited bandwidth, it will satisfy the Nyquist theory to guarantee no inter-symbol interference.

From Eqs. (146.5) to (146.8), we can calculate packet success rate Eq. (146.9), throughput expectation Eq. (146.10), packet delay expectation Eq. (146.11), and channel efficiency Eq. (146.12). Based on the targets, performance can be evaluated in a quantified way.

146.3 Simulation Based on VDL2

Asynchronous p-CSMA is typically applied in VDL2 air–ground communication which can provide reliable network services for future aviation network—ATN—and has been tested in Europe and identified as the main stream in the future system.

In VDL2 communication, there are two sides: ground station (GS) and aircraft (AC), by which a peer entity is composed. The communication can only be realized in half-duplex manner by peer entities. VDL2 applies hierarchical structure and conforms to ISO standard. From the bottom to the upper, there are three layers in VDL2 protocol stack [9]: physical layer, data link layer, and network layer. MAC, which is a sub-layer of data link layer, can realize Media access control and channel multiplexing with asynchronous p-CSMA. In order to simulate the performance of asynchronous p-CSMA, a model has been constructed and programmed for VDL2 protocol on OPNET 14.5 platform.

In the simulation, 16 ACs are equipped with scattered position around GS, which can closely meet the condition of the theory in the chapter, as shown in Fig. 146.1.

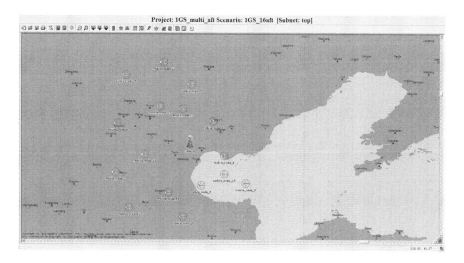

Fig. 146.1 Multi-ACs scenario

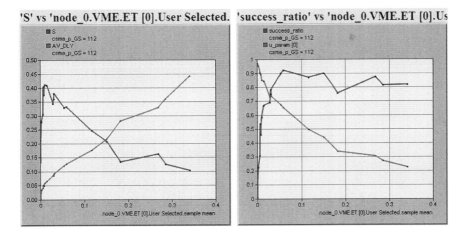

Fig. 146.2 E[T_0] trends with parameters

In simulation, parameter iat is defined as the mean of exponential distribution of the packet interval and we have iat $= 1/\lambda$ ground on Poisson theory. For packet length, we set the lower and upper limits to be $l_a = 128$ bits and $l_b = 8,320$ bits according to VDL2 protocol and assume that the random variable abides by uniform distribution.

From Eqs. (16), (18), and (19), we can see that the expectations of idle and busy period are keys to calculate the targets of both throughput and packet delay as well as channel efficiency. From Eqs. (146.7) and (13), we know that $E[T_0]$ and $E[T_1]$ change in opposite direction. Therefore, we can focus on $E[T_0]$, which is decided by multiple parameters. The simulation results are shown in Fig. 146.2. From the figure, we can see $E[T_0]$ has positive correlation with packet delay and has negative correlation with throughput and channel efficiency (u_param). There is no obvious correlation between success rate and $E[T_0]$. The results are in consistent with the theory we deduced.

With the method of numerical analysis, the simulation results of fixed position distribution have the good consistency with preemptive probability model we set up, which can verify the correctness of the mathematic model.

Furthermore, different scenarios are built by changing the distribution of nodes' positions. The results of simulation become complicated and variable. But with the multiple experiments, we can see that distribution pattern has a significant impact on the system performance. Specially, packet collisions will aggravate as the locations of stations become more uneven. The conclusion conforms to the sequitur from the probability model that success rate is negatively related to the sum of the absolute value of $d_{ij}(t) - d_{ik}(t)$, where $i \neq j \neq k$ and $i, j, k \in (0, N-1)$.

146.4 Conclusion

Taking propagation delay into account, a preemptive probability model is built for asynchronous p-CSMA. Network performance, e.g., throughput and packet delay, can be achieved with the idle and busy period expectation calculated in the model. Through simulation, the probability model is verified in scenarios with both fixed position distribution and varied position distributions.

The model can also be simplified by specialization. For instance, it can be specialized on Bianchi's critical saturation condition. Under the condition, the actor of packet rate is eliminated and P-CSMA performance Eqs. (146.9)–(146.12) can be expressed in a simplified form. Through specialization, further certain relationships between performances and parameters (especially position distribution of nodes) will be expected, which will benefit the network optimizations.

Acknowledgements This work is financially supported by the National Natural Science Foundation of China (No. 61170328) and Natural Science Foundation of Tianjin (No. 12JCZDJC20900).

References

1. Kleinrock, L., & Tobagi, F. A. (1975). Packet switching in radio channels: Part I-carrier sense multiple-access modes and their throughput-delay characteristics. *IEEE Transactions on Communications, 23*(12), 1400–1416.
2. Tobagi, F. A., & Kleinrock, L. (1977). Packet switching in radio channels: Part IV-stability considerations and dynamic control in carrier sense multiple accesses. *IEEE Transactions on Communications., 25*(10), 1103–1119.
3. Bianchi, G. (2000). Performance analysis of the IEEE 802.11 distributed coordination function. *IEEE Journal on Selected Areas in Communications., 18*(3), 535–547.
4. MacKenzie, R., & O'Farrell, T. (2010). Throughput and delay analysis for p-persistent CSMA with heterogeneous traffic. *IEEE Transactions on Communications., 58*(10), 2881–2891.
5. Abukharis, S., MacKenzie, R., & O'Farrell, T. (2011). Throughput and delay analysis for a differentiated p-persistent CSMA protocol with the capture effect. In *Proceedings of the 2011 I. E. 73rd vehicular technology conference* (pp. 1–5), Piscataway, NJ, USA.
6. Gao, Y., & Dai, L. (2011). On the throughput of CSMA, information sciences and systems. In *45th annual conference of IEEE Computer Society* (pp. 1–4), Piscataway, NJ, USA.
7. Wang, D., Hu, X., Xu, F., Chen, H., & Wu, Y. (2012). Performance analysis of P-CSMA for underwater acoustic sensor networks. In *IEEE OCEANS 2012 MTS* (pp. 1–6), Piscataway, NJ, USA.
8. Wargo, C. A., D'Arcy, J. F. (2011). Performance of data link communications in surface management operations. In *IEEE aerospace conference* (pp. 1–10), Piscataway, NJ, USA.
9. ICAO Annex 10 Aeronautical Telecommunications. (2007). Volume III part I, digital data communication systems.

Chapter 147
Extended Petri Net-Based Advanced Persistent Threat Analysis Model

Wentao Zhao, Pengfei Wang, and Fan Zhang

Abstract In order to display the attack scene in the description of the multistep process-oriented attack—advanced persistent threat, a specific model on advanced persistent threat behavior analysis—EPNAM is proposed, which is based on the Petri net and combined with the characteristics of APT. Firstly we carry out hierarchical analysis on the attack scene with AHP method to build the APT architecture and extract scene factors, then associate the attack scene with Petri net to construct extended Petri net, and finally, traverse the extended Petri net to generate the formal expression. The proposed model can achieve the combination of the attack scene, attack process, and state space, and its feasibility is proved by the application on actual case analysis of the RSA SecurID theft attack.

147.1 Introduction

APT (advanced persistent threat) is a kind of network attack launched by organizations (especially governments) or small groups, which aims at specific targets and works persistently by means of advanced techniques. Generally believed, APT is a particular type of attack on the network infrastructure, aiming to obtain the key information of a particular organization or government, including energy, power grid, financial organization, and national defense. APT utilizes a variety of approaches to get the authorization of an inner organization step by step, including advanced techniques and social engineering methods. Attackers launch a series of attacks to a specific target; what those attackers really want is usually not getting benefit in a short period of time but keeping on searching through the compromised host, until they thoroughly grasp the key information of the target person or thing [1–3].

W. Zhao • P. Wang (✉) • F. Zhang
School of Computer, National University of Defense Technology, Changsha 410073, China
e-mail: wpengfei_nudt@163.com

147.2 Related Works

APT brings the threat that continues to strengthen, but for the relative lack of research, systemic knowledge has not yet formed, nor has the practical research method been found.

Currently the most sophisticated models of network attack include the attack tree model, attack graph model, and attack net model based on Petri net. The attack tree model proposed by B. Schneier [4] is intuitive and hierarchical, but hard to distinguish attack state from attack behavior, and has poor scalability [5].

Swiler et al. proposed the attack graph model [6, 7] from the graph theory, but analysis of large-scale network by the attack graph model will lead the state space of the attack graph algorithms into exponential growth, which is difficult to search or has the flexibility to add and modify [8, 9].

McDermott proposed the attack net model based on Petri net [10], which can separate attack state from attack process, and has the hierarchical nature of the attack tree model. But as a general attack model, it is insufficient in the description of the characteristics of the APT, such as persistence and penetrability, especially it cannot reflect the scene features and milestones of APT.

Based on the above status, we propose an extended Petri net-based modeling approach, combined with persistence and penetrability of APT, as well as the depiction of the attack scene, which can completely include the key factors of APT, including the object, intention, preconditions, vulnerability, penetration, and state space.

147.3 APT Modeling Analysis Based on Extended Petri Net

The attack net model based on Petri net was first proposed by McDermott [10]. Places of Petri net represent the path attack stage or state, indicated as "O," transitions represent the attack behavior, indicated as "□"; paths represent the attack process, indicated as "→"; and Token represents the current state, indicated as "•."

The extended Petri net-based APT model proposed in this chapter combines with APT characteristics. Firstly do hierarchical analysis on the attack scene to extend the classical Petri net model according to the extracted scene factors. Then take advantage of the extended Petri net for modeling analysis to generate formal expression. And finally achieve the target of the attack scenes factors reflected in a multistep process-oriented attack description so as to complete the combination of attack scenes, attack process, and time states.

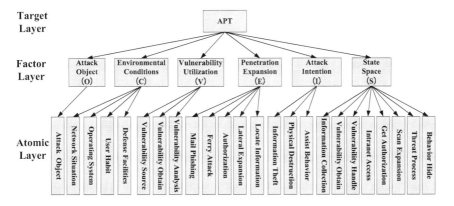

Fig. 147.1 Hierarchical analysis structure

147.3.1 Attack Scene Hierarchical Analysis

First analyze the attack scene of APT hierarchically by analytical hierarchy process. As shown in Fig. 147.1, target layer is defined as the top layer, then the factor layer forms 6-tuple {*O, C, V, E, I, S*} to characterize a specific APT scene, and the underlying atomic layer represents the specific operation.

1. *Attack object*: Potential target of an APT, such as the corporation intranet, mail server, or industrial control systems, denoted by *O*
2. *Environmental conditions*: Surrounding environment and preconditions to launch an APT, denoted by *C*
3. *Vulnerability utilization*: Vulnerability (0 day) obtaining and utilization is prerequisite for the implementation of APT, denoted by *V*
4. *Penetration expansion*: Locate important information by implementing lateral extension, which is the key step in APT implementation, denoted by *E*
5. *Attack intention*: The ultimate goal of the attacker, such as information theft, physical damage, or simply obtaining authorization to conceal, denoted by *I*
6. *State space*: Set of critical state in the procedure of APT, denoted by *S*

147.3.2 Extended Petri Net Modeling Principle

The proposed extended Petri net model is extended to a 6-tuple {*O, C, V, E, I, S*} on the basis of the original 3-tuple. State space set *S* comes from the place node set (*P* in the 3-tuple) of the classical Petri net, which describes the critical state in the procedure of APT. The first state from *S* is extracted to represent the attack object, denoted as *O*. Then further divide the original transition node set (*T* in the 3-tuple) into a set of environmental condition set, vulnerability utilization set, penetration

Table 147.1 Extended Petri net system

Scene factor	Atomic index	Meaning	Candidate value
O	object	Attack object	object_intranet, object_server, object_ctrlSystem
C	net	Network situation	net_connect, net_isolated
	os	Operating system	os_windows, os_linux, os_mac, os_embed
	habit	Habits characteristic	habit_office, habit_photo, habit_music, habit_movie
	defense	Defense facilities	defense _ids, defense _audit
V	source	Vulnerability source	source _office, source _os, source_app
	obtain	Vulnerability obtain	obtain_buy, obtain_dig, obtain_exchange
	analysis	Vulnerability analysis	analysis_0day, analysis_1day
E	mail	Mail phishing	mail_server, mail_pc, mail_social, mail_sqlInject
	ferry	Ferry attack	ferry_worm, ferry_Trojan
	authorize	Gain authorization	authorize_malWare, authorize_backDoor, authorize_bruteForce, authorize_phishing
	expand	Laterally expansion	expand _ scan User, expand_sendMsg
	locate	Information locating	locate_info, locate_asset, locate_person
I	info	Information theft	info_pack, info_compress, info_encrypt
	destroy	Physical destruction	destroy_modify, destroy_delete, destroy_cheat
	assist	Assist behavior	assist_conceal, assist_maintain
S	s1	Informationcollection	s_infoCollect
	s2	Vulnerability obtain	s_ vulnerabilityObtain
	s3	Vulnerability handle	s_ vulnerabilityHandle
	s4	Intranet entrance	s_intranetAccess
	s5	Get authorization	s_getAuthorization
	s6	Scan expansion	s_scanExpand
	s7	Threat process	s_threatProcess
	s8	Behavior conceal	s_behaviorConceal

expansion set, and attack intention set. In the need to characterize the dynamic attack scene, Token, which from the original Petri net is retained, the state node where Token at indicates the current state. The connection set (F in the 3-tuple) is omitted in the 6-tuple. Extended Petri net system is shown in Table 147.1.

Definition 1: Extended Petri net-based APT model. EPNAM = {$ob, co, vu, ex, it, st \mid ob \in O, co \in C, vu \in V, ex \in E, it \in I, st \in S$}. Where

O = {$object$}; C = {$net, os, habit, defense$}
V = {$source, obtain, analysis$}; E = {$mail, ferry, authorize, expand, locate$}
I = {$info, destroy, assist$}; S = {$s1, s2, s3, s4, s5, s6, s7, s8$}

Extended Petri net is generated, as shown in Fig. 147.2.

Definition 2: State change. State changes of EPNAM can be expressed as $O \times C \times V \times E \times I \times S \rightarrow S$. APT process can be classified into stages as environmental conditions, vulnerability utilization, penetration expansion, and attack intention; state change in each stage can be expressed as

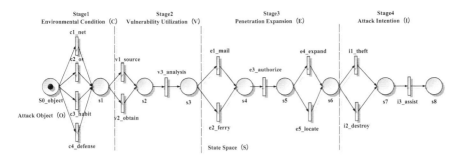

Fig. 147.2 Extended Petri net

Environmental conditions: $O \times C \rightarrow S$; Vulnerability utilization: $S \times V \rightarrow S$
Penetration expansion: $S \times E \rightarrow S$; Attacks intention: $S \times I \rightarrow S$

147.3.3 The EPNAM Generation Algorithm

Step 1: Scene hierarchical analysis
Analyze APT scene with AHP, extract each layer, and determine the 6-tuple $\{O, C, V, E, I, S\}$ of the factor layer, and the atomic indicators of each factor.

Step 2: Scenes association with Petri net
Extract key characteristics of the attack scene, then associate them with atomic layer indicators. Determine the value of each atom layer indicator from the existing values within the selecting domain to complete the 6-tuple.

Step 3: Construction of extended Petri net
The Petri net is constructed in accordance with the 6-tuple stage order with a serial structure. In each stage, the parallel structure is used between the atomic indicators and uses serial structure between the chosen values for each indicator.

Step 4: Formal expression generation
Traverse the constructed Petri net by the mobile of Token to generate formal expression. *attack_sqc* represents the attack sequence generated at the last stage, *stage_sqc* represents the attack subsequence produced at a certain stage in APT, and *atom_sqc* is the collection of value of a certain atomic indicator.

Algorithm 1 Generation of attack sequence
Input: $S = \{s1, s2, s3, s4, s5, s6, s7, s8\}$
Output: *attack_sqc*
PROCEDURE GENERATE_SEQUENCE(S)
BEGIN:
attack_sqc $\leftarrow \emptyset$; *section_sqc* $\leftarrow \emptyset$; *atom_sqc* $\leftarrow \emptyset$
token $\leftarrow$ s1 section $\leftarrow$ S1
FOR token $\leftarrow$ s1 TO s9 DO

(continued)

(continued)

```
    section_sqc ← ∅
    WHILE nextIndex(token) ≠ ∅
      atom_sqc ← ∅
      WHILE isSelected(atom) IN AtomDomain
        atom_sqc ← atom_sqc∧atom
        section_sqc ← section_sqc∨atom_sqc
      token ← token + 1
    IF changSection(token) THEN
      nextSection(section)
    attack_sqc ← (attack_sqc → section_sqc)
END
```

147.4 Case Study

In this chapter, we choose the famous RSA SecurID theft attack as the analysis case to prove the feasibility of the proposed model, which is the most representative one that contains the universal characteristics of APT.

RSA, which is the subsidiary company of EMC, suffered APT in March 2011, and part of the SecurID technology and customer information was stolen. Its consequences led to the theft of important information of the company which uses SecurID as the authentication credentials to establish a VPN network. The following analysis is implemented using the proposed EPNAM approach on the background of RSA SecurID theft attack.

1. Scene hierarchical analysis (Table 147.2)
2. Scene association with Petri net (Table 147.3)
3. Construction of extended Petri net (see Fig. 147.3)
4. Formal expression generation

Table 147.2 RSA SecurID theft attack scene analysis

Scene factor	Description
O	Intranet of RSA, mail server
C	Collect real-time information on RSA corporation. Imitate communication characteristics and habits of branch company. Intranet web server, windows operating system, and MS office, equipped with IDS and auditing system
V	Sending malicious e-mail to the chief manager in the name of the branch company, attachment (2011Recruitment Plan.xls) of which is embedded with a new 0-day vulnerability, Adobe Flash CVE-2011-0609
E	Trojan download from the C&C server; expand rapidly, get access to PC with high authority, raise authorization, locate important files
I	Gather the SecurID technique and customer information of RSA corporation, then pack compress, encrypt, and send out through FTP to a remote PC. Clear trace and maintain the current permission of the network
S	Information collection, vulnerability obtain, vulnerability utilization, mail penetration, lateral expansion, information theft, conceal maintain

147 Extended Petri Net-Based Advanced Persistent Threat Analysis Model

Table 147.3 Association

Scene factor	Atomic index	Candidate values
O	*object*	object_intranet, object_server
C	*net*	net_connect
	os	os_windows
	habit	habit_office, habit_photo
	defense	defense_ids, defense_audit
V	*source*	source_office, source_app
	obtain	obtain_dig
	analysis	analysis_0day
E	*mail*	mail_server, mail_pc, mail_social
	authorize	authorize_malWare, authorize_phishing
	expand	expand_scan User, expand_sendMsg
	locate	locate_info, locate_asset, locate_person
I	*info*	info_pack, info_compress, info_encrypt
	assist	assist_conceal, assist_maintain
S	s1	s_infoCollect
	s2	s_vulnerabilityObtain
	s3	s_vulnerabilityHandle
	s4	s_intranetAccess
	s5	s_getAuthorization
	s6	s_scanExpand
	s7	s_threatProcess
	s8	s_behaviorConceal

According to algorithm 1 to traverse the preconstructed extended Petri net, the generated formal expression is described as follows:

O(object_inNet∧object_server) → C(net_connect∨os_windows∨(habit_office∧habit_photo)∨(defense_ids∧defense_audit))→
V((source_office∧source_app)∨obtain_dig∨analyse_0day)→
E((mail_server∧mail_pc∧mail_social)∨
  (authorize_malWare∧authorize_phishing)∨
(expand_scanUser∧expand_sendMsg)
  ∨(locate_info∧locate_asset∧locate_person))→
I((info_pack∧info_compress∧info_encrypt∧)∨(assist_conceal∧assist_maintain))

Generation of attack states transformation is described as below:

s_infoCollect → s_vulnerability Obtain → s_vulnerability Handle → s_intranet Access →
s_getAuthorization → s_threatProcess → s_behaviorConceal

The above analysis proves the feasibility of the proposed model. Furthermore, the use of the model can do a better depiction on the attack scene, attack process as well as state change of APT, and the formal expression can be used as basis for further behavioral analysis and quantitative assessment.

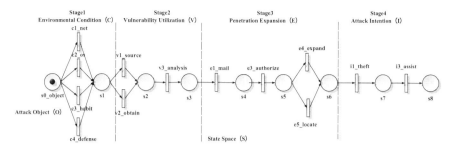

Fig. 147.3 Extended Petri net of RSA SecurID theft attack

147.5 Conclusion

In this chapter, an APT behavior analysis model based on extended Petri net—EPNAM model—is proposed, which can do hierarchical analysis on attack scene, refine the APT hierarchy and scene factors, associate the attack scene with Petri net to construct extended Petri net, and generate the formal expression by traversing the Petri net. EPNAM model can completely portray the characterization of an APT by embodying the attack scene in a multistep process-oriented attack description and finally complete the combination of the attack scene, attack process, and state space. The feasibility of the proposed model is proved by the analysis of actual case.

The following work focuses on refining the scene analysis and scene association, improving the adaptability of the proposed model on different APT cases, and implementing behavior analysis and quantitative assessment on the basis of the generation of formal expression.

Acknowledgements This work is supported by the National Natural Science Foundation of China (No. 61271252).

References

1. Huang, D., & Xue, Z. (2012). Research and analysis on advanced persistent threat behavior. *Information Security and Communication Secrecy, 1*(5), 94–96 (In Chinese).
2. Bo, N. (2012). APT: Conceal, aims at theft of enterprise secret. *Journal of China Computer, 1*(1), 1–3 (In Chinese).
3. Chen, J., Wang, Q., & Wu, M. (2012). Network APT attack and defense policy. *Information Security and Communication Secrecy, 1*(7), 30–33 (In Chinese).
4. Schneier, B. (1999). Attack trees-modeling security threats. *Journal of Software Tools, 24*(12), 21–29.
5. Li, W. (2010). Research on network attack modeling method. *Computer CD Software and Applications, 1*(13), 28–29.
6. Phillips, C., & Swiler, L. (1998). A graph-based system for network vulnerability analysis. In *Proceedings of the 1998 Workshop on New Security Paradigms* (pp. 71–79). New York, NY: ACM.

7. Swiler, L., Philips, C., Ellis, D., & Chakerian, S. (2001). Computer-attack graph generation tool. In *Proceedings of DARPA Information Survivability Conference and Exposition* (pp. 307–321). Los Alamitos, CA: IEEE.
8. Wang, G., Wang, H., Chen, Z., & Xian, M. (2009). An attack graph-based computer network attack modeling method. *Journal of National University of Defense Technology, 31*(4), 78–84.
9. Cheng, K. (2010). *Cyber attacks based on attack graphs and Petri nets model study*. Xi'an, China: Xi'an University of Architecture University of Science and Technology.
10. McDermott, J. (2000). Attack net penetration testing. In *Proceedings of the 2000 Workshop on New Security Paradigms* (pp. 15–21). New York, NY: ACM.

Chapter 148
Energy-Efficient Routing Protocol Based on Probability of Wireless Sensor Network

Kaiguo Qian

Abstract This chapter mainly discusses the problem of wireless sensor network routing protocols. Based on analysis of the disadvantages of information implosion and overlapping caused from implementation mechanisms of the flooding protocol, energy-efficient routing protocol based on probability of wireless sensor network (ERPBP) is proposed and evaluated. It uses the node distance and residual energy as the weights to calculate the forwarding probability of neighbor nodes and chooses some of maximum forwarding probability nodes as router. It saves the energy by avoiding redundancy packet copies produced and improves the disadvantage of flooding routing protocol. Performance analysis and simulation experiment show that the new protocol effectively reduces the data redundancy, reduces the energy consumption, and prolongs the network lifetime.

148.1 Introduction

Wireless sensor network (WSN) [1] is a self-organization network system that is widely used in environmental monitoring, medical care, urban traffic management, warehouse management, and military reconnaissance and is now becoming the hottest research field. Wireless sensor network (WSN) [2] has the following characteristics: node energy is limited and large-scale network of nodes is without uniform distribution, node mobility, and network topology dynamic change. Routing protocol [3] is designed to transmit the data from the source node to the base station by intermediate forwarding nodes. Primary goal of routing design is to meet the energy constraint of sensor node, to reduce energy consumption, and to prolong the network life cycle. Second, it has higher scalability because of the topological structure dynamic change. Third, routing protocol implementation is

K. Qian (✉)
Department of Physics Science and Technology, Kunming University,
Kunming 650031, China
e-mail: qiankaiguo@qq.com

simple on demands of weakness of computing power and storage capacity. Researchers have proposed hundreds of solutions, in which the flooding routing protocol originated in the ad hoc network is the simplest and reliable algorithm. Flooding does not need the information of the whole sensor network and has the advantages of high scalability, simple calculation, and high reliability. Whereas, it causes the problem of redundant packet copies, information implosion and overlap. With the analysis of Gossip [4] and comprehensively considering the influence of the node residual energy and communication distance, the chapter presents a new routing protocol base nodes forwarding probability calculated according to the node residual energy and communication distance. Performance analysis and simulation experiment show that the new protocol effectively reduces the data redundancy and the energy consumption and prolongs the network lifetime.

148.2 Related Work

In flood-routing protocol, source node sends collected data packet to all of the own neighbors. Every neighbor node receives data from other nodes and broadcasts to its neighbors, so that it goes on until the data transmitted to the destination sink. It produces a large amount of redundant information, causing an information implosion and overlap. Gossip [4] randomly chooses a neighbor node as the data receiver if a node has the data to transmit to sink, so do the neighbors. This approach decreases redundant data copies and relieves the information implosion and overlapping. It causes longer delay time for the packet transmission. SPIN [5] and DD [6] are two data-centric protocols. Clustering routing protocol is a kind of hierarchy that divides the nodes into the cluster head nodes and member nodes. Cluster heads manage member nodes in the native. Leach [7] and TEEN [8] are not reliable as flooding, which are representatives for the clustering routing protocol.

148.3 Problem Description

The energy-efficient routing protocol based on probability (ERPBP) routing protocol runs in the network application scenario as the following: N sensor node is randomly deployed in M × M square area. Each sensor node is stationary after completion of WSN deployment. All nodes in the network are homogeneous and energy constrained. We take the radio energy model [4] in order to verify energy efficiency of the ERPBP routing protocol:

$$E_{Tx}(k,d) = \begin{cases} kE_{elec} + k\varepsilon_{fs}d^2 : d < d_0 \\ kE_{elec} + k\varepsilon_{mp}d^4 : d > d_0 \end{cases} \quad (148.1)$$

Equation (148.1) is energy dissipation equation for transmitting k-bit message and for receiving k-bit message is shown Eq. (148.2):

$$E_{Rx}(k) = kE_{elec} \quad (148.2)$$

In above model, E_{elec} is the radio dissipate to run the transmitter or receiver circuitry.

148.4 Energy-Efficient Routing Protocol Based on Probability

There are two ways to improve energy efficiency for running routing protocol from the radio energy model. The first is decreasing packets on the data transmission process, that is, decrease k in the radio energy model. The second is shorting the communication distance. It rotates the routing node to realize balancing energy consumption. The ERPBP protocol calculates the neighbor node forwarding probability based on node residual energy and communication distance if nodes may transmit packets and selects a few nodes of the maximum probability as forwarding nodes.

148.4.1 The Principle of ERPBP Protocol

In flooding protocol, if the node S wants to transmit data to the sink, it sends the data copies to every neighbor node, and so does every neighbor except sending back to source node until that sink node receives the data. The data is discarded if the TTL (time to live) becomes 0. As can be seen from Fig. 148.1, it produces redundant data and information implosion.

In ERPBP protocol, Source node S_i calculates the forwarding probability p_i of neighbors according to Eq. (148.3) and chooses a few of nodes of the maximum probability to send packets. The forwarding probability is proportional to the node residual energy. The greater the residual energy of the node, the greater the forwarding probability. The forwarding probability is not a simple inverse ratio relationship with the communication distance between nodes. We assume that the forwarding probability will decrease quickly when node distance is greater than the threshold d_0. We put forward the probability calculation mathematical model as shown in Eq. (148.3):

$$p(e,d) = \begin{cases} ae + \dfrac{1}{(d+1)^\alpha} & (0 < d \leq d_0) \\ ae - bd + c & (d > d_0) \end{cases} \quad (148.3)$$

Fig. 148.1 Principle of flooding

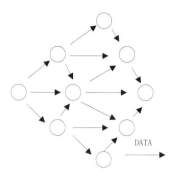

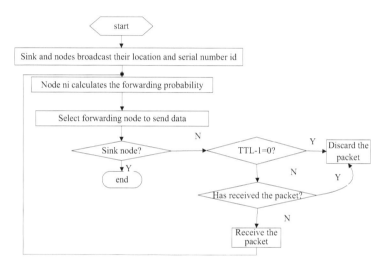

Fig. 148.2 Process of ERPBP protocol

148.4.2 The ERPBP Protocol Description

Data message contains TTL field and DATA field. TTL field in the message prevents unrestricted forwarding packet in the network. If a node n_i needs to send the data message, it does as shown in Fig. 148.2:

(a) Sink and nodes broadcast their location and serial number id.
(b) Node n_i calculates the forwarding probability of neighbors according to the node's residual energy and distance.
(c) Node n_i chooses a few nodes with maximum forwarding probability as next hops.

(d) If the received message node is sink node, the message has already arrived the destination; otherwise, jump to (e).
(e) If the TTL − 1 is equal to zero or the destination node has already received the packet, jump to (f); otherwise, turn to (b).
(f) Discard the packet.

148.5 The Performance Analysis and Performance

148.5.1 Performance Analysis

The ERPBP protocol carries the advantage that algorithm implementation mechanism is simple. It is suitable for topological dynamic change. In the process of packet routing, the neighbors of larger residual energy and near distance are preferentially selected for the forwarding nodes. Nodes, in which the distance is far and residual energy is small, do not participate in data forwarding. For this reason, it reduces energy consumption and realizes power load balance to prolong the system life. At the same time, the ERPBP protocol limits the number of nodes to avoid the generation of a large amount of redundant information and alleviates the information implosion in flooding algorithm.

148.5.2 Performance

An experiment is designed in MATlab to simulate and estimate the performance. 100 sensor nodes are randomly deployed in the area of 100 m × 100 m. Sink node is located at point 100 × 100.

1. Lifetime of system. Experiment running 5,000 rounds, the results of dead nodes for the traditional flooding protocol and the ERPBP algorithm is shown as Fig. 148.3. In the flooding protocol, the sensor nodes begin to die at about 250 rounds and the dead nodes reach to more than 70% at about 1,000 rounds. Ninety percent of nodes are dead at the end of running. When ERPBP protocol runs 1,500 rounds, the first dead node emerges. Eighty percent nodes die between 1,950th round and 2,000th round. The result proves that the energy consumption of the RPBEN protocol is balanced. At the same time, it shows that the RPBEN protocol has higher energy utilization rate than flooding and has longer life cycle.
2. Loss packets. Loss packet experimental results are as shown in Fig. 148.4. In traditional flooding protocol, the total number of lost packets is probably 4.4×10^5 when sensor network runs 1,000 rounds. The reason is that redundancy data copies are produced in the process of the operation of the network with flooding protocol. After 1,000 rounds, there is no data to produce because

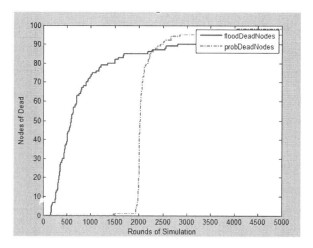

Fig. 148.3 Dead nodes with rounds

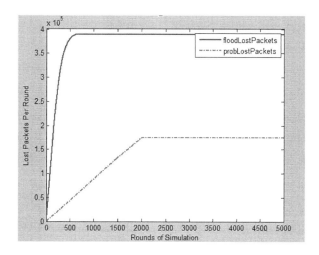

Fig. 148.4 Lost packets with rounds

sensor network is dead. In the ERPBP protocol, the total number of lost packets grows a lot less than the flooding protocol. Running at 1,950th round, loss packets achieve 1.8×10^5. There are no loss packets after 1,950th round because almost all nodes are dead.

Combining the experimental results of lifetime of system and loss packets show that the network life cycle of flooding lasted about 500 rounds and the ERPBP protocol life cycle lasted for 2,000 rounds. It is proved that the improved protocol is a more efficient flooding algorithm.

148.6 Conclusion

This chapter presents a new energy-efficient routing protocol based on neighbors' probability calculated for the weights of the node's residual energy and distance. Source node chooses a few neighbor as forwarding hop routing nodes according to probability. This mechanism balances energy consumption in sensor networks and reduces a large number of redundant packets of flood process. It improves the disadvantages of information implosion, overlapping and the blinding usage of resources in flooding algorithm. Performance analysis and simulation results show that the new protocol is more energy-efficient than the flooding protocol.

Acknowledgements This research was supported by Foundation of Yunnan Educational Committee (2011Y238), China. The authors thank the anonymous reviewers whose comments have significantly improved the quality of this chapter.

References

1. Akyildiz, L. F., Su, W. L., Sankarasubramaniam, Y., & Cayirci, E. (2002). A survey on sensor networks. *IEEE Communications Magazine., 40*(8), 102–114.
2. Ren, F. Y., Huang, H. N., & Lin, C. (2003). Wireless sensor networks. *Journal of Software., 14*(7), 1282–1291 (in Chinese).
3. Tang, Y., Zhou, M. T., & Zhang, X. (2006). Overview of routing protocols in wireless sensor networks. *Journal of Software., 17*(3), 410–421 (In Chinese).
4. Haas, Z., Joseph, Y., & Li, H. L. (2006). Gossip-based ad hoc routing. *IEEE/ACM Transactions on Networking (TON), 14*(3), 479–491.
5. Kulik, J., Rabiner, W., Balakrishnan, H. (1999). Adaptive protocols for information dissemination in wireless sensor networks. In Proceedings of the fifth annual ACM/IEEE international conference on mobile computing and networking (pp. 174–185). New York, NY: ACM.
6. Ntanagonwiwat, I. C., Govindan, R., & Estrin, D. (2000). Directed diffusion: A scalable and robust communication paradigm for sensor networks. In Proceedings of the sixth annual international conference on mobile computing and networking (pp. 56–57). New York: ACM.
7. Heinzelman, W. R., Anantha, C., & Hari, B. Energy-efficient communication protocol for wireless microsensor networks. In IEEE proceedings of the Hawaii international conference on system sciences (pp. 3005–3014), Maui, Hawaii.
8. Manjeshwar, A., & Agrawal, D. P. (2001). TEEN: A routing protocol for enhanced efficiency in wireless sensor networks. In Proceedings of the 15th parallel and distributed processing symposium (pp. 2009–2015), San Francisco, USA.

Chapter 149
A Dynamic Routing Protocols Switching Scheme in Wireless Sensor Networks

Zusheng Zhang, Tiezhu Zhao, and Huaqiang Yuan

Abstract Many sensor query processing systems have been developed to acquire, process, and aggregate data from wireless sensor networks. The energy consumption of query processing is significantly impacted by routing protocol. In this chapter, we propose a dynamic routing protocols switching scheme for query processing. The scheme supports multiple kinds of routing protocols coexisting in a single sensor node, and these protocols can be switched according to query tasks. Simulation results show that the dynamic scheme is more energy efficient than single routing protocol.

149.1 Introduction

The data query processing systems, such as TinyDB [1], are promising for wireless sensor networks. With these systems, user can inject SQL-style queries into a network through a PC. The networked sensor nodes then work together to process the queries and send results back to the PC. The performance of these sensor systems is greatly affected by the routing protocol, because routing protocol is responsible for result gathering.

Traditionally, in wireless sensor network each routing protocol is designed, developed, and evaluated separately, and the network layer uses a single routing protocol to deal with all kinds of queries in a query system [2]. However, a routing protocol is optimum for special task and network condition. For acquisition, such as "select temperature from sensors," a node doesn't aggregate data packets from neighbors, and all data packets are forwarded to BS through multi-hop. MintRoute [3], CTP [4], and other collection protocols [5] are energy efficient to establish flows up a tree and pull data out of a network. Another example, for aggregation,

Z. Zhang (✉) • T. Zhao • H. Yuan
Dongguan University of Technology, Dongguan 523808, China
e-mail: zushengzhang@126.com

such as "select average (temperature) from sensors," a node aggregates its sensed data with data packets received from neighbors into one packet. Clustering technique is energy efficient for aggregation. Many clustering protocols [6–8] that have been proposed for sensor networks, such as LEACH [6] and HEED [7], are the most popular. Jun and Julien [9] demonstrated that an optimal routing protocol can be selected for a particular application in ad hoc networks. He and Raghavendra [10] described a framework to build programmable routing services for sensor networks.

This chapter proposes the protocols switching scheme for reducing energy consumption of query processing. We decompose the network layer into shared modules and private modules. The design allows several routing protocols to coexist without burdensome memory requirement. BS decides protocols switching according to query tasks, attaches the information in the query, and disseminates the query to whole network. Then the network executes protocol switching. Simulation results show that dynamic switching scheme is more energy efficient than single routing protocol.

149.2 A Switching Scheme

149.2.1 Modular Design

Unlike traditional monolithic way, an overall modular design can increase code reuse and runtime sharing. Code reuse will foster more rapid protocol and application development. Runtime sharing refers to the sharing of code and resources such as memory and radio. Based on the TinyOS [11] code of MintRoute and the chapter of [12], we decompose the network layer into separate four parts: ForwardingEngine (FE), RoutingEngine (RE), MsgPool (MP), and NeighborMgt (NM), as shown in Fig. 149.1. The FE and MP modules support information exchanging for application and MAC layers by interfaces. The MP and NM are shared modules for all routing protocols. This is in contrast to the FE and RE modules; different routing protocols have different algorithms implemented in these modules.

The main function of an FE module is to obtain the next hop to which the packet is to be forwarded. Additional functions include detection of cycles, network level retransmission, and duplicate packet elimination. RE is responsible for creating and maintaining the routing topology. Examples of topologies are trees, geographic coordinates, or clusters. NM manages information about node's neighbors. MP is the buffer place for outgoing packets, and it supports packet scheduling for all network protocols.

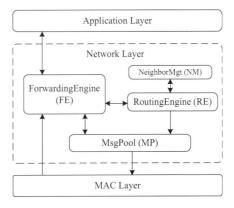

Fig. 149.1 Modular design of the network layer

Table 149.1 Polices of protocol selection

Keywords	Task	Routing protocol
SUM, AVE, MAX, MIN	Aggregation	HEED
WHEN	Event	CTP
Others	Collection	MintRoute

149.2.2 Protocol Selection

We adopt a centralized control mode for routing protocol selection. BS maintains a protocol selection table, as shown in Table 149.1. User defined the table entries, i.e., routing switching criteria, according to the overall knowledge of the network performance by theoretic analysis, simulation, and experiment results. Procedure at BS is as follows:

(a) Upon receiving a query from user, BS matches the query command to task. As shown in Table 149.1, for SQL language, BS checks the query task according to SQL keywords. (b) BS selects routing protocol according to query task. Then it attaches the routing protocol information on the query command. Such as "select temperature from sensors using MintRoute," it informs the network switching to MintRoute for the query. Finally, BS disseminates the query to the whole network.

149.2.3 Protocols Switching

To achieve the compatibility and expansibility, the dynamic scheme provides uniform interface for application and MAC layers. Small changes of the code inside of modules of routing protocols are required, while other layers' code need not be modified. The modular design of the dynamic scheme is shown in Fig. 149.2. Because FE and RE are private modules of each routing protocol, multiple

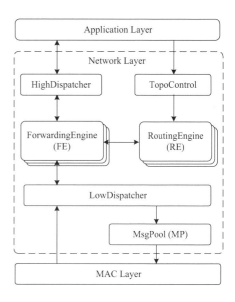

Fig. 149.2 Architecture of the dynamic scheme

Fig. 149.3 Network packet header format

instances of FEs and REs simultaneously exist on the single node. So correctly switching between FEs and REs modules is the key issue of the dynamic scheme. We add three modules: HighDispatcher, LowDispatcher, and TopoControl for routing protocols switching.

The TopoControl module controls the routing topology switching. When a node receives a query, explains the query, and starts the corresponding application program, TopoControl matches applications to REs. According to the routing protocol identifier (RpID for short) carried by the query command, when the query task starting, TopoControl starts the RpID's RE module. When the query task is finished, TopoControl stops the RpID's RE module. The HighDispatcher and LowDispatcher modules hide the differences of routing protocols for application and MAC layers. FE module is responsible for data packet sending and forwarding. As shown in Fig. 149.2, the dynamic scheme is competent for the switching between FEs by the HighDispatcher and LowDispatcher modules. The network header of a data packet is shown in Fig. 149.3. The dynamic scheme adds two fields: low header and high header. The low header is added and explained by HighDispatcher module. The HighDispatcher module provides a bridge for data packets correctly flowing between application layer programs and FEs. LowDispatcher manages the low header field in the packet header, and it provides a bridge for data packets correctly flowing between MAC layer and FEs.

149.3 Simulation

To evaluate the feasibility of our approach and to make the proposal concrete, we implemented the dynamic scheme in TinyOS [11]. We used two routing protocols: MintRoute and HEED for our implementation. Based on our previous work [13], BS selects routing protocol according to query tasks: for aggregation, BS chooses HEED as routing protocol; for acquisition, it chooses MintRoute. BS attaches the routing protocol information on the query command and sends it to the network using flooding algorithm.

We adopt the simple energy model [7]. The energy consumption for sending 1 bit data with distance d is E_{Tx}, and energy consumption for receiving 1 bit data is E_{Rx}. The energy model can be expressed as

$$E_{Rx}(l) = lE_{elec} \tag{149.1}$$

$$E_{Tx}(l,d) = \begin{cases} lE_{elec} + l\xi_{fs}d^2, & d < d_0 \\ lE_{elec} + l\xi_{mp}d^4, & d > d_0 \end{cases} \tag{149.2}$$

where E_{elec} is the circuitry power consumption, which depends on factors such as the digital coding, modulation, and filtering. $\xi_{fs}d^2$ and $\xi_{mp}d^4$ are the transmission power which depends on the distance to the receiver and acceptable bit-error rate. In our simulation, 100 nodes are randomly distributed in a 100 × 100 m area, node communication range is 50 m, and the sink node is located at (0, 0 m). Energy consumption of data fusion is 5 nJ/bit. Energy consumption for sleep and sense mode is ignored. Simulation parameters are shown in Table 149.2.

149.3.1 Code Size

One of the main objectives of creating a dynamic scheme for sensor networks is to increase code reuse, thereby allowing for multiple protocols to coexist cleanly and efficiently on a single node. Compared to the monolithic implementations, Fig. 149.4 compares code size between the different implementations. Modular is

Table 149.2 Simulation parameters

Parameter	Value
Threshold distance (d_0)	75 m
Data packet size	500 bytes
Control packet size	25 bytes
Initial energy	2 J
E_{elec}	50 nJ/bit
E_{fusion}	5 nJ/bit
E_{fs}	10 pJ/bit
ξ_{mp}	0.0013 pJ/bit

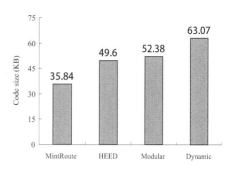

Fig. 149.4 Code size comparison

the modular design of coexist MintRoute and HEED, and Dynamic refers to dynamic scheme based on coexist of MintRoute and HEED. When we combine different protocols, we observe clear gains: the modular design use up 40 % less code size.

149.3.2 Continuous Query

BS sends an aggregation query to the network at first; when the aggregation query is finished, BS sends an acquisition query to the network continuously. The sample period of each query is 10 s and the duration is 30 min. We simulated the dynamic scheme, HEED, and MintRoute, respectively. The query system adopting dynamic scheme can switch routing protocols according to query tasks. As shown in Fig. 149.5a, we record the number of data packets received at BS and energy dissipation. When the number of data packets is about between 0 and 16,000, the network is running an aggregation query. In the interval, HEED is more energy-efficient than MintRoute, so the dynamic scheme selects HEED as routing protocol. When the number of data packets is about 16,000, the aggregation query is finished. And when the number of data packets is about between 16,000 and 32,000, the acquisition is executing. In the interval, the performance of MintRoute is better than HEED. The dynamic scheme can get the best of both worlds by switching from HEED to MintRoute. So it is more energy efficient than both MintRoute and HEED.

149.3.3 Random Query

The number of queries is randomly generated, query type is randomly selected from aggregation and acquisition, and query duration is a random integer between 1 and 30 min. Using these constraints, we generate the random queries scene, which contains 16 aggregation queries and 14 acquisition queries. Queries are injected into a network one by one, and there is only single query running in the network at a time.

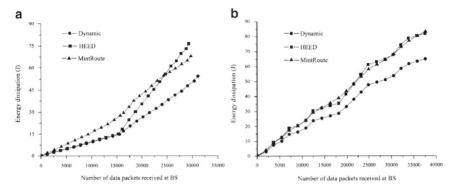

Fig. 149.5 Energy dissipation vs. the number of data packets received at BS (**a**) In the case of continuous query. (**b**) In the case of random query

We simulate MintRoute, HEED, and dynamic scheme using the same random queries scene. As shown in Fig. 149.5b, the dynamic scheme is more energy efficient than HEED and MintRoute.

149.4 Conclusion

This chapter describes the routing protocols switching scheme in sensor network that can switch between routing protocols according to query tasks. Simulation results show that it is more energy efficient than single routing protocol for query processing. The scheme decides protocols switching only based on query tasks. The performance of routing protocol also is significantly affected by network conditions. So as future work, we intend to let BS decide protocols switching based on knowledge by automatic learning.

References

1. Madden, S., Franklin, M. J., Hellerstein, J. M., & Hong, W. (2005). Tinydb: An acquisitional query processing system for sensor networks. *ACM Transactions on Database Systems, 30*(1), 122–173.
2. Luo, Q., & Wu, H. (2007). System design issues in sensor databases. In *Proceeding of the 2007 ACM SIGMOD international conference on management of data (SIGMOD)* (pp. 1182–1185). New York: ACM Press.
3. Woo, A., Tong, T., & Culler, D. (2003). Taming the underlying challenges of reliable multihop routing in sensor networks. In *Proceedings of the 1st international conference on embedded networked sensor systems (SenSys)* (pp. 14–17). New York: ACM Press.

4. Gnawali, O., Fonseca, R., Jamieson, K., Moss, D., & Levis, P. (2009). Collection tree protocol. In *Proceeding of the 7th ACM conference on embedded networked sensor systems (SenSys)* (pp. 1–14). New York: ACM Press.
5. Borsani, L., Guglielmi, S., Redondi, A., & Cesana, M. (2011). Tree-based routing protocol for wireless sensor networks. In *Proceedings of the IEEE international conference on Wireless On-Demand Network Systems and Services (WONS)* (pp. 158–163). Washington, DC: IEEE.
6. Heinzelman, W., Chandrakasan, A., & Balakrishnan, H. (2002). An application specific protocol architecture for wireless microsensor networks. *IEEE Transactions on Wireless Communications, 1*(2), 660–670.
7. Younis, O., & Fahmy, S. (2004). HEED: A hybrid, energy-efficient, distributed clustering approach for ad hoc sensor networks. *IEEE Transactions on Mobile Computing, 3*(4), 366–379.
8. Khan, A., Madani, S., Hayat, K., & Khan, S. (2012). Clustering-based power-controlled routing for mobile wireless sensor networks. *International Journal of Communication Systems, 25*(4), 529–542.
9. Jun, T., & Julien, C. (2007). Automated routing protocol selection in mobile ad hoc networks. In *Proceedings of ACM symposium on applied computing* (pp. 906–913). New York: ACM Press.
10. He, Y., & Raghavendra, C. S. (2005). Building programmable routing service for sensor networks. *Computer Communications, 28*(6), 664–675.
11. Levis, P., Madden, S., Polastre, J., Szewczyk, R., Whitehouse, K., Woo, A., et al. (2005). TinyOS: An operating system for sensor networks. In W. Weber, J. Rabaey, & E. Aarts (Eds.), *Ambient intelligence* (pp. 115–148). Berlin: Springer.
12. Cheng, T., Fonseca, R., Kim, S., Moon, D., Tavakoli, A., Culler, D., et al. (2006). A modular network layer for sensornets. In *Proceedings of the 7th USENIX symposium on operating systems design and implementation* (pp. 249–262). Berkeley, CA: USENIX Association.
13. Zhang, Z., & Yu, F. (2010). Performance analysis of cluster-based and tree-based routing protocols for wireless sensor networks. In *Proceedings of the international conference on Communications and Mobile Computing (CMC)* (pp. 418–422). Washington, DC: IEEE.

Chapter 150
Incipient Fault Diagnosis in the Distribution Network Based on S-Transform and Polarity of Magnitude Difference

Jinqian Zhai and Xin Chen

Abstract It is difficult for conventional relaying algorithms to detect incipient faults, such as insulator current leakage, electrical faults due to tree limbs, and transient or intermittent earth faults, which are frequent in distribution networks. With the time, they may lead to a catastrophic failure. In order to avoid this situation, S-transform technique is proposed to extract the suitable features of incipient fault in this chapter. A least square support vector machine (LS-SVM) classifier is developed utilizing the features so that incipient fault is distinguished from the normal disturbances. Then the polarity of magnitude difference of residual current is used to determine the fault section of distribution network. The proposed technique has been investigated by ATP/EMTP simulation software. Simulation results show that this technique is effective and robust.

150.1 Introduction

Incipient faults in power lines are normally characterized as the faulty phenomena with the relatively low fault currents, such as high-impedance faults, insulator leakage current faults, and intermittent/transient faults [1]. These faults represent little threat of damage to power system equipment. But with time, they may lead to a catastrophic failure (i.e., a permanent damage beyond repair). So, early detection of power line fault would undoubtedly be a great benefit to the utilities enabling them to avoid catastrophic failures, unscheduled outages, and thus loss of revenues.

Various methods have been proposed by researchers and protection engineers. Among them, harmonic analysis [2], randomness detection [3], artificial neural networks [4], Hilbert-transform-based [5], wavelet transform [6–9], etc. are used to extract the feature of incipient fault signals in the distribution line. But due to high

---

J. Zhai (✉) • X. Chen
Zhengzhou Power Supply Company, Zhengzhou 450000, China
e-mail: jinqianzhai@163.com

time resolution and low frequency resolution for high frequencies and high frequency resolution and low time resolution for low frequencies, wavelet transform can achieve a better solution. To overcome the shortcomings associated with the existing techniques, novel methods need to be developed. The purpose of this chapter is to develop an online system that uses voltage and current during a period to diagnosis incipient faults of power line; the proposed algorithms are as follows: The incipient fault features are extracted by S-transform. All these features are then used to train LS-SVM to enable it to make prediction of possible occurrences of incipient faults in the distribution network. Then the polarity of magnitude difference of residual current is used to determine the faulty section in the distribution network.

150.2 Incipient Fault Characterization and Simulation System

150.2.1 The Characterization of Incipient Fault

The term "incipient faults" refers to certain pre-fault "symptoms" or electrical activities taking place prior to a power system failure or blackout. Power line incipient faults are the primary causes of catastrophic failures in the distribution network. For underground cable, these faults develop in the extruded cables from gradual deterioration of the solid insulation due to the persisting stress factors. For overhead lines, incipient faults are associated with degraded equipment (insulators, arresters, transformer insulation and bushings, etc.) and the gradual intrusion of tree limbs as they grow into the overhead power line. Therefore, the present development aims at detecting these defects through prediction. The investigation searches out defect patterns or signatures that are still hidden, so they can be solved before they effectively interrupt the transmission system.

150.2.2 Simulation System

The 10-kV distribution system is supplied with power by a 110-kV grid via a 40-MVA transformer as shown in Fig. 150.1; the isolated neutral 10-kV system consists of several overhead lines and cables as radial feeders, which are simulated using ATP/EMTP, in which the processing is created by ATPDraw [10]. L2 and L3 are cables, 7 and 5 m, respectively; other feeders are overhead lines. L1 is 15 m, AB is 10 m, BC is 5 m, BD is 4 m, CE is 5 m, CF is 7 m, CG is 4 m, DH is 5 m, and DI is 6 m. The feeder overhead line and cable are represented using the frequency-dependent JMarti model. The high-impedance arc fault is simulated, and the several wireless sensors are installed in specific measuring sites [11, 12].

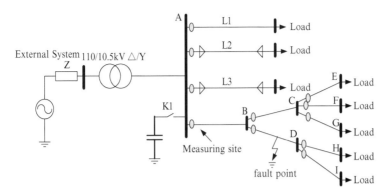

Fig. 150.1 Configuration of simulated system

150.3 S-Transform-Based Feature Extraction Method for Incipient Fault in the Distribution Network

150.3.1 The Principle of S-Transform

The S-transform, put forward by Stockwell in 1996, is considered as one of the most recent signal processing techniques, which is a kind of time-frequency analysis method based on continuous wavelet transformation and short-time Fourier transformation, and overcomes some of disadvantages of the wavelet transforms [13]. The basis function for the S-transform is the Gaussian modulation cosinusoids. The consinusoid frequencies are used for the interpretation of a signal that will result in the time-frequency spectrum. The output of the S-transform is an $N \times M$ matrix called the S-matrix whose row pertains to the frequency and columns to time [14]. Each element of the S-matrix is complex valued and is used for extracting features to classify the incipient faults. The S-transform performs multi-resolution analysis on a time-varying power signal as its window width varies inversely with the frequency. The S-transform for a function $x(t)$ is defined as [15]

$$S(\tau,f) = \int_{-\infty}^{\infty} x(t) \frac{|f|}{\sqrt{2\pi}} e^{-\frac{(\tau-t)^2 f^2}{2}} e^{-i2\pi ft} dt \qquad (150.1)$$

The S-transform will generate time-frequency contours; these contours can provide excellent features, which can be used by a pattern recognition system for classifying the incipient faults.

Table 150.1 Features from the S-transform of incipient fault signal

Features	Description
F1	The maximum of S-matrix for residual current
F2	The minimum of S-matrix for residual current
F3	Standard deviation of residual current
F4	Mean value of residual current
F5	The maximum of S-matrix for residual voltage
F6	The minimum of S-matrix for residual voltage
F7	Standard deviation of residual voltage
F8	Mean value of residual voltage

150.3.2 Feature Extraction from the S-Transform

Since the residual voltage and current are real-time summation of the three phase voltages and currents, respectively, any single-phase change is reflected in the residual voltage and current. Therefore, in this study, both residual voltage and current were analyzed. They are computed as

$$u_r = u_a + u_b + u_c \tag{150.2}$$

$$i_r = i_a + i_b + i_c \tag{150.3}$$

where u_r and i_r are the residual voltage and current, respectively. u_a, u_b, and u_c are the phase voltages. i_a, i_b, and i_c are the phase currents.

Characterization of the measurements of voltage and current is proposed to obtain features to be used as inputs of classifier. In this study, S-matrix is obtained by S-transform of residual voltage and current. According to S-matrix of incipient fault signal, several important information in terms of magnitude, phase, and frequency can be extracted. Many features such as amplitude, variance, mean, standard deviation, and energy of the transformed signal are widely used for proper classification. In this study, features are as follows.

The first set of features: The maximum value of the absolute S-matrix of residual current and voltage, which are F1 and F5. The second set of features: The minimum value of the absolute S-matrix of residual current and voltage, which are F2 and F6. The third set of features: Standard deviation of the data set comprising of the elements corresponding to maximum magnitude of each column of the S-matrix, which are F3 and F7. The fourth set of features: Mean value of the data set comprising of the elements corresponding to maximum magnitude of each column of the S-matrix, which are F4 and F8.

The overall features selected for characterizing the residual voltage and current waveforms are shown in Table 150.1.

150.4 Classification for Incipient Fault and Normal Switching in the Distribution Network

In this section, the LS-SVM is used to classify the incipient fault and normal disturbance, such as capacitor switching and load switching. The LS-SVM model is finally given as follows [16]:

$$y(x) = sign\left[\sum_{i=1}^{N} \alpha_i y_i K(x, x_i) + b\right] \qquad (150.4)$$

$K(x,x_i)$ is the kernel function. In our experiment, we choose the radial basis function (RBF) kernel where δ^2 is the bandwidth of the RBF kernel as our kernel, because it tends to achieve better performance. In determining the kernel bandwidth, δ^2 and the margin γ are set at 0.4 and 10, respectively. In order to fit the requirements of LS-SVM classifier, the training sample data must be collected and preprocessed before inputted into the classifier. By simulation using the ATP/EMTP, 600 numbers of residual voltage and current waveforms for incipient faults and normal disturbances were obtained for training LS-SVM. The features for all the waveforms were extracted from the S-transform. These features were used in developing the training database for the LS-SVM. Then, 300 numbers of residual voltage and current waveforms for incipient faults and normal disturbances were used for testing the training model. The results for test using LS-SVM classifier for input feature date from S-transform and wavelet transform are shown in Tables 150.2 and 150.3.

From Table 150.2, it is shown that LS-SVM classifier has good performance for these features extracted from S-transform; the correct classification rate is 90 %, which is higher than wavelet transform in Table 150.3. So S-transform-based incipient fault detection demonstrates the validity compared with wavelet transform widely used in power system.

Table 150.2 The classification result using features from S-transform

Disturbance type	No. of data	Correct classification	Accuracy (%)
Incipient fault	100	90	90
Capacitor switching	100	96	96
Load switching	100	93	93

Table 150.3 The classification result using features from wavelet transform

Disturbance type	No. of data	Correct classification	Accuracy (%)
Incipient fault	100	87	87
Capacitor switching	100	95	95
Load switching	100	94	94

150.5 Fault Section Discrimination for Incipient Fault in the Distribution Network

Once incipient fault is detected, it is necessary to find the position in power distribution network. In this section, a location method for incipient fault is proposed. Suppose the incipient fault occurs in the section DI. When the incipient fault occurs at 0.26 s, the corresponding residual current waveforms for these branches are shown in Figs. 150.2 and 150.3. We know that pre-fault residual current magnitude of some sections is lower than that during the fault, which is drawn in Fig. 150.2, and pre-fault residual current magnitude of other sections is larger than that during the fault, which is drawn in Fig. 150.3.

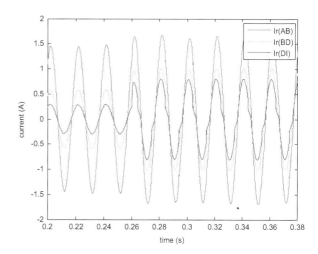

Fig. 150.2 The section of residual current magnitude during fault large than pre-fault when fault in DI

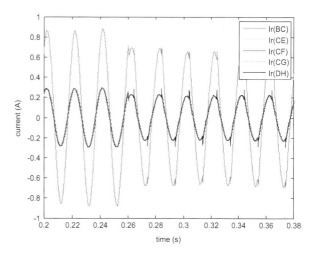

Fig. 150.3 The section of residual current magnitude during fault lower than pre-fault

Fig. 150.4 The section of residual current magnitude during fault large than pre-fault when fault in CF

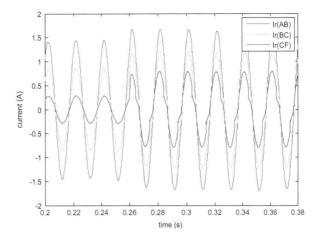

When the incipient fault occurred at 0.26 s, from Figs. 150.2 to 150.3, it is obvious that the residual current magnitude of every section has changed so much. Some magnitudes are bigger than pre-fault, and others are smaller than pre-fault. So this change is the difference between the residual current magnitude during and pre-fault measurements, which is described as

$$\Delta i_r = i_{r\_during} - i_{r\_pre} \tag{150.5}$$

Through the simulation in Fig. 150.1, it is well known that if $\Delta i_r > 0$, like the case of Fig. 150.2, it is the case with faulty section, and otherwise, if $\Delta i_r < 0$, like the case of Fig. 150.3, it is the case of healthy section. So, the positive or negative of Δi_r is considered as the discrimination rule to the status of power line.

The case was studied for incipient fault occurring at section CF. The corresponding residual current magnitude difference between during and pre-fault higher zero are shown in Fig. 150.4. By analyzing the Fig. 150.4, when the incipient fault occurred in section CF, the residual current magnitude of section AB is higher, and it is slightly reduced for each downstream section BC and CF. But for the healthy section, the residual current magnitude during the fault is lower than that of pre-fault.

150.6 Conclusion

In this chapter, a complete incipient fault diagnosis scheme is proposed. A novel technique based on S-transform of residual voltage and current is proposed to extract the features of incipient faults. According to these features, LS-SVM classifier is used to distinguish between fault status and normal disturbance. By analyzing the magnitude difference between during the fault and pre-fault, an estimation rule is proposed to determine the faulty section in the distribution network. The proposed scheme performance showed its high reliability and robustness to a wide variety of simulated tests.

References

1. Sidhu, T. S., & Xu, Z. H. (2010). Detection of incipient faults in distribution underground cables. *IEEE Transactions on Power Delivery, 25*(3), 1363–1371.
2. Kim, C. J., Jeong Hoon Shin, Myeong-Ho Yoo, & Gi Won Lee. (1999). A study on the characterization of the incipient failure behavior of insulators in power distribution line. *IEEE Transactions on Power Delivery, 14*(2), 519–524.
3. Benner, C. L., & Russell, B. D. (1997). Practical high-impedance fault detection on distribution feeders. *IEEE Transactions on Industry Applications, 33*(3), 635–640.
4. Al-Dabbagh, M., & Al-Dabbagh, L. (1999). Neural networks based algorithm for detecting high impedance faults on power distribution lines. *Proceedings of the International Joint Conference on Neural Networks, 5*(1), 3386–3390.
5. Cui, T., Dong, X. Z., Bo, Z. Q., & Andrzej Juszczyk (2011). Hilbert transform based transient/intermittent earth fault detection in noneffectively grounded distribution systems. *IEEE Transactions on Power Delivery, 26*(1), 143–151.
6. Butler, K. L. (1996). An expert system based framework for an incipient failure detection and predictive maintenance system. In *Proceedings of the 1996 intelligent systems application to power systems conference* (pp. 321–326). Orlando, FL.
7. Miri, S. M., & Privette, A. (1994). A survey of incipient fault detection and location techniques for extruded shielded power cables. In *26th annual Southeastern symposium on system theory* (pp. 402–405). Athens, OH.
8. Kim, C.J., Seung-Jae Lee, & Sang-Hee Kang (2004). Evaluation of feeder monitoring parameters for incipient fault detection using Laplace trend statistic. *IEEE Transactions on Industry Applications, 40*(6), 1718–1724.
9. Huang, S. J., & Hsieh, C. T. (1999). High-impedance fault detection utilizing a Morlet wavelet transform approach. *IEEE Transactions on Power Delivery, 14*(4), 1401–1410.
10. Prikler, L., & Hoildalen, H. (1998). *ATPDraw users' manual*. SINTEF TR A4790.
11. Darwish, H., & Elkalashy, N. (2005). Universal arc representation using EMTP. *IEEE Transactions on Power Delivery, 20*(2), 774–779.
12. Kizilcay, M., & Pniok, T. (1991). Digital simulation of fault arcs in power systems. *European Transactions on Electrical Power, 1*(1), 55–59.
13. Stockwell, R. G., Mansinha, L., & Lowe, R. P. (1996). Localization of the complex spectrum: The S-transform. *IEEE Transactions on Signal Processing, 44*(4), 998–1001.
14. Faisal, M.F., Azah Mohamed, & Aini Hussain (2009). S-transform based support vector regression for detection of incipient faults and voltage disturbances in power distribution networks. In *Proceedings of the 11th WSEAS international conference on mathematical methods, computational techniques and intelligent systems* (pp. 139–145). La Laguna.
15. Li, B., Zhang, P.L., Tian, H., Mi, S.S., Liu, D.S., & Ren, G.Q. (2011). A new feature extraction and selection scheme for hybrid fault diagnosis of gearbox. *Expert Systems with Applications, 38*(8), 10000–10009.
16. Suykens, J. A. K. (2002). *Least square support vector machines* (pp. 64–75). Singapore: World Scientific.

Chapter 151
Network Communication Forming Coalition S4n-Knowledge Model Case

Takashi Matsuhisa

Abstract This paper is to introduce the new concept of coalition Nash equilibrium of a strategic game. A coalition Nash equilibrium for a strategic game consists of (1) a subset S of players, (2) independent mixed strategies for each member of S, and (3) the conjecture of the actions for the other players not in S with the condition that each member of S maximizes his/her expected payoff according to the product of all mixed strategies for S and the other players' conjecture. Let us consider that each player communicates privately not only his/her belief about the others' actions but also his/her rationality as messages according to a protocol and then the recipient updates their private information and revises her/his prediction. Then we show that the conjectures of the players in a coalition S regarding the future beliefs converge in the long run communication, which lead to a coalition Nash equilibrium for the strategic game.

151.1 Introduction

For few decades, researchers in economics, AI, and computer science become entertained lively concerns about relationships between knowledge and actions. At what point does an economic agent sufficiently know to stop gathering information and make decisions? There are also concerns about cooperation and knowledge. What is the role of sharing knowledge to making cooperation among agents?

Considering a coalition among agents, we tacitly understand that each agents in the coalition shares their individual information and so they commonly know each other. In mathematical point of view yet a little is known what structure they have to know commonly. The aim of this paper is to fill the gap. Our point is that in a

T. Matsuhisa (✉)
Department of Natural Sciences, Ibaraki National College of Technology, Nakane 866, Hitachinaka-shi, Ibaraki, 312-8508, Japan
e-mail: mathisa@ge.ibaraki-ct.ac.jp

coalition, the members do not necessarily have common-knowledge about each other but they communicate his/her own beliefs on the others to each other through messages.

The purposes of this paper are to introduce the concept of coalition Nash equilibrium of a strategic game and to show that a communication among the players in a coalition leads to the equilibrium through messages. A coalition Nash equilibrium for a strategic game consists of (1) a subset S of players, (2) independent mixed strategies for each member of S, and (3) the conjecture about the actions for the players outside of S maximizes his/her expected payoff according to all mixed strategies for all members in S together with the conjectures of all the players outside of S.

This paper analyzes the solution concept from the Bayesian point of view: The players start with the same prior distribution on a state-space. In addition they have private information which is given by a reflexive and transitive binary relation on the state-space. Each player in a coalition S predicts the other players' actions as the posterior of the others' actions given his/her information. He/she communicates privately their beliefs about the other players' actions through messages among all members in S according to the communication network in S, which message is information about his/her individual conjecture about the others' actions. The recipients update their belief by the messages. Precisely, at every stage each player communicates privately not only his/her belief about the others' actions but also his/her rationality as messages according to a protocol, and then the recipient updates their privateinformation and revises her/his prediction. In this circumstance, we shall show that

Main theorem *In a communication process of the game according to a protocol with revisions of their beliefs about the other players' actions, the profile of their future predictions converges to a coalition Nash equilibrium of the game in the long run.*

This paper is organized as follows. In Sect. 151.2, after recalling the knowledge structure associated with an RT-information structure, we present a game on knowledge structure. The communication process for the game is introduced where the players send messages about their conjectures about the other players' action. In Sect. 151.3 we give the formal statement of the main theorem (Theorem 1) with an illustrated example, and we conclude with some remarks.

151.2 The Model

Let Ω be a *state-space* which is a nonempty *finite* set, N a set of finitely many *players* $\{1, 2, \ldots n\}$ at least two ($n \geq 2$), and let 2^Ω be the family of all subsets of Ω. Each member of 2^Ω is called an *event* and each element of Ω called a *state*. Let μ be a common probability measure on Ω for all players. For simplicity it is assumed that (Ω, μ) is a *finite* probability space with μ *full support*.[1]

151.2.1 Information and Knowledge[2]

By an *RT-information* structure[3] we mean $\langle \Omega, (\Pi_i)_{i \in \bar{N}} \rangle$ in which $\Pi_i : \Omega \to 2^\Omega$ satisfies the three postulates: For each $i \in \bar{N}$ and for any $\omega \in \Omega$,

Ref $\omega \in \Pi_i(\omega)$; **Trn** $\xi \in \Pi_i(\omega)$ implies $\Pi_i(\xi) \subseteq \Pi_i(\omega)$;

This structure is equivalent to a Kripke semantics for the multi-modal logic **S4n**. The set $\Pi_i(\omega)$ will be interpreted as the set of all the states of nature that i knows to be possible at ω, or as the set of the states that i cannot distinguish from ω. We will therefore call Π_i i's *possibility operator* on Ω and also will call $\Pi_i(\omega)$ i's *information set* at ω.

If the RT-information structure satisfies the below postulate

Sym If $\xi \in \Pi_i(\omega)$, then $\omega \in \Pi_i(\xi)$,

it is called an *information partition*, which is equivalent to a Kripke semantics for the multi-modal logic **S5n**.

Our interpretation is given as a player i for whom $\Pi_i(\omega) \subseteq E$ knows, in the state ω, that some state in the event E has occurred, and then we say that in the state ω the player i knows E. It follows

Definition 1 The *knowledge structure* $\langle \Omega, (\Pi_i)_{i \in N}, (K_i)_{i \in N} \rangle$ consists of an RT-information structure $\langle \Omega, (\Pi_i)_{i \in N} \rangle$ and a class of i's *knowledge operator* K_i on 2^Ω such that $K_i E = \{\omega \in \Omega \mid \Pi_i(\omega) \subseteq E\}$.

The set $K_i E$ will be interpreted as the set of states of nature for which i knows E to be possible.

We record the properties of i's knowledge operator[4]: For every E, F of 2^Ω,

N $K_i \Omega = \Omega$ and $K_i \emptyset = \emptyset$; **K** $K_i(E \cap F) = K_i E \cap K_i F$;
T $K_i F \subseteq F$; **4** $K_i F \subseteq K_i K_i F$;

i's possibility operator Π_i is uniquely determined by i's knowledge operator K_i satisfying the above four properties as $\Pi_i(\omega) = \bigcap_{\omega \in K_i E} E$. If Π_i satisfies further **Sym**, then the below property is also true:

5 $\Omega \setminus K_i(E) \subseteq K_i(\Omega \setminus K_i(E))$.

---

[1] That is, $\mu(\omega) \neq 0$ for every $\omega \in \Omega$.
[2] C.f.; Binmore [2] for information and knowledge.
[3] RT-information stands for a reflexive and transitive information.
[4] According to these we can say the structure $\langle \Omega, (K_i)_{i \in N} \rangle$ is a model for the multi-modal logic **S4n**.

151.2.2 Game and Knowledge[5]

In this paper, by a *game* G we always mean a *finite* strategic form game $\langle N, (A_i)_{i \in N}, (g_i)_{i \in N} \rangle$ with the following structure and interpretations: N is a finite set of players $\{1, 2, \ldots, i, \ldots n\}$ with $n \geq 2$, A_i is a finite set of i's *actions* (or i's pure strategies), and g_i is an i's *payoff function* of A into $\mathbb{R}$, where A denotes the product $A_1 \times A_2 \times \cdots \times A_n$, A_{-i} the product $A_1 \times A_2 \times \cdots \times A_{i-1} \times A_{i+1} \times \cdots \times A_n$. We denote by g the n-tuple $(g_1, g_2, \ldots g_n)$ and by a_{-i} the $(n-1)$-tuple $(a_1, \ldots, a_{i-1}, a_{i+1}, \ldots, a_n)$ for a of A. Furthermore we denote $a_{-I} = (a_i)_{i \in N \setminus I}$ for each $I \subset N$. A probability distribution σ_i on A_i is called an i's *mixed strategy* for a game G. We denote by $\Delta(A_i)$ the set of all i's mixed strategies, so we will denote $\Delta(A) = \prod_{i=1}^{n} \Delta(A_i)$ and $\Delta(A_I) = \prod_{i \in I} \Delta(A_i)$.

A profile $(\sigma_i)_{i \in N}$ of mixed strategies is called a *Nash equilibrium* if for each $i \in N$ and for every $b_i \in A_i$, we have

$$\sum_{a_{-i} \in A_{-i}} g_i(a_i, a_{-i}) \prod_{j \in N \setminus \{i\}} \sigma_i(a_j) \geq \sum_{a_{-i} \in A_{-i}} g_i(b_i, a_{-i}) \prod_{j \in N \setminus \{i\}} \sigma_i(a_j)$$

By i's *overall conjecture* (or simply i's *conjecture*) we mean a probability distribution $\varphi_i \in \Delta(A_{-i}) = \Delta(A_{N \setminus \{i\}})$. For each player j other than i, the marginal distribution on j's actions is called i's *individual conjecture* about j (or simply i's conjecture *about j*.) Functions on Ω are viewed like random variables in the probability space (Ω, μ). If $\mathbf{x}$ is a such function and x is a value of it, we denote by $[\mathbf{x} = x]$ (or simply by $[x]$) the set $\{\omega \in \Omega \mid \mathbf{x}(\omega) = x\}$.

The information structure (Π_i) with a common prior μ yields the distribution on $A \times \Omega$ defined by $\mathbf{q}_i(a, \omega) = \mu([\mathbf{a} = a] | \Pi_i(\omega))$; and the i's overall conjecture defined by the marginal distribution $\mathbf{q}_i(a_{-i}, \omega) = \mu([\mathbf{a}_{-i} = a_{-i}] | \Pi_i(\omega))$ which is viewed as a random variable of φ_i. We denote by $[\mathbf{q}_i = \varphi_i]$ the intersection $\bigcap_{a_{-i} \in A_{-i}} [\mathbf{q}_i(a_{-i}) = \varphi_i(a_{-i})]$ and denote by $[\varphi]$ the intersection $\bigcap_{i \in N} [\mathbf{q}_i = \varphi_i]$. Let $\mathbf{g}_i$ be a random variable of i's payoff function g_i and $\mathbf{a}_i$ a random variable of an i's action a_i.

According to the Bayesian decision theoretical point of view, we assume that each player i absolutely knows his/her own actions; i.e., letting $[a_i] := [\mathbf{a}_i = a_i]$, $[a_i] = K_i([a_i])$ (or equivalently, $\Pi_i(\omega) \subseteq [a_i]$ for all $\omega \in [a_i]$ and for every a_i of A_i.) i's action a_i is said to be *actual* at a state ω if $\omega \in [\mathbf{a}_i = a_i]$; and the profile a_I is said to be actually played at ω if $\omega \in [\mathbf{a}_I = a_I] := \bigcap_{i \in I} [\mathbf{a}_i = a_i]$ for $I \subset N$. The payoff functions $g = (g_1, g_2, \ldots, g_n)$ is said to be *actually played* at a state ω if $\omega \in [\mathbf{g} = g] := \bigcap_{i \in N} [\mathbf{g}_i = g_i]$. Let **Exp** denote the expectation defined by $\mathbf{Exp}(g_i(b_i, \mathbf{a}_{-i}); \omega) := \sum_{a_{-i} \in A_{-i}} g_i(b_i, a_{-i}) \, \mathbf{q}_i(a_{-i}, \omega)$.

---

[5] C.f., Aumann and Brandenburger [1].

By a *coalition* S we mean S is a non empty subset of N. Let $(\sigma_i)_{i \in S}$ be the profiles of mixed strategies of G for a coalition S. By S-*exectation* of i's payoff function g_i at ω we mean

$$\mathbf{Exp}_S(g_i(a_S, \mathbf{a}_{-S}); \omega) := \sum_{a_{-S} \in A_{-S}} g_i(a_S, a_{-S}) \left(\prod_{i \in S} \sigma_i \right) \mathbf{q}_i(a_{-S}, \omega).$$

Definition 2 A profile $(\sigma_i)_{i \in S}$ is called a *coalition S-Nash equilibrium* of G if each member i in S maximizes his/her $\mathbf{Exp}_S(g_i(a_S, \mathbf{a}_{-S}); \omega)$ for every $\omega \in \Omega$; i.e.: $\mathbf{Exp}_S(g_i(a_S, \mathbf{a}_{-S}); \omega) \geq \mathbf{Exp}_S(g_i(b_S, \mathbf{a}_{-S}); \omega)$ for every b_S in A_S.

A coalition S is called *rational* at ω if for every $i \in S$, each i's actual action a_i maximizes the expectation of his actually played payoff function g_i at ω when the other players' actions are distributed according to his conjecture $\mathbf{q}_i(\cdot\ ; \omega)$. Formally, letting $g_i = \mathbf{g}_i(\omega)$ and $a_i = \mathbf{a}_i(\omega)$, $\mathbf{Exp}(g_i(a_i, \mathbf{a}_{-i}); \omega) \geq \mathbf{Exp}(g_i(b_i, \mathbf{a}_{-i}); \omega)$ for every b_i in A_i. Let R_i denote the set of all of the states at which i is rational.

151.2.3 Protocol[6]

We assume that the players communicate by sending *messages*. Let T be the time horizontal line $\{0, 1, 2, \cdots t, \cdots\}$. A *protocol* on a coalition S of a game G is a mapping $\mathrm{Pr}_S : T \to S \times S$, $t \mapsto (s(t), r(t))$ such that $s(t) \neq r(t)$. Here t stands for *time* and $s(t)$ and $r(t)$ are, respectively, the *sender* and the *recipient* of the communication which takes place at time t. Simply we call it an *S-protocol*. We consider the protocol as the directed graph whose vertices are the set of all members in S and such that there is an edge (or an arc) from i to j if and only if there are infinitely many t such that $s(t) = i$ and $r(t) = j$.

In this paper a protocol Pr_S is assumed to be *fair*; that is, the graph is strongly connected: In words, every player in this protocol communicates directly or indirectly with every other player infinitely often. It is said to contain a *cycle* if there are players $i_1, i_2, \ldots, i_k$ with $k \geq 3$ such that for all $m < k$, i_m communicates directly with i_{m+1}, and such that i_k communicates directly with i_1. The communications are assumed to proceed in *rounds*.

151.2.4 Communication on Coalition

Let S be a coalition of G. A *coalition S-communication process* $\pi_S(G)$ with revisions of players' conjectures $(\varphi_i^t)_{(i,\ t) \in S \times T}$ according to a protocol for a game G is a tuple

---

[6] C.f.: Parikh and Krasucki [4].

$$\pi_S(G) = \langle G, (\Omega, \mu), \mathrm{Pr}_S, (\Pi_i^t)_{i \in S}, (K_i^t)_{i \in S}, (\varphi_i^t)_{(i,t) \in S \times T} \rangle$$

with the following structures: the players have a common prior μ on Ω, the protocol Pr_S among N, $\mathrm{Pr}_S(t) = (s(t), r(t))$, is fair and it satisfies the conditions that $r(t) = s(t+1)$ for every t and that the communications proceed in rounds. The revised information structure Π_i^t at time t is the mapping of Ω into 2^Ω for player $i \in S$. If $i = s(t)$ is a sender at t, the *message* sent by i to $j = r(t)$ is M_i^t. An n-tuple $(\varphi_i^t)_{i \in S}$ is a revision process of individual conjectures. These structures are inductively defined as follows:

- Set $\Pi_i^0(\omega) = \Pi_i(\omega)$.
- Assume that Π_i^t is defined. It yields the distribution $\mathbf{q}_i^t(a, \omega) = \mu([\mathbf{a} = a] | \Pi_i^t(\omega))$. Whence
  - R_i^t denotes the set of all the state ω at which i is *rational* according to his conjecture $\mathbf{q}_i^t(\cdot\,;\,\omega)$[7];
  - The message $M_i^t : \Omega \to 2^\Omega$ sent by the sender i at time t is defined by

$$M_i^t(\omega) = \bigcap_{a_{-i} \in A_{-i}} \{ \xi \in \Omega \,|\, \mathbf{q}_i^t(a_{-i}, \xi) = \mathbf{q}_i^t(a_{-i}, \omega) | \}.$$

  Then:
- The revised knowledge operator $K_i^t : 2^\Omega \to 2^\Omega$ is defined by $K_i^t(E) = \{\omega \in \Omega \mid \Pi_i^t(\omega) \subseteq E\}$.
- The revised partition Π_i^{t+1} at time $t+1$ is defined as follows: $\Pi_i^{t+1}(\omega) = \Pi_i^t(\omega) \cap M_{s(t)}^t(\omega)$ if $i = r(t)$, and $\Pi_i^{t+1}(\omega) = \Pi_i^t(\omega)$ otherwise,
- The revision process $(\varphi_i^t)_{(i,\,t) \in S \times T}$ of conjectures is inductively defined as follows: Let $\omega_0 \in \Omega$, and set $\varphi_{s(0)}^0(a_{-s(0)}) := \mathbf{q}_{s(0)}^0(a_{-s(0)}, \omega_0)$. Take $\omega_1 \in M_{s(0)}^0(\omega_0) \cap K_{r(0)}([g_{s(0)}] \cap R_{s(0)}^0)$,[8] and set $\varphi_{s(1)}^1(a_{-s(1)}) := \mathbf{q}_{s(1)}^1(a_{-s(1)}, \omega_1)$. Take $\omega_{t+1} \in M_{s(t)}^t(\omega_t) \cap K_{r(t)}^t([g_{s(t)}] \cap R_{s(t)}^t)$,[9] and set $\varphi_{s(t+1)}^{t+1}(a_{-s(t+1)}) := \mathbf{q}_i^{t+1}(a_{-s(t+1)}, \omega_{t+1})$.

The specification is that a sender $s(t)$ at time t informs the recipient $r(t)$ his/her prediction about the other players' actions as information of his/her individual conjecture. The recipient revises her/his information structure under the information. She/he predicts the other players' action at the state where the player knows

---

[7] That is, each i's actual action a_i maximizes the expectation of his payoff function g_i being actually played at ω when the other players actions are distributed according to his conjecture $\mathbf{q}_i^t(\cdot\,;\,\omega)$ at time t. Formally, letting $g_i = \mathbf{g}_i(\omega)$, $a_i = \mathbf{a}_i(\omega)$, the expectation at time t, $\mathbf{Exp}^t$, is defined by $\mathbf{Exp}^t(g_i(a_i, \mathbf{a}_{-i}); \omega) := \sum_{a_{-i} \in A_{-i}} g_i(a_i, a_{-i}) \, \mathbf{q}_i^t(a_{-i}, \omega)$. A player $i \in S$ is said to be S-rational according to his conjecture $\mathbf{q}_i^t(\cdot\,,\,\omega)$ at ω if for all b_i in A_i, $\mathbf{Exp}^t(g_i(a_i, \mathbf{a}_{-i}); \omega) \geq \mathbf{Exp}^t(g_i(b_i, \mathbf{a}_{-i}); \omega)$.
[8] We denote $[g_i] := [\mathbf{g}_i = g_i]$.
[9] It is noted that $\omega_{t+1} \in M_{s(t)}^t(\omega_t) \cap K_{r(t)}^t([g_{s(t)}] \cap R_{s(t)}^t) \neq$ for every $t = 0, 1, 2, \cdots$.

that the sender $s(t)$ is rational, and she/he informs her/his the predictions to the other player $r(t+1)$.

We denote by ∞ a sufficiently large $\tau \in T$ such that for all $\omega \in \Omega, \mathbf{q}_i^\tau(\cdot;\omega) = \mathbf{q}_i^{\tau+1}(\cdot;\omega) = \mathbf{q}_i^{\tau+2}(\cdot;\omega) = \cdots$. Hence we can write $\mathbf{q}_i^\tau$ by $\mathbf{q}_i^\infty$ and φ_i^τ by φ_i^∞.

151.3 The Result

We can now state the main theorem, and we will give an example to illustrate it.

Theorem 1 *Let G be a strategic form game. Suppose that the players in G have the knowledge structure with μ a common prior. Let S be a coalition in a game G. In a coalition S-communication process $\pi_S(G)$ according to an S-protocol Pr_S among all members in S, the $|S|$-tuple of their conjectures $(\varphi_i^t)_{(i,t) \in S \times T}$ converges to a coalition S-Nash equilibrium of the game in finitely many rounds.*

Proof will be omitted, and it will be appeared elsewhere.

Example 1 Let us consider the game $G = \langle N, (A_i)_{i \in N}, (g_i)_{i \in N} \rangle$ as follows:

- The set of players $N = \{1, 2, 3\}$:
- The action sets $A_1 = \{H, T\}, A_2 = \{H, T\}, A_3 = \{W, E\}$:
- The payoff functions g_1, g_2, g_3 are given in Table 151.1:

The game G has the unique Nash equilibrium $(\frac{1}{2}H + \frac{1}{2}T, \frac{1}{2}h + \frac{1}{2}t, W)$, and let us reconsider the situation as follows: *Each player knows his/her own actions, but he/she cannot know the other players' action.* To model the situation we introduce the game G as a Bayesian game equipped with the below information partition $(\Pi_i)_{i=1,2,3}$:

- $\Omega = \{\omega_1, \omega_2, \cdots, \omega_8\}$ with the equal probability μ; i.e., $\mu(\omega) = \frac{1}{8}$:
- The partitions $(\Pi_i)_{i=1,2,3}$ on Ω:
  - The partition Π_1: $\Pi_1(\omega) = \{\omega_1, \omega_2, \omega_5, \omega_6\}, \Pi_1(\omega) = \{\omega_3, \omega_4, \omega_7, \omega_8\}$,
  - The partition Π_2: $\Pi_2(\omega) = \{\omega_1, \omega_3, \omega_5, \omega_7\}, \Pi_2(\omega) = \{\omega_2, \omega_4, \omega_6, \omega_8\}$.
  - The partition Π_3: $\Pi_3(\omega) = \{\omega_1, \omega_2, \omega_3, \omega_4\}, \Pi_3(\omega) = \{\omega_5, \omega_6, \omega_7, \omega_8\}$.
- $\mathbf{a}_i$ is defined by

Table 151.1 Game G in Example 1

W	h	t
H	1, 0, 2	0, 1, 2
T	0, 1, 2	1, 0, 2

E	h	t
H	1, 0, 3	0, 1, 0
T	0, 1, 0	1, 0, 3

$\mathbf{a}_1(\omega) = H$ for $\omega = \omega_i (i = 1,2,5,6)$, $\quad \mathbf{a}_1(\omega) = T$ for $\omega = \omega_i (i = 3,4,7,8)$.
$\mathbf{a}_2(\omega) = h$ for $\omega = \omega_i (i = 1,3,5,7)$, $\quad \mathbf{a}_2(\omega) = t$ for $\omega = \omega_i (i = 2,4,6,8)$.
$\mathbf{a}_3(\omega) = W$ for $\omega = \omega_i (i = 1,2,3,4)$, $\quad \mathbf{a}_3(\omega) = E$ for $\omega = \omega_i (i = 5,6,7,8)$.

We can observe that the conjectures $\varphi_i(a_j) = \mathbf{q}_i(a_j; \omega_5)$ at ω_5 are: $\varphi_2(a_1) = \varphi_3(a_1) = \frac{1}{2}H + \frac{1}{2}T$, $\varphi_1(a_2) = \varphi_3(a_2) = \frac{1}{2}h + \frac{1}{2}t$, $\varphi_1(a_3) = \varphi_2(a_3) = \frac{1}{2}W + \frac{1}{2}E$. This shows that for each player i, any other players than i must agree on every i' actions, but these distributions $(\varphi_3(a_1), \varphi_1(a_2), \varphi_2(a_3))$ cannot form the Nash equilibrium for G, but $(\varphi_2(a_1), \varphi_1(a_2)) = (\varphi_3(a_1), \varphi_1(a_2))$ forms a coalition $\{1, 2\}$-Nash equilibrium. However $(\varphi_3(a_1), \varphi_1(a_3))$ is not a $\{1, 3\}$-Nash equilibrium. It should be noted that $(\varphi_3(a_1), \varphi_1(a_3))$ is commonly known among $\{1, 3\}$, and so the notion of common-knowledge cannot always yield a coalition Nash equilibrium.

Let us consider the coalition $\{1, 3\}$-communication process for game G with the protocol Pr : $T \to S = \{1, 3\}$. After two rounds in the communication, the below information partition can be obtained:

- The partitions $(\Pi_i^\infty)_{i \in N}$ on Ω:
  - The partition Π_1^∞ on Ω: $\Pi_1^\infty(\omega) = \{\omega_1, \omega_2\}, \Pi_1^\infty(\omega) = \{\omega_5, \omega_6\}, \Pi_1^\infty(\omega) = \{\omega_3, \omega_4\}, \Pi_1^\infty(\omega) = \{\omega_7, \omega_8\}$.
  - The partition Π_2^∞ on Ω is the same as the initial partition Π_2:
  - The partition Π_3^∞ on Ω is the same as the initial partition Π_3:

  We can observe that the conjectures $\varphi_i^\infty(a_j) = \mathbf{q}_i^\infty(a_j; \omega)$ at ω_5 are: $\varphi_2^\infty(a_1) = \varphi_3^\infty(a_1) = \frac{1}{2}H + \frac{1}{2}T$, $\varphi_1^\infty(a_2) = \varphi_3^\infty(a_2) = \frac{1}{2}h + \frac{1}{2}t$, $\varphi_1^\infty(a_3) = \varphi_2^\infty(a_3) = W$, and these distributions $(\varphi_3^\infty(a_1), \varphi_1^\infty(a_3))$ are a $\{1, 3\}$-Nash equilibrium. Furthermore, $(\varphi_2^\infty(a_1), \varphi_1^\infty(a_2), \varphi_1^\infty(a_3))$ forms the Nash equilibrium for G.

Remark 1 When a coalition S is the ground coalition N, the above theorem shows that the conjectures of the players lead to a Nash equilibrium through communication (Matsuhisa [3]). The notion of common-knowledge controls to form a Nash equilibrium for a game (Aumann and Brandenburger [1]), but it no longer play a crucial role to form a coalition Nash equilibrium.

151.4 Conclusion

This paper points out that for coalition Nash equilibrium, common-knowledge cannot play such role as for mixed strategy Nash equilibrium. In fact, the profile of conjectures of a coalition may not yield a coalition Nash equilibrium even when the profile is commonly known among all the members of the coalition. To improve the situation we adopt the knowledge revisions model as a communication model. The main theorem shows that communication instead of common-knowledge plays an essential role to form a coalition Nash equilibrium. We have observed that in a communication process with revisions of players' beliefs about the other actions among all the members in a coalition, their predictions induce a coalition Nash equilibrium of the game in the long run.

References

1. Aumann, R. J., & Brandenburger, A. (1995). Epistemic conditions for mixed strategy Nash equilibrium. *Econometrica, 63*, 1161–1180.
2. Binmore, K. (1992). *Fun and games.* xxx+642pp. Lexington, MA: D. C. Heath and Company.
3. Matsuhisa, T. (2000). Communication leading to mixed strategy Nash equilibrium I. In T. Maruyama (Ed.) *Mathematical economics*, Suri-Kaiseki-Kenkyusyo Kokyuroku, vol. 1165, (pp. 245–256).
4. Parikh, R., & Krasucki, P. (1990) Communication, consensus, and knowledge. *Journal of Economic Theory 52*, 178–189.

Chapter 152
An Optimization Model of the Layout of Public Bike Rental Stations Based on B+R Mode

Liu He, Xuhong Li, and Dawei Chen

Abstract In order to find out the optimal layout of bike rental stations for B+R mode, a bi-level programming model combined of genetic algorithm and the joint model with mode split and traffic assignment model is built. The optimal layout plan can minimize the total travelling cost and facility cost. A case is used to test and verify the practicability of the model. The result shows that the model can effectively solve the layout problem of bike rental stations for B+R mode and can offer suggestions for related planning.

152.1 Introduction

The connection modes of metro stations are mainly walking, bus, and car. As the trip distance of walking is short, it can be regarded as direct passenger flow. People prefer to reach the trip destination directly considering refueling and parking. For bus, the door-to-door service is restricted by lines and much time is wasted walking to the bus stop and waiting there. For B+R mode, public bus expands the covering area that metro stations radiate on the one hand and offer a second connection for bus on the other hand. As a result, the attraction is improved along metro line and more indirect passenger flow is brought in. So the study of the layout of bike rental stations has an essential meaning for metro passenger flows.

Current researches into the layout of bike rental stations are focused on the analysis, forecast, and principles. Li-hui Li et al. put forward the principles as "Control total quantity, divide and sort, balance the scale, and adjust flexibly" and divide rental stations into five types: bus stops, public buildings, blocks, resting places, and schools. They point out that the scale of blocks should be the same as the others [1]. Zhenghao Li works out the capacities that each bike rental station needs

---

L. He (✉) • X. Li • D. Chen
School of Transportation, Southeast University, Nanjing 210096, Jiangsu, China
e-mail: 124675282@qq.com

with Markov chain model [2]. Xue Geng et al. get the number and size of bike rental station in Paris after calculating the average number of trip [3]. Jenn-Rong Lin and Ta-Hui Yang choose level of service as the factor that affects the layout of public bike rental station [4]. Zhonghua Wei et al. analyze the road density that R+B model needs [5]. Karel takes three European cities in his study and shows how public bike's trip distance, purpose, and car affect each other [6].

The chapter considers what influence public bike has on the passenger flow attraction of metro stations. Based on bi-level planning, an optimization model of the layout of bike rental stations is built and solved to offer support for making such layout decisions.

152.2 Modeling

Imagining there exists a road network illustrated in Fig. 152.1, bold lines are metro and bus lines and the rest are roads; dots are metro stations, diamonds are bus stops, and triangles are residential areas. As is illustrated in Fig. 152.2, public bike raises passenger attractions from the original two to five, with direct two and indirect one.

When planning public bike rental stations in the region, residential areas and bus stops are both alternative, and public bike is only allowed within the region. Starting from residential areas, travellers choose to get to metro station by car, by bus, or on foot before reaching the destination by metro. The trip modes to be chosen are illustrated in Fig. 152.3.

Based on the above assumptions and analysis, the question is brought forward: How to determine the layout of public bike rental stations so as to maximize the passenger flow volume of metro to the destination?

The layout optimization of public bike rental stations is a discrete network design problem (DNDP). Based on the DNDP theory, the chapter builds a planning model that is suitable for discrete transportation network design problem (DTNDP).

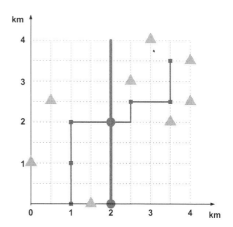

Fig. 152.1 A sketch map of a certain region

Fig. 152.2 The effect public bike has on the radiation

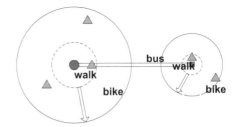

Fig. 152.3 Trip modes within the area

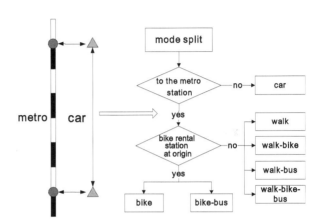

The optimization model of the layout of public bike rental stations based on B+R mode is a two-layer model, in which the upper model is the generation and screening of the plan set, and the lower model is the evaluation of every single plan. With the feedback between the two models, the best layout plan that satisfies both layers' targets can be found.

The building of the lower level should fully describe users' trip modes and route selection on the existing network, that is, mode spilt and traffic assignment.

In the whole system, every traveller needs to decide the trip mode and route that cost the least. Considering the IIA (independence from irrelevant) among different trip modes [7], selections among all modes use mixed logit model [8, 9]:

$$P_{in} = \int \frac{\exp(\lambda V_{in})}{\sum_j \exp(\lambda V_{jn})} g(\beta/\theta) d\beta \qquad (152.1)$$

$g(\beta/\theta)$ uses the normal distribution function [10] and the utility function uses linear form whose limbs including metro, walk, bicycle, public bicycle, car, and bus. The variables of trip mode are fare and travelling time while the variables of travellers are whether to have a car, whether to have a private bicycle, age, and trip purpose.

In the assignment model, every traveller selects the path whose travelling cost is the lowest. When the network reaches equilibrium, travellers cannot reduce their costs by changing the selected mode and path. Though user equilibrium model achieves theoretical optimization, users actually make the decision according to their expected cost. In that case, stochastic user equilibrium (SUE) model is more suitable.

When planning the layout of bike rental stations, the government should consider the characteristics of regional travel demand. Then the optimal layout can be chosen to minimize the total travelling cost and facility cost.

At the same time, according to the principles for the layout of public bike rental stations, four constraints should be satisfied. The first is the distance between bike rental stations should be reasonable considering travelling distance of bicycle; the second is the density of rental stations should be within a certain range; the third is the proportion of public bike travelling should reach a certain standard; the fourth is each rental station should have enough space. Then constraints (152.3), (152.4), and (152.5) of the model can be found and listed below.

From all the above, the model of the upper level can be expressed as follows:

$$\max z(x) = \sum_{i \in R} Q_i - \sum_{i \in R} Q_i P_i^c \qquad (152.2)$$

$$s.t. \quad d_{\min} \leq d_{ij} \leq d_{\max} \quad i,j \in T, \varsigma_i = 1, \varsigma_j = 1 \qquad (152.3)$$

$$\rho_{\min} \leq \sum_{i \in T} \varsigma_i / S \leq \rho_{\max} \qquad (152.4)$$

$$\min(C_i) \geq Q^b{}_{\min} \quad \varsigma_i = 1 \qquad (152.5)$$

In constraint (152.3), d stands for distance and is the set of alternative public bike rental stations; ς means whether to build a rental station there and 1 means yes, while 0 no. In constraint (152.4), ρ is the density and S is the area of the region. C in constraint (152.5) means the capacity of the rental station.

The upper level is a system optimization problem whose objective function can get the best solution of the lower level.

152.3 Model Solving

The bi-level planning model is combined of the upper level from the aspect of the government and the lower level from the users. The government hopes that public bike can help improve the proportion of metro travelling while users want to minimize their broad travelling expenses, the average of time, and narrow travelling expenses. The final result is a best layout plan found to optimize the two objective functions.

When solving the lower level, a super network should be created to do mode spilt and traffic assignment in consideration of multi-modes.

When solving the upper level, an effective plan should be built for the lower layer according to the constraints. Genetic algorithm has an advantage in solving the upper level.

When doing mode spilt with mixed logit model, the difficulty lies in building utility matrix of different modes, whose solving depends on creating a super network. In addition, traffic assignment is needed to get each rental station's size. So the lower level is combined of two steps: mode spilt and traffic assignment. The detailed steps are listed below:

Step 1: Creating the network. A super network is created according to the layout of public bike rental stations.

Step 2: Calculating utility. The shortest paths for each OD of different modes are found and then used to calculate the utility matrix. Dijkstra algorithm can be used here. Penalty fees increase when transferring.

Step 3: Mode spilt. The travelling proportion of each mode is calculated with (152.1).

Step 4: Traffic assignment. Trace back with Dijkstra algorithm in Step 2 and distribute OD volume onto the super network. The scale of each rental station is obtained at the same time.

The location of public bike rental stations needs the support of modeling and quantitative method, and the bi-level planning model has the character of NP (nondeterministic polynomial time complete, NP-complete). The layout plan of the rental stations can be described with a series of discrete genetic data and then explored by crossover and mutation. Due to its outstanding advantage in solving such problems, genetic algorithm is applied to the upper level [11].

One: Construction of Solutions A n-piece gene cluster is used to stand for a set of layout plans of public bike rental stations. Each gene is composed an n-unit 0–1 variable. Each unit means whether to build a rental station in the location it stands for, and 0 means no, 1 means yes.

Two: Generating Original Cluster L individuals are generated randomly to compose the original cluster on condition that the coding plan is satisfied. The cluster is named $G_0 = \{g_1, g_2, \cdots, g_L\}$, and its feasibility should be verified.

Three: Fitness Function Fitness function is one that measures the degree individuals approach the optimal solution. In the problem here, it measures each plan. The higher the value is, the closer it is to the optimal solution.

$$Fit\widetilde{z}(x) = f - (f)^k_{\min} + \xi^k \tag{152.6}$$

where $\widetilde{z}(x)$ is the objective function of SO problem; f is the objective function value of the lower level; $f^k_{\min}$ is the minimized objective function value of individuals in the kth generation; and ξ^k is the adjusted value of selection pressure, which decreases as k increases. It's set as follows:

$$\begin{cases} \xi^0 = M \\ \xi^k = c \cdot \xi^{k-1} \end{cases} \qquad (152.7)$$

where f_{min}^k is the smallest individual of the kth generation and M, c are constants, $c \in [0.9, 0.999]$.

Four: Genetic Manipulation There are three simple genetic manipulations of genetic algorithm: selection, crossover, and mutation. Selection is a process during which individuals with strong vitality produce new clusters; crossover means two homologous chromosomes recombine through mating to form a new chromosome during the evolution; and mutation means new chromosome is produced due to gene mutation and shows some new biological traits.

152.4 Case Study

Imagine the travel demand of all settlements in a region during a certain period as illustrated in Fig. 152.4. The travelling distance obeys uniform distribution between 4 and 12 km. The bi-level model is used to solve the optimization problem.

Constraints in the upper level are processed with penalty function methods. Those that don't satisfy constraints (152.3), (152.4), and (152.5) are punished to one-third of the original value.

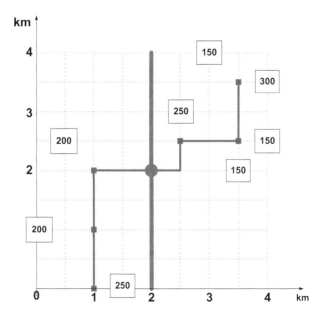

Fig. 152.4 The production volume in settlements

Fig. 152.5 The optimal layout plan

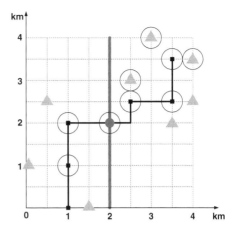

Table 152.1 Scale of public bike rental stations

No.	Volume of picking up	Volume of dropping off	Recommended capacity
21	45	28	50
39	64	22	70
41	0	116	120
51	0	39	40
53	55	39	60
60	36	0	40
71	0	76	80
72	83	0	90
79	37	0	40
Total	320	320	590

The selection of genes uses roulette strategy. That is to say, the selection probability is connected with fitness and individuals are selected at random probability to the next generation. The crossover probability is 0.9 and mutation probability is 0.04.

Matlab programming can help realize the two levels' algorithm of bi-level planning and be applied to the case. After 60 generations' evolution, the optimization plan is found. The optimal layout plan of station is illustrated in Fig. 152.5.

The final recommended capacity of each rental station is listed in Table 152.1.

As is shown in Table 152.1, 9 public bike rental stations need to be built in the region with a total capacity of 320; thus, the recommended capacity is 590.

152.5 Conclusion

The optimal layout of public bike rental stations based on B+R mode is an essential way to increase the attraction of metro's passenger flow. Considering the characteristics of the layout of public bike rental stations, discrete transportation network design method and bi-level planning model are chosen to describe and solve the problem. For the upper level, to maximize the metro trip volume, from the aspect of travel planners, the model designs a reasonable layout plan under all constraints. And the lower level uses a combined model of mode spilt and traffic assignment to describe users' travelling habits. To validate the bi-level model, a case is taken as an example. The result shows bi-level model can not only well describe the optimal layout problem of public bike but also offer technical and scientific support for decisions of layout of public bike rental stations around metro stations.

References

1. Li, L.-h., Chen, H., & Sun, X.-l. (2009). Bike rental station deployment planning in Wuhan [J]. *Urban Transport of China, 7*(4), 39–44 (In Chinese).
2. Li, Z. (2010). Analysis of the size of public bike rental stations development [J]. *Jiaotong Jieneng Yu Huanbao, 2*, 44–46 (In Chinese).
3. Geng, X., Kai Tian, Y., & Zhang, Q. L. (2009). Bike rental station planning and design in Paris [J]. *Urban Transport of China, 7*(4), 21–29 (In Chinese).
4. Lin, J.-R., & Yang, T.-H. (2010). Strategic design of public bicycle sharing systems with service level constraints [J]. *Transportation Research Part E, 47*, 284–294.
5. Wei, Z., Huabing, D., & Ren, F. (2005). A research into R_B tripping in large cities of China [J]. *Road Traffic & Safety, 5*(4), 1–4 (In Chinese).
6. Martens, K. (2004). The bicycle as a feedering mode: Experiences from three European countries [J]. *Transportation Research Part D., 9*, 281–294.
7. Huapu, L. (2006). *Theory and method in transportation planning (second edition) [M]* (pp. 24–28). Beijing: Qsinghua University Press (In Chinese).
8. Jia, W. (2011). *Mixed logit model and its application research [D]*. Ningbo: Ningbo University (In Chinese).
9. Wang, S.-s., Huang, W., & Zhen-bo, L. (2006). Study on mixed logit model and its application in traffic mode split [J]. *Journal of Highway and Transportation Research and Development, 23*(5), 88–91 (In Chinese).
10. Li, H.-m., Huang, H.-j., & Liu, J.-f. (2009). Parameter estimation of the mixed logit model and its application [J]. *Journal of Transportation Systems Engineering and Information Technology, 7*(4), 39–46 (In Chinese).
11. Chen, D.-w., Zhong, X., & Li, X.-h. (2010). Layout optimizing model of alternative interchange in highway network [J]. *Journal of Traffic and Transportation Engineering, 10*(3), 72–76 (In Chinese).

Chapter 153
Modeling of Train Control Systems Using Formal Techniques

Bingqing Xu and Lichen Zhang

Abstract Train control systems must guarantee a very high level of safety because their incorrect functioning may have very serious consequences such as loss of human life, large-scale environmental damages, or considerable economical penalties. The software reliability is related to several factors, such as completeness, consistency, and lack of ambiguity. Formal methods are widely recognized as fault avoidance techniques that can increase dependability by removing errors during the specification of requirements and during the design stages of development. In this chapter, a brief overview of existing results on formal specification of train control systems is first presented. Then we propose an integrated formal approach to specify train control systems; this integrated approach combines CSP and Object-Z with Clock theory to specify the Railway Control System concerning both the linear track and crossing area, especially the time delay between any two aspects of the railway system.

153.1 Introduction

Train control systems must guarantee a very high level of safety because their incorrect functioning may have very serious consequences such as loss of human life, large-scale environmental damages, or considerable economical penalties [1]. To meet safety and reliability requirements, the relative international standards recommend the application of formal methods in specifying development specifications and design for train control systems, the approaches using formal methods which can eliminate ambiguities of specification, and specify and prove system specification in a rigorous mathematical way, were highly recommended for the railway systems which belong to software safety integrity level 4 [2]. Although the

B. Xu • L. Zhang (✉)
Shanghai Key Laboratory of Trustworthy Computing, East China Normal University, Shanghai 200062, China
e-mail: zhanglichen1962@163.com

train control system safety and reliability depend not only on software but also on hardware, without a proper software system support, the system cannot work perfectly. In addition, compared with hardware faults which are mostly physical, the detecting and correcting of faults in software systems are usually more abstract and more troublesome. The software reliability is related to several factors, such as completeness, consistency, and lack of ambiguity [3]. Formal methods are widely recognized as fault avoidance techniques that can increase dependability by removing errors during the specification of requirements and during the design stages of development. Formal methods can be used to increase the safety of systems by formally verifying that certain safety properties hold on a model of system. In addition most formal method approaches are well suitable to be mechanized and a great variety of tools for automatic validation are nowadays available. Complex system such as train control system is a system with many different aspects, and the mechanism of communication between different aspects is hard to define. With the help of formal methods, we can find a way to construct a detailed specification of each aspect and the link mechanism among various aspects, while a communication mechanism is not enough to describe the state change and data change in the system. Above all, the author tends to use Communicating Sequential Processes (CSP) [4] to specify the communication part of the train control system. Concerning the time characteristics in the system, clock [5] specifies the system time requirements better [6]. For the state and data changes, Object-Z [7] is ideal for analysis in data change in a schema box form.

In this chapter, we propose an integrated formal approach to specify train control systems; this integrated approach combines CSP and Object-Z with Clock theory to specify the Railway Control System concerning both the linear track and crossing area, especially the time delay between any two aspects of the railway system.

153.2 Related Works

Formal methods are approaches, based on the use of mathematical techniques and notations, for describing and analyzing properties of software systems. That is, descriptions of a system are written using notations which are based on mathematical expressions rather than informal notations. These mathematical notations are typically drawn from areas of discrete mathematics, such as logic, set theory, or graph theory. There exists a lot of work on applications of formal methods in train control system.

Modeling the controller of the railway network, having resource sharing based on mutual exclusion constraints, is an important problem. Ahmad, Farooq and Khan, and Sher Afzal firstly address the specification of safety properties for the model of a complex railway crossing. The operations, i.e., occupied, free, and block, are formalized to describe the safety properties along railway crossing. Second, to develop the control model of the crossing system, they construct the subnet representing the train flow along the tracks in the crossing region and the set of monitors or supervisors are also modeled as subnets. Arc-constant colored Petri net (ac-CPN) is used to construct the train flow subnet while the monitors are modeled using the place/transition net. Arc-constant colored Petri net enforces the specification of not to shift the train from a track to another one. Bottom-up approach is adopted to model the control for railway crossing as a synchronous synthesis of the subnets is applied to build the final model. Finally, to verify the safety properties in the developed controller, the coverability tree method is used for the analysis of the final model [8].

Defects in requirements specification of train control system may have fatal consequences. Zhao Lin, Tang Tao, Cheng Ruijun, and He Liyun present a property-based requirements analysis approach for train control system [9], which provides support for constructing precise, complete, and consistent requirements specification and analyzing them with formal techniques.

Level crossings (LCs) are considered to be a safety black spot for railway transportation since LC accidents/incidents dominate the railway accident landscape in Europe, thus considerably damaging the reputation of railway transportation. LC accidents cause more than 300 fatalities every year throughout Europe, which represents up to 50 % of all deaths for railway. That is why LC safety is a major concern for railway stakeholders in particular and transportation authorities in general. LCs with an important traffic moment $\$^1$ are generally equipped with automatic protection systems (APSs). Here, they focus on two main risky situations, which have caused several accidents at LCs. The first is the short opening duration between successive closure cycles relative to trains passing in opposite directions. The second is the long LC closure duration relative to slow trains. Mekki Ahmed, Ghaze Mohamed, and Toguyeni Armand suggest a new APS architecture that prevents these kinds of scenarios and therefore increases the global safety of LCs [10]. To validate the new architecture, a method based on well-formalized means has been developed, allowing one to obtain sound and trustworthy results. Their method uses a formal notation, i.e., timed automata (TA), for the specification phase and the model-checking formal technique for the verification process. All the steps are progressively discussed and illustrated.

Railway interlocking system is a distributed, safety, monetary, and environmentally critical system, and its failure may cause the loss of human life, severe injuries, loss of money, and environmental damages. The complexity of this system requires formal modeling and step-by-step refinement for its construction and development. The formal specification-based languages, such as VDM, Z-notation, and RAISE, have been used for its modeling using crisp (two-valued logic) theory. However, due to the continuous and inexact features, like speed, weight, and moving block (breaking distance including length of a train), fuzzy distributed multi-agent approaches are required to capture the inexactness and uncertainty present in the existing system.

---

[1] LCs with an important traffic moment are generally equipped with automatic protection systems (APSs).

The RBC (radio block center) handover is an important part of European Train Control System (ETCS) level 2 which is a typical safety-critical hybrid system. Liu et al. [11] build a formal model of RBC handover procedure using Differential Dynamic Logic, which is a first-order dynamic logic for specifying and verifying hybrid systems, and identify some constraints that are necessary for ensuring safety of train control, including collision avoidance as well as derailment avoidance. Moreover, they formally verify the safety-related properties of their model with deductive verification tool KeYmaera. The experimental results show the validity and feasibility of the method. Meanwhile, the safety constraints and safety-related properties verified in the chapter can be helpful to the practical application of train control.

Flaws in requirements may have unacceptable consequences in the development of safety-critical applications. Formal approaches may help with a deep analysis that takes care of the precise semantics of the requirements. However, the proposed solutions often disregard the problem of integrating the formalization with the analysis, and the underlying logical framework lacks either expressive power, or automation. We propose a new, comprehensive approach for the validation of functional requirements of hybrid systems, where discrete components and continuous components are tightly intertwined. The proposed solution allows to tackle problems of conversion from informal to formal, traceability, automation, user acceptance, and scalability.

Jochen Hoenicke uses a combination of three techniques for the specification of processes, data, and time: CSP, Object-Z, and Duration Calculus [12–16]. The basic building block in our combined formalism CSP-OZ-DC is a class. First, the communication channels of the class are declared. Every channel has a type which restricts the values that it can communicate. There are also local channels that are visible only inside the class and that are used by the CSP, Z, and DC parts for interaction. Second, the CSP part follows; it is given by a system of (recursive) process equations. Third, the Z part is given which itself consists of the state space, the Init schema and communication schemas. For each communication event a corresponding communication schema specifies in which way the state should be changed when the event occurs. Finally, below a horizontal line the DC part is stated. The combination is used to specify parts of a novel case study on radio-controlled railway crossings. Johannes Faber formally specifies a part of the ETCS with the specification language CSP-OZ-DC treating the handling of emergency messages.

153.3 Formal Specification of Train Control Systems

Train control system is composed of two parts [3]: onboard system and control center. The onboard system is composed of the following: (1) a basic state detection subsystem, including position and speed of train; (2) an ATP subsystem, which is used to monitor the speed and will produce appropriate actions if certain situations happen (unresponsive train operator, earthquake, disconnected rail, overrun of the authority, etc.) to prevent accidents from happening; (3) a communication subsystem, which is used to send and receive messages and commands to and from the control center when it is necessary, such as approaching the station and departing from the station; and (4) a record subsystem. The control center is mainly composed of the following: (1) a communication system, which is used to send and receive commands and messages to and from trains. The control center knows the position and station block where every train is, which is the basis of authorization generation; (2) an interlocking system, which is responsible for the set of points, and route set based on information received from the trains; and (3) a traffic operation management, which creates train timetables. All trains share their information with the control center periodically. In the event of a natural disaster, once the train loses communication with the control center for a certain time, it will stop automatically.

In order to keep the description focused, we concentrate on some particular points in train control systems rather than the detailed descriptions of all development process. The specification is made by integrating Object-Z, CSP, and Clock theory. The Object-Z is used to specify the data aspect and the operation on the data of the system; CSP is used to specify communication aspect of the system, and clock is used to specify the time aspect of the system.

Clock theory [5] puts forward the possibility to describe the event in physical world by using a clock and can analyze and record the event by clock. To use clock to specify cyber physical systems, the time description is clearer to every event and can link continuous world with discrete world better [6].

Let e be an event. Clock (e) denotes the clock that records the time instants when the event e occurs. And we use clock (event(c)) to denote the event that take place at every time instant c[i].

The controller should take care of trains running over the track. It should control the safety of the configuration, i.e., no two trains may enter the critical section. When one critical section is occupied, some others, which share some part of section with this one, should be locked. The controlled can control the status, speed, and position of trains as shown in Fig. 153.1.

We integrate CSP and Object-Z with Clock theory to specify real-time aspects as shown in Fig. 153.2.

Finally, woven model is shown as Fig. 153.3.

153.4 Conclusion

Train control systems must guarantee a very high level of safety because their incorrect functioning may have very serious consequences such as loss of human life, large-scale environmental damages, or considerable economical penalties. The software reliability is related to several factors, such as completeness, consistency, and lack of ambiguity. Formal methods are widely recognized as fault

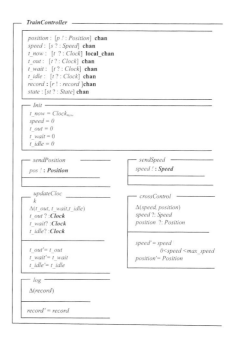

Fig. 153.1 Train control system

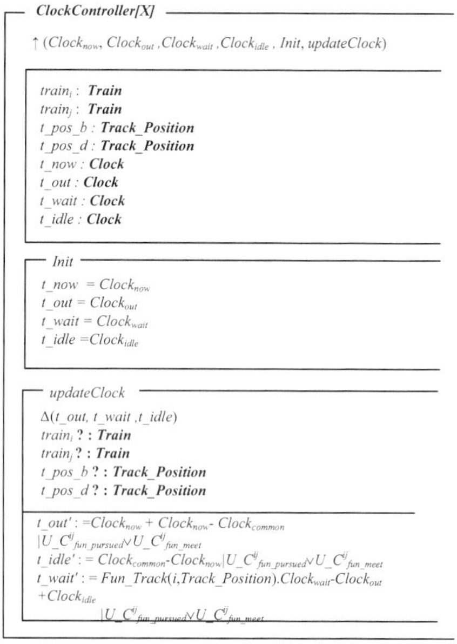

Fig. 153.2 Model of clock

avoidance techniques that can increase dependability by removing errors during the specification of requirements and during the design stages of development. In this chapter, we propose an integrated formal approach to specify train control system; this integrated approach combines CSP and Object-Z with Clock theory to specify

Fig. 153.3 Woven aspects of diagram

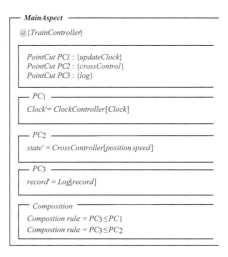

the train control system. In this chapter, we have applied the proposed approach to the specification of train control systems; the proposed approach is expressive enough to represent the functional requirements, behavior requirements, and real-time aspects of train control system, yet simple enough to allow for the use by nonexperts in formal methods.

Future work focuses on the verification tool development of our proposed method.

Acknowledgments This work is supported by Shanghai Knowledge Service Platform Project (No. ZF1213); National High Technology Research and Development Program of China (No. 2011AA010101); National Basic Research Program of China (No. 2011CB302904); the National Science Foundation of China under Grant Nos. 61173046, 61021004, 61061130541, and 91118008; Doctoral Program Foundation of Institutions of Higher Education of China (No. 20120076130003); and National Science Foundation of Guangdong Province under Grant No. S2011010004905.

References

1. Jo, H.-J., Yoon, Y.-K., & Hwang, J.-G. (2009). Analysis of the formal specification application for train control systems. *Journal of Electrical Engineering & Technology, 4*(1), 87–92.
2. IEC62278:2002. *Railway applications: Specification and demonstration of reliability, availability, maintainability and safety (RAMS)*.
3. Xie, G., Hei, X., Mochizuki, H., Takahashi, S., & Nakamura, H. (2013). Safety and Reliability Estimation of Automatic Train Protection and Block System. Quality and Reliability Engineering International. © John Wiley & Sons, Ltd.
4. Reed, G. M., & Roseoe, A. W. (1988). A timed model for communicating sequential processes. *Theoretical Computer Science, 58*, 249–261.
5. Xu, B. Q., He, J., & Zhang, L. C. (2013). Specification of cyber physical systems based on clock theory. *International Journal of Hybrid Information Technology, 6*(3), 45–54.

6. Xu, B. Q., et al. (2013). Specification of cyber physical systems by clock. In *AST2013, SERSC* (Vol. 20, pp. 111–114). SERSC (Science & Engineering Research Support Society) Korea
7. Najafi, M., & Haghighi, H. (2013). An integration of UML-B and object-Z in software development process. In K. Elleithy & T. Sobh (Eds.), *Innovations and advances in computer, information, systems sciences, and engineering* (pp. 633–648). New York: Springer.
8. Ahmad, F., & Khan, S. A. (2013). Specification and verification of safety properties along a crossing region in a railway network control. *Applied Mathematical Modelling, 37*(7), 5162–5170.
9. Zhao, L., Tang, T., Cheng, R., & He, L. (2013). Property based requirements analysis for train control system. *Journal of Computational Information Systems, 9*(3), 915–922.
10. Mekki, A., Ghaze, M., & Toguyeni, A. (2012). Validation of a new functional design of automatic protection systems at level crossings with model-checking techniques. *IEEE Transactions on Intelligent Transportation Systems, 13*(2), 714–723.
11. Dewang Chen, Rong Chen, & Yidong Li. Formal modeling and verification of RBC handover of ETCS using differential dynamic logic. In *Proceedings of 2011 10th international symposium on autonomous decentralized systems* (pp. 67–72).
12. Hoenicke, J. (1999). Specification of radio based railway crossings with the combination of CSP. In G. Smith & I. Hayes (Eds.), *Towards real-time object-Z, Lecture notes in computer science* (pp. 49–65). Berlin: Springer.
13. Hoenicke, J. (2006). *Combination of processes, data, and time*. PhD thesis, University of Oldenburg.
14. Hoenicke, J., & Maier, P. (2005). Model-checking of specifications integrating processes, data and time. In J. S. Fitzgerald, I. J. Hayes, & A. Tarlecki (Eds.), *FM 2005: Formal methods* (Vol. 3582, pp. 465–480). Berlin: Springer.
15. Hoenicke, J., & Olderog, E.-R. (2002). CSP-OZ-DC: A combination of specification techniques for processes, data and time. *Nordic Journal of Computing, 9*(4), 301–334.
16. Hoenicke, J., & Olderog, E.-R. (2002). Combining specification techniques for processes data and time. In M. Butler, L. Petre, & K. Sere (Eds.), *Integrated formal methods. Lecture notes in computer science* (Vol. 2335, pp. 245–266). Berlin: Springer.

Chapter 154
A Clock-Based Specification of Cyber-Physical Systems

Bingqing Xu and Lichen Zhang

Abstract In cyber-physical systems, the elapse of time becomes the most important property of system behavior, and time is central to predicting, measuring, and controlling properties of the physical world. A cyber-physical system is composed of two interacting subsystems: a cyber system and a physical system. The behavior of the cyber system is controlled by the execution of programs on a distributed digital computer system, while the laws of physics control the behavior of the physical system. The different models of time—continuous physical time in the physical system versus discrete execution time in the cyber system and the impossibility of perfect synchronization of the physical clocks of the nodes of a distributed computer system, lead to interesting phenomena concerning the joint behavior of these two subsystems. The chapter describes the case studies in applying clock theory to the production cell. The clock theory described is very simple, in that it models clocks as potentially infinite lists of reals. Xeno's paradox and similar problems are avoided by specifying limits on clock rates, which effectively means that the model sits somewhere between a discrete synchronous model and a fully dense continuous-time model as assumed by some other formalisms. The case study of the specification of the production cell shows that using clock theory to specify cyber-physical systems can give a more detailed description of the every subsystem and give a much more considerate observation of the time line and sequence of every event.

B. Xu • L. Zhang (✉)
Shanghai Key Laboratory of Trustworthy Computing, East China Normal University, Shanghai 200062, China
e-mail: zhanglichen1962@163.com

154.1 Introduction

Cyber-physical systems consist of the class of large-scale infrastructures that have significant cyber and physical components and have wide-ranging impact on society in their deployment [1]. Time is central to predicting, measuring, and controlling properties of the physical world: given a physical model, the initial state, the inputs, and the amount of time elapsed, one can compute the current state of the plant. A cyber-physical system consists of two interacting subsystems: a cyber system and a physical system. The behavior of the cyber system is controlled by the execution of programs on a distributed digital computer system, while the laws of physics control the behavior of the physical system. The different models of time—dense physical time in the physical system versus discrete execution time in the cyber system and the impossibility of perfect synchronization of the physical clocks of the nodes of a distributed computer system, lead to interesting phenomena concerning the joint behavior of these two subsystems.

Cyber-physical systems-related research is based on two, originally different worldviews: on the one hand, the dynamics and control (DC) worldview and on the other hand the computer science (CS) worldview. The DC worldview is that of a predominantly continuous-time system, which is modeled by means of differential (algebraic) equations or by means of a set of trajectories. The CS worldview is that of a predominantly discrete-event system. A well-known model is a (hybrid) automaton, which modeling of discrete-event systems is also based on. As new CPS applications start to interact with the physical world using sensors and actuators, there is a great need for ensuring that the actions initiated by the CPS is timely. Cyber-physical systems are characterized by their stringent requirements for time constraints such as predictable end-to-end latencies and timeliness.

Since cyber-physical systems are dynamic systems that exhibit both continuous and discrete dynamic behavior, the continuously bilateral interaction between discrete events and continuous time flow makes it hard to know the dynamic feature of the system. So specifying the timing issues is a really vital work in the early stage. This chapter describes the case study in applying clock theory to the production cell. The clock theory described is very simple, in that it models clocks as potentially infinite lists of reals. Xeno's paradox and similar problems are avoided by specifying limits on clock rates, which effectively means that the model sits somewhere between a discrete synchronous model and a fully dense continuous-time model as assumed by some other formalisms.

154.2 Related Works

Duggirala et al. present an algorithm for checking global predicates from distributed traces of cyber-physical systems for an individual agent [2]. Each observation has a possibly inaccurate timestamp from the agent's local clock. The challenge is to symbolically over-approximate the reachable states of the entire system from the unsynchronized traces of the individual agents. The presented algorithm first approximates the time of occurrence of each event, based on the synchronization errors of the local clock, and then over-approximates the reach sets of the continuous variables between consecutive observations. The algorithm is shown to be sound; it is also complete for a class of agents with restricted continuous dynamics and when the traces have precise information about timing synchronization inaccuracies. Experimental results illustrate that interesting properties like safe separation, correct geocast delivery, and distributed deadlocks can be checked for up to 20 agents in minutes.

The specification of modeling and analysis of real-time and embedded (MARTE) systems is an extension of the unified modeling language (UML) in the domain of real-time and embedded systems. Even though MARTE time model offers a support to describe both discrete and dense clocks, the biggest effort has been put so far on the specification and analysis of discrete MARTE models. To address hybrid real-time and embedded systems, Liu et al. propose to extend statecharts using both MARTE and the theory of hybrid automata [3]. As a case study, they model the behavior of a train control system with hybrid MARTE statecharts to demonstrate the benefit.

The Object Management Group (OMG) UML profile for MARTE aims at using the general-purpose modeling language UML in the domain of real-time and embedded (RTE) systems. We have also defined a non-normative concrete syntax called the CCSL to demonstrate what can be done based on this structure. Mallet Frédéric gives a brief overview of this syntax and its formal semantics and shows how existing UML model elements can be used to apply this syntax in a graphical way and benefit from the semantics [4].

The UML Profile for MARTE has been recently adopted. The CCSL allows the specification of causal, chronological, and timed properties of MARTE models. The IEEE Property Specification Language (PSL) provides a formal notation for expressing temporal logic properties that can be automatically verified on electronic system models [5]. We identify and restrict the CCSL constructs that cannot be expressed in temporal logics so that CCSL become tractable in temporal logics.

Event clock automata (ECA) are a model for timed languages that has been introduced by Alur, Fix, and Henzinger as an alternative to timed automata, with better theoretical properties (for instance, ECA are determinizable while timed automata are not). Gilles et al. revisit and extend the theory of ECA [6].

In the chapter of Lamport [7], the concept of "happening before" defines an invariant partial ordering of the events in a distributed multiprocess system. The representation of a closed finitary real-time system as a graph annotated with clock constraints is called a timed automaton [8], since clocks range over the nonnegative reals, every nontrivial timed automaton has infinitely many states. If the clocks of a finitary real-time system are permitted to drift with constant, rational drift bounds, one obtains a finitary drifting-clock system. The representation of a closed finitary drifting-clock system as a graph annotated with constraints on drifting clocks is called an initialized rectangular automaton [9]. Two popular specification languages for the algorithmic verification of untimed systems are finite automata and propositional temporal logics. In order to specify timing constraints, these languages can be extended by adding clock variables. If we judiciously add clocks to finite automata, one obtains the timed automata (TA); from propositional linear temporal logic, one obtains the real-time logic TPTL [10]; and from the propositional, branching-time logic CTL, one obtains the real-time logic TCTL [11].

Bujorianu et al. presented a multiclock model for real-time abstractions of hybrid systems [12]. They call hybrid time systems the resulting model, which is constructed using category theory. They define a timed (or clock) system as a functor from a category of states to a category of time values. They further define concurrent composition operators and bisimulation.

154.3 Clock Theory

Clock theory [13] puts forward the possibility to describe the event in physical world by using a clock, and can analyze and record the event by clock. To use clock to specify cyber-physical systems, the time description is clearer to every event and can link continuous world with discrete world better [14].

154.4 Case Study: Specification of the Production Cell Steam Based on Clock Theory

The production cell [15] is composed of two conveyor belts, a positioning table, a two-armed robot, a press, and a travelling crane. Metal plates inserted in the cell via the feed belt are moved to the press [16]. There, they are forged and then brought out of the cell via the other belt and the crane [17]. As is known to all, the production cell is widely used in factories, and it has brought a large amount of benefit to the manufacture [18]. Since its multifunction and high efficiency, metal blank can be made into various shapes to fit different requirements [19]. Figure 154.1 shows the typical type of the production cell; there are sensors named sensor1 and sensor2 which conveys information when the press is loaded or unloaded separately. First of all, the lower part of the press is raised to middle, and the metal blank is fed. When the press is loaded, sensor1 conveys the information [20]. When the lower part is raised to the top, the press moves down to the bottom position to place the metal blank to the lower convey belt. Now sensor 2 conveys the information when the press is unloaded. Then the lower part is raised to the middle position and waits for the press to be loaded again. The production cell is a kind of cyclical control system [21]. All the necessary parameters are listed in Table 154.1.

To guarantee the safety of the press, the two basic rules are as follows. The current height of press is always lower than or equal to the top position, and

Fig. 154.1 The production cell

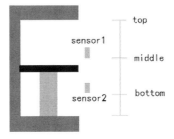

Table 154.1 Parameter list for the production cell

Parameter	Value
v	Current speed of press
h	Height of press
c	Height change of press
Bottom, middle, top	Position of press
unload, load, pressing	Operations of press
mov_unload, mov_load	Press moves to load and unload

it is always higher than or equal to the bottom position. Bottom position is always lower than middle position, and so is the middle position to top position.

Bottom $\leq$ h $\leq$ Top
Bottom < Middle < Top

First of all, we consider the process of press; h denotes the height of press; c denotes the height change of press; v denotes the current speed of press; and load and unload are events which control the convey belt to feed metal blanks. During the process, the press moves down and up. Suppose the initial situation is that the press is at the bottom position and is ready to be raised to the middle position.

```
climb(h,Middle - c) ≤ clock(load) ≤clock(pressing)
                  ≤climb(h, Top -c )
      ≤ drop(h, Bottom + c ) ≤ clock(unload)
Between different phases, there exists time latency.
      ρ(climb(h,Middle - c), clock(load)) ≤ c/v
         ρ(clock(load),clock(pressing)) ≤ c/v
       ρ(clock(pressing),climb(h, Top - c)) ≤ c/v
       ρ(climb(h, Top - c), drop(h, Bottom + c )) ≤ c/v
         (drop(h,Bottom + c), clock(unload)) ≤ c/v
```

From above, it is obvious that some couple of events in the equations above have noninterference, since mov_load and mov_unload cannot happen at the same time, and so do load and unload.

```
            clock(mov_load)[1]) >0
clock(mov_load) ≤ clock(mov_unload) ≤ clock(mov_load)'
```

$$\text{clock(mov\_unload)} \wedge \text{clock(mov\_load)} = \emptyset$$
$$\text{clock(load)[1]} > 0$$
$$\text{clock(load)} \leq \text{clock(unload)} \leq \text{clock(load)}'$$
$$\text{clock(load)} \wedge \text{clock(unload)} = \emptyset$$

In the equations, **V** denotes continuous speed change, and **v** denotes discrete speed at each clock unit. The continuous variable and discrete variable can be linked as below:

$$\mathbf{V} = \mathbf{v} \text{ init } \mathbf{v}_0$$
$$\mathbf{v} = \mathbf{V} \text{ every c init } \mathbf{v}_0$$

154.5 Conclusion

In cyber-physical systems, the elapse of time becomes the most important property of system behavior and time is central to predicting, measuring, and controlling properties of the physical world. A cyber-physical system is composed of two interacting subsystems: a cyber system and a physical system. The behavior of the cyber system is controlled by the execution of programs on a distributed digital computer system, while the laws of physics control the behavior of the physical system. The different models of time—continuous physical time in the physical system versus discrete execution time in the cyber system and the impossibility of perfect synchronization of the physical clocks of the nodes of a distributed computer system, lead to interesting phenomena concerning the joint behavior of these two subsystems. This chapter presented the case study in applying clock theory to the production cell. Case study shows that using clock theory to specify cyber-physical systems can give a more detailed description of every subsystem and give a much more considerate observation of the time line and sequence of every event.

It is brilliant to connect the event with clock, while it is so difficult to handle so many local clocks with the global clock. In my point of view, it is always hard work to make local clocks keeping consistent with the global clock, and the verification of the security and accuracy of the synchronization is very complicated, and we need more ideas to do this work.

Acknowledgements This work is supported by Shanghai Knowledge Service Platform Project (No. ZF1213), National High Technology Research and Development Program of China (No. 2011AA010101), National Basic Research Program of China (No.2011CB302904), the National Science Foundation of China under grants (No.61173046, No. 61021004, No. 61061130541, No.91118008), Doctoral Program Foundation of Institutions of Higher Education of China (No. 20120076130003), National Science Foundation of Guangdong Province under grant (No. S2011010004905).

References

1. Eidson, J., Lee, E. A., Matic, S., Seshia, S. A., & Zou, J. (2012). Distributed real-time software for cyber-physical systems. *Proceedings of the IEEE (Special Issue on CPS), 100*(1), 45–59.
2. Duggirala, P. S. Johnson, T. T, Zimmerman, A., et al. (2012). Static and dynamic analysis of timed distributed traces. *Proceedings of the 2012 I.E. 33rd Real-Time Systems Symposium* (pp. 173–182). IEEE.
3. Liu, J., Liu, Z., He, J., Frédéric, M., & Ding, Z. (2013). Hybrid MARTE statecharts. *Frontiers of Computer Science, 7*(1), 95–108.
4. Frédéric, M. (2008). Clock constraint specification language: Specifying clock constraints with UML/MARTE. *Innovations in Systems and Software Engineering, 4*(3), 309–314.
5. Regis, G., Frederic, M., & Julien, D. (2011). Logical time and temporal logics: Comparing UML MARTE/CCSL and PSL. In *Proceedings of the International Workshop on Temporal Representation and Reasoning* (pp. 141–148). IEEE.
6. Gilles, G., Jean-François, R., & Nathalie, S. (2011). Event clock automata: From theory to practice. In U. Fahrenberg & S. Tripakis (Eds.), *Formal modeling and analysis of timed systems*: Vol. 6919. *Lecture notes in computer science* (pp. 209–224). Berlin: Springer.
7. Lamport, L. (1978). Time, clocks, and the ordering of events in a distributed system. *Communications of the ACM, 21*(7), 558–565.
8. Rajeev, A., & Dill, D. L. (1994). A theory of timed automata. *Theoretical Computer Science, 126*(2), 183–235.
9. Henzinger, T.A., Kopke, P.W., Puri, A. & Varaiya, P. (1995). What's decidable about hybrid automata? In *Proceedings of the 27th Annual Symposium on Theory of Computing* (pp. 373–382). ACM Press.
10. Alur, R., & Henzinger, T. A. (1994). A really temporal logic. *Journal of the ACM, 41*(1), 181–204.
11. Alur, R., Courcoubetis, C., & Dill, D. L. (1993). Model checking in dense real time. *Information and Computation, 104*(1), 2–34.
12. Bujorianu, M. C. Bujorianu, L. M., & Langerak, R. (2008). An interpretation of concurrent hybrid time systems over multi-clock systems. In *Proceedings of the 17th IFAC World Congress* (Vol. 17, Part 1, pp. 3635–3640). IFAC
13. Xu, B., He, J., & Zhang, L. (2013). Specification of cyber physical systems based on clock theory. *International Journal of Hybrid Information Technology, 6*(3), 45–54.
14. Xu, B., et al. (2013). Specification of cyber physical systems by clock. In *AST 2013, ASTL 20*:111–114, 2013 © SERSC 2013.
15. Back, R. J., Petre, L., & Porres, I. (2000). Generalizing action systems to hybrid systems. *Lecture Notes in Computer Science, 1926*, 202–213.
16. Aboutrab, M. S., Brockway, M., Counsell, S., & Hierons, R. M. (2013). Testing real-time embedded systems using timed automata based approaches. *Journal of Systems and Software, 86*(5), 1209–1223.
17. Budde, R. (1995). ESTEREL applied to the case study production cell. In: C. Lewerentz and T. Lindner eds. Formal Development of Reactive Systems-Case Study Production Cell, Springer, Berlin, 75–100
18. Lewerentz, C., & Lindner, T. (1995). Case study 'production cell': A comparative study in formal specification and verification. In: Manfred Broy, Stefan Jähnichen. Berlin. *Lecture Notes in Computer Science, 1009*, Springer Berlin Heidelberg, 388–416.
19. Burns, A. (2003). How to verify a safe real-time system: The application of model checking and timed automata to the production cell case study. *Real-Time Systems, 24*(2), 135–151.
20. Benghazi Akhlaki, K., Capel Tuñón, M. I., Holgado Terriza, J. A., & Mendoza Morales, L. E. (2007). A methodological approach to the formal specification of real-time systems by transformation of UML-RT design models. *Science of Computer Programming, 65*(1), 41–56.
21. El-Maddah Islam, A. M. (2005). Component-based development of process control systems. In *3rd ACS/IEEE International Conference on Computer Systems and Applications* (pp. 797–804).

Chapter 155
Polymorphic Worm Detection Using Position-Relation Signature

Huihui Liang, Jiwen Chai, and Yong Tang

Abstract This chapter proposes a novel worm signature that is appropriate for the polymorphic worm detection. Most of the recent worm signatures are constructed based on worm bytes themselves or relationships between worm bytes. In this case, most of these signatures cannot detect the polymorphic worms successfully. Our worm signature takes the worm bytes themselves and the relationships between worm bytes into consideration. So, it is called position-relation signature (PRS). The new signature is capable of handling certain polymorphic worms. The experiments show that the algorithm could be used as a basis to implement a worm detection system.

155.1 Introduction

In recent years, Internet worms have proliferated with the development of computer hardware and software. When a host is found to be worm infected, a large number of hosts may have been infected. So, fast spreading worms have presented a huge threat to the security of the Internet. Whatever in academia or in industry, there are two major problems that must be solved, the method [1] to detect worms in time and how to curb worm propagation effectively [2, 3]. In this chapter, we mainly focus on extracting the signatures of the worms to do a basis for detection.

With the development of polymorphism technology, the next generation of Internet worms is likely to be polymorphic. Polymorphic worms can be changed in their spreading process. As a consequence, copies of a polymorphic worm might no longer share a common invariant substring of sufficient length, and the existing systems will not recognize the network streams containing the worm copies as the manifestation of a worm outbreak. Therefore, the polymorphic worms that have

H. Liang (✉) • J. Chai • Y. Tang
Sichuan Electric Power Research Institute, Chengdu 610072, China
e-mail: liang_huihui@sohu.com

evolved to the network have brought great threat. Defense of the damage effectiveness mainly depends on the quality of extracted signatures of worms. Most of the recent worm signatures are constructed based on worm bytes themselves or relationships between worm bytes. Polymorphic worms can automatically change its content, so to detect them is a challenging task. To detect polymorphic worm, extracting the invariant signatures of the same polymorphic worm in different situations is the key to success.

This chapter presents a novel worm signature based on signature fusion. This worm signature combines the worm bytes themselves and the relationships between worm bytes-neighborhood-relation signature. It is called position-relation signature (PRS). And this chapter is structured as follows. Section 155.2 discusses related work. Section 155.3 presents the description of the new signature. In Sect. 155.4, we present the experimental results using different signatures to evaluate our signature. Section 155.5 briefly concludes and points out limitations of the current signatures.

155.2 Related Works

Signature is the most fundamental factors in worm detection system that is based on the signatures. The effectiveness of the polymorphic worm defense system depends on signature describing the ability of worms. The complexity of worm technology that is increasing puts forward higher request for the ability of detection and defense about worms. A lot of research is dedicated to produce worm signatures.

In recent years, most worm signatures mainly divided into two categories are as follows:

155.2.1 Signatures Based on Worm Bytes Themselves

In such signatures, the longest common substring (LCS) is a typical signature [4, 5]. Many systems use LCS. However, it cannot detect polymorphic worms. For example, when the samples of polymorphic worms added instructions, LCS cannot detect the samples after deformation. Newsome, Karp, and Song [6] put forward a polymorphic worm signature generation system—polygraph—and present a kind of suitable system for matching the characteristics of the polymorphic worms. Polygraph system extracts many tokens from suspicious traffic pool; these tokens referring to independent substring have emerged at least k times in the suspicious pool n series.

Position-aware distribution signature (PADS) is proposed by Tang and Chen [7]. It is a collection of position-aware byte-frequency distribution function. Compared with tokens signature, PADS is more flexible. When the polymorphic worm uses encryption technology and uses limited decryption routines, the decryption routines can be identified if we get enough samples of worms. However, we find it difficult to get enough polymorphic worm samples in a short period of time, so suspicious traffic pool cannot contain all polymorphic worm samples produced by each routine. Therefore, the PADS extracted from suspicious traffic pool cannot detect polymorphic worm which includes many decryption routines successfully.

155.2.2 Signatures Based on Relationships Between Worm Bytes

Most of recent worm signatures are constructed based on worm bytes themselves. They can be used to detect one pattern of worms successfully, but are not appropriate when treating on polymorphic worms since these worms can change their patterns dynamically. A class of neighborhood-relation signatures (NRSs) [8] is proposed based on neighborhood relationship between worm bytes. The NRS can exhibit characteristics of polymorphic worm and can be used to detect polymorphic worm with no noise efficiently. However, NRS is difficult to have a good performance when the polymorphic worm has noise.

Each of these two category signatures has its own advantages and disadvantages. In this chapter, we combine these two kinds of signatures, combine these advantages, and gain a better performance.

155.3 Position-Relation Signature Description

Before introducing the signatures of worms, we give an introduction on how to determine the significant regions of the worms. Given a set S of worm variants, at first, we know either the significant regions of the worm variants or the signature of this worm. If we know the signature of a worm category, we can compute the significant region of a worm variant; in the same way, given the significant region, we can compute the signature of the worm category. This is a "missing data problem" in statistics; in this chapter, we solve it using expectation-maximization (EM) algorithm.

Expectation-maximization (EM) algorithm can estimate the maximum-likelihood parameter by using an iterative procedure. In this chapter, we use it to determine the signature and the significant regions. Given a set $S = S_1, S_2, \cdots, S_n$ of collected worm byte sequence, let $(a_1, a_2, \cdots, a_n)$ represent the start positions of the significant regions and parameter F represent the signature.

At first, we initialize the start positions $(a_1, a_2, \cdots, a_n)$ of the significant regions for the worm byte sequences $(S_1, S_2, \cdots, S_n)$ and compute the maximum-likelihood estimate of the signature F. Then, we calculate the new locations of the significant regions based on the computed signature F. In this algorithm, the new start location

$a_j (j = 1, 2, \cdots, n)$ of the significant region of the worm variant $S_j(j = 1, 2, \cdots, n)$ is the position that the worm byte sequence has the maximum match sore with the current signature F. The formulation of the start position of the significant region is shown as follows:

$$a_i = \arg \max_{a_i} Score(F, S_i, a_i), \qquad (155.1)$$

where $Score(F, S_i, a_i)$ is the matching score function; it is formulated in Function 1.5. This is the expectation step. And then, based on the current locations of the significant regions, we compute the maximum-likelihood estimate of signature; this is called a maximization step.

The EM algorithm iterates between the expectation step and the maximization step. It terminates while the difference of the current average matching score and the score of the previous iteration is less than ϵ. In this chapter, the signature width is denoted by w. The EM algorithm decides the significant regions and the signature. In the following, we introduce three kinds of signatures.

155.3.1 Position-Aware Distribution Signature

The PADS signature is a frequency distribution in different byte positions. We use $(f_1, f_2, \cdots, f_w)$ to denote the worm signature or the byte-frequency distribution of the worm, and let f_0 represent the byte-frequency distribution of the normal traffic or the normal signature. The PADS signature is $F_{pads} = (f_0, f_1, f_2, \cdots, f_w)$.

Then we give detailed information on how to compute the signature while the significant regions of the worm sequence are known. At first, we compute the byte-frequency distribution for each significant region position. For position $p (p = 1, 2, \cdots, w)$, the maximum-likelihood estimation of frequency $f_p(x)$ is defined as follows:

$$f_p(x) = \frac{c_{p,x}}{n} \qquad (155.2)$$

where $c_{p,x}$ is the number of times x appears at position p in the significant regions $x \in [0, \cdots, 255]$. The above function is constrained by $\sum_{b=0}^{255} f_p(b) = 1$.

155.3.2 Neighborhood-Relation Signature

NRS takes the frequency of the neighbor distance of the bytes of the significant regions as the feature. The worm variant of each kind of worm contains at least one significant region to complete the function of the worm. Most of the worm detection method detects worms directly based on the worm payload bytes or subsequences. Because the load contents of the polymorphic worm are often changed, such methods cannot effectively detect polymorphic worms. However, NRS can be used to detect polymorphic worms effectively. This signature combines the characteristics of the polymorphic worms and takes the relationship of the payload bytes and the subsequence into important consideration. NRS has a good performance on polymorphic worm detection.

For convenience, at first, we give the definition of neighbor distance. Given a sequence $S_j = c_1 c_2 \cdots c_n$, assume the byte distance between c_i and c_{i+1} is $d_{i,i+1}$, and $d_{i,i+1}$ is referred to as the neighbor distance of location i of the sequence S_j. The formulation of the neighbor distance $d_{i,i+1}$ is as follows:

$$d_{i,i+1} = |c_{i+1} - c_i| (i = 1, 2, \cdots, n-1). \tag{155.3}$$

The sequence set $S = S_1, S_2, \cdots, S_n$ is a kind of collected worm variants. Assume the length of the significant regions of such kind of worm is w. The start locations of the significant regions of the worm sequences are $a_i (i = 1, 2, \cdots, n)$. Then we introduce how to compute NRS of this worm.

Given a value $p(p = 1, 2, \cdots, w-1)$, the neighbor distance of the location index $a_i + p$ of sequence S_i is d_{a_i+p, a_i+p+1}; $count(p,d)$ represents the occurrence number of value d of $d_{a_i+p, a_i+p+1}, (i = 1, 2, \cdots, n)$ for the sequences of the sequence set S. Define the distribution function of the neighbor distance as follows:

$$f'_p(d) = \frac{count(p,d)}{n} \tag{155.4}$$

with the constraint, $\sum_{d=0}^{255} f'_p(d) = 1, \quad p = 1, 2, \cdots, w-1$.

$(f'_1, f'_2, \cdots, f'_{w-1})$ is the NRS of the n sequences of the sequence set S. The signature length is $w-1$. f'_0 denotes the normal signature. The NRS signature of a worm category is denoted as $F_{nrs} = (f'_0, f'_1, f'_2, \cdots, f'_{w-1})$.

155.3.3 Position-Relation Signature

PRS is a novel signature that combines the byte-frequency information and the neighbor distance information. The PRS F is represented as (F_{pads}, F_{nrs}). This signature combines the merits of PADS and NRS. In the following experiments, it has a good performance on worm detection.

For an incoming connection S_i, we want to decide whether S_i is a worm variant. Assume the length of S_i is l. $S_{i,j}(j = 1, 2, \cdots l)$ is the byte of S_i at position j.

$seg(S_i,a_i)$ denotes the w-byte segment of S_i with the start position a_i. The matching score of $seg(S_i,a_i)$ with the PRS is defined as follows:

$$Score(F, S_i, a_i) = \Pi_{p=1}^{w} \frac{f_p(S_{i,a_i+p-1})}{f_0(S_{i,a_i+p-1})} + \Pi_{p=1}^{w-1} \frac{f'_p(d_{a_i+p,a_i+p+1})}{f'_0(d_{a_i+p,a_i+p+1})}. \quad (155.5)$$

We define the maximum of $Score(F,S_i,a_i)$ among all possible position a_i as the matching score of the byte sequence S_i with the PRS. The formulation is shown as follows:

$$Score(F, S_i) = \max_{a_i=1}^{l-w+1} Score(F, S_i, a_i). \quad (155.6)$$

Alternatively, we reformulate the score function as the logarithmic form. The final matching score of S_i with the PRS F as shown is

$$\Theta(F, S_i) = \max_{a_i=1}^{l-w+1} \left(\sum_{p=1}^{w} \frac{1}{w} \log \left(\frac{f_p(S_{i,a_i+p-1})}{f_0(S_{i,a_i+p-1})} \right) + \sum_{p=1}^{w-1} \frac{1}{w-1} \frac{f'_p(d_{a_i+p,a_i+p+1})}{f'_0(d_{a_i+p,a_i+p+1})} \right). \quad (155.7)$$

The w-byte segment that maximizes $\Theta(F,S_i)$ is called the significant region of S_i. The matching score of the whole byte sequence is the matching score of the significant region.

For the incoming byte sequence S_i, if $\Theta(F,S_i)$ is greater than a threshold (here, the threshold is set as 0), we take S_i as a worm byte sequence. If the value of $\Theta(F,S_i)$ is above 0, it means S_i is closer to the worm signature; else if the value is below 0, it means S_i is closer to the normal signature.

155.4 Experiments

This experiment adopted Blaster worm and SQL Slammer worm as the test cases. Blaster worm spreads through the leak of Windows RPC DCOM. Blaster worm will get a copy of the infected files in the system, when the attack after successful execution. SQL Slammer worm uses a leak of SQL Server to attack during buffer. In the experiment, we use the polymorphic technology to generate Blaster worm sample and SQL Slammer worm sample.

Generate data sets. First of all, to generate 1,000 Blaster worm sequences, the number of samples in the pool is 1,000, and then, replace worm sequences with noise sequence. In this experiment, we consider five kinds of situations. The details are in Table 155.1 to extract PADS [8], NRS, and PRS, respectively, and to use them to detect worms. The results of detection are expressed in the missing report rate (MRR) and error report rate (ERR) in Table 155.1.

155 Polymorphic Worm Detection Using Position-Relation Signature

Table 155.1 Missing report rate and error report rate about Blaster worm

Number of noise	MMR (%) NRS	MMR (%) PADS	MMR (%) PRS	ERR (%) NRS	ERR (%) PADS	ERR (%) PRS
0	0.0	0.0	0.0	0.0	0.0	0.0
100	67.2	0.0	0.0	0.0	80.8	0.0
200	0.0	0.0	0.0	0.0	92.4	0.0
300	64.8	0.0	19.2	0.0	96.8	49.2
400	63.7	0.0	0.0	0.0	97.9	0.0
500	0.0	0.0	0.0	0.0	99.9	0.0

Table 155.2 Missing report rate and error report rate about SQL Slammer worm

Signature length	MMR (%) NRS	MMR (%) PADS	MMR (%) PRS	ERR (%) PADS	ERR (%) PADS	ERR (%) PRS
10	0.0	12.5	0.0	0.0	0.0	0.0
20	0.0	30.6	0.0	0.0	0.0	0.0
30	5.5	30.2	3.2	0.0	0.0	0.0
40	5.7	30.1	4.8	0.0	0.0	0.0
50	14.5	30.4	14.5	0.0	0.0	0.0
60	17.6	30.9	16.8	0.0	0.0	0.0
70	19.3	31.2	10.3	0.0	0.0	0.0
80	26.2	31.3	20.3	0.0	0.0	0.0
90	27.3	31.7	28.9	0.0	0.0	0.0
100	29.8	31.3	25.6	0.0	0.0	0.0

Experiment results. As seen in Table 155.1, when suspicious pool contained in article 200 and 500 of the noise sequences, the MRR of NRS is 0; however, it has higher MRR in other situations. The MRR of PADS and PRS is 0. That means the NRS have unstable performance. In this case, the performance of PRS is similar with PADS. In Table 155.1, PADS perform higher ERR than NRS and PRS mainly because PADS does not exclude noise jamming. The PADS not only include worm features but also include features about noise sequences.

Table 155.2 shows the MRR and ERR about SQL Slammer worm signatures. PADS, NRS, and PRS have different performance when the signature has different length. From the result, the PRS has better performance than others in MRR and ERR.

155.5 Conclusion

To detect polymorphic worm, this chapter presented a novel worm signature: PRS which is based on signature fusion. This worm signature combines the worm bytes themselves—PADS and the relationships between worm bytes, NRS. In this chapter, a large number of experiments are completed. We get the following conclusion:

PRS is a combination of two kinds of signatures. It is better than PADS and NRS. It is more suitable for the polymorphic worm detection, which is complicated and changing.

References

1. Korczyński, M. (2012). Classifying application flows and intrusion detection in the internet traffic. PhD thesis, UNIVERSITÉ DE GRENOBLE.
2. Fan, W. K. G. (2012). An adaptive anomaly detection of WEB-based attacks. *International Conference on Computer Science & Education (ICCSE)* (pp. 690–694). Melbourne, VIC: IEEE.
3. Magkos, E., Avlonitis, M., Kotzanikolaou, P., & Stefanidakis, M. (2013). Toward early warning against Internet worms based on critical-sized networks. *Security and Communication Networks, 6*(1), 78–88.
4. Cai, M., Hwang, K., & Pan, J. (2007). WormShield: Fast worm signature generation with distributed fingerprint aggregation. *IEEE Transactions on Dependable and Secure Computing, 4*(2), 88–104.
5. Portokalidis, G., & Bos, H. (2007). SweetBait: Zero-hour worm detection and containment using low- and high-interaction honeypots. *Computer Networks, 51*(11), 1256–1274.
6. Newsome, J, Karp, B, Song, D. (2005). Polygraph: Automatically generating signatures for polymorphic worms. *IEEE symposium on Security and Privacy Symposium* (pp. 226–241). Washington, DC: IEEE Computer Society.
7. Tang, Y., & Chen, S. (2007). An automated signature-based approach against polymorphic internet worms. *IEEE Transactions on Parallel and Distributed Systems, 18*(7), 879–892.
8. Wang, J, Wang, J, Sheng, Y, Chen, J. (2009). Polymorphic worm detection using signatures based on neighbourhood relation. *IEEE International Conference on High Performance Computing and Communications* (pp. 347–353). Seoul: IEEE.

Chapter 156
Application of the Wavelet-ANFIS Model

Rijun Zhang, Caishui Hou, Hui Lin, Meiyan Zhuo, Meixin Zhang, Zhongsheng Li, Liwu Sun, and Fengqin Lin

Abstract Since many predicting methods, such as CS and PP are not very precise, Wavelet-ANFIS with high estimation precision is always used to model the decomposed series recently. This chapter uses wavelet analysis to decompose water level series and then uses ANFIS to model the decomposed series; in the end, it combined these series and predicted Lingxi Reservoir's runoff. The runoff forecast of reservoir is essential for its flood control safety. The forecast result shows that the prediction accuracy of Wavelet-ANFIS is very high and the model is quite fit to use in daily runoff and water level prediction.

156.1 Introduction

Lingxi Reservoir was built and put into operation in June 1956, whose total capacity is 30.6 million m^3. It is a comprehensive utilization of medium-sized reservoir, which is combined with functions of water supply, irrigation, and flood protection. Currently, the reservoir is responsible for industries and domestic water supply of Fujian Refinery and Hui'an county, and its downstream is an important transport area. Therefore, the runoff and water level forecast of the reservoir is essential for its flood control safety [1].

This chapter predicts Lingxi Reservoir's runoff; it is the basic of influence research of the dam project. Once there were many predicting methods, such as CS and PP [2], but these methods are not very precise. This chapter uses wavelet analysis to decompose water level series and then it uses ANFIS to model the decomposed series; in the end, it combined these series.

Wavelet analysis is a good multi-resolution frequency method; by the multi-scale sequence analysis, it can effectively recognize the main frequency

R. Zhang (✉) • C. Hou • H. Lin • M. Zhuo • M. Zhang • Z. Li • L. Sun • F. Lin
Fujian College of Water Conservancy and Electric Power, Yong'An 366000, China
e-mail: Zhangrj_vip@163.com

components and local information. ANFIS has strong nonlinear approximation functions and self-learning, adaptive characteristics. So they can be combined and give full play to the advantage of both [3].

156.2 Wavelet Analysis

Wavelet analysis is a window fixed but the shape can variable (variable bandwidth and when wide) changing in the time-frequency analysis method. It has adaptive time-frequency window: high frequency, frequency-domain window increased, the time window reduced, the time window expands, and the frequency domain window reduced. Wavelet analysis is the key to satisfy certain conditions, the introduction of the basic wavelet function ψ(t) to replace the Fourier transform of basic functions $e^{-i\omega t}$. The next is stretching and translation functions:

$$\psi_{a,b}(t) = |a|^{-1/2} \psi\left(\frac{t-b}{a}\right) \quad a,b \in R, a \neq 0 \quad (156.1)$$

in which $\psi_{a,b}$ is called the analysis wavelet or continuous wavelet; a is measurements (telescopic) factor, in a sense, is corresponding to frequency ω; b is the time (translation) factor, and it reacts to time's translation.

156.3 ANFIS

Artificial Fuzzy Neural Network (ANFIS) [4, 5] is combined by Artificial Neural Network and fuzzy theory. It uses ANN to construct the fuzzy system. According to input and output sample, it can automatically design and adjust the design parameters of the fuzzy system, and then it can achieve fuzzy system's self-learning and adaptive function. It can fit to complex input and output's linear and nonlinear mapping relations. So it is especially applicable to complex nonlinear hydrological system [6].

The network structure of Artificial Fuzzy Neural Network (ANFIS) is shown in Fig. 156.1.

In Fig. 156.1, the connecting line between the nodes only shows the flow of signal; it doesn't associate with weight. Square nodes represent nodes which have adjustable parameters; circular nodes represent nodes which don't have adjustable parameters. ANFIS's structure can be divided into five levels:

Level 1: Membership $z_{ij} = \mu_i(x_j, \theta_i)$, $i = 1, 2, L, M_{num}$, $j = 1, 2, L, I_{num}$, M_{num} is the number of membership functions; I_{num} is the number of input variable; μ(.) is the generalized membership function; and the commonly used membership

Fig. 156.1 Structure of ANFIS

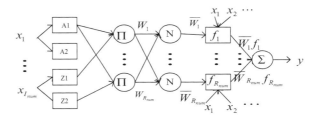

functions are triangle membership function, trapezoidal membership function, Gaussian membership function, and bell-shaped membership function. The form of triangle and trapezoidal membership functions is simple, and their computing efficiency is high. However, because their membership function is constructed by linear line, the corner points of some specified parameters are not smooth enough. Gaussian function and bell-shaped membership function have smooth and simple representation, so they are the most commonly used form for definite fuzzy sets. Formula 1 is the bell-shaped membership function:

$$\mu_i(x, \theta_i) = 1 / \left[1 + \left[\frac{x - c_i}{a_i} \right]^{2b_i} \right] \quad (156.2)$$

x and $\theta_i = [a_i \ b_i \ c_i]$ are respectively I the membership function's input and original reasoning parameter set.

Level 2: k the incentive intensity, in which R_{num} is the number of fuzzy rules.

$$W_k = \prod_{}^{I_{num}} Z_{ij}, \quad i \in [1, 2, \cdots, M_{num}], \quad i = 1, 2, \cdots, R_{num} \quad (156.3)$$

Level 3: Normalized incentive intensity is the ratio of the rule's incentive intensity and the sum of all of the rule's incentive intensity.

$$\overline{W}_k = \frac{W_k}{\sum_{j=1}^{R_{num}} W_j}, \quad k = 1, 2, \cdots, R_{num} \quad (156.4)$$

Its vector form is:

$$\overline{W} = \left[\overline{W}_1, \overline{W}_2, \cdots, \overline{W}_{R_{num}} \right]^T \quad (156.5)$$

Level 4: Fuzzy rule's conclusion, which is accurate output.

$$f_i = p_{i1} x_1 + p_{i2} x_2 + \cdots + p_{ij} x_i + r_i \quad i = 1, 2, \cdots, R_{num}, \quad j = I_{num} \quad (156.6)$$

The parameter set, which is composed by all of $\{p_{ij}, r_i\}$, is called consequent parameter set.

Level 5: After weight-average, the overall output of the net can get:

$$y = \sum_{i=1}^{R_{num}} \overline{W}^T F \qquad (156.7)$$

in which $F = [f_1, f_2, L, f_{R_{num}}]$.

This chapter uses hybrid learning algorithm to optimally select the parameter of ANFIS; this gets the smallest sum of square error between the final output result and the goal. The core idea of this algorithm is that in the forward calculation, it keeps the value of all of the original reasoning parameter unchanged and improves the value of consequent parameter by recursive least squares; then, it keeps the value of all of the improved consequent parameter unchanged and improves the value of original reasoning parameter by error back-propagation.

156.4 Wavelet-ANFIS Rainoff Forecast Model

Since wavelet analysis is a good multi-resolution frequency method and ANFIS has strong nonlinear approximation functions and self-learning, adaptive characteristics, the result will work better and give full play to the advantage of both if they are combined. It is named Wavelet-ANFIS rainfall forecast model.

In the model, first, data sequence is decomposed and reconstructed J times by Mallat wavelet, by which a low frequency signal C_J and the high frequency signals $(D_1, D_2,,, D_J)$ in J scale can be obtained; then ANFIS models are established for each decomposed signal$(C_J, D_1, D_2,,, D_J)$; after that it is time to determine the model parameters and to start prediction. Finally, the ultimate forecast result is obtained by synthesis of forecast results of the decomposed signals, and accuracy of the results is analyzed. The combined model flowchart is shown in Fig. 156.2.

156.5 Calculation Example

This chapter uses Wavelet-ANFIS model to find the suitable level of the runoff series; by some trying, it finds level 2 as the best level. The decomposed result is delayed as shown in Fig. 156.3.

In the figure we can clearly see that wavelet analysis can well separate the trend item and the wave item.

Next we use ANFIS to separately predict the trend item and the wave item.

We use runoff historical data from 1954 to 1993 as the modeling series and 1994–1996 as the predict series. In order to combine them in the end, we select the

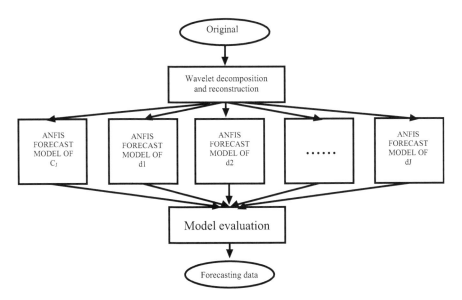

Fig. 156.2 Wavelet-ANFIS model flow diagram

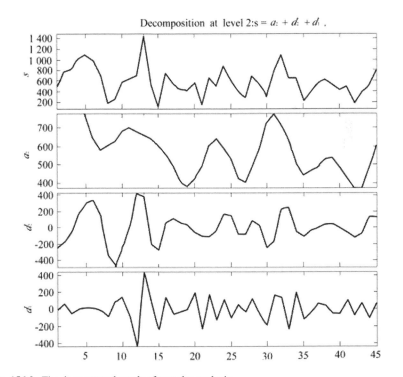

Fig. 156.3 The decomposed result of wavelet analysis

Fig. 156.4 Comparison of original sequence and the final result

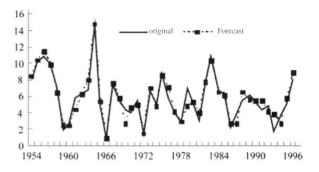

Table 156.1 Compared result

Relative error (%)	<5	5–20	>20
Proportion (%)	94.97	5.03	0.00

same function for these series. This chapter adopts ANFIS function of MATLAB's toolbox to predict these series; the type of function is bell-shaped membership function. After the 20th iteration, we can get a good simulating result. Then this chapter combined the result of this series to make the final result. The contrast result of the forecast result and the original sequence is delayed as shown in Fig. 156.4.

The evaluation of this model is delayed as shown in Table 156.1.

From Fig. 156.3 and Table 156.1, we can see that the result is very good. The relative error is nearly all less than 5 %. So we can use Wavelet-ANFIS model to forecast Lingxi Reservoir's runoff. The precision is very high. This is also the basic research of other water resources' researches.

156.6 Conclusion

This chapter combined wavelet analysis and ANFIS to predict Lingxi Reservoir's runoff, which is called Wavelet-ANFIS model with the advantage of both; through series of analysis, we can find that Wavelet-ANFIS model is very fit to predict water level. The forecast result is very close to the original sequence and can be used to evaluate the influence of Lingxi Reservoir.

References

1. Ma, X., Mu, X., & Guo, H. (2012). Reservoir monthly runoff forecast model based on wavelet-ANFIS analysis. *Hydroelectric Energy, 1*(1), 12.
2. Ma, X., He, X., & Zhao, D. (2012). BP network hidden layer on the water quality impact analysis results of the evaluation. *Hydroelectric Energy., 20*(3), 121.

3. Wang, W., Ding, J., & Li, Y. (2005). *Hydrological wavelet analysis (pp. 32–36)*. Beijing: Chemical Industry Publishing House.
4. Jang, J. S. R. (2011). ANFIS: Adaptive-network-based fuzzy inference system. *IEEE Transactions on System, Man and Cybernetics, 2*, 235–238.
5. Zhang, Z., & Sunetc, C. (2010). *Neuron fuzzy and soft computing* (pp. 86–92). Xi'an: Xi'an Communication University Press.
6. Zhang, B., et al. (2011). *Research on Poyang Lake (pp. 22–28)*. Shanghai: Shanghai Technology Press.

Chapter 157
Visualization of Clustered Network Graphs Based on Constrained Optimization Partition Layout

Fang Huang, Wenjie Xiao, and Hao Zhang

Abstract Hybrid layout is a common visualization technique for clustered network graphs. Since most previous hybrid layout methods do not consider a reasonable balance between screen utilization and layout aesthetics of the network graphs, the inappropriate partition of the display region may result in unpleasant display effect of network graphs. This chapter proposes to address this problem with nonlinear constrained optimization techniques. This chapter analyzes why the circular algorithm would fail in region partition. To ensure that every subgroup of network nodes can be assigned to a rectangular region, the maximal utilization of the display area is taken as an objective function and the rectangular ratio is taken as constraints. The constrained optimization layout model leads to efficient balance between regional utilization and layout aesthetic. Experimental results show that the constrained optimal partition layout generates more balanced relation network graphs with better visual effects.

157.1 Introduction

Visualization of clustered network graphs is an important research topic in social relations modeling and analysis based on network information technology. To clearly display relation networks, it is necessary to construct a clustered graph by dividing all nodes into clusters according to network closeness degrees. In the process of clustered graphs visualization, it is crucial to decide how to display all the nodes in various clusters in the screen. Besides, in order to achieve a balanced aesthetic effect, we need to spread nodes to the whole area. Integrating the two issues together, we can obtain efficient and coordinated picture by optimizing

---

F. Huang (✉) • W. Xiao • H. Zhang
School of Information Science and Engineering, Central South University, Changsha 410083, China
e-mail: hfang@mail.csu.edu.cn

layout of nodes for clustered network graphs. The optimal layout can present the nodes and associations of network diagram more clearly and evenly.

In 2008, Huang proposed a hybrid layout method [1] which is commonly used in visualization of clustered graphs. This algorithm firstly divides all the nodes into subgroups according to the tightness of contact in relation network structure, then partitions the screen into several rectangular regions in accordance with the number of subgroups, and finally layouts nodes in the rectangular regions. In the hybrid layout, it is supposed that the display area is fully utilized, and a reasonable length-to-width ratio is required so that the layout is aesthetic. If only one of the two constraints are not satisfied, the program executing would be terminated, which will cause the subgroups to be not incompletely distributed to the rectangular region and sequentially impacts on the effect of network layout. Therefore, this chapter proposes an optimization region partition method for the hybrid layout, which merges utilization rate of display area, aesthetic proportion of rectangle as well as uniform density of nodes. We employ the constrained nonlinear optimization model to conduct region partition, which can ensure that every subgroup can occupy a rectangular region by dividing the screen completely and meet the conditions of aesthetic layout. As a result, the effect of visualization of relation network graphs can be improved.

So far, the visualization of clustered graphs is still an open topic in social relations modeling and analysis. In 2006, Eades proposed that the planar clustered graphs can be converted into convex polygons and can be drawn by a straight line, which is named as straight-line algorithm [2]. Since confusion visualizations are usually caused by many cross-edges in the clustered graphs, in 2007, Omote provided a novel force-directed algorithm, which reduces the number of the cross-edges with clearer graphs [3]. In 2008, Huang made a synthesis process for the network graphs visualization, which integrates partition of clusters and layout of nodes [1]. In the scheme, the hierarchy-clustering algorithm is employed to build a node-clustered tree, then utilize the region partition algorithm to assign the display area for every cluster and, lastly, take force-directed algorithm to layout the nodes in each area. In 2009, Battisia raised a directly visualized method that organizes all nodes by clustered trees and the network graph shows the shape of the tree [4]. In 2012, Liu achieved more effective clustering while considering the relationship strength between nodes and more attributes of nodes [5]. Since the region partition algorithm plays an important role in clustered graphs visualization, it has been paid much attention. In 2005, Nguyen investigated the region partition algorithms, such as circular region partition, Squarified treemaps, and Sunburst, and compared the balance of nodes and orderliness of the edges [6]. The conclusion is that the circular region partition is better in effect and aesthetic. So, in 2008, Huang adopted circular region partition as a part of his hybrid layout process [1]. However, in the method, the unbalanced node of clustered trees easily leads to an incomplete partition process. The purpose of the study is to explore an optimal region partition to improve visual effect of clustered graphs.

157.2 Analysis of Circular Region Partition

Hybrid layout is composed of nodes clustering [7], circular region partition [6], and force-directed layout [8]. Firstly, nodes closely related to each other are classified as various subgroups by nodes clustering. Then, the circular region partition is adopted to divide the display screen into several rectangular regions roughly so that each subgroup is positioned in their respective rectangle. Finally, the force-directed layout is applied to spread all the nodes of the subgroup in the rectangle in order to achieve balanced effect. In the hybrid layout, the circular region partition plays a decisive role for the overall effect of the relation networks visualization. The circular region partition assigns a rectangular region to each subgroup one by one clockwise along the four edges of the display screen, which is a key factor for aesthetics and integrity of the circular region partition.

- Calculating the Size of Rectangles

In Fig. 157.1, the width of display area in the screen is denoted by W, and the height is denoted by H. Assume there are n subgroups to be assigned, and we denote the number of nodes in subgroup i by wg_i to serve as the subgroup weight. For example, m subgroups are divided into rectangles in the left part of the display area as shown in Fig. 157.1. Since the area of rectangle R_i is S_i, the width w_i, height h_i, and S_i of each rectangle can be calculated, respectively, by the following:

$$Wp_i = wg_i / \sum_{i=1}^{n} wg_i \qquad (157.1)$$

$$S_i = W * H * Wp_i. \qquad (157.2)$$

In formulas (157.1) and (157.2), Wp_i represents the percentage of the weight of subgroup i in all subgroups. Expression (157.2) means the area S_i of the rectangle R_i is a percentage of the total area, which corresponds to the ratio between the number of subgroup nodes and the total number of all nodes in the network.

Expression (157.3) shows that h_i is the percentage of the vertical height of the display screen, which is the ratio between the number of subgroup nodes and the total number of nodes in the m subgroups along the vertical edge.

$$h_i = H \left(wg_i / \sum_{i=1}^{m} wg_i \right) \qquad (157.3)$$

$$w_i = W \left(\sum_{i=1}^{m} wg_i / \sum_{i=1}^{n} wg_i \right) \qquad (157.4)$$

Fig. 157.1 Circular region partition

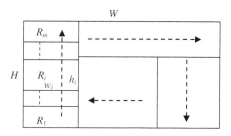

From expression (157.4), w_i is decided by the ratio of the number of m subgroup nodes to the total number of nodes. According to the above formulas, it is sure that the sum of all rectangular areas is equal to the display area. In other words, the circular partition algorithm takes full utilization of display area as a precondition.

- Constraining the Proportion of Width and Height

In addition to the width and height of a rectangle, the ratio between them must also meet the aesthetic criteria of a rectangle in the circular partition. Otherwise, the algorithm will be terminated. That is to say, not only the ratio of rectangles decides the aesthetics but also the probability of partition success. Here, we use $\lambda_i = w_i/h_i$ to represent the ratio between width and height. The ratio is set to be larger than 0.67 by trial and error and it is manually set in order to achieve the best rectangular layout [6].

- Process of Circular Region Partition

In Fig. 157.1, the partition divides the first rectangle on the bottom-left corner along the direction of the vertical dashed arrow. First, the area, height and width of the rectangle is calculated according to expression (157.2) (157.3) and (157.4). Next, the width/height ratio of the rectangle is checked. If it meets the constraints, try to assign two subgroups in the direction and calculate the height, area, and width of the two rectangles successively. Similarly, check whether the ratio of each rectangle meets the requirements. Trying and computing repeatedly, an extra subgroup is added in each iteration until one of these rectangles doesn't meet the proportional constraints. At this point, the partition in the current direction will be ended. The partition continues in the next direction clockwise. If the attempts in all direction are finished but the remaining subgroups cannot be assigned to any suitable region, the circular region partition fails. The number of rectangles arranged on one edge of the display screen is m, which is in accordance with the ratio constraints [6].

In the circular region partition, dividing rectangles is based on assuming rectangular seamless arrangement. If the premise does not hold, the rectangular formula is not valid, and the segmentation process cannot be carried out. Thus, the first rule of the circular partition algorithm is the full utilization of display area. Secondly, the aspects ratios of the rectangle should satisfy the constraint condition to ensure the aesthetics of networks. Regardless of whether the other subgroups are partitioned to the region, the algorithm will terminate if the constraint cannot be established. In other words, the circular partition has the effectiveness and aesthetics as two absolutely independent conditions. However, in applications, the aesthetics should be on the basis of the validity. It is first to assure that all subgroups can be assigned to a rectangular region, then to consider the aesthetics. Meanwhile, the aesthetic criteria can be flexible, if only a good visual effect is kept. Therefore, to take the above factors into consideration, we propose a nonlinear constraint optimization for the region partition layout.

157.3 Constrained Optimization Partition Layout

Constrained optimization layout model includes an objective function and constraints [9]. For the region partition of subgroups after nodes clustering, it is the main goal to guarantee that all subgroups can get the right rectangle from the display area. On this basis, adjusting the size and position of the rectangle leads to maximum utilization of the display screen. We designed the optimization model which aims to maximize the sum area of all rectangles and takes rectangular aesthetics and reasonable position as constraints. A set of the rectangles with optimal size and position are found by solving the model so as to achieve the optimal layout.

157.3.1 The Effect of Rectangular Aspects Ratio on Aesthetics

The aspects ratio decides the shape and size of rectangles in the process of region partition. Its purpose is to layout nodes of the subgroup in the region, namely, the regional division of the subgroups. We should conduct the quantitative evaluation for the aesthetic effect of display area objectively. Since visualization should meet the aesthetic standards of people, the feasible solution of constrained optimization model must be limited in the acceptable range. Therefore, the quantitative range of aesthetics can be derived to be the model constraints.

Battista proposed a method to measure the effect of visualization according to aesthetic criteria which include the overall edge lengths d_i, variance e_i of edge lengths, the number of cross-edges c_i, and variance a_i of an included angle between associate edges [10]. If the overall edge lengths are longer, the layout of nodes is more dispersed. Minimizing the variance of edge lengths can result in more balanced layout of nodes to a certain extent. The less cross-edges show more reasonable layout of nodes and smaller variance of angles makes network diagrams clearer. We build a sample dataset of relationship networks to evaluate the effect of rectangle aspects ratios on aesthetics. For k relationship networks, the comprehensive assessment for the above four rules can be defined as following formulas:

$$D_{ave} = \frac{1}{k}\sum_{i=1}^{k} d_i, E_{ave} = \frac{1}{k}\sum_{i=1}^{k} e_i, C_{ave} = \frac{1}{k}\sum_{i=1}^{k} c_i, A_{ave} = \frac{1}{k}\sum_{i=1}^{k} a_i. \quad (157.5)$$

Here D_{ave}, E_{ave}, C_{ave}, and A_{ave} are the average values of total edge lengths, variance of edge lengths, the number of cross-edges, and variance of angle, respectively. In addition, DL_i, DE_i, CE_i, and AN_i defined below are the proportion of deviation from i network graph to the average value of all networks on the above four rules, respectively.

$$DL_i = \frac{d_i - D_{ave}}{D_{ave}}, DE_i = \frac{e_i - E_{ave}}{E_{ave}}, CE_i = \frac{c_i - C_{ave}}{C_{ave}}, AN_i = \frac{a_i - A_{ave}}{A_{ave}} \quad (157.6)$$

We also define NET_i as a comprehensive estimation for network i in formula (157.7).

$$NET_i = DL_i - DE_i - CE_i - AN_i \quad (157.7)$$

In the current relation network, when d_i gets larger, the layout of nodes is more dispersed. If d_i-D_{ave} is positive, it shows that the overall edge lengths in the current network will be larger than the average value. That is, DL_i contributes to the comprehensive assessment, so it is a positive sign. For smaller e_i, the layout of nodes is more uniform. If e_i-E_{ave} is positive, the variance of edge lengths will be greater than the average; at the same time, the DE_i has a negative impact on the comprehensive assessment, so we can use minus sign for the item. CE_i and AN_i are similar to DE_i. In other words, when DL_i is positive and DE_i, CE_i, and AN_i are negative, the NET_i is positive. That means that the aesthetics of current network is better than the average effect of all networks. The comprehensive assessment reflects the relative value of visualization between the current network and all networks.

In order to determine the effect of rectangular aspects ratios on the aesthetics of the display area, take the network containing 21 nodes as inputs, set the rectangular aspects ratio λ from 0.1 to 0.7 for constraint layout, and then get the quantitative evaluations, respectively.

In Table 157.1, when the rectangular aspects ratio equals 0.1, 0.2, 0.3, 0.4, and 0.5, respectively, the comprehensive evaluations are negative. That indicates the aesthetics of these networks is poorer than the average effect. When the ratio is 0.6 or 0.7 and the assessments are positive, it means a better effect. So we set the rectangular aspects ratio greater than 0.6 as the constraint of rectangular aesthetics.

157 Visualization of Clustered Network Graphs Based on Constrained... 1387

Table 157.1 Quantitative evaluations of visual networks

Ratio λ	d_i	e_i	c_i	a_i	NET_i
0.1	1,235	14.08	58	11.08	−0.4789
0.2	1,220	12.9	44	15	−0.15
0.3	1,204	16.65	38	18	−0.34
0.4	1,180	16.21	32	16	−0.03
0.5	1,157	12.35	26	23.71	−0.13
0.6	1,162	13.73	12	16	0.64
0.7	844	12.81	8	17	1.1

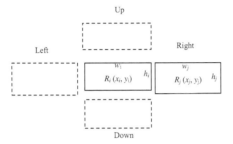

Fig. 157.2 Nonoverlapping between rectangles

157.3.2 Nonoverlapping Criteria

All rectangles in the display area should not be overlapping, which is a rigorous constraint in the process of region partition. In Fig. 157.3, there are two rectangles: R_i and R_j. Supposing R_i is fixed, there are four situations that R_i is not overlapping with R_j: the R_j is on the right, left, up, and down of the R_i, respectively.

In Fig. 157.2, x_i and y_i are the center coordinate of R_i, w_i and h_i is the width and height, and x_j and y_j are the center of the R_j. For example, when the R_j locates on the right of the R_i, the nonoverlap rule is that the abscissa of R_j left edge must be greater than the abscissa of R_i right edge. When the R_j is on the left of R_i, the abscissa of the R_j right edge is smaller than the abscissa of the R_i left edge. Similarly, R_j stands above R_i; the ordinate of R_j bottom must be greater than the ordinate of R_i top, and R_j is below R_i; the ordinate of R_j top is smaller than the ordinate of R_i bottom.

157.3.3 Nonlinear Constrained Optimization Layout Model

In optimization layout, the objective function and constraints are defined by the following.

- Objective Function for the Maximizing Utilization of Display Area

In layout partitioning, every subgroup must correspond with a rectangle in display area and minimize the remaining area. In other words, we maximize the utilization to reduce center-of-gravity shift and unbalance of networks in the display area, which can cause too much blank areas. Therefore, the utilization of display screen is taken as the objective function:

$$\max \sum_{i=1}^{n} w_i h_i \qquad (157.8)$$

Here, w_i and h_i are the width and height of R_i; n is the number of all rectangles.

- Constraints of Rectangular Aesthetics

$$0.6 \leq \lambda_i = \frac{w_i}{h_i} \leq 1.67 \; i = 1, 2, \ldots, n \qquad (157.9)$$

Here λ_i is the ratio between width and height of R_i.

- Nonoverlapping Constraints

$$\left(x_i + \frac{w_i}{2}\right) - \left(x_j - \frac{w_j}{2}\right) \leq 0, i = 1, 2, \cdots n, j = 1, 2, \cdots, n, i \neq j \qquad (157.10)$$

or

$$\left(x_i - \frac{w_i}{2}\right) - \left(x_j + \frac{w_j}{2}\right) \geq 0 \; i = 1, 2, \cdots, n, j = 1, 2, \cdots, n, i \neq j \qquad (157.11)$$

or

$$\left(y_i + \frac{h_i}{2}\right) - \left(y_j - \frac{h_j}{2}\right) \leq 0 \; i = 1, 2, \cdots, n, j = 1, 2, \cdots, n, i \neq j \qquad (157.12)$$

or

$$\left(y_i - \frac{h_i}{2}\right) - \left(y_j + \frac{h_j}{2}\right) \geq 0 \; i = 1, 2, \cdots, n, j = 1, 2, \cdots, n, i \neq j \qquad (157.13)$$

Formulas (157.10)–(157.13) show the constraints of rectangular center coordinate to ensure nonoverlapping and just located on right, left, up, and down.

- Constraints for Exceeding Boundary

$$h_i \leq Hi = 1, 2, \ldots, n \quad (157.14)$$

$$w_i \leq Wi = 1, 2, \ldots, n \quad (157.15)$$

$$\frac{w_i}{2} \leq x_i \leq W - \frac{w_i}{2} i = 1, 2, \ldots, n \quad (157.16)$$

$$\frac{h_i}{2} \leq y_i \leq H - \frac{h_i}{2} i = 1, 2, \ldots, n \quad (157.17)$$

The constraints limit of the width and height of rectangles should be smaller than the width and height of display area, respectively.

- Relaxed Constraint for the Rectangular Area

$$H \times W \times Wp_i \times (1 - 5\%) \leq S_i \leq H \times W \times Wp_i \times (1 + 5\%) \quad (157.18)$$

The constraint of rectangular areas, formula (157.18), is a relaxation condition based on the rectangular area accounted for the proportion of total area which is equal to the subgroup weight to total weight. Since small changes of the rectangular area bring about little impact on the visual aesthetics, the relaxation can help to expand the range of feasible solutions. The relaxed range of rectangular area is 5 %.

The best partition layout is determined by calculating the width and height of rectangles and the abscissa and ordinate of centers in constrained optimization model.

157.4 Comparison and Analysis of Visualization

Since the objective of the proposed optimal region partition is to avoid layout failures of the circular region partition, it is necessary to compare the visual effect for circular region partition and optimal region partition. Four relationship networks containing 21, 40, 80, and 160 nodes, respectively, are chosen to implement the experiments.

157.4.1 Visual Comparison

Figure 157.3 is for the circular region partition; since the ratio of the rectangle in the middle of display area does not satisfy the aesthetic constraint, it leads to partition failure. However, the process of layout of nodes has to be implemented by the rectangle with inappropriate proportion so that it is more crowding and has many cross-edges in the middle. For optimal region partition, Fig. 157.4 shows more scattered nodes, fewer cross-edges, and clearer visualization.

Fig. 157.3 21 nodes by circular partition layout

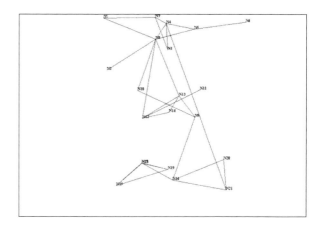

Fig. 157.4 21 nodes by optimizing partition layout

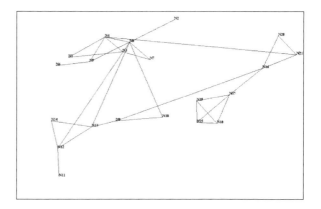

In Fig. 157.5, since the rectangle of inappropriate ratio stands on the left space of display area, it results in nonuniform distribution of nodes on the left. After optimizing the partition layout, nodes in Fig. 157.6 are distributed in a more uniform way.

Figure 157.7 is a network diagram by the circular region partition for 80 nodes, and all the nodes are concentrated at the middle of the display area. The visualization result is poor because of small distances among the nodes. After optimal region partition, the nodes in Fig. 157.8 are more uniformly and clearly distributed.

Figures 157.9 and 157.10 show the visualization results for a network with 160 nodes using circular and optimization partition layout, respectively. Comparing with Fig. 157.9, Fig. 157.10 presents the network distribution with better visual effect. In particular, we choose the network with fewer connections, in order to better contrast on vision.

Fig. 157.5 40 nodes by circular partition layout

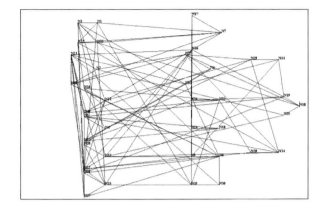

Fig. 157.6 40 nodes by optimizing region layout

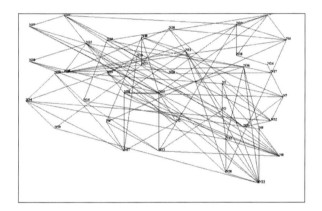

Fig. 157.7 80 nodes by circular partition layout

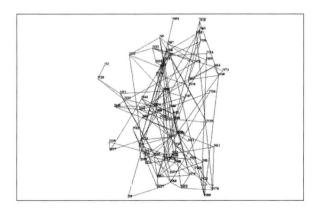

1392

Fig. 157.8 80 nodes by optimizing partition layout

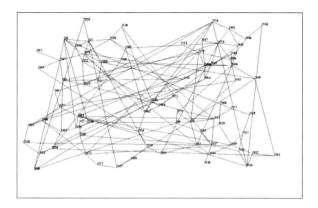

Fig. 157.9 160 nodes by circular partition layout

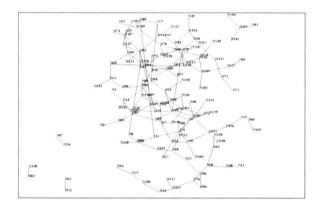

Fig. 157.10 160 nodes by optimizing partition layout

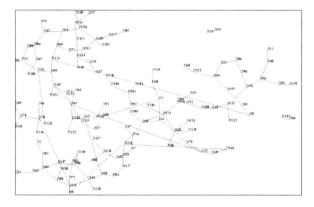

Table 157.2 Aesthetics evaluations of 21 nodes

Aesthetics criteria	d	e	c	a
Circular partition	2,030	125.75	26	21.3
Optimizing partition	2,046	162	20	18.6

Table 157.3 Aesthetics evaluations of 40 nodes

Aesthetics criteria	d	e	c	a
Circular partition	11,232	115	985	24.2
Optimizing partition	16,287	210	754	29.3

Table 157.4 Aesthetics evaluations of 80 nodes

Aesthetics criteria	d	e	c	a
Circular partition	22,500	48.5	3,048	29.1
Optimizing partition	34,596	62.7	2,169	25.3

Table 157.5 Aesthetics evaluations of 160 nodes

Aesthetics criteria	d	e	c	a
Circular partition	2,076	65.5	385	101
Optimizing partition	2,112	76.4	283	95

157.4.2 Assessment on Aesthetics of Visualization

We elaborate the quantitative evaluation of visualization for the above four networks including 21, 40, 80, and 160 nodes, respectively. Aesthetics criteria include the total edge lengths d, edge length variance e, the number of cross-edges c, and variance of angle a. The statistical results are shown in Tables 157.2, 157.3, 157.4, and 157.5 separately.

From the above four tables, it can be concluded that optimization partition layout can produce greater total length of edges, fewer cross-edges, and smaller angle variance, which result in wider distribution of nodes and more uniform and clearer network. The most critical point is the cross-edge by using optimization layout less than the number generated by circular region partition. It makes network connections clearer. In the four criteria, only the edge length variance is larger by circular region partition, which indicates that the distance between nodes should not be more uniform, but a smaller variance of edge length cannot absolutely guarantee a uniform distribution of nodes. The optimal partition performs better results on the three aesthetic criteria. Synthetically, the relation network graphs produced from optimization layout is superior to circular partition layout.

157.5 Conclusion

In the optimization region partition, the nonlinear constrained optimization model is employed to avoid the failure partition in circular region layout. It effectively balances the contradictions of the aesthetics and utilization by taking the utilization of the display area as the objective function and putting the rectangular ratio as constraints. It ensures that all subgroups of nodes can be assigned to a rectangular region. We randomly generated 60 relationship networks as the experimental inputs, then ran programs with the following three steps: nodes clustering, optimizing region partition, and force-directed node layout. Since the optimization region partition is implemented on LINGO 13 and other two parts are programmed by Java, the total running time is the sum of the three parts. The experiment results showed that the average running time of circular region partition and optimal region partition are 1.23 and 1.8 s, respectively. However, the latter has a 92 % average utilization of the display screen in all experiment networks and effectively avoids the layout failures. The results showed that the optimization partition layout not only achieved better quantitative evaluations on aesthetic criteria but also made the visual effects of relationship networks more balanced and clearer. In future work, we will further improve the algorithm in order to display scalable large-scale relation networks.

Acknowledgements This work was supported by Project 61073105 of National Natural Science Foundation of China.

References

1. Huang, M. L., & Nguyen, Q. V. (2008). Large graph visualization by hierarchical clustering. *Journal of Software, China, 9*(8), 1933–1946.
2. Eades, P., Feng, Q. W., Lin, X. M., & Nagamochi, H. (2006). Straight-line drawing algorithms for hierarchical graphs and clustered graphs. *Algorithmca, 44*(1), 1–32.
3. Omote, H., & Sugiyama, K. (2007). Force-directed drawing method for intersecting clustered graphs. *Proceedings of Asia-Pacific Symposium on Visualization 2007 (APVIS2007)* (pp. 85–92). Sydney: IEEE.
4. Battista, G. D., Drovandi, G., & Frati, F. (2009). How to draw a clustered tree. *Journal of Discrete Algorithms, 7*(4), 479–499.
5. Liu, X., Glänzel, W., & Moor, B. D. (2012). Optimal and hierarchical clustering of large-scale hybrid networks for scientific mapping. *Scientometrics, 91*(2), 473–493.
6. Nguyen, Q. V., & Huang, M. L. (2005). EncCon: An approach to constructing interactive visualization of large hierarchical data. *Information Visualization, 4*(1), 1–21.
7. Huang, M. L., & Nguyen, Q. V. (2007). A fast algorithm for balanced graph clustering. *Proceedings of IEEE 11th International Conference Information Visualization (IV'07)* (pp. 46–52). Zurich: IEEE.
8. Eades, P. (1984). A heuristic for graph drawing. *Utilitas Mathematica, 42*(11), 149–160.
9. Yuan, Y. X., & Sun, W. Y. (1999). *Optimization theory and methods* (pp. 35–75). Beijing, China: Science Press.
10. Battista, G. D., Eades, P., Tamassia, R., & Tollis, I. G. (1999). *Graph drawing: Algorithms for the visualization of graphs* (pp. 187–234). Upper Saddle River, NJ: Prentice-Hall.

Chapter 158
An Ultra-Wideband Cooperative Communication Method Based on Transmitted Cooperative Reference

Tiefeng Li, Ou Li, and Zewen Zhou

Abstract In order to decrease the power waste of relay node, the paper presents a novel ultra-wideband cooperative communication method that uses two relay nodes to transmit reference impulses and data impulses separately. A transmitted cooperative-reference UWB cooperative communication model is developed in this paper. Based on the model and sampling expansion approach, a closed-form SER expression was deduced for delay-hopped transmitted-reference UWB systems which use cooperation strategy of decode and forward relaying and equal-gain combining. Simulation results show that the transmitted cooperative-reference method can obtain multi-order diversity gains.

158.1 Introduction

It becomes the current research hotspot to involve cooperative diversity technique in the ultra-wideband communication system because of the additional spatial diversity. By doing this, both transmission reliability and communication coverage will be substantially enhanced. At present, studies of cooperative UWB technology are mainly focused on multiband orthogonal frequency division multiplexing ultra-wideband (MB-OFDM-UWB). There are a few essays which refer to the issue that combines impulse radio ultra-wideband with cooperative communication technique and few researches on cooperative communication that are based on the transmitted reference (TR) [1]. The paper just simply combines two techniques and chooses a relay node with best SNR to forward. In fact, it had developed a relay-forward

T. Li (✉) • O. Li • Z. Zhou
China National Digital Switching System Engineering and Technological Research Center, Zhengzhou 450002, China
e-mail: 13838146019@126.com

channel without diversity gain [1]. As well known, TR techniques can implement reliable communication with low complexity in the random or unknown channel [2]. Therefore, it is necessary to research the performance of cooperative TR-UWB system.

The transmitted-reference UWB can be classified into three main categories: time-domain transmitted reference [3], frequency-domain transmitted reference [4], and code-domain transmitted reference [5]. In the time-domain transmitted-reference system, since the nodes not only transmit data impulse but also send reference impulses that independent of the data, so it wastes half of the total energy. If all cooperative nodes are using this modulation to participate in cooperative, apparently the relay nodes have to pay half power as much as the source node. This cooperative approach can be named "transmitted-reference UWB cooperative." This paper presents a new method that is named "transmitted cooperative-reference UWB cooperative." It uses two relay cooperative nodes to send cooperative-reference impulses and cooperative-data impulse separately. Obviously, it is effective to save transmitting power by using this cooperation strategy.

The paper is organized as follows: In Sect. 158.2, we build transmitted cooperative-reference UWB communication system model. Next, in Sect. 158.3, based on the sampling expansion approach, a closed-form SER formulation was deduced from the delay-hopped transmitted-reference UWB systems which use cooperation strategy of decode and forward relaying, equal-gain combining. Then, Sect. 158.4 details the experiment results and discussions.

158.2 Transmit Cooperative Reference UWB Cooperative Communication System Model

We consider a cooperation strategy with two phases and three nodes in the cooperative model, which can be extended to the multi-node. Figure 158.1 shows the specific three-node cooperative model:

Compared to the three-node model, the transmitted cooperative-reference UWB cooperative communication model added an additional relay node as shown in Fig. 158.2. The model includes a source node (S), two relay nodes ($R1$ and $R2$), and a destination node (D). Firstly, S sends a signal; $R1$, $R2$, and D receive it. Secondly, $R1$ and $R2$ forward the decoded reference impulses and data impulses separately, and then D receives them.

Assume that the system can always find two relay nodes $R1$ and $R2$, which can establish a good time-synchronous accordance with the destination node through the upper layer cooperate protocol. Assume $R1$ only sends the cooperative-reference impulse, and $R2$ only sends cooperative-data impulse. Good synchronization means that when the impulse transmitted from each relay nodes arrived to the destination node, the time delay between the reference impulse and data impulse sent separately by $R1$ and $R2$ has the same effect as R in Fig. 158.1. So, the model is

Fig. 158.1 Transmitted-reference UWB cooperative model

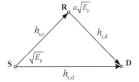

Fig. 158.2 Transmitted cooperative-reference UWB cooperative model

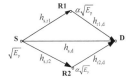

equivalent to the classic three-node model. R in Fig. 158.1 can replace the role of R1 and R2 in Fig. 158.2. In cooperative communication network with multi-nodes, it is possible to satisfy such conditions.

Figure 158.2 depicts the transmitted cooperative-reference UWB cooperative model; in stage one, the source node broadcasts to the destination node and the relay nodes. The signals received by destination and relay, respectively, are defined as:

$$\begin{aligned} y_{s,d} &= \sqrt{E_p} h_{s,d} s + n_{s,d} \\ y_{s,r1} &= \sqrt{E_p} h_{s,r1} s + n_{s,r1} \\ y_{s,r2} &= \sqrt{E_p} h_{s,r2} s + n_{s,r2} \end{aligned} \quad (158.1)$$

where E_p is transmission energy carried by single impulse and S is the symbol signal calculated as:

$$s(t) = \sum_{n=0}^{\frac{N_s}{2}-1} \left(p(t - 2nT_f) + d_0 p(t - nT_f - T_r) \right) \quad (158.2)$$

$h_{s,d}, h_{s,r1}, h_{s,r2}, h_{r2,d}$, and $h_{r1,d}$ are the channel coefficients correspond to S to D or S to R, and R to D, respectively, are modeled as zero mean, complex Gaussian random variables with variances $\delta_{s,d}^2, \delta_{s,r1}^2, \delta_{s,r2}^2$, etc. $n_{s,d}, n_{s,r1}$, and $n_{s,r2}$ are the additive white Gaussian noise. T_f is the delay time between the reference impulse and data impulse. A frame consisted of a pair of impulse, that is, reference impulse and data impulse. T_r is the time delay between the reference impulse and data impulse. N_s is the number of impulse in one symbol signal. d_0 is the binary data bit.

In stage two, the relay nodes separately decode and only forward reference impulse or data impulse. For simplicity to derivate and analyze, the role of *R1* and *R2* in Fig. 158.2 can be replaced by *R* in Fig. 158.1.

After two periods of this stage, the destination received three copies of signals via source and relay channel. The received *SNR* of the destination is compounded with the *SNR* of S and R. The optimal method of maximizing the total *SNR* is maximum-ratio combining (*MRC*). The output *SNR* via *MRC* is equal to the summation of *SNR* of each branch. However, it needs to know the real-time *SNR* of each branch. It implies that we need to obtain the real-time channel estimation. On the contrary, the advantages of transmitting-reference scheme can implement reliable receive without channel estimation. It is more suited for transmitted-reference scheme with the gain-equal-combination strategy. Based on the gain-equal combination, the output r_{rec} at destination is given by

$$\begin{aligned} r_{rec} &= y_{s,d} + y_{r,d} \\ &= \left(\sqrt{E_p}h_{s,d} + a\sqrt{E_p}h_{s,r}h_{r,d}\right)s + \left(n_{s,d} + ah_{r,d}n_{s,r} + n_{r,d}\right) \\ &= s' + n' \end{aligned} \qquad (158.3)$$

where

$$s' = \left(\sqrt{E_p}h_{s,d} + a\sqrt{E_p}h_{s,r}h_{r,d}\right)s = \sqrt{E'_p}\alpha_l s$$

$$E'_p = E_p(1 + ah_{s,r})^2, \quad n' = (n_{s,d} + ah_{r,d}n_{s,r} + n_{r,d}).$$

We assume that three transmission channels are independent and identically obey the Rayleigh distribution. The fading coefficient here is represented by the variable h. Since all signals transmitted in the channel of S to R have been processed by relay, we employed a constant $\overline{h}_{s,r}$ to represent the statistical average of the fading coefficient of the channel. Assuming that the components of channels of S to R have been combined by relay, we extend the above model to the multipath case. Let

$$c = \sum_{l=1}^{L_{CAP}} h^{(l)} = \sum_{l=1}^{L_{CAP}} \alpha_l \qquad (158.4)$$

$$\begin{aligned} r_{rec\_multi} &= \sqrt{E'_p}cs + n'' \\ &= \sum_{m=0}^{\frac{N_s}{2}-1} b'_r(t - 2mT_f) + d_0 b'_d(t - 2mT_f - T_r) + n \end{aligned} \qquad (158.5)$$

where $n'' = (n_{s,d} + ah_{r,d}n_{s,r} + n_{r,d})$; $b'_r(t) = b'_d(t) \triangleq \sum_{l=1}^{L_{CAP}} \sqrt{E'_p}\alpha_l p(t - 2mT_f)$. L_{CAP} is the number of maximum multipath that can be captured by the receiver.

158.3 Performance Analysis of DHTR-UWB System

In this paper, we focus on delay-hopped transmitted-reference ultra-wideband (DHTR-UWB). The structure of DHTR-UWB receiver is shown in Fig. 158.3.

The decision signal Z_{TR} can be given as

$$Z_{TR} \triangleq \sum_{m=0}^{\frac{N_s}{2}-1} \int_0^T \left[\tilde{b}'_r(t+2mT_f) + n''(t+2mT_f) \right]$$
$$\cdot \left[d_0 \tilde{b}'_d(t+2mT_f+T_r) + n''(t+2mT_f+T_r) \right] dt$$
$$= \sum_{m=0}^{\frac{N_s}{2}-1} \int_0^T \left(w'_m(t) + \eta'_m(t) \right) \left(d_0 w'_m(t) + \xi'_m(t) \right) dt$$
$$= \sum_{m=0}^{\frac{N_s}{2}-1} U_m \qquad (158.6)$$

where $U_m \triangleq \int_0^T (w'_m(t) + \eta'_m(t))(d_0 w'_m(t) + \xi'_m(t)) dt$, $\tilde{b}'_r(t+2mT_f)$, and $\tilde{b}'_d(t+2mT_f+T_r)$ are the impulse response of the BPZF, denoted as reference impulse and data impulse, respectively. $\eta'_m(t) = n''(t+mT_f)$; $\xi'_m(t) = n''(t+2mT_f+T_r)$. m is the frame number in a symbol.

We assumed that all impulses experience the same channel, implying that

$$w'_m(t) \triangleq \tilde{b}'_r(t+2mT_f) = \tilde{b}'_d(t+2mT_f+T_r). \qquad (158.7)$$

Though it is easy to solve SER with traditional Gaussian approximation approach, some restrictions must be taken into account [6]. In addition, the analytical solution cannot be obtained by using Gaussian approximation approach, thus making it difficult to carry out further research, such as power allocation and relay selection. An analytical method is introduced that combines sampling expansion approach with time transmitted reference and solved for the analytical solution of the system [6–8]. In this paper, we extend it to DHTR scheme to analyze the performance of cooperative DHTR-UWB system.

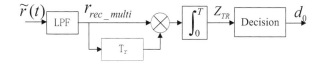

Fig. 158.3 DHTR-UWB receiving model

According to the sampling theorem, we consider to use 2WT-dimensional discrete signal to project the received continuous waveform U_m without losing:

$$U'_m = \frac{1}{2W} \sum_{l=1}^{2WT} \left(w'_{m,l} + \eta'_{m,l} \right) \left(d_0 w'_{m,l} + \xi'_{m,l} \right)$$

$$= \frac{1}{2W} \sum_{l=1}^{2WT} \left(d_0 w'^2_{m,l} + w'_{m,l} \xi'_{m,l} + d_0 w'_{m,l} \eta'_{m,l} + \eta'_{m,l} \xi'_{m,l} \right) \quad (158.8)$$

where the lth sample of $w'_m(t)$, $d_m(t)$, $\eta'_m(t)$ are $w'_{m,l}(t)$, $d_{m,l}(t)$, $\eta'_{m,l}(t)$, respectively. In the condition on d_0, we can rewrite U'_m as follows:

$$U'_{m|d_0=+1} = \frac{1}{2W} \sum_{l=1}^{2WT} \left(w'^2_{m,l} + w'_{m,l} \xi'_{m,l} + w'_{m,l} \eta'_{m,l} + \eta'_{m,l} \xi'_{m,l} \right) \quad (158.9)$$

$$U'_{m|d_0=-1} = \frac{1}{2W} \sum_{l=1}^{2WT} \left(-w'^2_{m,l} + w'_{m,l} \xi'_{m,l} - w'_{m,l} \eta'_{m,l} + \eta'_{m,l} \xi'_{m,l} \right) \quad (158.10)$$

We obtain the following simplification:

$$U'_{m|d_0=+1} = \sum_{l=1}^{2WT} \left[\left(\frac{1}{\sqrt{2W}} w'_{m,l} + \chi_{1,m,l} \right)^2 - \chi^2_{2,m,l} \right] \quad (158.11)$$

$$U'_{m|d_0=-1} = \sum_{l=1}^{2WT} \left[-\left(\frac{1}{\sqrt{2W}} w'_{m,l} - \chi_{2,m,l} \right)^2 + \chi^2_{1,m,l} \right] \quad (158.12)$$

where $\chi_{1,m,l}$ and $\chi_{2,m,l}$ are given by Eqs. 158.13 and 158.14

$$\chi_{1,m,l} = \frac{1}{2\sqrt{2W}} \left(\xi'_{m,l} + \eta'_{m,l} \right)$$

$$= \frac{1}{2\sqrt{2W}} \left[\left(n^{(l)}_{1,s,d} + ah^{(l)}_{1,r,d} n_{1,s,r} + n^{(l)}_{1,r,d} \right) + \left(n^{(l)}_{2,s,d} + ah^{(l)}_{2,r,d} n_{2,s,r} + n^{(l)}_{2,r,d} \right) \right]$$

$$(158.13)$$

$$\chi_{2,m,l} = \frac{1}{2\sqrt{2W}} \left(\xi'_{m,l} - \eta'_{m,l} \right)$$

$$= \frac{1}{2\sqrt{2W}} \left[\left(n^{(l)}_{1,s,d} + ah^{(l)}_{1,r,d} n_{1,s,r} + n^{(l)}_{1,r,d} \right) - \left(n^{(l)}_{2,s,d} + ah^{(l)}_{2,r,d} n_{2,s,r} + n^{(l)}_{2,r,d} \right) \right]$$

$$(158.14)$$

$\xi'_{m,l}$ and $\eta'_{m,l}$ are white Gaussian noises, with the following mean and variance:

$$E\{\xi'_{m,l}\} = E\{\eta'_{m,l}\} = 0. \tag{158.15}$$

The variance of DHTR-UWB can be defined as

$$\sigma'^2_{TR} = E\left\{\left[\xi'_{m,l} \pm \eta'_{m,l}\right]^2\right\} = \frac{N_0}{4}\left(2 + a^2\delta^2_{r,d}\right) \tag{158.16}$$

We can define the four normalized random variables as shown below:

$$Y'_1 \triangleq \frac{1}{2\sigma'^2_{TR}} \sum_{m=0}^{\frac{N_s}{2}-1} \sum_{l=1}^{2WT} \left(\frac{1}{\sqrt{2W}} w'_{m,l} + \chi_{1,m,l}\right)^2 \tag{158.17}$$

$$Y'_2 \triangleq \frac{1}{2\sigma'^2_{TR}} \sum_{m=0}^{\frac{N_s}{2}-1} \sum_{l=1}^{2WT} \chi^2_{2,m,l} \tag{158.18}$$

$$Y'_3 \triangleq \frac{1}{2\sigma'^2_{TR}} \sum_{m=0}^{\frac{N_s}{2}-1} \sum_{l=1}^{2WT} \left(\frac{1}{\sqrt{2W}} w'_{m,l} - \chi_{2,m,l}\right)^2 \tag{158.19}$$

$$Y'_4 \triangleq \frac{1}{2\sigma'^2_{TR}} \sum_{m=0}^{\frac{N_s}{2}-1} \sum_{l=1}^{2WT} \chi^2_{1,m,l} \tag{158.20}$$

where Y'_2 and Y'_4 are central chi-squared random variables with $N_s WT$ degrees. Y'_1 and Y'_3 are noncentral chi-squared random variables with same degrees. The noncentrality parameter is given by

$$\mu'_{TR} \triangleq \frac{1}{2\sigma'^2_{TR}} \sum_{m=0}^{\frac{N_s}{2}-1} \sum_{l=1}^{2WT} \frac{1}{2W} w'^2_{m,l}. \tag{158.21}$$

So, we obtain as

$$\mu'_{TR} = \frac{1}{2\sigma'^2_{TR}} \sum_{m=0}^{\frac{N_s}{2}-1} \int_0^T w'^2_m(t)dt = \frac{(1+a\bar{h}_{s,r})^2}{(2+a^2\delta^2_{r,d})} \frac{E'_s}{N_0} \sum_{l=1}^{L_{CAP}} \alpha_l^2 \tag{158.22}$$

where $E'_s = E'_p N_s = (1+a\bar{h}_{s,r})^2 E_p N_s$, $\alpha_l^2 = h^{(l)2}$.

We assume that $q_{TR} = N_s WT/2$, $\gamma'_{TR} = \mu'_{TR}/2$, and the conditional *SER* of cooperative DHTR-UWB system are as follows [6]:

$$P\{e|\gamma'_{TR}\} = P\{Y_2 > Y_1|d_0 = +1\}$$

$$= \frac{e^{-\gamma'_{TR}}}{2^{q_{TR}}} \sum_{i=0}^{q_{TR}-1} \frac{(\gamma'_{TR})^i}{i!} \sum_{k=i}^{q_{TR}-1} \frac{1}{2^k} \frac{(k+q_{TR}-1)!}{(k-i)!(q_{TR}+i-1)!} \quad (158.23)$$

Considering the uncertainty of channel coefficient h^2, we define the expectation of $P\{e|\gamma'_{TR}\}$ as follows [7, 9]:

$$P_{e,TR} = E\{P\{e|\gamma'_{TR}\}\}$$

$$= \frac{1}{2^{q_{TR}}} \sum_{i=0}^{q_{TR}-1} \frac{E\{(\gamma'_{TR})^i e^{-\gamma'_{TR}}\}}{i!} \sum_{k=i}^{q_{TR}-1} \frac{1}{2^k} \frac{(k+q_{TR}-1)!}{(k-i)!(q_{TR}+i-1)!}$$

$$= \frac{1}{2^{q_{TR}}} \sum_{i=0}^{q_{TR}-1} \frac{(-j)^i}{i!} \frac{d^i}{dv^i} \psi'_{\gamma'_{TR}}(jv) \bigg|_{jv=-1} \cdot \sum_{k=i}^{q_{TR}-1} \frac{1}{2^k} \frac{(k+q_{TR}-1)!}{(k-i)!(q_{TR}+i-1)!} \quad (158.24)$$

$$\triangleq P_e\left(\psi'_{\gamma'_{TR}}(jv), q_{TR}\right)$$

where $\psi'_{\gamma'_{TR}}(jv)$ is the CF of γ'_{TR} [9]:

$$\psi'_{\gamma'_{TR}}(jv) = \prod_{l=1}^{L_{CAP}} \psi'_{l,\gamma'_{TR}}(jv) = \prod_{l=1}^{L_{CAP}} \frac{1}{K(1-jv\overline{\gamma}'_{TR})} \quad (158.25)$$

where $\overline{\gamma}'_{TR} = \left[\frac{(1+a\overline{h}_{s,r})^2}{2(2+a^2\delta^2_{r,d})}\right] \frac{\Omega E_s}{N_0}$, $\Omega = \frac{1}{L}$.

158.4 Results and Discussion

In this section, we evaluate the performance of DHTR-UWB system. We consider a dense multipath channel, where each path's signal experiences flat Rayleigh fading. $h_{s,d}$, $h_{s,r}$, $h_{r,d}$ are the channel fading coefficient which are defined as complex Gaussian random variables with zero mean and normalized square deviation $\delta^2_{s,d} = \delta^2_{s,r} = \delta^2_{r,d} = 1$, respectively. The actual number of multipath components is $L = 30$, and the number of multipath components captured by the autocorrelation receiver (AcR) is denoted by $L_{CAP}=TW$. We also discuss the effect of time-bandwidth product, number of impulse, and instantaneously received SNR on the cooperative DHTR-UWB system as shown in Figs. 158.4, 158.5 and 158.6.

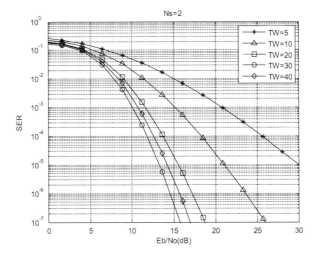

Fig. 158.4 DHTR-UWB's SER performance with $Ns = 2$

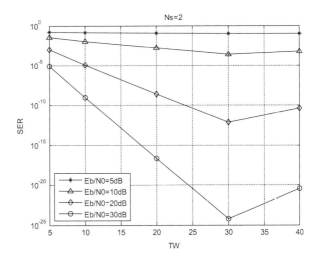

Fig. 158.5 Effect of TW on the performance of DHTR-UWB

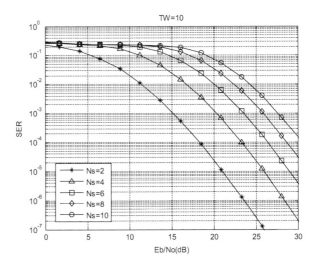

Fig. 158.6 SER performance of DHTR-UWB with Ns

In Fig. 158.4, we note that the tendency of *SER* performance is from bad to good then turns into bad again as *TW* increases. This is due to the fact that time-bandwidth product *TW* represents for the integral duration. $L < TW$ implies that multipath components cannot be captured completely, thus severely degrades the *SNR* at the receiver termination and drops the *SER* performance of the cooperative system. On the contrary, increasing *TW* beyond *L* will only accumulate more noise energy, and it degrades performance as well.

We assumed that the actual number of multipath components is $L = 30$ in this simulation. For the case $TW = 30$, at which point integral interval equals to the multipath delay, the energy of multipath components is fully captured and the noise is maximally restrained. As we can see from Fig. 158.3, the cooperative DHTR-UWB system gains five-order diversity under the condition of optimum integration interval. This embodies the advantage of impulse scheme in dense multipath channel transmission.

Figure 158.5 details the effect of time-bandwidth product *TW* on the *SER* performance of DHTR-UWB cooperative system. It can be observed that the *SER* decreases with *TW* until it rises to a value of 30, which is equal to *L*. We obtain the best *SER* performance at the inflection point $TW = 30$ for each fixed E_b/N_0. Due to the fixed bandwidth *W* at the transmitting terminal, we can easily deduce that the optimum integral interval is $30/W$.

In order to analyze the number of impulse Ns impact on *SER* performance, we fixed time-bandwidth product and specify the optimum integral interval $T = 10/W$. We observe that *SER* decreases with N_s that correspond to Fig. 158.5. Increasing N_s implies less energy per impulse and leads to more noise energy accumulation. This can be seen as the reason why performance degradation as Ns increases in Fig. 158.6.

Obviously, the transmitted cooperative-reference UWB cooperative approach is not suitable for simply amplify and forward (AF) because in AF the relay node do not decode. They cannot separate and regenerate the reference impulses or data impulses to forward further alternatively. Only decode and forward supports transmitted cooperative-reference cooperative approach.

Further, the transmitted cooperative-reference UWB cooperative approach is only applicable to the transmitted-reference system and applies only to time-domain transmitted-reference system. For the frequency-domain and code-domain transmitted-reference system, the node finally transmits the composite impulse of the reference impulse and the data impulse [4, 5]. It is not possible that the relay node decodes and forwards only one of the reference impulse and the data impulse.

Furthermore, in the TR-UWB cooperative communication network, it can be the degradation or evolution between of three cooperative communication modes: two-relay transmitted cooperative-reference mode, single-relay transmitted cooperative-reference mode, and direct transmission mode. System will convert to two-relay transmitted cooperative-reference mode when we got two ideal relays, and the power of each relay will reduce by half. Similarly, single-relay mode will be chosen if there is only one proper relay in the system. Otherwise, the system will degenerate into direct transmission mode.

158.5 Conclusion

The paper proposes a novel transmitted cooperative-reference UWB cooperative communication method that uses two relay nodes to forward reference impulses and data impulses separately. For each relay node is concerned, it can save half of the energy and help to encourage nodes to participate in cooperation. It also provides a new relay strategy for cooperative communication. We established a cooperation model based on this method and analyzes the performance of it. Simulation result shows that the DHTR-UWB cooperative system has favorable SER performance. System can obtain SER $= 10^{-3}$ performance under optimal integration time with receiving SNR is 10 dB. Cooperative system can gain approximately five-order diversity in dense multipath channels. Future research will focus on the effect of distance and power distribution on DHTR-UWB system.

References

1. Shen, Q., Wu, X., Lin, D., & Qiu, X. (2010). Performance analysis of cooperative ultra-wideband communication system. *2010 International Conference on Communications and Mobile Computing (CMC), IEEE* (pp. 217–220).
2. Chao, Y. L., & Scholtz, R. (2003). Optimal and suboptimal receivers for ultra-wideband transmitted reference systems. *IEEE Global Telecommunications Conference, 2003. GLOBECOM'03* (pp. 759–763).
3. Hoctor, R., & Tomlinson, H. (2002). Delay-hopped transmitted-reference RF communications. *Digest of Papers, 2002 I.E. Conference on Ultra Wideband Systems and Technologies* (pp. 265–269).
4. Goecke, D. L., & Zhang, Q. (2007). Slightly frequency-shifted reference ultra-wideband (UWB) radio. *IEEE Transactions on Communications, 55*(3), 508–519.
5. Jian, Z., Han-Ying, H., Luo-Kun, L., & Tie-Feng, L. (2007). Code orthogonalized transmitted reference ultra-wideband wireless communication system. *International Conference on Wireless Communications, Networking and Mobile Computing, 2007. WiCom 2007* (pp. 528–532).
6. Quek, T. Q. S., & Win, M. Z. (2005). Analysis of UWB transmitted-reference communication systems in dense multipath channels. *IEEE Journal on Selected Areas in Communications., 23*(9), 1863–1874.
7. Quek, T. Q. S., Win, M. Z., & Dardari, D. (2007). Unified analysis of UWB transmitted-reference schemes in the presence of narrowband interference. *IEEE Transactions on Wireless Communications, 6*(6), 2126–2139.
8. Ngo, H. Q., Quek, T. Q. S., & Shin, H. (2010). Amplify-and-forward two-way relay networks: error exponents and resource allocation. *IEEE Transactions on Communications, 58*(9), 2653–2666.
9. Simon, M. K., & Alouini, M. S. (1998). A unified approach to the performance analysis of digital communication over generalized fading channels. *Proceedings of the IEEE, 86*(9), 1860–1877.

Author Index for Volume 1

A
Ao, X., 421
Arshad, M.R.H.M., 355

B
Bai, C., 385
Bai, J., 337, 577
Bao, W., 577

C
Cai, W., 107
Cao, G., 671
Cao, L., 635
Cao, S., 671
Cao, Y., 549
Chai, J., 643
Chen, C., 559, 609
Chen, G., 219
Chen, J., 365
Chen, K., 671
Chen, L., 481
Chen, M., 163
Chen, Q., 55
Chen, X., 243, 515, 635
Chen, Y., 47
Chen, Z., 171, 307
Cheng, W., 481
Cheng, Y., 541

D
Dai, M., 577
Dong, J., 99
Duerling, B., 329

F
Fang, P., 153
Fang, X., 9
Fard, S.P., 125, 355
Fei, S., 209
Feng, W., 577
Feng, X., 265

G
Gai, S., 625
Gao, J., 171
Gao, S., 609
Gao, Z., 31
Geng, X., 395
Gu, C., 91, 145
Gu, H., 429
Gu, Y., 375
Guan, H., 345
Guan, Y., 653
Guo, C., 21
Guo, J., 291

H
Han, W., 255
Hao, X., 577
He, H., 489
Hong, L., 541
Hong, Z., 499
Hou, C., 321
Hou, L., 489
Hu, H., 437
Hu, S., 345
Huang, X., 329, 661
Huang, Y., 507
Huang, Z., 91, 375

J
Jiang, C., 345
Jiang, M., 593
Jiao, Z., 365
Jin, Z., 209

K
Kang, Y., 73, 83

L
Li, D., 411, 635
Li, G., 65, 193
Li, H., 177, 465, 593, 679
Li, J., 47
Li, K., 177, 679
Li, L., 201, 321, 585
Li, M., 643
Li, Q., 3
Li, T., 39
Li, X., 55
Li, Y., 115
Li, Z., 193, 201
Liang, H., 643
Liang, T., 533
Lin, A., 525
Lin, M., 617
Lin, Y., 73, 83
Liu, J., 107
Liu, Q., 617
Liu, W., 65
Liu, X., 115, 255
Liu, Y., 185
Lu, H., 73, 83
Luo, P., 153
Luo, S., 177, 679
Luo, X., 533
Luo, Y., 617

M
Ma, J., 291
Ma, L., 403
Ma, R., 307
Ma, Y., 337, 585
Meng, Q., 177, 679
Ming, J., 695
Mohamed, H.H., 355
Mousavi, S.A., 355
Mu, F., 201
Mu, S., 337

P
Peng, X., 489

Q
Qiu, L., 445
Qiu, Q., 153

S
Sedaghat, L., 329
Shen, H., 559
Shen, J., 55
Shen, Z., 403
Shi, M., 255
Shu, S., 281
Sumari, P., 355
Sun, J., 365
Sun, S., 3
Sun, Z., 299

T
Tan, X., 163
Tang, J., 243
Tang, Z., 329
Tao, Q., 91, 145

W
Wang, B., 219
Wang, F., 47, 201, 411
Wang, L., 585
Wang, P., 429
Wang, Q., 177, 403
Wang, R., 695
Wang, S., 653
Wang, W., 533
Wang, X., 273, 661
Wang, Y., 365, 395, 455
Wang, Z., 411
Wen, D., 437
Wen, Q., 209
Wu, C., 671
Wu, Y., 321

X
Xia, C., 99
Xia, X., 625
Xiao, B., 525
Xiao, Y., 185, 601

Xing, Y., 567
Xu, D., 465
Xu, H.-Y., 9
Xu, J., 201
Xu, R., 653
Xu, Z., 201, 307

Y

Yan, X., 365
Yang, B., 411, 679
Yang, F., 567
Yang, G., 473
Yang, H., 227
Yang, J., 265, 687
Yang, T., 515
Yang, Y., 201, 445
Yang, Z., 385
Yao, C., 541
Yin, L., 39
Yin, S., 403
Ying, S., 345
Yu, B., 567
Yu, F., 65
Yu, P., 465
Yu, Q., 617
Yu, X., 429
Yu, Z., 385
Yuan, F., 39
Yuan, H., 235, 421
Yuan, J., 671

Z

Zainuddin, Z., 125
Zeng, H., 55
Zeng, W., 695
Zhang, B., 609
Zhang, C., 91, 473, 577
Zhang, H., 219
Zhang, J., 235, 291
Zhang, M., 107
Zhang, S., 307
Zhang, T., 671
Zhang, W., 99
Zhang, X., 193, 473
Zhang, Y., 73, 83, 635
Zhang, Z., 39, 135
Zhao, D., 375
Zhao, J., 3
Zhao, L., 507
Zhao, T., 421
Zhao, Y., 429
Zhao, Z., 321
Zheng, S., 489
Zhou, D., 21
Zhou, H., 465
Zhou, L., 107
Zhou, W., 153
Zhu, S., 687
Zhu, X.-f., 687
Zhu, Y., 525
Zou, G., 337

Subject Index for Volume 1

A

Active decoy, 437–440, 442, 443
Adaptive, 3–8, 65, 107–114, 171–176, 292, 379, 386, 457, 478, 507–510, 512, 525–532, 580, 631, 701
  filter, 171–176
  sampling, 107–114
AF. *See* Array factor (AF)
Affinity propagation, 177–183
Android, 154, 155, 161, 371
ANNs. *See* Artificial neural networks (ANNs)
Ant colony algorithm, 227–233
Ant colony system, 73–80, 83–90
Anti-radiation missile (ARM), 437–443
Apache, 210, 265
Apriori algorithm, 281–285
Architectural space, 549–558
ARM. *See* Anti-radiation missile (ARM)
Array factor (AF), 40, 41, 43
Artificial neural networks (ANNs), 22, 48, 50–52, 100, 102, 125, 126, 220, 467, 672
Association rule, 281–288, 458

B

Bayesian models, 385–392
Bayesian networks (BNs), 386, 391, 392, 525–532, 649, 650
BCI. *See* Brain–computer interface (BCI)
Beam position, 5, 6, 8
Binary image, 478, 479, 483, 688, 691, 692
Binary phase coded, 445–451
Biometrics, 465, 466, 469
BNs. *See* Bayesian networks (BNs)

BP neural network (BPNN), 21–30, 243–250, 671–677
Brain–computer interface (BCI), 355–362
Business process execution language (BPEL) process, 345–352

C

Camera calibration, 484–486
Centroid, 122, 499, 515–522, 561, 562, 657
Certainty-based active learning, 235–242
CH. *See* Cluster head (CH)
Chaos perturbation, 230, 233
Classification
  accuracy, 51, 321, 322, 326, 327, 500, 503, 504, 578, 583, 584, 586, 591, 661, 666–668, 676, 698, 701
  and recognition, 672–674, 676
Classify, 50, 322, 326, 357, 468, 500, 502, 594, 596, 598, 600, 602, 644–648, 650, 661, 662
Cloud computing, 202, 207, 209, 210, 425, 429
Cluster head (CH), 65–71, 110, 111, 163–170
Clustering, 48, 65, 66, 69, 85, 108, 110, 163–170, 177–183, 221–223, 301, 322, 339, 455–462, 501, 635–641, 680, 697
Collection sites, 395–402
Color management, 617–619
Component docking, 473–480
Compression matrix, 281–288
Computer technology, 48, 300, 549–558
Computer vision (CV), 223, 224, 301, 507, 533, 609, 654
Conditional independence tests, 526
Connected-element interferometry, 411–418

Contour features, 597, 599
Convergence, 22–26, 28–30, 43, 44, 65, 80, 97, 104, 150, 172, 174, 175, 185–187, 193, 194, 229, 230, 233, 526, 531, 544, 546, 585–592, 664
Copy-move forgery, 680, 684
Corner extraction, 482, 485, 486
Curve on surface, 10, 11, 14, 16–19
CV. See Computer vision (CV)
Cycle analysis, 56, 69

D

Data center network, 429–435
Data intensive, 375–383
Data mining, 209, 210, 281, 300, 458, 499, 644, 645, 649
DBN. See Deep belief network (DBN)
Decision degree, 277, 278
Decision-theoretic rough set model (DTRS), 273–279
Decoupling interrupts, 291–296
Deep belief network (DBN), 661–668
Dense matching, 625–633
Density-sensitive, 177–183
Descriptive, 339, 508
Device driver, 329–336
Dictionary learning, 541, 578, 580–581
Difference received signal strength (DRSS), 115–123
Differential geometric characteristic, 13
Dijkstra algorithm, 153, 155–157, 227, 233
Distributed computing, 228
Distributed system, 31, 74, 75, 80, 83–90, 404
Document object model (DOM), 368–370
Domain ontology learning, 489–496
Double approximate identity functions, 126, 127
Double flexible approximate identity functions, 126, 132
Double flexible approximate identity neural networks, 125–132
DRSS. See Difference received signal strength (DRSS)
DTRS. See Decision-theoretic rough set model (DTRS)
Dynamic route guidance system (DRGS), 227–233
Dynamic voltage scaling (DVS), 73–80

E

ECFG. See Exception control flow graph (ECFG)
Electroencephalographic (EEG) signal classification, 355–362

Energy consumption, 65, 66, 69, 71, 74–77, 79, 80, 108, 110, 112, 113, 164–166, 168–170, 558
Energy effective, 107–114
ENN. See Extension neural network (ENN)
Exception control flow graph (ECFG), 345–352
Exception handling, 345, 346, 349, 351
Extended Kalman filter, 610, 611, 613–614
Extension neural network (ENN), 219–225

F

Factorization, 255, 258, 263, 321–327
Fall detection, 653–659
FAS algorithm. See Frequency based sampling (FAS) algorithm
FCA. See Formal concept analysis (FCA)
Feature extraction, 48, 49, 243, 245, 322, 355, 356, 358, 362, 466, 469, 471, 476, 516, 594, 597, 600, 662
Feature selection, 299–306, 361, 635–641, 645–646
Flexible approximate identity functions, 126, 132
Forecasting model, 101–102, 104, 248
Forgery detection, 680, 683
Formal concept analysis (FCA), 490–493, 495, 496
Formal method, 307
Forward-error correction, 291
Fragments, 507–512, 534
Frequency based sampling (FAS) algorithm, 107–114
Frequency identification, 243–250
2FSK. See Pseudorandom binary frequency shift keying (2FSK)
Fuzzy system, 99–106

G

GA. See Genetic algorithm (GA)
Gabor wavelet, 662, 663, 668
Gait feature, 515–522
Gait recognition, 516
Gauss-Jacobi iterative, 542, 543, 546
Genetic algorithm (GA), 21–30, 39–46, 79, 88, 89, 92, 146, 457
Geographic information system (GIS), 338, 339, 342–344
Gesture manipulation, 135–142
GLCM. See Gray-Level Co-occurrence Matrix (GLCM)
Global orientation, 653–659

Graph cut, 538
Graphical user interface (GUI) testing, 385–392
Gray-Level Co-occurrence Matrix (GLCM), 662, 663, 665, 666, 668
Green supply chain management (GSCM), 395, 396, 402
Grey model, 220, 224, 225
GUI testing. *See* Graphical user interface (GUI) testing

H
Hadoop, 209–216, 265–269, 300, 376, 425, 591
Hand tracking, 135–142
Harmonic function (HF), 321–327
HBase, 265–271, 376
HCI. *See* Human–computer interaction (HCI)
Heterogeneous, 73–80, 83–90, 164, 186, 191, 375, 379
Heterogeneous system, 73–80, 84, 85
Heuristic, 75, 77–80, 84–89, 111, 229, 230, 258, 291, 293, 294, 395–402, 526, 676, 700
HF. *See* Harmonic function (HF)
Hough transformation, 571–573, 675
HowNet, 635–641
Human–computer interaction (HCI), 135–142, 559, 675
Human shape, 653–659
Hybrid particle swarm optimization, 185–191
Hyperbola model, 609–616

I
ICC profile. *See* International Color Consortium (ICC) profile
Identifying, 343, 490, 538, 567–575, 581
Image
  forensics, 679, 680
  monitoring, 475
  processing, 322, 475, 476, 481–486, 568, 594, 668
  restoration, 541–546
  skeleton, 687–691
  softproofing, 617–624
  tampering, 601
Implicit information, 255, 260–262
Improved tolerance relation, 274, 279
IMRT. *See* Intensity modulated radiation therapy (IMRT)
Incomplete decision table, 274–279
Indoor three-dimensional positioning, 115–123
Information extraction, 202, 366, 489, 579
Information service, 455–462
Infrared (IR), 695–697
Integral sliding mode controller, 31–37
Intelligent visual surveillance, 653
Intense illumination, 567–575
Intensity modulated radiation therapy (IMRT), 193–198
Interlocking, 307–319
International Color Consortium (ICC) profile, 617, 622
IR. *See* Infrared (IR)

J
Jsoup, 368, 369

K
Kernel-nearest neighbor (kNN), 48, 52, 326, 499–505, 680, 683
K-optimal path, 231
K–T transform, 578, 579, 582, 583

L
Lane detection, 609–616
Lane mark, 567–575, 615
Lattice reduction, 262
LEACH-SC, 66–71
Least mean square error (LMS) algorithm, 171–176
Least squares method, 615
Lifetime, 65, 70, 71, 169, 170
Linear antenna array, 39–46
Linear matrix inequality (LMI), 34–37
Link16, 55–63
LLL algorithm, 256, 258
LMS algorithm. *See* Least mean square error (LMS) algorithm
Loading data, 265–271

M
Machine learning, 240, 241, 299–301, 305, 321, 500, 580, 585, 608
MapReduce, 210, 214, 215, 265–269, 299–306, 376, 429, 585–592
MapReduce-support vector machine (MR-SVM), 585–592
Markov models, 291–296
Mass data, 403–409
Maude, 307–319

Mean comparison, 412, 416, 417
Mean shift, 136–142
Mel Frequency Cepstrum Coefficient (MFCC), 244–247, 249, 250
Mellin approximate identity functions, 126
Mismatch eliminating, 629
Missile-borne phased array radar, 3–8
MongoDB, 376, 377, 379, 403–409
MR-SVM. *See* MapReduce-support vector machine (MR–SVM)
Multihop communication, 164
Multi-objective optimization, 193
Multi-sensor fusion, 559–565
Multispectral remote sensing images, 578, 581, 584
Mutual information, 300–301, 303–305, 526, 527, 529

N
NAR. *See* Nonparametric auto-regression analysis (NAR)
Navigation software, 153–161
NDVI. *See* Normalized Difference Vegetation Index (NDVI)
Network control system (NCS), 31–37
NMF. *See* Non-negative matrix factorization (NMF)
Non-dominated neighbor-based immune algorithm (NNIA), 193–198
Non-local, 541–546
Non-negative matrix factorization (NMF), 321–327
Nonparametric auto-regression analysis (NAR), 219–225
Non-relational databases, 404
Normalized Difference Vegetation Index (NDVI), 578–579, 582, 583
Not Only SQL (NoSQL), 266, 376, 377, 404

O
On-line, 108, 109, 136, 201, 204, 376, 456, 458–460, 462, 507–512, 559–565, 637
Online gesture recognition, 559–565
Optimization, 6, 22, 27–28, 30, 40, 41, 43, 48, 50, 75, 76, 84–88, 91–98, 103, 108, 138, 145–151, 183, 185–191, 193–198, 228, 230, 233, 248, 281, 324, 343, 422, 457, 538, 580, 586, 589, 590, 592, 595, 604, 612, 647, 697

P
PageRank algorithm, 204–206
Pairwise constraints, 178, 180–182
Palmprint, 465–471
Paper submission, 201–208
Parallel
  processing framework, 375–383
  projection, 9–19
Parallel computing, 186–189, 191, 210, 376, 377
Parallelize, 188
Parameter optimization, 50, 592
Partial code replicated, 451
Particle swarm optimization (PSO), 39, 50, 91–98, 185–191
PCA. *See* Principal components analysis (PCA)
PCA-SIFT, 679–685
PC radar. *See* Pulse compression (PC) radar
P300 detection, 355–362
Performance analysis, 57–62, 294, 421, 422
Performance evaluation, 112–113, 376, 422, 470, 640–641
Piano, 243–250
Polar space, 533–540
Predict interaction between proteins, 48
Prediction model, 100, 219–225, 426, 697
Principal components analysis (PCA), 466, 467, 471, 682–684, 695–701
Processing time, 21–30, 75, 76, 84, 88, 379–381
Process quality monitoring, 473–480
Pseudorandom binary frequency shift keying (2FSK), 171–176
PSO. *See* Particle swarm optimization (PSO)
Pulse compression (PC) radar, 445–451

Q
Quantum-inspired genetic algorithm (QGA), 145–151

R
Range deception, 449
Range-gate, 450, 451
RBF neural network, 99–106, 126
Recommendation, 201–208, 455–462
Region growing, 625–633
Regularization, 103, 541–546, 647
Remote sensing image, 577–584, 671–677
Rendering intent, 617–624
Repositioning, 505

Subject Index for Volume 1   1415

Restricted searching area, 153–159, 161
Reverse engineering, 329–336
Reverse logistics, 395–402
Rewriting logic, 307–309, 318
Road impedance factor, 229–233
Routing protocol, 66, 69, 163

S

SA. *See* Simulated annealing (SA)
Sag measurement, 481–486
Sample selection, 236–241, 674–676
Scheduling algorithm, 74–76, 84, 85, 89
SDA. *See* Subclass discriminant analysis (SDA)
Search, 3–7, 23, 40, 41, 50, 76, 77, 80, 86, 95, 153, 157, 161, 172, 181, 186, 188, 194, 195, 201–208, 210, 228, 291, 307, 378, 455, 457, 460, 474, 511, 526, 612, 613, 626, 630, 631, 633, 636, 680, 683, 700
Search engine, 201–208, 636
Segmentation, 204, 208, 240, 248, 250, 477–479, 490, 491, 520, 533–540, 638
Semantic dependency, 489–496
Semantics, 206, 309, 348, 376–378, 382, 459, 489–496, 635–641
Semi-supervised clustering, 177–183
Semi-supervised learning, 178, 179, 321–327
Server-centric architecture, 433, 435
Shortest path, 153–161, 228
Short-term load forecasting, 99–106
Short texts clustering, 635–641
Side lobe level, 39–46
Similarity, 177–183, 202, 206, 207, 222, 239, 303, 322, 372, 460, 461, 466, 469, 502, 508–510, 512, 541–544, 560–563, 580, 635–640, 682, 683
Simulated annealing (SA), 186, 525–532
Simulation, 3–8, 36, 37, 48, 69–70, 74, 104, 112–114, 167–176, 194, 196, 232, 233, 244, 292, 300, 307, 412, 416–418, 433, 434, 437–443, 445–451, 474, 479, 550–557, 563–564, 614, 615
Simulation analysis, 168–170, 418, 550–557
Slot allocation, 55–63
Sparse principal component analysis (SPCA), 695–701
Sparse representations, 577–584
Spatial analysis, 337–344
Spatial data, 343, 375–383
Spectral analysis, 243, 518–520

Speeded-up robust features (SURF), 594, 597–599
Spontaneous speech summarization, 235
SPRINT, 209–216
Standard processing time table (SPTT), 22–30
State feedback, 33, 37
Statistics, 21, 154, 159, 300, 301, 338, 339, 341, 342, 490, 529, 593, 602, 667, 681
Structure learning, 525–532
Subclass discriminant analysis (SDA), 465–471
Supply chain network, 91–98, 145–151
Support vector machine (SVM), 49, 52, 236, 240–242, 300, 304, 305, 467, 499, 500, 502–505, 582–600, 602–605, 607, 608, 643–650, 695–701
SURF. *See* Speeded-up robust features (SURF)
Survival time, 168, 170
Switching function, 32–34, 37
Synthetic aperture radar (SAR) oil spill image, 661–668

T

Taylor approximation, 12, 19
TB. *See* Tuberculosis (TB)
TDMA, 55, 57–62
Template matching method, 559
Text classification, 222, 499–505
Thinning procedure, 687–692
Time delay, 411–418
TOLU, 291–296
Topological properties, 430, 431, 435
Topology, 125, 220, 232, 375, 387–389, 392, 430, 586, 689, 691, 692
Tracing, 10, 11, 534, 551, 555, 655
Tracking, 3, 135–142, 411, 412, 439, 440, 448–450, 474, 507–512, 567–575, 610, 611, 613, 614, 635, 654
Transforming, 145–151, 179, 536–537
Transforming operator, 92–94, 96, 98, 148, 149, 151
Transmission delay, 58–61
Tuberculosis (TB), 337–344, 380, 381, 383

U

Uncertainty, 31–37, 102, 237, 300, 357, 502, 626, 629, 676
Uneven clustering, 170
Universal approximation, 125–132

USB, 295, 296, 329–336
User model, 456–459, 462

V

Variable precision rough set (VPRS), 500–503, 505
Vehicle classification, 593–600
Vertical search, 201–208
Visual design, 549–558

W

WarpingLCSS, 559–565
Wavelet analysis, 47–52
Wavelet transform, 48–50, 99–106
Web sentiment, 219–225
WiFi, 115–123
Wind turbine, 403–409
Wireless sensor networks (WSN), 65–71, 107–114, 163–170
Worm detection, 643–650

Author Index for Volume 2

A
Alahmadi, A., 731

C
Cao, B., 799
Chai, J., 1365
Chen, B., 723
Chen, D., 1341
Chen, G., 755
Chen, K.-L., 1045
Chen, Q., 783, 825
Chen, X., 1323
Chen, Y., 1209
Chen, Y.-W., 713
Chen, Z., 799
Chi, T., 739
Chou, W., 955
Chu, Y.-Y., 713
Collados, K., 1191
Cui, Y., 905

D
Di, Y., 1099
Dong, M., 955
Dong, W., 1201
Dong, Y., 783
Dou, L., 1233
Du, J., 755
Du, Y., 755

F
Feng, S., 1117, 1125
Fu, T., 747

G
Gao, L., 1289
Gao, W., 971
Gao, Y., 747
Gorricho, J.-L., 1191
Guan, H., 1173
Guo, W., 971
Guo, X., 1245

H
He, L., 1341
He, Q., 963
He, Y., 913
He, Z., 997
Hou, C., 979, 1373
Hu, D., 809
Hu, P., 1109
Hu, Q., 883
Hu, W., 1017
Hu, X., 855
Huang, F., 1381
Huang, J.-F., 1035, 1045
Huang, T., 705
Huang, Y., 883

J
Jiang, C., 1173
Jiang, D., 1007
Jiang, F., 855
Jiang, Z., 971
Jiangn, Y.-J., 1035
Jiao, J., 1091

L

Li, B., 791, 865
Li, F., 1073, 1083
Li, G., 835
Li, J., 1225
Li, M., 971
Li, O., 1395
Li, T., 1395
Li, W., 905, 1073, 1083, 1143
Li, X., 929, 939, 1341
Li, Y., 1217, 1263
Li, Z., 963, 1373
Liang, H., 1365
Lin, F., 1373
Lin, H., 1373
Liu, D., 835
Liu, F., 1181
Liu, H., 1165
Liu, J., 997
Liu, L., 835, 1063
Liu, T., 739
Liu, Y., 799, 955, 1055, 1253, 1281
Lu, L., 847, 1233
Luan, D., 921
Lv, S., 791, 865

M

Man, Q., 763
Mao, T., 775
Matsuhisa, T., 1331
Miao, Z., 1263

N

Nie, W., 809

P

Peng, I.-H., 713
Peng, Y., 893

Q

Qian, K., 1307
Qin, P., 755
Qin, Z., 1099

R

Ren, Y., 1055
Rui, P., 763

S

Serrat, J., 1191
Shao, Y., 929, 939
Shi, B., 1271
Shi, F., 929, 939
Shuo, C., 1201
Soh, B., 731
Song, B., 971
Su, A.Y.S., 713
Su, D., 1073, 1083
Su, Q., 921
Sun, J., 963
Sun, L., 1373
Suo, Y., 739

T

Tang, Y., 1365
Tang, Y.H., 873
Tao, H., 979
Tian, M., 791, 865
Tian, Y., 913

W

Wang, H., 963
Wang, K., 783
Wang, L., 947, 1109
Wang, P., 971, 1297
Wang, S., 1263
Wang, Y., 723, 893, 1055, 1209
Wang, Z., 723, 847, 1217
Wei, S., 835, 1063
Wen, J.-Y., 1035
Wu, C., 989
Wu, G., 1181
Wu, H., 1181
Wu, J., 997
Wu, L., 1165, 1209
Wu, R., 783
Wu, Y., 809
Wu, Z., 1289

X

Xi, J., 1281
Xia, M., 783
Xiao, J., 1143, 1263
Xiao, T., 947
Xiao, W., 1281, 1381
Xie, L., 913
Xu, B., 1349, 1357

Xu, C., 1253
Xu, K., 1253
Xu, M., 1201, 1245
Xu, P., 1217
Xu, W., 1263
Xu, Y., 775
Xu, Z., 775
Xue, Y., 825
Xue, Z., 1017

Y
Yang, B., 775
Yang, F., 1027
Yang, J., 747
Yang, J.-L., 1045
Yang, L., 1165
Yang, T., 913
Yang, Y., 763
Yang, Z., 1099, 1233, 1245
Yao, J., 1181
Yi, C.-F., 713
Yi, D., 979
Yin, B., 1027
Yin, S., 835, 1063
Yin, Y., 1007
Ying, S., 1173
Yu, G., 747
Yu, Q., 1133
Yuan, H., 817, 1315

Z
Zha, F., 971
Zhai, J., 1073, 1083, 1323
Zhang, C.N., 1133
Zhang, F., 1297
Zhang, G., 1073, 1083
Zhang, H., 1381
Zhang, L., 755, 1117, 1125, 1349, 1357
Zhang, M., 1217, 1373
Zhang, P., 921
Zhang, Q., 1099, 1263
Zhang, R., 1373
Zhang, X., 847, 905, 1099
Zhang, Y., 791, 865, 1017, 1165
Zhang, Z., 817, 1099, 1315
Zhao, H., 1271
Zhao, S., 847
Zhao, T., 817, 1315
Zhao, W., 1297
Zhao, X., 963
Zheng, H., 1191
Zheng, Z., 847
Zhou, F., 1155
Zhou, X., 1225
Zhou, Y., 1017
Zhou, Z., 1395
Zhu, M., 1063
Zhuo, M., 1373
Zou, Y., 1063
Zuo, Y., 1233

Subject Index for Volume 2

A

AADL. *See* Architecture Analysis and Design Language (AADL)
Access category (AC), 1165, 1171
Access control, 784, 791–797, 1203–1206, 1294
Advanced persistent threat (APT), 1297–1304
Advertising delivery strategy, 856
Aesthetic criterion, 1384, 1385, 1393, 1394
Android, 740, 749, 788, 837, 1099–1106, 1112–1116
ANFIS, 1373–1378
APT. *See* Advanced persistent threat (APT)
Architecture Analysis and Design Language (AADL), 1117–1123, 1125–1131
Area efficiency, 1064, 1069, 1071
Aspect-oriented, 1125–1131
Asynchronous mode, 1001, 1002, 1290

B

Basic Detection Strategy, 802–805, 807
Bike rental station, 1341–1348
Bi-level model, 1346, 1348
Bipartite graph, 826, 829, 830, 832
Bounded retransmission protocol (BRP), 1245–1251
B+R mode, 1341–1348

C

Cache, 764–767, 786, 900, 1064, 1143–1151
Carrier aggregation, 714, 715, 786, 990–992, 1212, 1291
Chain network, 1209–1216
Channel reservation, 1166, 1169, 1199

Churn, 1263, 1264, 1267–1269
Clock, 787, 840, 843, 906, 907, 910, 989, 993, 1050, 1290, 1350, 1353–1355, 1357–1362
Clock theory, 1350, 1353, 1354, 1358, 1360–1362
Cloud computing, 723–727, 747–754, 791–797, 799–801, 805, 807, 817–823, 865, 866, 893–902, 1112
Cloud GIS, 893, 894, 896–902
Cloud security, 759, 865–871, 894
Cloud storage, 847–853, 895, 898
Cluster-based, 1253–1261
CMOS, 843, 906, 909, 912, 1036, 1040, 1042, 1043, 1046, 1047, 1051, 1091–1095, 1097
Collaborative filtering, 1017–1024
Communicating Sequential Processes (CSP), 1350, 1352–1354
Communication, 724, 775–782, 784–787, 790, 813, 835, 836, 843, 844, 868, 875, 886, 905, 963, 997, 1005, 1007, 1008, 1012, 1035–1043, 1056, 1063, 1073, 1091, 1109–1116, 1118, 1120, 1128, 1165, 1182, 1183, 1191, 1202, 1204, 1206, 1209, 1223, 1246, 1264, 1271, 1294, 1302, 1308, 1309, 1319, 1331–1339, 1350, 1352, 1353, 1395–1405
Complex network, 1226, 1228–1231, 1281, 1282
Computer graphics, 947
Computer power management system, 913–920
Configurability, 998, 999, 1005
Conjecture, 1332, 1334–1338
Constrained optimizing layout, 1381–1394

Context-triggered piecewise hash (CTPH), 1100, 1102, 1103, 1105, 1106
Continuous and discrete, 1128
Control mode, 929–936
Cooperative communication, 1395–1405
Coq, 1233–1243
Cost optimization, 1343, 1344
Co-training, 979–986
CPS. *See* Cyber physical systems (CPS)
Cross-domain, 791–797
CSP. *See* Communicating Sequential Processes (CSP)
Cyber counterwork, 1155–1162
Cyber physical systems (CPS), 1117–1119, 1125–1130, 1353, 1357–1362

D

D/A converter, 1091–1097
Data analysis, 810, 815, 870, 1254
Decision-making, 767, 855–862, 885, 886, 1157
Decision support system, 825–832
Dedicated short-range communication (DSRC), 1035–1043
Delay-hopped transmitted reference UWB, 1396, 1399
Denotational semantics, 1240
Desktop virtualization, 751–752, 754
Digital library, 747–754
Digital publishing, 888, 889
Discrete-time Markov chain (DTMC), 1246–1248
Dispatching, 723–730
Distributed file system, 817–823, 847, 898
Distribution network, 1077, 1083–1089, 1323–1330
Distribution system, 1073–1080, 1084, 1086, 1088, 1324
Domain specific languages, 1217–1224
Drowsy, 1143–1151
DSRC. *See* Dedicated short-range communication (DSRC)
DTMC. *See* Discrete-time Markov chain (DTMC)
DVB-RCS, 776–777, 782
Dynamic adaptive, 1156
Dynamic routing switching, 1317
Dynamic voltage scaling, 989–996

E

Earliest deadline first, 989–996
E-commerce application level, 873–881
EDCA. *See* Enhanced distributor channel access (EDCA)

Edge marking algorithm, 947–952
Educational Technology, 871
e-Health monitoring, 731–738
Electronic product code (EPC), 1271–1278
Electronic vehicle (EV), 1055–1060
Embedded Linux, 956–959, 993
Embedded processor, 1143, 1150, 1151
Embedded system, 947, 990, 1007–1015, 1117, 1130, 1217, 1218, 1359
Energy conservation, 913, 914
Energy consumption, 920, 963–969, 990, 995, 996, 1144, 1191, 1210, 1211, 1214, 1215, 1307–1309, 1311, 1313, 1316, 1319
Enhanced distributor channel access (EDCA), 1166, 1169–1171
EPC. *See* Electronic Product Code (EPC)
Exception handling framework, 1173–1179
Expert system, 809–816
Extended Petri net, 1297–1304

F

Face detection, 913–920
Fault detection, 911, 1083–1089, 1327
Fault diagnosis, 737, 825–832, 1323–1330
Fault-tolerance design, 731, 732, 735–738
Fault-tolerant requirements, 732
Field authority, 1017–1024
Filtering, 803, 804, 971–977, 1017–1024, 1103, 1203, 1204, 1240
Fire-fighting, 741–743
Forecasting model, 1377
Formal expression, 1301–1304
Formal specification, 1128, 1242, 1351–1353
Format-compliant, 778–781
Forwarding probability, 1308–1310
Frequency synthesizer, 1045–1053
Fuzzy logic, 1192, 1193, 1199

G

GA. *See* Genetic algorithm (GA)
General interface system, 998
Genetic algorithm (GA), 1060, 1345, 1346
Genre, 1018–1021, 1024
GTIN-716, 1272–1276
Guest OS, 1202, 1204–1206

H

Hadoop, 817–823, 847, 848, 851, 898
Hadoop Distributed File System (HDFS), 817–823, 847–850, 853, 898
Handheld controller, 955–961

HDFS. *See* Hadoop Distributed File System (HDFS)
Healthcare, 733, 1126
Hierarchical local-interconnection, 1063–1071
Hierarchical mobile IPv6 network, 763–773
Hierarchical scene analysis, 1298, 1299, 1301, 1302
High-dimensional data, 980, 981, 986
High impedance fault, 1073–1080, 1083, 1323
Host OS, 1203
Human role, 1155–1162
Hybrid layout, 1382, 1383

I
IaaS. *See* Infrastructure as a Service (IaaS)
Identity federation, 756, 759–761
IEEE802.15.4, 1210, 1213–1216
IEEE 802.11p, 1165–1172
Incipient fault, 1083–1087, 1089, 1323–1330
Indoor GIS, 739–745
Indoor map, 739–745
InfiniBand switch, 963–969
Influence diagram, 855–862
Information dissemination, 886, 1225–1231
Infrastructure as a Service (IaaS), 895, 896, 899, 900
Instruction, 815, 1101, 1102, 1143–1151, 1205, 1275, 1366
Integration, 744, 750, 751, 896, 897, 905, 1005, 1056, 1119, 1123, 1130, 1404, 1405
Integrity measurements, 869
Intelligent transportation system (ITS), 1035, 1165
Interface bridge, 835–844
Internal resistance, 1028–1030, 1033
Inter-turn short circuit, 940
IPv6, 763–773, 1007–1015
Isolated neutral system, 1087, 1324
ITS. *See* Intelligent transportation system (ITS)
IVI, 1008, 1009

K
Key management, 1115, 1134, 1138, 1141
Knowledge, 710, 810–814, 816, 856, 857, 860, 874, 983, 1002, 1056, 1134, 1158, 1166, 1177, 1178, 1298, 1317, 1321, 1331–1339

L
Laser bending, 921–927
Learning system, 1198

Light protocol, 1209–1216
Load control scheme, 763–773
Localization, 1067, 1253–1261
Long-Term Evolution Advanced (LTE-A), 713–720

M
Mac protocol, 784–786, 790
Magnetic resonance, 1056
Map updating, 744
MDA. *See* Model-driven architecture (MDA)
MDS-MAP, 1253–1261
Measurement indicators, 875
Media selection, 855–862
Memory controller, 837, 905–912
Message, 715, 724, 764–767, 770, 771, 801, 1110, 1111, 1113, 1133, 1135–1137, 1140, 1166, 1168, 1173–1179, 1205, 1219, 1225, 1234, 1241, 1245, 1255, 1256, 1275, 1290–1292, 1309–1311, 1332, 1335, 1336, 1352, 1353
Method of moments (MoM), 1056–1058, 1060
Method of Poynting vector, 1028, 1033
Mirror-role, 797
Mobile payment, 706–711
Model-based development, 1130, 1243
Model-driven architecture (MDA), 1118
Modelica, 1117–1123
Model transformation, 1118, 1123, 1127
MoM. *See* Method of moments (MoM)
Multi-bank, 705, 710, 711
Multi-loop theory, 939–945
Multiple-target, 835–844
Multiple view, 979–986

N
NAND flash, 836, 841–844
Nash equilibrium, 1332, 1334, 1335, 1337–1339
NAT-PT, 1008, 1009
Nearest neighbor, 1017, 1020–1022
Near field communication (NFC), 1109–1116
Network design, 783–790, 1264, 1342, 1348
NFC. *See* Near field communication (NFC)
Nonlinear frequency analysis, 1073–1080

O
OAuth, 756–759, 761
Object Linking and Embedding for Process Control (OPC), 997–1006

Object-Z, 1350, 1352–1354
Off-line payment, 706, 709
One-way hash chain, 1133–1141
On-line payment, 708–709
OODA loop, 1156–1157, 1162
OPC. *See* Object Linking and Embedding for Process Control (OPC)
Operating system (OS), 749, 750, 752, 821, 837, 851, 867, 869–871, 894, 897, 994, 1013, 1112, 1156, 1158, 1202–1206, 1223, 1299, 1300, 1302, 1303
OPNET, 1210, 1213, 1214, 1294
OS. *See* Operating system (OS)

P
PaaS. *See* Platform as a Service (PaaS)
Packet delivery, 764–768, 770, 772
Packing algorithm, 847–853
Parking guidance, 783–790
p-CSMA, Persistent carrier sense multiple access (p–CSMA)
Performance, 714, 723, 732, 750, 764, 776, 805, 810, 817, 836, 847, 870, 874, 886, 907, 919, 922, 963, 976, 980, 990, 1000, 1018, 1028, 1041, 1048, 1056, 1089, 1091, 1112, 1127, 1140, 1143, 1162, 1169, 1182, 1193, 1215, 1247, 1263, 1290, 1308, 1315, 1327, 1367, 1396
Peripheral environment simulation, 1217–1224
Persistent carrier sense multiple access (p-CSMA), 1290, 1291, 1293, 1294, 1296
Phase-locked loop, 1048
Plasma column antenna, 1027–1029, 1032
Platform as a Service (PaaS), 895, 896, 899, 900
PLL, 1046, 1048, 1049, 1052, 1053
Poisson distribution, 726, 1283, 1290, 1292
Polygon filling, 947, 951
Polymorphic, 1365–1372
Position-relation signature (PRS), 1365–1372
Power, 711, 714, 737, 785, 810, 844, 858, 912, 924, 929, 939, 955, 963, 989, 1010, 1028, 1035, 1045, 1055, 1073, 1083, 1095, 1109, 1118, 1126, 1134, 1143, 1192, 1229, 1235, 1271, 1282, 1297, 1308, 1319, 1323, 1352, 1396
Power amplifier, 1039
Power-law distribution, 1282, 1283, 1285–1287
Power saving, 713–720, 914–916, 919, 920
Prediction, 875, 974, 980, 1017, 1018, 1020–1023, 1084, 1143–1151, 1159, 1324, 1332, 1336, 1337, 1339, 1376

PRISM, 1246, 1247, 1250, 1251
Probabilistic model checking, 1246, 1247, 1251
Probability, 727, 728, 770, 803, 825–832, 856, 858, 860–862, 870, 906, 934, 942, 1166, 1168–1171, 1225–1231, 1247, 1249–1251, 1290, 1291, 1295, 1296, 1307–1313, 1333, 1334, 1337, 1347, 1384
Process control, 898, 997–1005
Propagation delay, 1168, 1169, 1171, 1291, 1296
Protocol, 708, 756, 763, 775, 784, 836, 956, 963, 997, 1008, 1035, 1110, 1120, 1134, 1158, 1166, 1174, 1203, 1209–1245, 1263, 1290, 1307, 1315, 1332, 1396
PRS. *See* Position-relation signature (PRS)
Public bike, 1341–1348

Q
QMR. *See* Quadruple Modular Redundant (QMR)
QoS. *See* Quality of services (QoS)
Qt GUI, 959, 960
Quadruped robot, 972, 975–977
Quadruple Modular Redundant (QMR), 905–912
Quality of services (QoS), 714–716, 732, 1125–1131, 1166, 1171, 1192–1198
Query processing, 1315, 1316, 1321

R
Radiation resistance, 1028, 1031–1033
Radio-frequency identification (RFID), 733, 1110, 1271
Radio resource management, 1191–1199
Radio resource scheduling, 713–720
Range-free, 1253–1261
Rating prediction, 1020–1023
RCAA. *See* Resource congestion avoidance algorithm (RCAA)
RC4 Based Hash Function (RC4 BHF), 1133–1141
RC4 BHF. *See* RC4 Based Hash Function (RC4 BHF)
RC4 stream cipher, 1134–1136
Real-time, 741, 743, 776, 782, 826, 898, 955–961, 971–977, 989–991, 995, 996, 1126–1128, 1157, 1173, 1175, 1181, 1215, 1218, 1246, 1302, 1326, 1353, 1354, 1359, 1360, 1398
Recommender system, 1017

Subject Index for Volume 2   1425

Reconfigurable processing unit (RPUs), 1063–1071
Regional registration, 764, 766–768, 770, 772
Region partition, 1382–1385, 1387, 1389, 1390, 1393, 1394
Reinforcement learning, 1191–1199
Reliability, 710, 731, 735, 736, 751, 755, 870, 871, 905–912, 1005, 1089, 1118, 1159, 1210, 1211, 1234, 1243, 1245–1247, 1308, 1330, 1349, 1350, 1353, 1395
Repackage, 1099, 1106
Resonance, 972, 1032, 1033
Resource allocation, 714–720, 724, 902, 1128, 1181–1189
Resource congestion avoidance algorithm (RCAA), 1183–1189
Resource discovery, 1263, 1264
Resource information organization, 1263–1270
Resoure management, 723–725, 728, 729, 1191–1199
RFID. *See* Radio-frequency identification (RFID)
Routing protocol, 1172, 1307–1313, 1315–1321
RPUs. *See* Reconfigurable processing unit (RPUs)
R-2R ladder, 1091–1097
Ruby, 1218, 1220–1222
Runoff forecast, 1373

S

SaaS. *See* Software as a Service (SaaS)
Sampling expansion, 1396, 1399
Satellite communication, 775–782
Scale-free network, 1230, 1281–1287
Scoring system, 809–816
SD. *See* Security Digital (SD)
SDH network, 1181–1189
SEC. *See* Spectral embedded clustering (SEC)
Security, 709–710, 724, 740, 745, 748, 753, 756, 758–761, 775, 792–797, 836, 865–871, 894, 897, 898, 914, 1109–1112, 1115, 1118, 1119, 1123, 1126–1128, 1130, 1133–1136, 1138–1141, 1159, 1201–1206, 1278, 1362, 1365
Security Digital (SD), 835–844
Selective content encryption, 775–782
Server virtualization, 749–751, 754
Service channel, 1165–1172, 1196
SET. *See* Single event transient (SET)
SEU. *See* Single event upset (SEU)
SGTIN-798, 1272–1277
Sheet, 921–927

Short circuit fault, 932, 1084
Similarity, 984, 1020, 1021, 1100, 1101, 1103–1106, 1121, 1282
Simulation testing environment, 769–770, 827, 1217–1224
Simulation training platform, 809–816
Single event transient (SET), 906, 907
Single event upset (SEU), 905–909, 912
Single sign-on, 755–761
SIR model, 1226–1228, 1231
Small file, 847–853
S4n-logic, 1333
Software as a Service (SaaS), 895, 896, 899–900
Software engineering, 1119, 1233
Software testing, 1217
Spectral embedded clustering (SEC), 979–986
Stateful translation, 1007–1015
Stateless translation, 1007–1015
Stator, 931, 939–945
Storage efficiency, 847, 848
Storage system, 725, 817, 818, 847, 848, 898, 963, 964, 969
S-transform, 1323–1330
Structure, 715, 744, 747, 776, 788, 810, 826, 837, 868, 894, 906, 914, 929, 939, 948, 957, 979, 990, 999, 1029, 1036, 1056, 1063, 1121, 1128, 1134, 1145, 1156, 1176, 1209, 1219, 1225, 1234, 1263, 1273, 1283, 1294, 1299, 1307, 1331, 1359, 1366, 1374, 1382, 1399
Structured P2P, 1263, 1264, 1266, 1269
Synchronous generator, 939, 940, 942

T

Tag, 1001, 1004, 1005, 1018, 1019, 1112, 1143–1151, 1174, 1271–1273, 1276, 1277
TCP/IP, 776, 777, 821, 836, 955–961, 1007, 1010–1013, 1166–1168, 1170, 1171, 1203–1206
Technical support, 901
Thermal stress, 921, 922
Threshold Strategy, 804, 805
Time analysis, 1325
Titanium, 921–927
Tracking, 976, 977, 998–1002, 1005, 1159, 1350, 1351, 1353
Traffic, 717, 719, 727, 763, 770, 772, 773, 964, 1014, 1035, 1105, 1112, 1166, 1171, 1172, 1183, 1264, 1267–1269, 1291, 1307, 1343–1345, 1348, 1351, 1353, 1366–1368
Train control systems, 1349–1355

Transceiver front-end, 1035–1043
Transmitted cooperative reference, 1395–1405
Trust, 760, 791–797, 865–871, 1138, 1139
Trusted cloud, 865–871
Trusted computing, 865–871
T-type network, 1091–1097
Twist, 921–927

U

Ubiquitous computing (UBICOMP), 835, 836, 844
UML sequence diagrams, 1233–1243
Unhealthy cloud system status, 799–807

V

VANET. *See* Vehicular ad hoc network (VANET)
VCO. *See* Voltage-Controlled Oscillator (VCO)
VDL2, 1294–1295
Vehicular ad hoc network (VANET), 1130, 1165–1172
Verification, 745, 842, 843, 959–961, 1130, 1206, 1233–1243, 1245–1251, 1281–1287, 1351, 1352, 1355, 1360, 1362
Vertex-degree sequence, 1281–1287
Video transmission, 955–961

Virtualization, 747–754, 820, 897, 898, 1201
Virtual machine monitor, 820, 1202
Visualization of clustered network graph, 1381–1394
VMware vSphere, 748, 749, 751, 752
Voltage-Controlled Oscillator (VCO), 1046–1052
Voltage transient stability, 929–936

W

Wavelet analysis, 1087, 1373, 1374, 1376–1378
Wavelet transform, 1083–1085, 1323–1325, 1327
Wavelet-ANFIS, 1373–1378
Web Service, 899, 900, 1173–1179
Weighted current, 1091–1097
Wind generator, 929–936
Wireless Local Area Network (WLAN), 1109–1116
Wireless Power Transmission (WPT), 1055–1058, 1060
Wireless sensor network (WSN), 784, 843, 1209, 1254, 1307–1313, 1315–1321
WLAN. *See* Wireless Local Area Network (WLAN)
Worm detection, 1365–1372
WPT. *See* Wireless Power Transmission (WPT)
WSN. *See* Wireless sensor network (WSN)

Printed by Publishers' Graphics LLC